The Physics and Technology of Ion Sources

The Physics and Technology of Ion Sources

Edited by

IAN G. BROWN

*Lawrence Berkeley Laboratory
University of California
Berkeley, California*

A WILEY-INTERSCIENCE PUBLICATION
JOHN WILEY & SONS
New York • Chichester • Brisbane • Toronto • Singapore

Copyright © 1989 by John Wiley & Sons, Inc.

All rights reserved. Published simultaneously in Canada.

Reproduction or translation of any part of this work beyond that permitted by Section 107 or 108 of the 1976 United States Copyright Act without the permission of the copyright owner is unlawful. Requests for permission or further information should be addressed to the Permissions Department, John Wiley & Sons, Inc.

Library of Congress Cataloging in Publication Data:

The Physics and technology of ion sources.

    Includes bibliographies and indexes.
    1. Ion sources. I. Brown, Ian G.
QC702.3.P48 1989   541.3'72   88-20507
ISBN 0-471-85708-4

Printed in the United States of America

10 9 8 7 6 5 4 3 2 1

. . . If I had waited long enough I probably never would have written anything at all since there is a tendency when you really begin to learn something about a thing not to want to write about it but rather to keep on learning about it always and at no time, unless you are very egotistical, which, of course, accounts for many books, will you be able to say: now I know all about this and will write about it. Certainly I do not say that now; every year I know there is more to learn, but I know some things which may be interesting now, and I may be away from the bullfights for a long time and I might as well write what I know about them now.
—ERNEST HEMINGWAY

*Ernest Hemingway, excerpted from DEATH IN THE AFTERNOON. Copyright 1932 Charles Scribner's Sons; copyright renewed © 1960 Ernest Hemingway. Reprinted with the permission of Charles Scribner's Sons, an imprint of Macmillan Publishing Company.

Contributors

D. Aitken*
Applied Materials Implant Division
Horsham, England

R. J. Anderson
Physics Department
University of Arkansas
Fayetteville, Arkansas

A. E. Bell
Oregon Graduate Center
Beaverton, Oregon

I. G. Brown
Lawrence Berkeley Laboratory
University of California
Berkeley, California

C. Burkhart
Institute for Accelerator and Plasma
　Beam Technology
University of New Mexico
Albuquerque, New Mexico

E. D. Donets
Laboratory of High Energies
Joint Institute for Nuclear Research
Dubna, USSR

B. F. Gavin
Lawrence Berkeley Laboratory
University of California
Berkeley, California

R. A. Gerber
Sandia National Laboratories
Albuquerque, New Mexico

A. J. T. Holmes
UKAEA Culham Laboratory
Culham, England

R. H. Hughes
Physics Department
University of Arkansas
Fayetteville, Arkansas

S. Humphries, Jr.
Institute for Accelerator and Plasma
　Beam Technology
University of New Mexico
Albuquerque, New Mexico

J. Ishikawa
Kyoto University
Kyoto, Japan

*Current affiliation: Superion, Surbiton, Surrey, England.

Contributors

Y. Jongen
Laboratoire du Cyclotron
Universite Catholique de Louvain
Louvain-la-Neuve, Belgium

R. Keller*
GSI, Gesellschaft fur
  Schwerionenforschung
Darmstadt, Federal Republic of
  Germany

L. K. Len
Institute for Accelerator and Plasma
  Beam Technology
University of New Mexico
Albuquerque, New Mexico

K. N. Leung
Lawrence Berkeley Laboratory
University of California
Berkeley, California

C. M. Lyneis
Lawrence Berkeley Laboratory
University of California
Berkeley, California

N. Sakudo
Hitachi Central Research Laboratory
Tokyo, Japan

P. Spädtke
GSI, Gesellschaft fur
  Schwerionenforschung
Darmstadt, Federal Republic of
  Germany

L. W. Swanson
Oregon Graduate Center
Beaverton, Oregon

*Current affiliation: Lawrence Berkeley Laboratory, University of California, Berkeley, California.

Preface

This book is designed to serve as a review and reference for the rapidly evolving field of ion source technology. The target reader is the physicist who is actively engaged in ion source development or related work. The object of the book is to provide a comprehensive, easily understood survey of the field in sufficient depth to be useful to the laboratory worker, and of sufficient scope to elucidate the forest as well as the individual trees.

The field of ion source development has grown dramatically in recent years. New areas of application have evolved, calling for ion beams of unprecedented energy, current, species, focus, uniformity, size, and charge states, among other parameters. New kinds of ion sources have been developed and existing kinds have been improved greatly. Along with the emergence of new fields of ion source application, new laboratories and research groups have formed. The curious situation has come about in which workers in a given subarea of ion source development share backgrounds from one specific scientific field, mostly attend scientific conferences within that same field, and mostly publish in journals more-or-less specific to that field. There is not a great cross-field communication between ion source workers in different fields. Thus, the wide mix of kinds of sources that are addressed in detail in this book is quite purposeful. The intent has been to present the field of ion source physics and technology from as global a perspective as can sensibly be done. While global, however, the treatment is not encyclopedic. The work presented here in no way preempts the usual necessity for a good literature search; hopefully it will help that chore.

The different chapters have been written by researchers who are expert in the topic discussed. The chapters are independent and self-contained. The terminology, notation, and units used are as is conventional within the subfield under discussion and vary a little from chapter to chapter. The reader should expect consistent use of symbols and units only within a given chapter. So long as this is recognized, no confusion should follow.

Ion source research and development is certainly one of the more empirical of scientific endeavors, and for the most part the emphasis presented in this book is experimental. It has overwhelmingly been the case that theoretical understanding of an ion source has followed after its construction and experimental performance. This situation is reflected in the various presentations here.

The academic background assumed of the reader is roughly physics graduate level. A working knowledge of plasma physics, atomic physics, and electromagnetic theory would aid the digestion.

I thank my colleagues in ion sourcery and plasma physics at the Lawrence Berkeley Laboratory and elsewhere for their encouragement in this endeavor, and some for reading and commenting on several of the manuscripts. I wish most especially to express my deep gratitude to Wulf Kunkel, whose support has been abundant and important. I am also most grateful to my long-suffering contributors, who persevered a plethora of persnickety proofing; my reward on this score has been in getting to know them.

IAN G. BROWN

Berkeley, California
March, 1988

Contents

1 Introduction   1
I. G. Brown

2 The Plasma Physics of Ion Sources   7
I. G. Brown

  1 Density, Temperature, and Distribution Functions, 7
      1.1 Percentage Ionization, 8
      1.2 Distribution Functions, 8
  2 Sheaths and Electric Fields, 11
      2.1 Debye Length, 12
      2.2 Charge Neutrality, 12
  3 Collisions, 13
  4 The Plasma Frequency, 14
  5 Magnetic Field Effects, 15
      5.1 Cyclotron Radius, 15
      5.2 Cyclotron Frequency, 16
  6 Plasma Pressure, 16
  7 Particle Flow, 17
  8 Ionization, 17
  9 Multiple Ionization, 19
      References, 21

3 Ion Extraction   23
R. Keller

  1 The Physics of Ion Extraction, 23
  2 First-Order Treatment of Ion Extraction, 27
  3 Beam Quality Parameters, 30

    4 Refined Treatment of Ion Extraction, 35
    5 Voltage Scaling Laws for Beam Current and Brightness, 39
    6 Special Extraction Systems, 45
       References, 50

4 Beam Transport 53
A. J. T. Holmes

1 Beam Optics in the Absence of Collective Effects, 51
   1.1 Busch's Theorem and the Paraxial Ray Equation, 55
   1.2 Solutions of the Paraxial Ray Equation and the Matrix Method, 58
   1.3 Focusing of Ion Beams, 59
       1.3.1 Electrostatic Lenses, 60
       1.3.2 Magnetic Lenses, 62
           1.3.2.1 The Solenoid, 62
           1.3.2.2 The Quadrupole, 63
           1.3.2.3 Combination of Quadrupole Lenses, 65
           1.3.2.4 The Magnetic Sector Lens, 66
   1.4 Beam Transport Channels, 69
2 The Transport of Laminar Beams with Self Fields, 70
   2.1 Motion of an Ion Beam without External Fields, 71
   2.2 Self Fields and the Paraxial Ray Equation, 73
   2.3 Beam Propagation Using an Axial Magnetic Field, 74
3 Space-Charge Neutralization of Ion Beams, 75
   3.1 Neutralization of Positive Ion Beam Charge, 76
   3.2 A Simple Model for a Positive Ion Beam Plasma, 77
   3.3 Neutralization of Negative Ion Beams, 80
       3.3.1 A Simple Model of Negative Ion Beam Neutralization, 86
   3.4 The Effect of Magnetic Fields, 88
4 Nonlaminar Beams, 91
   4.1 Liouville's Theorem, 91
   4.2 Emittance and Brightness, 92
   4.3 Measurement of Beam Emittance, 95
   4.4 Emittance and Beam Propagation, 95
   4.5 The Kapchinskij-Vladimirskij Distribution, 99
   4.6 Emittance Growth and Nonlinear Optics, 100
5 The Beam Envelope of a Drifting Beam, 101
   5.1 Solutions of the K-V Equation, 103
       References, 105

5 Computer Modeling 107
P. Spädtke

1. History, 110
2. Techniques of Ray Tracing Programs, 112
   - 2.1 Poisson Equation, 115
   - 2.2 Cathode Simulation, 116
   - 2.3 Plasma Simulation, 118
   - 2.4 Ray Tracing, 119
3. Representation of Results, 120
4. A 2D Example, 120
   - 4.1 Ion Beam Extraction, 121
   - 4.2 Drift Space, 121
   - 4.3 Acceleration, 122
5. A 3D Example, 122
   - Appendix, 132
   - References, 133

6 Some Basic Ion Sources 137
I. G. Brown

1. Sources Made from Bayard-Alpert Gauges, 137
2. Other Electron-Bombardment Sources, 139
3. Metal Ion Sources, 142
4. Microwave Ion Sources, 145
5. Miscellaneous Sources, 148
   - References, 149

7 High-Current Gaseous Ion Sources 151
R. Keller

1. Sources with Single-Stage Discharge, 153
2. Sources with Two-Stage Discharges, 158
3. Multiply Charged Ions, 161
4. Ions from Elements with Low Vapor Pressure, 162
5. Outlook, 164
   - References, 164

8 PIG Ion Sources   167
B. F. Gavin

1 Background, 167
2 Characteristics of Cold Cathode PIGs, 169
3 Characteristics of Hot Cathode PIGs, 171
4 PIG Discharge Mechanisms, 171
5 High Charge State Ion Production, 175
6 Operating Parameters, 176
7 Metallic Ion Production, 179
8 PIG Ion Beam Characteristics, 182
9 Conclusion, 183
   References, 183

9 The Freeman Ion Source   187
D. Aitken

1 The Early Freeman Source, 188
2 Freeman Sources in Laboratory Scale Separators and Early Implanters, 190
   2.1 The Original Harwell Design, 190
   2.2 Source Variants, 190
   2.3 The Harwell Isotope Separator as an Implanter, 190
   2.4 The Freeman Source in Commercial Ion Implanters, 194
3 Source Characteristics, 194
   3.1 Arc Current Control, 194
   3.2 Nonuniformity of the Arc Discharge along the Length of the Filament, 196
4 Freeman Source Variants, 199
   4.1 Magnetic Bottle, 199
   4.2 Electrostatic Electron Reflectors, 199
   4.3 AC Filament, 199
   4.4 Source Magnet Control, 200
   4.5 Other Variants, 201
5 Ion Beam Extraction, 201
6 Arc Control, 202
7 Feed Materials, 203
8 Summary, 205
   References, 205

10 Electron Cyclotron Resonance/Ion Sources 207
Y. Jongen and C. M. Lyneis

1 A Brief History of the Development of ECR Sources, 208
2 Basic Principles of ECR Sources, 211
   2.1 Ionization by Electron Impact, 211
   2.2 Magnetic Confinement, 212
   2.3 ECR Heating, 213
   2.4 Losses by Charge Exchange, 214
   2.5 Ion Confinement, 215
   2.6 The Plasma Injector Stage, 215
   2.7 Gas Recirculation in ECR Sources, 216
3 World ECR Sources, 216
4 Illustration of a Typical ECR Source: The LBL ECR, 219
   4.1 Source Design, 219
   4.2 LBL ECR Performance, 222
   4.3 ECR Source Beams from Solids, 222
   References, 227

11 Microwave Ion Sources 229
N. Sakudo

1 Plasma Generation by Microwave Discharge in a Magnetic Field, 229
2 Some Practical Considerations, 230
3 Versatility of Beam Extraction, 233
   3.1 Large Cross-Sectional Beam Extracted through a Multiaperture Lens, 233
   3.2 Slit-Shaped Ion Beam for Ion Implanters, 235
   3.3 Further Improvements in Slit-Shaped Ion Beams, 237
      3.3.1 Increasing Ion Extraction Voltages, 237
      3.3.2 Increasing Extracted Currents, 238
   3.4 Compact Microwave Ion Sources, 239
4 Diversity of Available Ion Species, 240
5 Summary, 243
   References, 243

12 Electron Beam Ion Sources 245
E. D. Donets

1 Physical Basis and Mode of Operation of EBIS, 245
   1.1 Physical Basis, 246

    1.2 Basic Operation of the Electron Beam Method of Producing Highly Charged Ions, 250
        1.2.1 Injection of Ions into the Trap, 251
        1.2.2 Containment of the Ions: Ionization, 254
        1.2.3 Ion Extraction, 256
        1.2.4 Some Remarks about the Collective Model, 257
2 Types of EBIS Construction, 258
    2.1 Construction of KRION-2, 259
    2.2 Construction of CRYEBIS, 260
3 Experimental Development of EBIS, 262
    3.1 Capacity of the Ion Trap, 262
    3.2 Containment of Ions in the Beam, 265
    3.3 Production of Highly Charged Ions, 266
    3.4 Other Experimental Results, 267
4 Short Review of Projects and Devices, 269
5 Some Special Applications, 271
    5.1 Ionization of Positive Ions by Electron Impact, 272
    5.2 Dependence of Nuclear Properties on the Ion Charge State, 274
    5.3 Spectroscopy of Slow Highly Charged Ions on Solid Surfaces, 276
6 Problems and Prospects of EBIS Development, 278
References, 278

13 Beam-Plasma Ion Sources   281
J. Ishikawa

1 Ion Source Characteristics, 282
2 Source Construction, 283
    2.1 Ion Extraction and Electron Gun Region, 286
    2.2 Drift Space, 286
    2.3 Electron Beam Collector Region, 286
3 Beam–Plasma Discharge, 287
    3.1 Beam–Plasma Discharge Due to Hydrodynamic Instability, 288
    3.2 Quasilinear Relaxation, 290
4 Ion Extraction with Ion Space-Charge Compensation, 291
    4.1 Relaxation of the Space-Charge Limitation of Extractable Ion Current, 292
    4.2 Extracted Ion Current, 294
5 Summary, 296
References, 296

14 Laser Ion Sources — 299
R. H. Hughes and R. J. Anderson

1. Background, 299
2. Laser Plasmas Used as an Ion Source, 301
3. Performance of Laser Ion Sources, 303
   - 3.1 The Dubna Laser Sources, 303
   - 3.2 The Arkansas Experiments, 306
4. Future of Laser Ion Sources, 309
   References, 310

15 Liquid Metal Ion Sources — 313
L. W. Swanson and A. E. Bell

1. Principle of Operation, 314
2. Source Characteristics, 318
   - 2.1 Total Energy Distribution and Angular Intensity of the LMIS, 320
   - 2.2 Virtual Source Size of the LMIS, 324
3. Alloy Sources, 325
4. Applications of Liquid Metal Ion Sources, 326
5. Summary, 328
   References, 329

16 The Metal Vapor Vacuum Arc Ion Source — 331
I. G. Brown

1. The Metal Vapor Vacuum Arc Plasma, 332
2. Source Design, 334
3. Embodiments, 339
   - 3.1 MicroMEVVA, 339
   - 3.2 MEVVA II, 339
   - 3.3 MEVVA III, 343
   - 3.4 MEVVA IV, 343
4. Source Performance, 343
   - 4.1 Ion Species, 343
   - 4.2 Beam Current, 345
   - 4.3 Extraction Voltage, 347
   - 4.4 Beam Noise, 346
   - 4.5 Duty Cycle, 346
   - 4.6 Source Lifetime, 348
   - 4.7 Charge State Distribution, 349

5 Other Versions, 351
6 Future Directions, 353
   References, 353

17 Negative Ion Sources 355
K. N. Leung

1 Surface-Produced Negative Ion Sources, 355
   1.1 Sputtering-type Negative Ion Sources, 355
   1.2 Plasma Surface Conversion Negative Ion Sources, 356
2 Volume-Produced Negative Ion Sources, 359
   2.1 Duoplasmatron, Magnetron, and Penning-type H$^-$ Sources, 359
   2.2 Multicusp Negative Ion Sources, 362
   2.3 Volume-Produced Heavy Negative Ion Sources, 364
   References, 368

18 Light-Ion Sources for Inertial Confinement Fusion 371
R. A. Gerber

1 Background, 371
2 Passive Sources, 373
   2.1 Flashover Sources, 373
   2.2 Cryogenic-Anode Sources, 377
   2.3 EHD-Driven Liquid Sources, 378
3 Active Sources, 383
   3.1 XUV-Driven Sources, 384
   3.2 Gas-Breakdown Plasma Anode, 385
   3.3 BOLVAPS/LIBORS Lithium Source, 387
   3.4 Other Research on Active Lithium Sources, 390
4 Summary, 391
   References, 393

19 Ion Sources for Pulsed High-Brightness Beams 397
S. Humphries, Jr., C. Burkhart, and L. K. Len

1 Emittance and Brightness, 398
2 Principles of Grid-Controlled Ion Extractors, 400
3 Limiting Factors, 403
4 Arc Plasma Sources for Pulsed Ion Beams, 406
5 Observation of Grid-Controlled Extraction, 409
6 Flux Measurements, 412

7 Measurements of Beam Species and Emittance, 414
 8 Conclusions, 417
    References, 418

APPENDIXES

Appendix 1 Physical Constants   **421**

Appendix 2 Some Plasma Parameters   **422**

Appendix 3 Table of the Elements   **423**

Appendix 4 Ionization Potentials for Multiply Charged Ions of All the Elements   **425**

Appendix 5 Work Functions of the Elements   **426**

Appendix 6 Vapor Pressure Curves of the Elements   **427**

Name Index   **431**
Subject Index   **441**

1

Introduction

I. G. Brown
*Lawrence Berkeley Laboratory
University of California
Berkeley, California*

This book is an introduction to the physical principles of ion sources and a detailed review of a number of different kinds of sources. It is intended to be a guide for ion source researchers and users, and an overview of the field for new entrants to the discipline. It is not encyclopedic. The principles covered and source variants discussed form a representative assemblage of the field, that in toto should serve to convey a picture of the state of the art at the present time.

There has been a rapid growth in activity in ion source research and development in the last decade or two. This development has assuredly been symbiotic with the growth of fields of ion source applications. Thus, for example, the fields of particle accelerator physics, ion implantation, and controlled fusion research have contributed greatly to the growth in ion sourcery, to use a term that seems to be making a slow transition toward respectability. It is an interesting situation that workers in ion source R&D have in the main been drawn from the particular application field: ion sourcerers (to push the limits of verbal respectability yet further) involved in development of sources for accelerator application have mostly been of nuclear or accelerator physics background, those involved in development of sources of ion implantation application have mostly been of solid state or materials background, and those involved in development of sources for controlled fusion research have mostly been of plasma physics background. This is not a universal rule, of course, simply a generality. A lot of important developments have come about as a result of researchers in one field becoming aware of an application in another. A nice example of this cross-field technology transfer is provided by the ECR ion source, where the initial research and development of hot electron, microwave-driven plasmas was done within the controlled fusion research community and the techniques were recognized as being of value for the production of highly stripped ions for particle accelerator (mainly cyclotron) injection; this particular ion source development subfield was given much impetus by the pioneering work of Geller. But on the whole, there has not been a great deal of intercommunication between the different ion sourcery fields of application. This observation has formed the core of the philosophy with which this book has been tailored. It is quite deliberate that a wide range of superficially unrelated ion source types have been gathered together here for detailed description. If this circumstance is evident to the reader, then an aim of the book has at least in part been accomplished—to help provide an awareness of the diversity of phenomena available to the ion source experimentalist.

The first part of the book is concentrated on a number of fundamental aspects of ion sources: some very basic plasma physics, discussion of ion extraction, a review of beam transport, computer modeling, and a simple survey of some fairly basic kinds of ions sources. The second part of the book consists of detailed descriptions of specific ion sources: high-current gaseous sources, PIGs, Freeman sources, ECR sources, microwave sources, EBISes, beam-plasma sources, laser ion sources, liquid metal ion sources, metal vapor vacuum arc ion sources, negative-ion sources, light ion sources of the kind used for inertial confinement fusion research, and pulsed high-brightness sources. Each of these chapters has been written by researchers actively engaged in, and at the forefront of, the specific ion source subfield addressed.

The chapter on the plasma physics of ion sources (Chapter 2, by I. G. Brown) serves as a brief review of some of the elements of plasma behavior that are valuable to be aware of when considering the plasmas that one meets in ion sources. The approach is simple; more elegant and more complete treatments abound, and references to a number of such texts are given. Furthermore, many of the later chapters dedicated to specific types of ion sources provide discussion of those aspects of plasma physics required for a good understanding of that ion source type.

Essentially all ion sources have two basic ingredients: a plasma source and a means of forming the ion beam. The ion beam formation system is conventionally called the ion extractor, even though from a strict plasma physics point of view the term is a little misleading for positive ions, which usually fall out through the plasma sheath at relatively low energy and are then accelerated, rather than being energetically extracted from the interior of the plasma region. It is usual that the extractor system is made from a number of elements, for example, multiaperture grids or slots, and the detailed design of the extractor so as to optimize the extracted ion beam is a complicated subject. Chapter 3 (by R. Keller, GSI/Darmstadt, FRG) addresses the concern of ion extraction.

Subsequent to beam formation, any real-world application of ion beams will necessarily involve some form of beam transport. As just one example, it would not be atypical for the beam to be focused using magnetic quadrupole or electrostatic lenses, then to be mass and/or charge state analyzed by passing through a dipole analysis magnet, then to be further focused, then perhaps transported through a series of focusing and bending elements spaced along a beam transport pipe. The complexity of the situation might be further complicated by the presence of space

Introduction

charge forces tending to disassemble the beam; neutralizing electrons (for the case of positive ion transport) may be provided via collisions of the beam with the background gas, or from elsewhere. Finally, the beam must be transported through the array of optical elements without loss of ions from the beam and without spoiling the beam quality. Beam transport is treated in considerable detail in Chapter 4 (by A. J. T. Holmes, Culham Laboratory, England).

The rapid growth of computing power has brought with it the availability of computational techniques for the calculation of ion trajectories from their origin at the plasma side of the ion extractor, through the beam formation and acceleration region, and into and through the ensuing ion optical system, whatever it may be. Significant progress has been made in the capacity of the computer programs to treat 3D geometries and to include effects due to space charge. Computer modeling of ion extraction and beam transport is reviewed in Chapter 5 (this chapter was written by P. Spaedtke, GSI/Darmstadt, FRG).

The rest of the book is dedicated to fairly detailed reviews of different kinds of ion sources. First, in Chapter 6 (by I. G. Brown), a number of fairly simple kinds of sources that have been described in the literature are discussed. This serves two purposes: It provides a handy guide for the experimentalist faced with the need of setting up a basic ion source facility, and it provides an introduction to the detail and sometimes complexity of the ion sources described in the following chapters.

The class of high-current gaseous ion sources contains a variety of orders. Here we have chosen to consider several different kinds of related sources under one heading. There are many similarities of construction and also of problems encountered in these different kinds of high-current sources. For the most part they are used in the same fields of application, and meet with similar experimental demands and restrictions. This class of sources has been reviewed in Chapter 7 (written by R. Keller, GSI/Darmstadt, FRG).

PIG ion sources are discussed in Chapter 8 (by B. F. Gavin, Lawrence Berkeley Laboratory, Berkeley, California). These sources are very closely related to those discussed in Chapter 7, and have found widespread application in the injectors of the large particle accelerators used for nuclear and high-energy physics research. The term "PIG" is eminently conventional in the ion source community, and we have refrained from using the unabbreviated name.

The Freeman ion source is the source of choice of the semiconductor ion implantation community. This kind of source is widely used in the commercially available ion implantation equipment. Much of the pioneering work in ion implantation was done using the Freeman source. This situation has occurred not without reason, since the beam produced by the Freeman source is a particular high-quality beam measured in a number of ways that are important to this application. These sources are described in Chapter 9 (D. Aitken, Applied Materials Implant Division, Horsham, England).

Microwave-produced plasmas have been used to advantage in ion sources. There are two rather different parameter regimes. In one, the microwave coupling into the plasma occurs at the electron cyclotron frequency, the gas pressure is low and the plasma is largely collisionless. In the second, the coupling does not depend on a resonance with the magnetic field, the gas pressure is higher and the plasma is largely collisional. Going along with these plasma preparation conditions, the parameters of the ion beams produced by ion sources utilizing these two kinds of plasmas differ significantly also. In the first kind the charge states of the ions produced are impressively high while the ion beam current is generally low; in the second kind of source the ion beam current is impressively high while the charge states are mostly singly ionized. The first kind of source has become well known as the electron cyclotron resonance ion source, or more commonly, *the ECR ion source* or *ECRIS*; a universally accepted name for the second kind seems to be still in evolution; here we have used the term *microwave ion source*. ECR sources have found important application as high-charge state ion injectors for nuclear physics particle accelerators, especially cyclotrons, and microwave ion sources for the filamentless production of high-current beams for high-dose ion implantation into semiconductors. ECR sources are described in Chapter 10 (by C. Lyneis, Lawrence Berkeley Laboratory, and Y. Jongen, Louvain-la-Neuve, Belgium); microwave ion sources, in Chapter 11 (by N. Sakudo, Hitachi, Tokyo).

The record for high-charge state ion production from an ion source (i.e., not including ions that are accelerated to very high energy and stripped with a stripper foil) is easily held by the electron beam ion source, or EBIS. In the EBIS the ions are contained within a high-energy electron beam in a strong magnetic field and in a very high vacuum ambient, for a sufficiently long time for very high charge state ions to be produced. The ion output is relatively low. This kind of ion source is particularly suited to particle accelerator injection, especially heavy ion synchrotrons, as well as to basic atomic physics. EBIS devices are described in Chapter 12 (by E. D. Donets, Joint Institute for Nuclear Research, Dubna, USSR).

The beam–plasma ion source is a relatively new kind of source in which ionization occurs as a result of the interaction of an intense injected electron beam with the background plasma. A beam–plasma instability occurs, and beam energy is efficiently transferred into the plasma. Another novel feature of this kind of ion source is that the extracted ion beam current density can significantly exceed the classical Child-Langmuir limit. This unique ion source is described in Chapter 13 (by J. Ishikawa, Kyoto University, Kyoto, Japan).

Chapter 14 (by R. H. Hughes and R. J. Anderson, University of Arkansas) describes laser ion sources. The very high energy fluxes that can be delivered onto solid targets using readily available high-power pulsed lasers can be used as a means for producing dense pulsed plasmas with a high fraction of highly stripped ion species. A suitable extractor system can be used to form an ion beam from the laser-produced plasma, thus completing the basic laser ion source configuration. These kinds of sources are in a relatively early research phase, and show promise of being of value for applications such as heavy ion synchrotron injection.

Liquid metal ion sources stand alone in their ability to produce finely focused ion beams. The ions are formed by ion field emission from a tiny point onto which a film of liquid metal flows. The beam is of very low current, but the current density at focus can be respectable because of the submicron spot size. This kind of ion source has application to ion beam lithography and to ion microscopy.

Chapter 15 (by L. W. Swanson and A. E. Bell, Oregon Graduate Center, Beaverton, Oregon) is devoted to liquid metal ion sources.

In the metal vapor vacuum arc (MEVVA) ion source, a dense metal plasma is created at the cathode spots formed on the surface of a solid metallic cathode by an arc discharge that is initiated at high vacuum. The plasma plumes away from the cathode in a manner reminiscent of laser-produced plasmas, and an intense metallic ion beam can be formed from the plasma plume. Very high current beams of virtually all the solid metals have been produced. This kind of source has application for particle accelerator injection and for broad-beam ion implantation. MEVVA ion sources are described in Chapter 16 (by I. G. Brown, Lawrence Berkeley Laboratory, Berkeley, California).

Negative-ion sources are of importance for particle accelerator injection and for the production of very high energy neutral beams for fusion plasma heating. Techniques for the production of negative ions are quite different from positive ion production techniques, and there has been a steady and impressive growth in the beam currents available from negative ion sources. Negative-ion sources are discussed in Chapter 17 (by K. N. Leung, Lawrence Berkeley Laboratory, Berkeley, California).

In the inertial confinement fusion (ICF) research program there is a need for short-pulse, high-current beams of light ions. In an ICF process, beams of this kind would bombard a spherical fuel pellet target, compressing it to very high density and temperature, and part of the target material would undergo a fusion burn. In a reactor, the process would be repeated continuously. Because of the very high beam currents required, albeit at low duty cycle, progress in the development of these kinds of sources has led to a number of quite novel designs and ion production concepts. Light-ion sources for inertial confinement fusion are discussed in Chapter 18 (by R. A. Gerber, Sandia National Laboratories, Albuquerque, New Mexico).

Finally, some unique concepts and new developments related to the production of pulsed high-brightness ion beams are described in Chapter 19 (by S. Humphries, Jr., C. Burkhart, and L. K. Len, University of New Mexico, Albuquerque, New Mexico). A novel and effective means of producing a very quiet beam, even though the source plasma has a high noise level, via a technique called grid-controlled extraction, is described.

As mentioned above, the variety of ion sources discussed here is not comprehensive. There are a great many other kinds of ion sources that could have been discussed. But the range covered does span a divers spectrum of beam parameters and source techniques. Perhaps, with a little serendipitous fortune, some of the ideas presented here in description of one or another kind of source might prove to be fertile material for new kinds of ion sources and ion beam devices.

2

The Plasma Physics of Ion Sources

I. G. Brown
Lawrence Berkeley Laboratory
University of California
Berkeley, California

The characteristics of an ion beam are determined by the plasma and the extractor. Thus, for example, the ion beam current is determined by the plasma density, the plasma electron temperature, the extraction voltage, and the extractor geometry. The beam emittance is determined by the plasma ion temperature and the extractor geometry; and, clearly, the beam composition is determined by the composition of the plasma. The physics of the ion source is thus largely *plasma* physics. Quite a few texts have been written that provide an excellent coverage of the fields of basic plasma physics and ionization phenomena that are important to ion sources; see, for example, Refs. 1-9. In this chapter we review those principles of plasma physics required for an understanding of ion source performance and behavior.

1 DENSITY, TEMPERATURE, AND DISTRIBUTION FUNCTIONS

The most basic parameter used to describe a plasma is the plasma density. The constituents of the plasma are ions and electrons, as well as un-ionized neutrals, and so there is a plasma electron density n_e, a plasma ion density n_i, and a neutral particle density n_n. It is usual to express the density in units of particles per cm^3 or perhaps particles per m^3. For most ion sources the plasma density—meaning the electron density—is in the very broad range 10^{10} to 10^{16} cm^{-3}. As a comparison recall that gas at STP has a particle density of 2.5×10^{19} molecules/cm^3, and solids have atomic densities near 10^{23} cm^{-3}; a vacuum vessel at a pressure of 10^{-6} Torr contains gas molecules at a density of 3.3×10^{10} molecules/cm^3.

1.1 Percentage Ionization

The concept of percentage ionization follows naturally,

$$\text{percentage ionization} = n_i/(n_i + n_n) \times 100\% \qquad (1)$$

When the percentage ionization is greater than about 10% or so, the plasma is generally said to be highly ionized, and the physics of the medium is dominated by plasma effects. If, on the other hand, the percentage ionization is low, say less than 1%, then the interaction with neutrals must be considered.

1.2 Distribution Functions

The plasma particles have kinetic energy of motion, and the ensemble of particles can be represented by a velocity distribution function. The velocity distribution function $f(\mathbf{v})$ describes the number of particles in a given velocity interval,

$$\begin{aligned} dn &= f(\mathbf{v})\, d\mathbf{v} \\ &= f(v_x, v_y, v_z)\, dv_x\, dv_y\, dv_z \\ &= f(v_x) f(v_y) f(v_z)\, dv_x\, dv_y\, dv_z \end{aligned}$$

where $\mathbf{v}$ is the vector velocity (v_x, v_y, v_z) and $d\mathbf{v}$ is the velocity space element $dv_x\, dv_y\, dv_z$. It is usual that $f(\mathbf{v})$ is normalized via the particle density,

$$n = \int_{-\infty}^{\infty} f(\mathbf{v})\, d\mathbf{v} \qquad (2)$$

In the absence of external forces the plasma will tend toward thermal equilibrium, and the distribution is then Maxwellian,

$$f(v_x, v_y, v_z) = n(m/2\pi kT)^{3/2} \exp\left[-\tfrac{1}{2}m(v_x^2 + v_y^2 + v_z^2)/kT\right] \qquad (3)$$

where m is the particle mass, k is Boltzmann's constant,

$$k = 1.38 \times 10^{-23}\ \text{J/K}$$

and T is the plasma temperature. It is usual to refer to the temperature in units of electron volts (eV), and

$$1\ \text{eV} = 11{,}600\ \text{K}$$

An arc plasma for example might have a temperature of 5 eV, or about 50,000 K.

The electron volt is an energy unit and

$$1 \text{ eV} = 1.6 \times 10^{-19} \text{ J}$$

It may be that all the ions in the plasma are in equilibrium with each other and that it is thus meaningful to talk of an ion temperature T_i; the electron component of the plasma, however, may be in a different equilibrium and thus described by a different temperature T_e. The neutral component is usually at a much lower temperature than the ions or electrons. The term *plasma temperature* is ill-defined unless equilibrium between the ion and electron components is assumed, when $T_i = T_e$.

The particle distribution functions sometimes come in handy: One-dimensional Maxwellian velocity distribution:

$$f(v_x) = n\left(\frac{m}{2\pi kT}\right)^{1/2} e^{-mv_x^2/2kT} \tag{4}$$

Maxwellian distribution of speed, $F(v)$, where now $F(v)\,dv$ is the mean number of particles per unit volume with speed $v = |v|$ in the range v to $v + dv$:

$$\begin{aligned} F(v) &= 4\pi v^2 f(v) \\ &= 4\pi v^2 n \left(\frac{m}{2\pi kT}\right)^{3/2} e^{-mv^2/2kT} \end{aligned} \tag{5}$$

Maxwellian distribution of energy, $f(E)$:

$$\begin{aligned} f(E) &= F(v)\frac{dv}{dE} \\ &= n(4/\pi)^{1/2}(kT)^{-3/2} E^{1/2} e^{-E/kT} \end{aligned} \tag{6}$$

The mean energy of a particle is

$$\overline{E} = \tfrac{3}{2}kT \tag{7}$$

and in an isotropic plasma this is divided up equally between the three orthogonal degrees of freedom—the x, y, and z coordinates,

$$\overline{E}_x = \overline{E}_y = \overline{E}_z = \tfrac{1}{2}kT \tag{8}$$

The mean speed of a particle is

$$\bar{v} = \left(\frac{8kT}{\pi m}\right)^{1/2} \quad (9)$$

and the rms particle speed is

$$v_{\text{rms}} = \left(\frac{3kT}{m}\right)^{1/2} \quad (10)$$

In a Maxwellian hydrogen plasma at a temperature of 1 eV, the mean speed of positive ions (i.e., protons) is 1.56 cm/μs, and of electrons 67 cm/μs.

The three kinds of distribution functions referred to above and the parameters that can be derived from them are shown in Figures 2.1–2.3.

A *plasma map* can be drawn with density and temperature as axes, on which

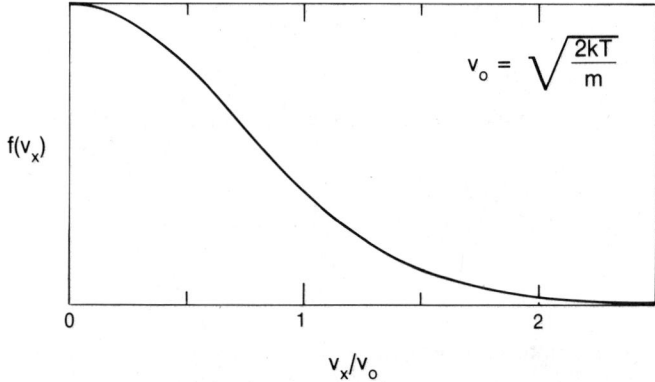

Figure 2.1 One-dimensional Maxwellian velocity distribution, $f(v_x)$.

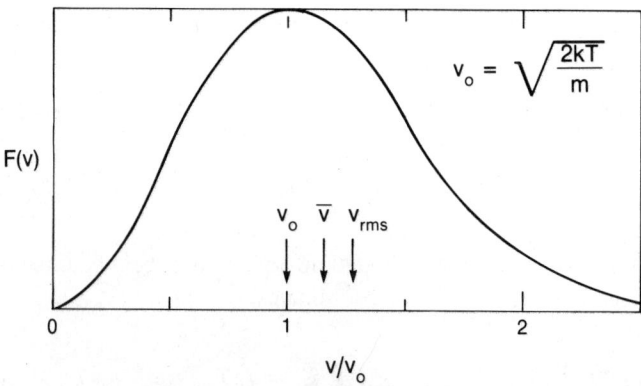

Figure 2.2 Maxwellian distribution of speed, $F(v)$.

2 Sheaths and Electric Fields

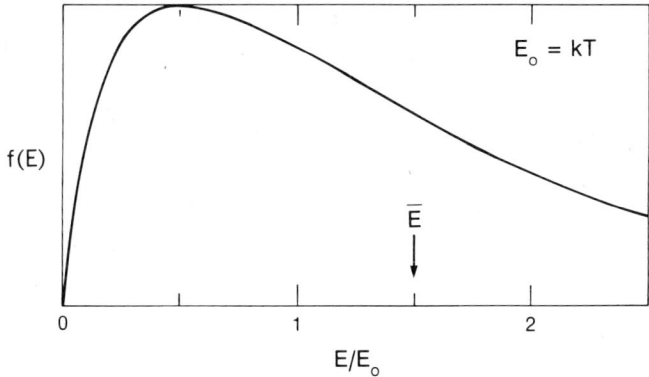

Figure 2.3 Maxwellian energy distribution, $f(E)$.

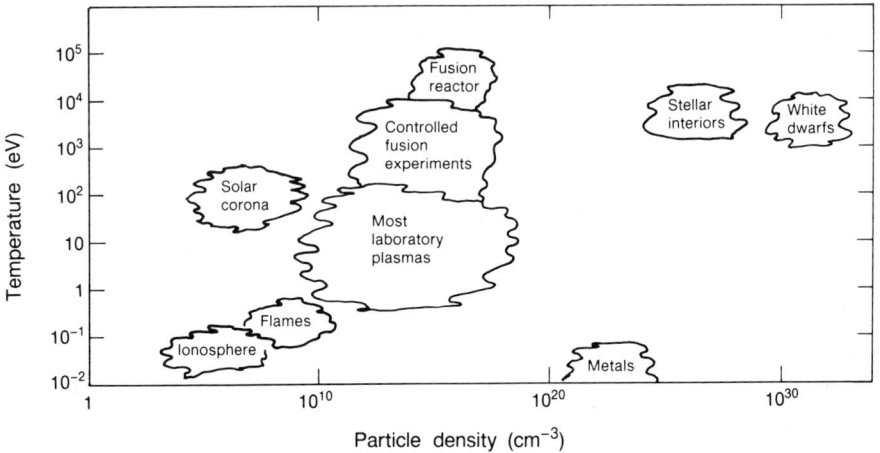

Figure 2.4 n-T plasma map.

we can indicate all the different kinds of plasmas, from ionospheric through "worldly" to white dwarfs, and including ion source plasmas and the plasmas of controlled thermonuclear research. It's an instructive way of looking at the universe of plasmas, and it helps to engender perspective. Such a map is shown in Figure 2.4. In this very general representation the term plasma temperature is used loosely, since the ion temperature is not always equal to the electron temperature for all the many different kinds of plasmas encompassed by Figure 2.4.

2 SHEATHS AND ELECTRIC FIELDS

The plasma particles—the ions and electrons—are charged particles, and therefore interact with one another at a distance via their electric and (if in relative motion) magnetic fields. There is a scale length for the plasma which defines the distance over which the electric field of a test particle extends; field particles closer than this distance know of the test particle, and those beyond it do not. Thus, this is a *screening distance*—the distance it takes for the plasma to effectively screen out, or shield itself from, an applied electric field. The concept also holds for electric fields that are applied to the plasma from an external source: the field penetrates into the plasma only for a short distance. More precisely, the local density of field particles is redistributed so as to cancel the applied field, which then is attenuated exponentially over a distance determined by the plasma parameters.

2.1 Debye Length

The distance within the plasma over which electric fields are effectively excluded is called the *screening* or *shielding distance*, but it is better known as the *Debye length*, after Peter Debye, who first recognized and studied the phenomenon [10]. The Debye length is denoted by the symbol λ_D, and is given by

$$\lambda_D^2 = \frac{\epsilon_0 k T_e}{e^2 n_e} \tag{11}$$

or, in more practical units,

$$\lambda_D = 743 \sqrt{\frac{T_e}{n_e}} \tag{12}$$

where in the latter expression T_e is the plasma electron temperature in electron volts, n_e is the plasma electron density in particles per cm^3, and λ_D is in cm.

This plasma parameter is very basic and makes an appearance throughout virtually the entire field of plasma physics. It is especially important in the consideration of beam formation through the use of a set of biased grids, where a strong electric field is applied to the plasma from outside (the extractor).

That part of the plasma that acts to shield out the externally applied electric field is called the *plasma sheath*. Within the sheath the electric field is not zero and the plasma is not charge neutral. Beyond the sheath, the plasma is considered to be essentially unperturbed by the external field. This is depicted in Figure 2.5.

2.2 Charge Neutrality

The concept of charge neutrality—or, more correctly, quasineutrality—is another fundamental concept that is related to Debye length. Within the plasma there is approximate charge neutrality to a very high degree,

$$e \sum Q n_i - e n_e = 0 \tag{13}$$

where Q is the ion charge state and the sum is taken over all such states. Note that here we consider only positive ions. In the simple case where the only ions are

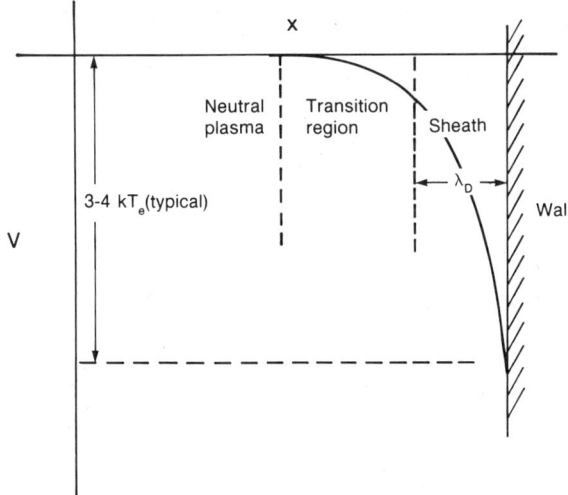

Figure 2.5 Schematic of the plasma sheath region.

singly charged positive ions, then $n_e = n_i$. The plasma cannot depart from the quasineutrality condition, Eq. (13), by very much at all, because very intense electric fields would then occur to reestablish the quasineutrality. This situation prevails throughout the plasma, so long as the dimensions of the sampled plasma volume are large compared to the Debye length; if we consider dimensions small compared to the Debye length, quasineutrality is not required. Random fluctuations of electron density lead to fluctuating electric fields—plasma noise. As implied by Eq. (11), potentials of magnitude up to the electron temperature can exist over lengths up to the Debye length.

3 COLLISIONS

Collisions between the charged particles in a plasma are fundamentally different from collisions between the neutral particles in an ordinary gas. Energy and momentum are exchanged mostly via a large number of distant encounters, rather than by single close encounters. Thus, for example, the velocity vector of a test particle is altered in direction through a random walk process of many minute angular steps. In this situation the concepts of collision time and mean free path, which have a clear intuitive meaning for the case of neutral particle collisions, need a new definition. This is the *relaxation time*. The relaxation time is the time required for collisions to make a major change in the distribution function. Thus, one speaks of the *energy relaxation time*, meaning the time to exchange energy from one component of the plasma to another; for example, if there are hot electrons in a cold ion background then the ions will be heated by electron–ion energy transfer collisions and the ion temperature will rise logarithmically toward a new equilibrium temperature at a rate characterized by the electron–ion energy relaxation time τ_{ei}. Similarly, the angular relaxation time τ_θ is the mean time for a particle of a particular species to be deflected through, say, 90° as a result of many small-angle collisions. Notwithstanding these things, it is usual to speak loosely of the *energy exchange time*, or the *ion–ion collision time*; the imprecision involved usually doesn't matter. Note that the time for energy exchange from electrons to ions is much greater than the electron scattering time.

The mean-free path λ is related to the cross section σ for a given process by

$$\lambda = 1/n\sigma \tag{14}$$

where n is the appropriate particle density. The collision time τ is related to the mean free path and the cross section by

$$\tau = \lambda/v = 1/n\sigma v \tag{15}$$

where v is the appropriate mean particle velocity. The collision frequency is the reciprocal of the collision time,

$$\nu = 1/\tau = n\sigma v \tag{16}$$

In these expressions the concept of cross section is a little artificial, and is not easily reconciled with the small-angle scattering picture described above. However, it is sometimes convenient.

There are a number of different energy exchange times and collision times, corresponding to interactions between both like and unlike particles and in various parameter ranges. Usually these collision times lie in the broad range of nanoseconds to milliseconds.

4 THE PLASMA FREQUENCY

There are a number of natural modes of oscillation of a plasma, the most fundamental and important of which are the electron plasma oscillations and ion plasma oscillations. Any small departures from complete charge neutrality with the plasma give rise to a restoring force and the plasma oscillates at a frequency called the electron plasma frequency. This parameter is given the symbol ω_{pe},

$$\omega_{pe}^2 = \frac{e^2 n_e}{\epsilon_0 m_e} \tag{17}$$

or

$$f_{pe} = 8980\sqrt{n_e} \quad (\text{Hz}) \tag{18}$$

where in the latter expression n_e is in cm^{-3}.

Under some circumstances the ion component can oscillate also (e.g., as ion acoustic waves—sound waves in the plasma), and the natural frequency with which the ions want to oscillate is called the *ion plasma frequency* ω_{pi},

$$\omega_{pi}^2 = \frac{Q^2 e^2 n_i}{\epsilon_0 m_i} \tag{19}$$

where Q is the ion charge state (i.e., ion charge $= eQ$), or

$$f_{pi} = 210 Q \sqrt{\frac{n_i}{A}} \quad (\text{Hz}) \tag{20}$$

where A is the ion mass in amu.

Often the electron plasma frequency is referred to simply as *the plasma frequency*. Usually the electron plasma frequency is in the microwave band, some GHz; the ion plasma frequency is in the rf band, low MHz.

5 MAGNETIC FIELD EFFECTS

A particle with charge q in motion with velocity $v_\perp$ perpendicular to a magnetic field of strength B experiences a force on it that is perpendicular to both the particle velocity and the magnetic field. This force is given by

$$F = Bqv_\perp \tag{21}$$

5.1 Cyclotron Radius

Because the force is transverse both to the velocity and to the field, the particle moves in a circular orbit, as if "tied" to a magnetic field line. The radius of this gyro-orbit is given by

$$\rho = \frac{mv_\perp}{qB} \tag{22}$$

This is called the *cyclotron radius* or the *gyroradius*; a plasma permeated by a magnetic field is sometimes called *gyrotropic*. The mean electron cyclotron radius of the plasma electrons of temperature T_e (eV) is conveniently expressed as

$$\rho_e = 0.0038 \frac{\sqrt{T_e}}{B} \text{ cm} \tag{23}$$

where the magnetic field B is given in kG. Similarly, the mean ion cyclotron radius for ions of charge state Q, mass number A and temperature T_i (eV) can be written as

$$\rho_i = 0.16 \frac{\sqrt{AT_i}}{QB} \text{ cm} \qquad (24)$$

(B in kG). For the same temperature and field strength, the electron cyclotron radius is smaller than the ion cyclotron radius by the square root of the mass ratio. For temperatures of a few electron volts and magnetic fields of a few kilogauss, ρ_e is about 0.1 mm and ρ_i is millimeters to centimeters.

Note that the magnetic field is experienced by the particle motion only via their mutually perpendicular components. Thus, in the expressions above the particle temperatures are referred to the energy components in the direction perpendicular to the magnetic field. The particle velocity parallel to the magnetic field is unaffected, and the resultant particle motion is a helix centered about the field line to which the particle is tied. Because the electrons and ions carry opposite electric charges, the sense of their gyromotion is opposite; electrons rotate in a right-handed sense with respect to the direction of the magnetic field, and ions are left-handed.

5.2 Cyclotron Frequency

The particle cyclotron motion occurs at a very well-defined frequency, called the *ion* or *electron cyclotron frequency*, which is given by

$$\omega_c = \frac{eB}{m} \qquad (25)$$

This can be expressed conveniently as

$$f_{ci} = 1.52 QB/A \text{ MHz} \qquad (26)$$

and

$$f_{ce} = 2.8B \text{ GHz} \qquad (27)$$

where B is again in kilogauss.

For magnetic fields of a few kilogauss the electron cyclotron frequency is a few to a few tens of GHz—the low microwave region of the spectrum—and the ion cyclotron frequency is typically a few hundred kHz to a few MHz.

6 PLASMA PRESSURE

Like any gas, a plasma exerts a pressure. Exactly as per gaseous kinetic theory, the pressure is given by

$$P = nkT \qquad (28)$$

Contributions to the plasma pressure are made by the different plasma components; the electrons provide a pressure $P_e = n_e k T_e$ and the ions a pressure $P_i = n_i k T_i$. When the plasma resides in a magnetic field, an anisotropy is introduced and temperatures may not be the same in different directions. Thus, one speaks of $T_{i\perp}$, $T_{i\|}$, $T_{e\perp}$, and $T_{e\|}$, where the subscripts $\perp$ and $\|$ refer to directions perpendicular to and parallel to the magnetic field, and similarly the plasma pressures $P_\perp$ and $P_\|$ can be distinguished.

7 PARTICLE FLOW

The plasma particles can flow, or stream, in a manner similar to an un-ionized gas. The Maxwellian velocity distribution function of Eq. (4) is symmetric about $v = 0$; that is, the mean velocity is zero. The distribution can be shifted by an amount $v = u$, and now the new distribution is symmetric about u. In this case the distribution is referred to as a *shifted Maxwellian*, and is described by Eq. (4) with v_x replaced by the shifted velocity $v_x - u$. The velocity u is the macroscopic flow velocity of the ensemble of particles, and is also commonly referred to as the drift velocity or the streaming velocity.

Because the plasma particles are charged, a macroscopic flow can give rise to an electric current. In a solid conductor, electric current is a phenomenon due to electron flow alone—the ions constituting the metallic lattice remain stationary. In a plasma, however, the ions are free to move also, and both can contribute quite independently to the net current. The current density j is given by

$$j = j_i - j_e$$
$$= n_e e (u_i - u_e) \qquad (29)$$

when the positive ions are singly charged and thus $n_i = n_e$, where j_i, j_e and u_i, u_e are the ion and electron current densities and drift velocities. The two components add, since under the influence of an external electric field the two velocities u_i and u_e are oppositely directed.

8 IONIZATION

Ionization of neutral atoms to form the plasma state can be accomplished through a number of different processes, including, for example, surface ionization, photoionization, and field ionization. Ion sources that make use of these ionization phenomena are surface ionization sources, some kinds of UV-excited ion sources, and liquid metal ion sources. Another ionization process, more widely used in the technology of plasma preparation, is ionization from the gaseous state by electron impact [1,2,4,5,8]. A gaseous discharge can be brought about under a wide range of experimental conditions, and is the dominant process in virtually all plasma or ion sources in which the feed material is introduced as a gas or vapor. In this case the energy input into the plasma from the external power supplies goes mostly into the electron component of the plasma; the energetic electrons then further ionize the background gas by electron–neutral ionizing collisions.

A fundamental condition that must be satisfied for ionization to occur is that the energy of the electron impact be greater than the energy needed to remove the bound electron,

$$E_e > e\phi_i \qquad (30)$$

where ϕ_i is the first ionization potential of the atomic species involved. This criterion oversimplifies the picture, however:

1. In a plasma there is a distribution of electron energies. A more appropriate parameter is the electron temperature T_e; the mean electron energy is $\frac{3}{2}kT_e$.

2. The probability of ionization of a neutral atom by electron impact varies with electron energy. The cross section in fact rises from zero at an electron energy equal to the ionization potential up to a maximum at an energy equal to about 3–4 times the ionization potential.

Thus, for efficient ionization, the electron density should be high, so that many electron–neutral collisions can occur. Also, the electron temperature should be sufficiently high that collisions result in ionizing events, but not so high that the probability of ionization falls off significantly. Ideally, the electron energy should be several times greater than that corresponding to the ionization potential—a condition that is in reality very rarely met.

These conditions will maximize the ionization of the gas, but to efficiently form a high-density plasma, other conditions must be met also. Thus, for example, it is considerably more efficient to reuse electrons for ionization many times, rather than having to create them afresh from neutral atoms and then heat them starting from low energy each time. It is conventional to refer to the hot (=energetic) electrons that carry out the ionization as the primary electrons, and the cold electrons that are formed along with the positive ions as a consequence of collisions as the secondary electrons. Primary electron confinement can be assisted in a number of ways:

1. The electrons can be caused to reflect backward and forward between negatively biased electrodes while tied to an axial magnetic field. This kind of discharge is called a reflex discharge. The term *pigging* is also used, after the effect that occurs in the PIG (Penning, or Philips, *I*onization *G*auge; see Chapter 8 on PIG ion sources). The effect is simply the oscillation of the electrons in the potential well formed along the magnetic field between the electrodes. Reflexing electron current, or pigging, is an ionization-enhancing effect that is often used in ion sources. It also sometimes occurs as an unwanted cause of electrical breakdown in other plasma or vacuum devices

2. The electrons can be confined in a magnetic mirror, preferably a stabilized magnetic mirror. This is the configuration used in electron cyclotron resonance

(ECR) ion sources, where the magnetic field serves both to provide a resonance (electron cyclotron resonance) for coupling microwave power into the electrons, and also to enhance the confinement of the energetic electrons so formed. This scheme is discussed in Chapter 10, on ECR ion sources.

3. A magnetic multipole field—really a series of mirrorlike field regions—can be used to form a magnetic barrier surrounding the plasma discharge. This configuration is called a *magnetic bucket*, and is described in various places throughout the book.

Another condition that helps in the production of a dense plasma is, reasonably enough, provision of some form of plasma confinement, so that ions formed by electron impact are not lost from the discharge. For the most part, the same configuration that is used to confine the primary electrons, for example, as above, also provides plasma confinement. Thus, the stabilized magnetic mirror and multipole magnetic bucket configurations work well as plasma confinement geometries. Ideally, the only ion loss from the plasma would be via the ion extractor, with no ions lost to the walls or by ion–electron recombination. In practice, this condition can be only quite poorly approached.

9 MULTIPLE IONIZATION

The same kinds of considerations as described above, which are required for efficient ionization of the working gas to provide a dense plasma from which the ion beam is to be extracted, can be extended to cover the case of removal of more than one electron from the positive ion. In this case the ion is said to be *multiply ionized*, *multiply stripped*, or *highly charged*. The usual process (but not the only possible one) whereby multiply stripped ions are produced is that of stepwise ionization by successive electron impact. Clearly, the electrons must now be more energetic (hotter) than needed for singly ionized plasmas, since the ionization energies for successively higher-stripped species become progressively greater [11]; as an extreme example, the threshold energy needed to produce a fully stripped uranium ion, U^{92+}, is about 130 keV [12].

This process, multiple ionization via successive electron impact, is considered in several chapters in this book, most especially in the chapter on electron beam ion sources (Chapter 12). The situation can be summarized as follows. For the case of ionization of a neutral gas into a singly ionized plasma there are two main parameters that determine the process—the electron temperature T_e and the electron density n_e. For the case of the evolution of a multiply stripped plasma there is a third important parameter that enters in—the ion confinement time τ_i. The electrons must be sufficiently energetic to strip to the desired charge state, the electron density must be sufficiently high for there to be many ionizing collisions, and the ions must remain within the ionization region for a sufficiently long time to allow the stripping to proceed. It turns out that the electron density and the ion confinement time enter via their product $n_e \tau_i$. Calculations of the parameters necessary to achieve different charge states from a variety of elements have been carried out by a number of authors [13,14], and involve evaluating expressions of the type

$$n_e \tau_i(Q) = \sum_{k=0}^{Q-1} \frac{1}{\langle \sigma_{k,k+1} v_e \rangle} \quad (31)$$

where $\tau_i(Q)$ is the time needed to produce ions of charge state Q, $\sigma_{k,k+1}$ is the cross section for ionization from charge state k to charge state $k + 1$, v_e is the electron velocity, and the average $\langle \sigma v \rangle$ is taken over the distribution of electron velocities, usually assumed to be Maxwellian. The cross sections can be taken as given by the semiempirical formulae of Lotz [15]. One can alternatively consider the parameter $j\tau$ rather than $n\tau$, where j is the electron current density, and

$$n\tau = j\tau/v_e \quad (32)$$

and it is understood that j is measured as a particle flux, electrons/cm²s.

An example of the predictions of calculations of this type is shown in Figure 2.6, which is a plot of the $j\tau$ necessary to produce various charge states of uranium

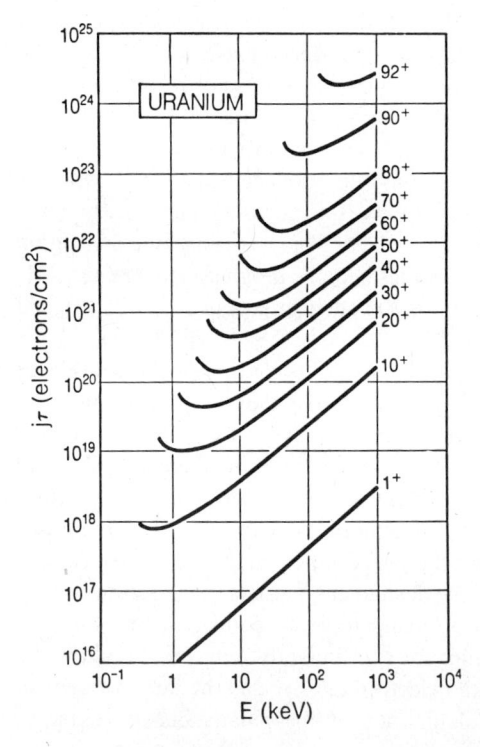

Figure 2.6 $j\tau$ needed to produce various charge states of uranium as a function of electron energy. Courtesy Lawrence Berkeley Laboratory, University of California. Redrawn from data presented in Ref. 14.

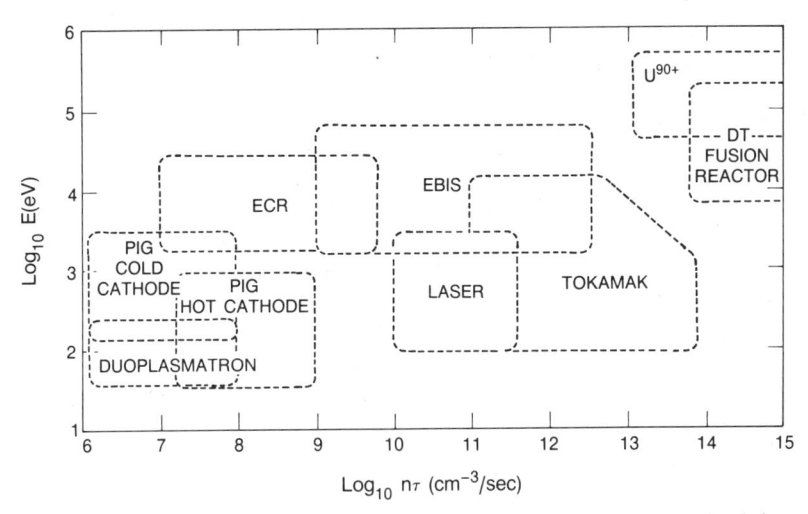

Figure 2.7 E-$n\tau$ space for various kinds of ion sources. Courtesy Lawrence Berkeley Laboratory, University of California.

as a function of the electron energy. It is interesting to note that the performance factor $j\tau$ spans eight orders of magnitude. Further results of this type have been calculated by Donets and are shown in Chapter 12 on electron beam ion sources.

Thus, the parameters $n_e \tau_i$ and T_e (or equivalently E_e, depending on whether one is considering a plasma whose electron energy is described by a temperature or a monoenergetic electron beam) provide a means of comparison of different plasmas and ion sources with respect to their suitability for the production of highly stripped ions. A comparison of this kind is shown in Figure 2.7. Note here that whereas the electron density and electron energy are pertinent to high-charged state ion production, it is the ion density and ion energy that are pertinent to the case of fusion plasmas; thus, the parameters n and E are used somewhat loosely in the figure. Note also that the region boundaries are approximate only, and are being steadily expanded. It is an active and exciting field of research to discover and develop new methods for extending the achievable charge states of multiply charged ions produced in ion sources.

REFERENCES

1. S. C. Haydon (Ed.), *An Introduction to Discharge and Plasma Physics*, University of New England, Armidale, Australia, 1964.
2. S. C. Brown, *Introduction to Electrical Discharge in Gases*, Wiley, New York, 1966.
3. J. L. Shohet, *The Plasma State*, Academic, New York, 1971.
4. E. Nasser, *Fundamentals of Gaseous Ionization and Plasma Electronics*, Wiley, New York, 1971.
5. M. Mitchner and C. H. Kruger, Jr., *Partially Ionized Gases*, Wiley, New York, 1973.
6. F. F. Chen, *Introduction to Plasma Physics*, Plenum, New York, 1974.

7. M. N. Hirsh and H. J. Oskam, *Gaseous Electronics*, Academic, New York, 1978.
8. E. E. Kunhardt and L. H. Luessen (Eds.), *Electrical Breakdown and Discharges in Gases*, Plenum, New York, 1981.
9. J. A. Bittencourt, *Fundamentals of Plasma Physics*, Pergamon, New York, 1986.
10. P. Debye and W. Huckel, *Physikal. Z.* **24,** 183, 305 (1923).
11. See, e.g., *CRC Handbook of Chemistry and Physics*, 57th ed., CRC Press, Cleveland, 1976, p. E-68; or the appendix to this book.
12. T. A. Carlson, C. W. Nestor, Jr., N. Wasserman, and J. D. McDowell, *At. Data* **2,** 63 (1970).
13. H. Winter and B. Wolf, in *Proceedings of the Second Symposium on Ion Sources and Formation of Ion Beams*, Berkeley, CA, October 1974, Lawrence Berkeley Laboratory, Report LBL-3399.
14. Laboratoire National Saturne, Report SFEC T 10—Cryebis II, Center for Nuclear Studies, Saclay, France, 1981.
15. W. Lotz, *Z. Phys.* **216,** 241 (1968).

3

Ion Extraction

R. Keller*
GSI, Gesellschaft für Schwerionenforschung
Darmstadt, Federal Republic of Germany

Understanding of the basic rules governing the process of ion extraction has been extended considerably in the last two decades, mostly due to developments in the area of high-current ion sources where high brightness and high beam current values are of vital importance for many different applications. Ideally, the extraction system can be examined independently of the ion generator, and the only requirement imposed on the latter is that it must provide ions of the needed current density over the area to be exploited by the extraction system. Many ion source types do not in general fulfil this condition, thereby setting quite stringent conditions for the layout of the extraction system. For this reason, in the present chapter extraction systems for high-current ion sources will be treated in general, and only for a few cases will other special extraction systems be discussed.

For most of the chapter, ion extraction through one round aperture will be treated, with less but still finite emphasis on slits or multiple apertures. Other implicit premises applying unless otherwise stated are that the ions are positive, generated within a quasineutral plasma, and extracted from a discharge region with no magnetic field of any regard. Good examples for plasma generators fulfilling these premises are the multicusp sources, as presented in Chapter 7, on gaseous high-current sources.

Apart from the design and performance characteristics of various kinds of extraction systems, some more general aspects of extraction are also discussed in this chapter, and in discussing them a scaling model for beam current and brightness as a function of extraction voltage will be derived.

1 THE PHYSICS OF ION EXTRACTION

The extraction process basically consists of applying a high voltage between an ion reservoir and a perforated acceleration electrode. The trajectories of the accel-

---

Current affiliation: Lawrence Berkeley Laboratory, University of California, Berkeley, California.

erated ions, which immediately determine the maximum beam quality, are influenced by several factors, such as the applied field strength, the shape of the emitting surface which may be solid (field and surface ionization sources) or flexible (plasma sources), and also on the space-charge density of the resulting beam itself. In the case of plasma sources, the emitting surface is commonly termed the *meniscus*. Its detailed shape again depends on the electrical field distribution due to the applied boundary conditions and the local densities of plasma ions, electrons, and accelerated ions. With H^- sources, electrons are accelerated together with the ion beams, leading to even more complicated conditions than found with positive ions.

For sources with a solid emitting surface, there is an important difference between surface ionization sources where the emitter can be shaped at will, following analytical considerations or computer simulations, and field emission sources where the tip of the emitting needle or the tiny plasma cone in front of it do not permit any reasonable shaping from the viewpoint of ideal beam formation.

With plasma sources the meniscus acts as the boundary layer between the discharge plasma and the accelerated beam particles. The depth and position of this layer, with respect to the surrounding electrodes, depends on the densities of plasma electrons and ions and their mobilities which can be expressed in terms of temperatures. Such temperatures are not frequently measured in ion source plasmas, but there are some data available, and from ion optical considerations one can deduce that the ion temperature for some sources might typically be around 0.2 eV [1], whereas a typical electron energy would be a few eV [2]. A one-dimensional theory of the meniscus as a "classical sheath" was developed by Self [3], and is incorporated into several simulation codes that permit the treatment of a variable meniscus shape (see the chapter on simulation codes).

From a practical standpoint, a detailed knowledge of the behavior of the classical sheath is not required, as long as the density of the discharge plasma can be varied over a broad range and only first-order effects are considered. The meniscus (see Figure 3.1) will be approximately spherical with its center of curvature inside the plasma when the plasma density is relatively high, about plane under intermediate conditions, and again nearly spherical but with the center of curvature outside the plasma when the plasma density is still lower or, correspondingly, the extraction field higher.

Within the extraction gap, the created beam is usually far from being neutral because any electrons that could compensate the positive space charge of the beam particles are accelerated (backward) by the extraction field to much higher velocities than the ions; thereby their own space charge is severely diluted. Only in some cases, for example, with the beam-plasma source [4] (see the dedicated chapter in this book), have attempts been undertaken to create enough electrons and inject them into the extraction gap, to preserve space-charge neutralization. After an intense beam has passed the extraction electrode, however, its space charge has to be compensated; otherwise severe blowup of the beam envelope would occur. An argon beam of a few mA current at 30 keV energy already suffers from space-charge effects [5]. In many cases some of the residual gas particles that are present within the beam line will be ionized by impinging beam ions, and this process can generate a sufficient number of compensating electrons (see the chapter

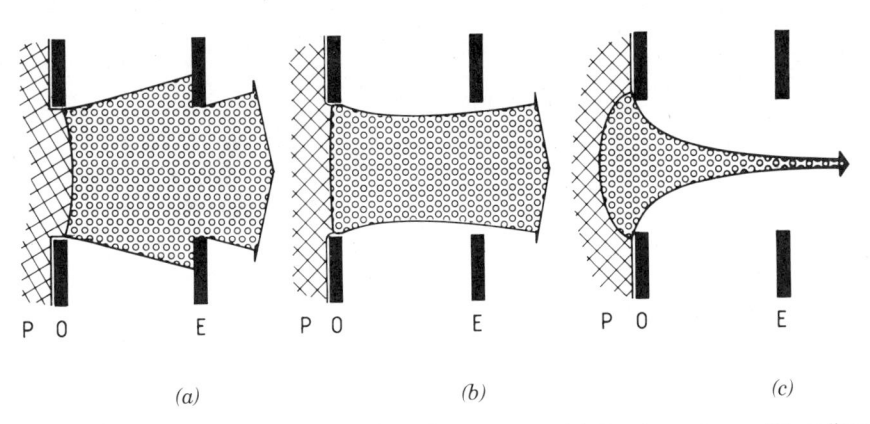

Figure 3.1 Three cases of ion extraction from plasma sources. (a) Overdense plasma, (b) medium density plasma, (c) underdense plasma, always in comparison to the applied extraction field strength. P, plasma; O, outlet electrode; E, extraction electrode. Plasma boundary (meniscus) and beam envelopes were drawn according to simulation runs by the author, using the code AXCEL-GSI [Ref. 20].

on beam transport); but the electrons must be kept from being accelerated back into the source by the extraction field. This shielding can be achieved by means of a so-called accel/decel system where a screening, or suppressor, electrode is introduced into the main extraction gap and biased to a sufficiently low potential so as to create a negative potential well and form an electron trap. An example of such an accel/decel system is shown in Figure 3.2.

So far, only those aspects of ion extraction have been discussed that determine the beam divergence. But the beam quality also depends on the absolute current value and the size of the emitting area, that is, on the current density. The maximum current density that can possibly be expected for any charged particle species accelerated by an electric field is obtained under space-charge limited conditions and follows the Child-Langmuir law [6, 7],

$$j = \tfrac{4}{9}\epsilon_0 (2q/m_i)^{1/2} U^{3/2}/d^2 \tag{1}$$

where j is the current density, q is the ion charge ($q = \zeta e$, $e =$ electron charge), m_i is the ion mass, U is the applied voltage, and d is the extraction gap width. This can be put into the more useful form,

$$j = 1.72 (\zeta/A)^{1/2} U^{3/2}/d^2 \tag{2}$$

where now j is in mA/cm^2, ζ is the ion charge state, A is the ion mass (amu), U is the extraction voltage in kV, and d is the gap width in cm.

These equations hold true under space-charge limited conditions, that is, when the plasma generator is producing more ions than are removed in any given time. Another condition is much harder to satisfy: The emitting surface must be exactly

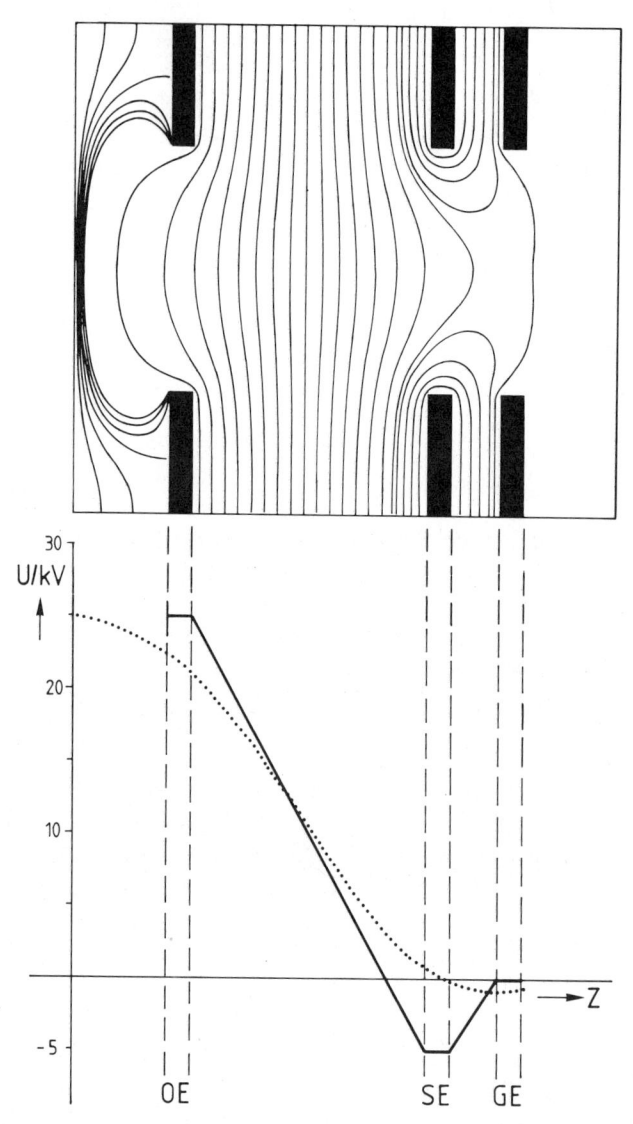

Figure 3.2 *Top*: Example of an unsophisticated accel/decel extraction system with equipotentials shown for pure vacuum conditions, without plasma or beam. The voltage differences between negative equipotential lines, near the screening electrode, are less than those between positive equipotentials. *Bottom*: Schematic drawing of the potential distribution U along the system axis z, without beam, far from the axis (solid line) and with beam, on axis (broken line). OE, outlet electrode; SE, screening electrode; GE, ground electrode. Calculation by the author using the simulation code AXCEL-GSI [Ref. 20].

plane. Equation (1) gives rise to two definitions: the perveance P of an ion gun,

$$P = I \times U^{-3/2} \times (A/\zeta)^{1/2} \tag{3}$$

and the normalized, or proton equivalent, current I_n,

$$I_n = I \times (A/\zeta)^{1/2} \tag{4}$$

where I is the ion current, U is the extraction voltage (exactly: the voltage that determines the beam energy), A is the atomic mass number, and ζ the ion charge state.

Definition (3) does not require the units of the involved quantities to be defined. To emphasize the distinction from absolute currents, the units for normalized currents, usually (mA), will be put in round brackets throughout this chapter. The use of these two quantities allows comparisons of extraction systems working under completely different conditions. With the quantity perveance, however, caution is recommended because by employing multiaperture systems one can easily obtain huge beam currents and correspondingly high perveance values for the system in total. To judge the ion optical performance of extraction systems, one would better compare perveance values of single apertures only, not of the entire ion gun.

2 FIRST-ORDER TREATMENT OF ION EXTRACTION

The problem of ion extraction would be solved if one could create an absolutely parallel beam. This task was attemptd by Pierce [8] for electrons extracted by either round holes or straight slits. Pierce derived electrode contours that should produce the required parallel beams, see Figure 3.3, but due to several reasons extraction systems with these contours do not precisely yield the intended result, especially in the case of ions being extracted from a plasma. The reasons are as follows: The numerical results for the potential distributions outside the beam are only approximations to the exact solutions; the outlet electrode shaping with a 67.5° angle against the beam axis is a good approximation only if the plasma surface has the same potential as this electrode, which is more often not true in reality; all electrodes should be infinitely thin, at least in the immediate vicinity of the beam, but this cannot be fulfilled, due to mechanical and power load reasons; the ion emission is required to be uniform over the entire emitting area, but a plasma is always less dense near solid electrodes; and, finally, the equipotential surfaces in the zone close to the extraction aperture (the opening in the second electrode) are curved outward and not plane as required for an ideal Pierce geometry (this is termed the *electrode-* or *aperture-lens effect*).

Even if all premises for Pierce's theory were exactly fulfilled there still remains the finite ion temperature as one cause for a residual beam divergence. A quanti-

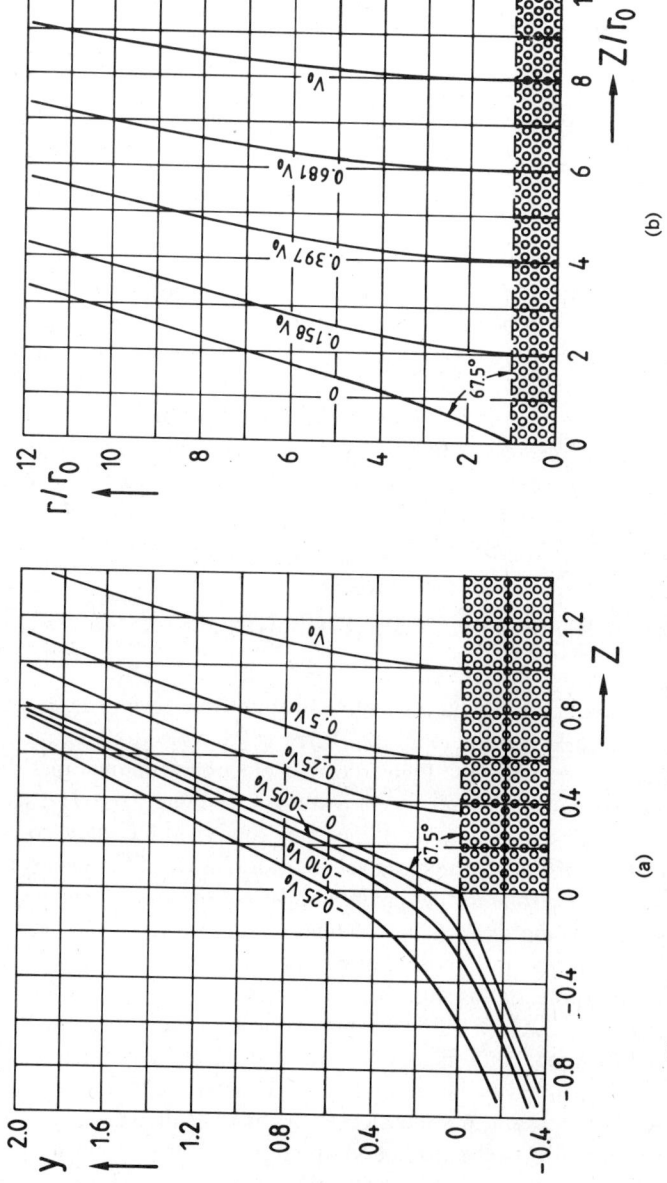

Figure 3.3 Pierce contours for ribbon (*a*) and cylindrical (*b*) beams, after Ref. 8. The extraction electrodes must coincide with the equipotentials at zero (outlet electrode) and full (extraction electrode) potential. V_0 must have a negative value in this illustration if positive ions are extracted. Pierce does not treat accel/decel systems.

28

tative expression for this ultimate lower divergence limit can be derived from Eq. (13) or (14), below, dividing a given emittance value by the known aperture width. An example of a multielectrode "exact" Pierce system is described in Ref. 9. "Exact" here means that the theoretical electrode contours are approximated mechanically as closely as possible.

In view of the electrode–lens effect it is better to make an initially convergent beam, to counteract the divergence that the beam would attain after flowing in a perfectly parallel pattern between outlet and extraction aperture. However, the perveance of a convergent beam is lower than the maximum perveance of a system with plane electrodes as given by the Child-Langmuir limit, Eq. (1). Taking into account a quantitative expression for the lens effect [10], one can derive that an exactly parallel beam should be created at a perveance value lower by a factor of 0.47 than the Child-Langmuir limit [11].

An example of a "quasi-Pierce" extraction system that has actually been used with a high-current ion source is shown in Figure 3.4. The detailed contours of the electrodes were determined by a computer code that is able to automatically optimize such shapes [12].

It should be clear that no real extraction system can deliver an absolutely parallel beam with zero divergence because the electrode lens effect represents only one of the violations of Pierce's premises. Even if this effect could be totally suppressed, aberrations and finite ion temperature would still be present. Before individual features in the performance of extraction systems can be discussed in detail, however, definitions of some beam quality parameters are given in the following section.

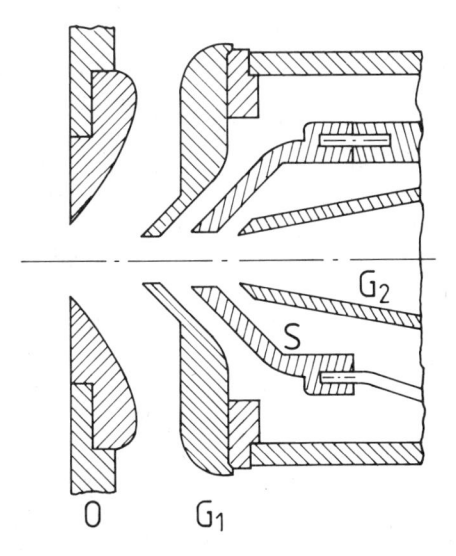

Figure 3.4 Quasi-Pierce extraction system with round apertures [Ref. 12]. O, outlet electrode; G_1, first ground electrode; S, screening electrode; G_2, second ground electrode. For reliability of operation the screening electrode is enclosed between two ground electrodes, resulting in a tetrode structure. Such a tetrode must be well distinguished from a two-gap tetrode, see below. Note that the curvatures of the outlet and first ground electrodes are convex toward each other. © 1983 IEEE.

3 BEAM QUALITY PARAMETERS

The definitions of perveance P, beam current I, and normalized beam current I_n have already been given in Section 1, see Eqs. (3) and (4). If not mentioned otherwise, here the perveance will always be related to a single emitting aperture, not to an entire multiaperture extraction system.

The definitions of these electrical parameters given above are generally accepted, but with the following definitions related to optical beam qualities extreme care is needed when comparing numerical values quoted from different publications. These qualities are: the beam radius r, and r_0 at the beam waist; the divergence half-angle α, or α_0 for trajectories crossing the beam axis at the plane under investigation; the emittance ϵ; and the brightness B. Even without yet going into the detailed definitions of such quantities, one should make clear whether the absolute limit values are considered for a given beam or 'reasonable' border values, cutting off the weaker beam halo. In this case the threshold beam density value should be specified, where the cutoff occurs. Or, lastly, one can also use statistical mean (rms = root-mean-square) values of the pertinent quantities. This is common practice in the particle accelerator community where single, scalar values for the mentioned quantities are essential to get insight into quite complicated transport phenomena.

Another approach consists of examining only those parts of the beam that are actually transported in the beam line, without taking into account the remaining halo of the beam. This way is similar to the one where a density threshold is imposed, but here the cutoff criterion is a purely geometrical condition, not an electrical one, and this criterion could be either a divergence limit, or a maximum radius, or an acceptance. The acceptance of a beam line is the largest emittance a beam can have that still passes through the line without any losses.

By strict definition the beam emittance is related to the pattern that the beam particles occupy within the six-dimensional phase space. In many practical cases the three coordinate pairs within the entire phase space are completely decoupled, and the longitudinal projection of the actual pattern does not have any meaning for quasistationary beams; thus, the two remaining transverse projections only are of importance in the present context. Assuming further that transverse motions are slow compared to the velocity in the beam direction, and nonrelativistic conditions are fulfilled throughout, one can substitute the transverse linear momenta $m\,dx/dt$ and $m\,dy/dt$ by the tangent values $x' = dx/dz$ and $y' = dy/dz$ of the divergence angles for all individual trajectories. Thus, the commonly used two-dimensional emittance definitions regard the patterns that the trajectories independently occupy in the (x, x') and (y, y') planes.

In the standard case that will be treated here, one round emitting aperture will be assumed; then the cylindrical symmetry of the problem allows one radial section only, that is, on the (r, r') half-plane. The difference between either (x, x') and (y, y') emittance patterns or (r, r') patterns is that in the former two cases all trajectories of a beam are projected into the plane considered while an (r, r') pattern represents a radial half-section through the beam only. The 100%-border lines of both patterns are identical (apart from the trivial consideration that a true (r, r') pattern exists only for positive r values), but a huge difference arises from the fact that the projections into the (x, x') and (y, y') planes artificially accumulate trajectory points near the origin, thereby converting a uniform density distribution into an elliptical one. A radial section leaves the density distribution unchanged, of course. For the more realistic truncated Gaussian distributions, this geometry effect leads to underestimating the density of the beam in its outer zones, and if the threshold of the emittance measurement is set to values believed absolutely reasonable (a few percent of the peak density at the center) then a serious misjudgment of the true size of the beam emittance will occur for (x, x') and (y, y') measurements. A thorough discussion of various emittance definitions and measurement techniques is given in Ref. 13.

As a gross classification, a beam is divergent if its emittance pattern mostly extends from the third into the first quadrant of the coordinate plane, convergent if its major extension runs from the second into the fourth one, roughly parallel if it is extended along the positional coordinate $(x, y,$ or $r)$, and it has a focus if the emittance pattern runs along the angle coordinate $(x', y',$ or $r')$. But for real beams the emittance pattern always has a finite width, and therefore there are always some divergent trajectories in an overall convergent beam, and so on; see Figure 3.5.

To quantify the size of an emittance two different conventions are in use; One can either directly take the area occupied by the emittance pattern and express its size, for example, in units of [mm mrad] or of [m rad] (the latter unit is larger by a factor of 10^6), or simply in [m] because [rad] is dimensionless. But in many cases the other convention is followed, that is, to define the size of an emittance by the value of the area of its pattern, divided by π. The reason for this second convention is that quite frequently the emittance patterns are of elliptical shape, and then one can directly deduce the extension of the second semiaxis of this ellipse when the emittance size and the extension of one semiaxis are numerically known. The trouble arises because in the majority of publications it is not explicitly mentioned which one of these conventions is being adhered to. Therefore, a third convention is being used more and more and will be adopted throughout this present chapter: One takes the actual size as size of the emittance but writes the π as a distinct factor to it, practically giving it the meaning of a measuring unit. In these cases the understanding is that the product π times the numerical size value gives

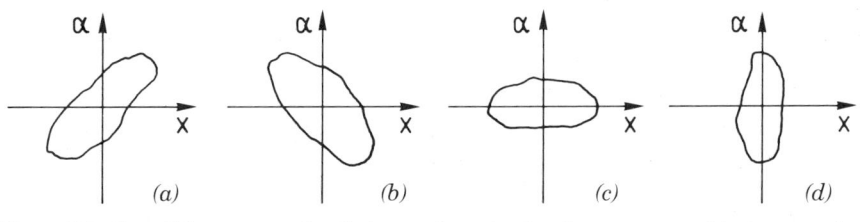

Figure 3.5 Four different cases of realistic two-dimensional emittance patterns. The corresponding beams are roughly divergent (a), convergent (b), parallel (c), and focused (d), but there are always some trajectories deviating from the gross behavior of the beams.

the area size, and still the advantages of the second convention just mentioned can be used. The units to measure emittances according to this third convention are then, for example, [π mm mrad].

The confusion brought about by these three conventions is increased further when statistical (rms) emittances are being treated. They are defined as the second moments of the distributions that represent the beam in each one of the two transverse, two-dimensional subspaces of the phase space, or commonly their equivalents in the angle/position planes, as explained above. But two definitions for the quantitative expression of rms emittances are commonly used, and they differ by a factor of 4 for the same pattern:

$$\epsilon_{rms} = \left(\overline{x^2 x'^2} - \overline{(xx')^2}\right)^{1/2} \tag{5}$$

according to Ref. 14 and

$$\epsilon_{4\,rms} = 4\left(\overline{x^2 x'^2} - \overline{(xx')^2}\right)^{1/2} \tag{6}$$

according to Ref. 15.

Definition (6) has the advantage that for a Kapchinskij-Vladimirskij distribution [16] the so-defined 4 rms-emittance exactly covers the real emittance pattern. The distinction between rms and 4 rms emittances is very helpful in avoiding some of the many possible misunderstandings regarding quantitative emittance values, but it is clearly not yet generally accepted.

Statistical emittances, no matter according to which definition, have the advantage that they are expressed by one scalar number. By similar formulae to (5) and (6) the other parameters of the corresponding ellipses can also be derived [16]. The disadvantage associated with these statistical computations is that the actual beam fraction contained within the ellipses is not generally known and depends on the distribution of beam particles over the plane under investigation.

In a different approach [17], an algorithm has been derived by which truly encompassing ellipses of minimum size are calculated for arbitrary distribution types. After identifying and eliminating those particles that give rise to a certain ellipse size, the distribution is gradually reduced, and smaller ellipses are drawn around the remaining beam fractions. In this way, the contained beam fractions are always exactly known, but the mathematical effort is considerably higher than in the case of rms emittances.

Plotting beam fraction against emittance size usually allows a distinction between a beam halo, where the density of the occupied area is low, and the beam core with higher density, where a further reduction of the remaining beam fraction no longer leads to still higher particle density; see Figure 3.6.

The concept of emittance offers yet another source for confusion that needs to be clarified to avoid any misunderstanding when reading published emittance values. The emittance of a beam will shrink if the beam is further accelerated, because for given transverse velocities the longitudinal velocity has increased. To

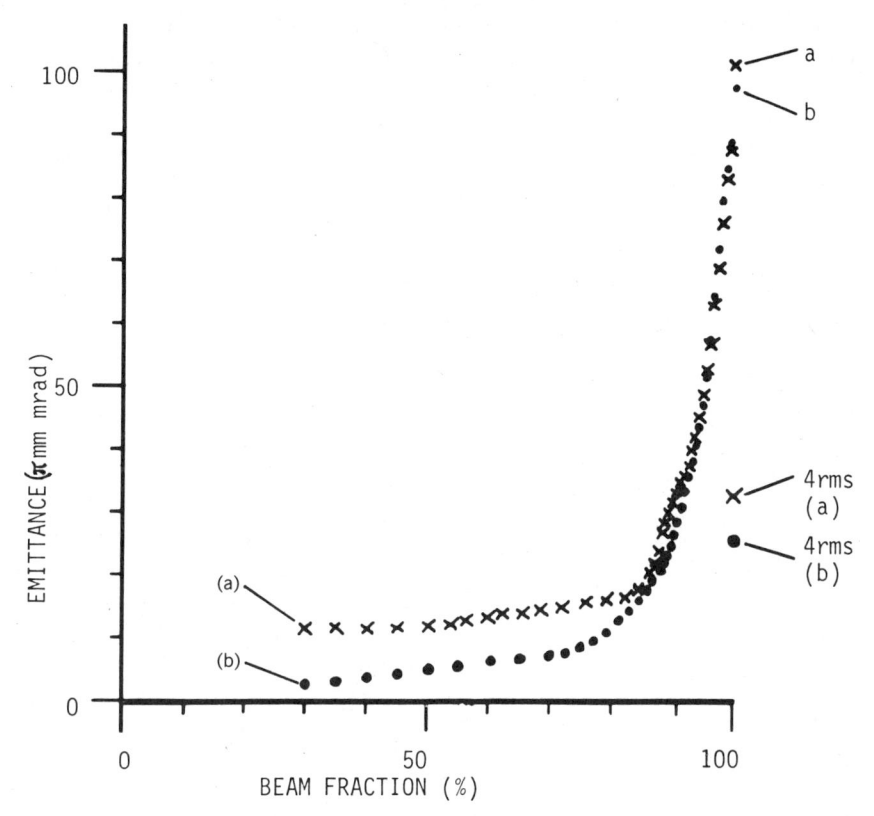

Figure 3.6 Plot of emittance size versus beam fraction for simulated particle distributions representing beams created by two similar quasi-Pierce extraction systems [17]. Note that the beam cores are quite different in emittance size whereas the total sizes of the two distributions are nearly equal. The 4 rms values for the entire distributions are marked by two data points on the 100% abscissa. © 1985 IEEE.

get rid of this effect, "normalized" emittances are defined as follows:

$$\epsilon_n = \beta\gamma\epsilon \tag{7}$$

where ϵ_n is the normalized emittance; $\beta = v/c$, ratio of particle velocity to the velocity of light in vacuum; and $\gamma = (1 - \beta^2)^{-1/2}$.

The relativistic parameter β can easily be calculated from the known beam parameters according to

$$\beta = 1.46 \times 10^{-3}(\zeta U/A)^{1/2} \tag{8}$$

where ζ is the charge state of the ions, U is the acceleration voltage of the beam measured in kV, and A is the atomic mass number of the ions.

The reason for introducing the definition (7) lies in the fact that the normalized emittance of a beam is constant, according to Liouville's theorem, as long as only conservative forces are acting and the two planes of observation are truly decoupled. One should be careful in making use of this theorem because it strictly applies to the actual emittance pattern, not to encompassing convex geometrical shapes like ellipses by which the so-called *effective emittances* are defined. For the sake of clarity, one should speak of *absolute* emittances whenever one wants to emphasize the distinction against normalized ones.

Hardly any other physical quantity gives rise to as much misunderstanding as does the brightness of an ion beam. In principle, this quantity simply relates beam current and emittance in the appropriate way, that is, via the product of the two transverse emittance values (or the square of the radial emittance in the case of cylindrical symmetry):

$$B = I/(\epsilon_x \epsilon_y) \qquad (9)$$

but very rarely indeed does an author disclose which of the many existing definitions for the emittance is being used, and even the beam current I is open to different interpretations, depending on whether the total extracted current is meant or only the transported part of it (here termed I_{tr}) or, finally, only the denser core of a transported beam. Further, it is common practice to use the term *particle current* (electrical beam current divided by the particle charge) whenever the number of transported particles is more important than the electrical quantity, such as in nuclear physics experiments. And for the sake of a smooth normalization, a second definition for beam brightness is widely used [18], reading:

$$B = 2I/(\pi^2 \epsilon_x \epsilon_y) \qquad (10)$$

Brightness values according to (10) are smaller by a factor of $\pi^2/2 \cong 5$ than those calculated according to (9) with identical beam parameters. It should be clear that in any brightness determination the current as well as the emittance values must be taken for identical beam fractions or for an entire beam, in one well-defined cross-sectional plane along the beam path.

In the following, brightness values will be used following the former definition (9) only. But for the sake of completeness two more conventions are introduced that will greatly facilitate the comparison of beam brightness values for different experimental conditions. Whenever the normalized emittances are taken for the calculation the resulting brightness is called *emittance-normalized*, with the definition

$$B_{en} = I/(\epsilon_{n,x} \epsilon_{n,y}) \qquad (11)$$

The Child-Langmuir equation (1) suggests the comparison of brightness values for different ion species by using the normalized currents I_n, according to definition

(4), rather than the absolute ones, but the absolute emittances. Therefore, such a brightness is defined as

$$B_{cn} = I_n/(\epsilon_x \epsilon_y) \qquad (12)$$

and is called *current-normalized brightness*.

4 REFINED TREATMENT OF ION EXTRACTION

The definitions and explanations presented in the preceding section may appear somewhat cumbersome, but they are essential for developing a full understanding of the detailed effects that determine the formation of ion beams. As a practical example for the following discussion, a high-brightness extraction system is shown in Figure 3.7, and simulated emittance patterns of the beams to be obtained from this system are shown in Figure 3.8.

Even though the trajectories shown in Figure 3.7 appear to be nearly laminar, the emittance patterns reveal the actual quality limits of the generated beams: Pattern 3.8a is strongly S-shaped, with pronounced curved tails that indicate the aberrations of the extraction system. This means that in the plane of observation the beam core is divergent while its halo is less divergent or even convergent. The total normalized beam current of 283 (mA) can be transported within an acceptance of 55π mm mrad. It is difficult to decide for this case where the beam halo (that in configuration space corresponds to the emittance tails) should be cut off to enhance the beam brightness, all the more because every trajectory (marked by a single spot on the diagram) represents an annular zone, with proportionally more current carried by the outer zones that originate at larger radii. Note that in this current-optimized case the beam is compressed at its waist to slightly less than half the width of the outlet aperture.

Pattern 3.8b exhibits a straighter shape and fewer trajectories within the aberration tails. It was obtained after greatly reducing the current density, as compared to case *a*, and widening the aperture of the screening and ground electrodes as indicated in Figure 3.7. Now a sharp distinction between core and halo can be seen, and the acceptance necessary to transport the core alone amounts to 7.5π mm mrad only. The price for this improvement is a large reduction of beam current: Within the core only 100 (mA) normalized current is contained. But due to the gain in emittance and the fact that the emittance determines the brightness by its square in our present case [see the text above Definition (9)], the brightness of this reduced beam is greater by a factor of 19 than that of the entire beam in case *a*.

This example, more than theoretical explanations, illustrates how carefully one should judge the performance of given extraction systems and their quoted brightness values. Even after ascertaining which units of measurement and definitions are being used, the development goal for the system is an important piece of information, too.

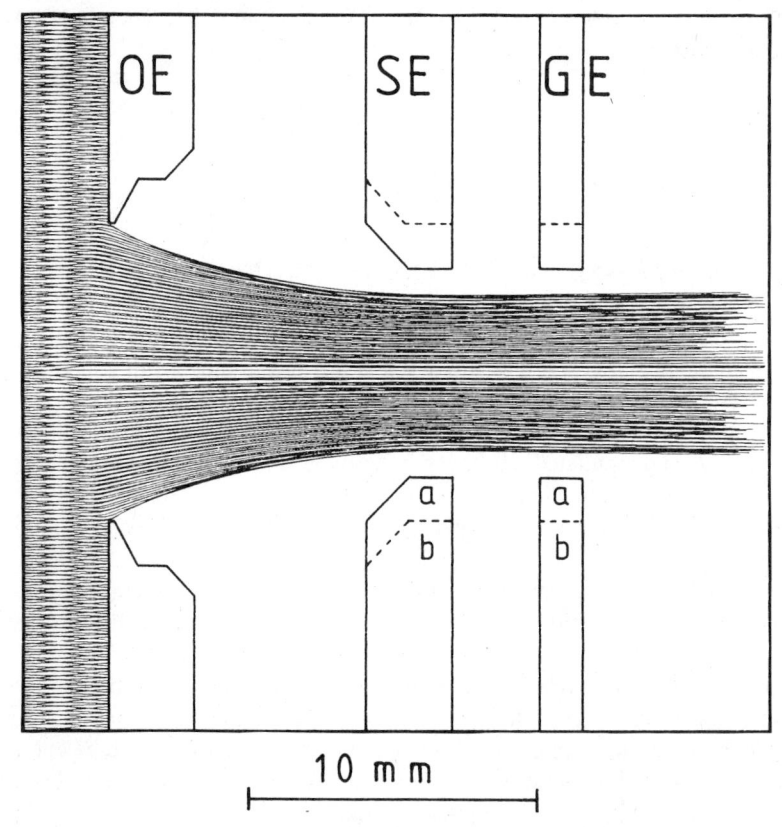

Figure 3.7 Round single-aperture extraction system [19], designed with the aid of the simulation code AXCEL-GSI [20]. The broken lines indicate a geometry variation that leads to higher beam brightness, at the expense of lower beam current, see Fig. 3.8. OE, outlet electrode, 50 kV; SE, screening electrode, −4 kV; GE, ground electrode, 0 kV. Reprinted with permission by Pergamon Journals, Ltd.

Looking at the electrode shapes of the system presented in Figure 3.7, the complicated contour of the outlet aperture is most striking. It represents a compromise between best performance and ease of fabrication, particularly in view of building multiaperture systems. Similar systems with different contours have been published in Ref. 21 and 22 for circular apertures and in Ref. 23 for a slit geometry (see Figure 3.16, Section 6).

The detailed design of high-performance extraction systems is now always based on simulation runs by computer codes; see the special chapter dedicated to this topic. The analytical guide lines of Sections 1 and 2 of the present chapter are useful as a rough check on the numerical results of such calculations, without being able, however, to furnish any detailed solutions. As we just pointed out, the goals of optimizations may be quite different: highest current, highest brightness, least intensity in the beam halo, just to mention a few. And correspondingly the solutions

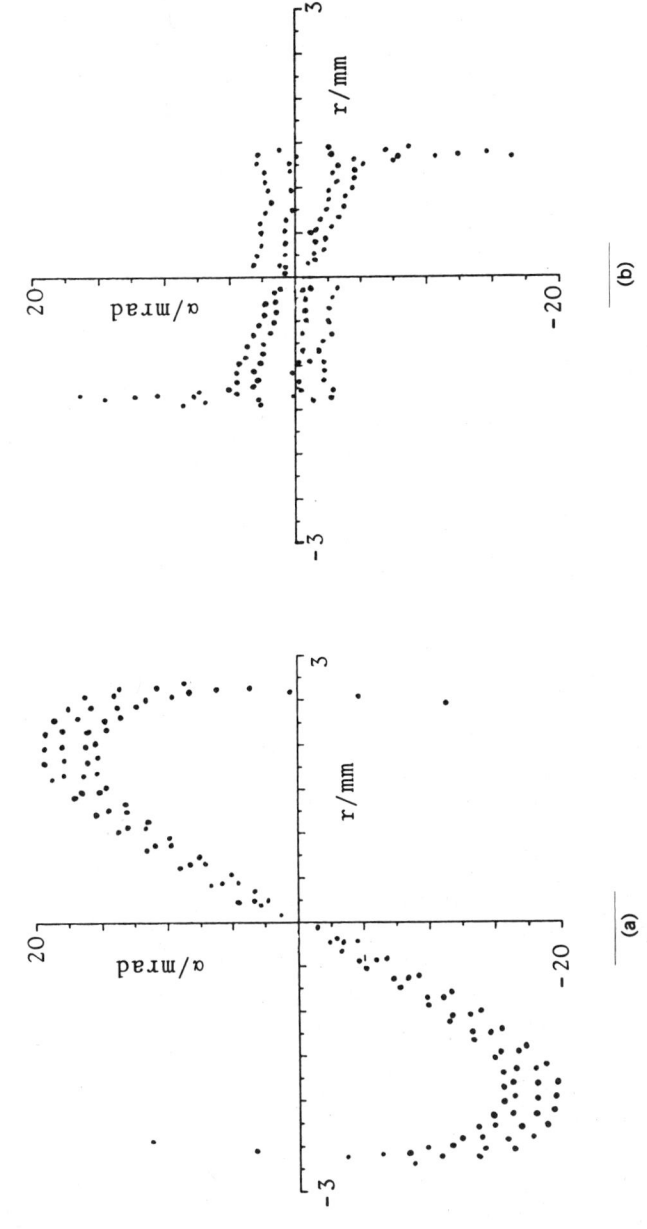

Figure 3.8 Emittance patterns of the beams generated by the extraction system shown in Fig. 3.7, calculated for a plane located 2 mm downstream of the ground electrode by the code AXCEL-GSI [20]. (a) System with contours shown by solid lines, with ion current density chosen so as to yield a maximum transported ion current within 20-mrad divergence half-angle, amounting to 283 (mA) normalized current. (b) System with changed contours for screening and ground electrodes, as shown by the broken lines. The ion current density in this case was chosen so as to yield a maximum brightness value for the beam core, totally neglecting the tails of the emittance pattern. The beam core then contains 100 (mA) normalized current, with 5 mrad divergence half-angle. After Ref. 19, reprinted with permission by Pergamon Journals, Ltd.

will differ substantially. Thus, there is still room for improvement in this field. What a user of a simulation code must try to learn through systematic case studies is to identify which part of a system affects a certain portion of the resulting emittance pattern. With this knowledge it will become easy enough to improve a given system up to the possible limit under the imposed boundary conditions.

Any optimization procedure can at best lead to exactly straight emittance patterns, suppressing all possible causes of aberrations. But even then the resulting emittances are finite, because due to the actual ion temperature there is always some amount of transverse momentum left for some trajectories. The resulting minimum emittance size depends on the type of velocity distribution of the ions, and cannot be expressed in general. For a Maxwellian energy distribution, the dependence of the normalized 4 rms emittance sizes from the ion temperature has been derived in Ref. 24 for circular apertures as well as for slits and, with the fundamental constants filled in, reads for a circular aperture

$$\epsilon_{n,4\,\text{rms}} = 0.0653 r \left(kT/A\right)^{1/2} \quad [\pi \text{ mm mrad}] \qquad (13)$$

where r is the aperture radius (mm), A is the mass number of the ions, and kT is the ion temperature (eV), and for a slit aperture

$$\epsilon_{n,4\,\text{rms}} = 0.0377 s \left(kT/A\right)^{1/2} \quad [\pi \text{ mm mrad}] \qquad (14)$$

where s is the slit width (mm), A is the atomic mass number of the ions, and kT is the ion temperature (eV). This formula holds for the direction transverse to the slit axis. These 4 rms emittances include about 89% of the total beam for circular apertures and 93% for slits.

As an example, the finite divergence of the beam core in Figure 3.8b is caused by the ion temperature of 0.1 eV assumed for the calculation. Application of Eq. (13) with $A = 1$ leads to a 4 rms emittance size of 0.10π mm mrad, while the value 7.5π mm mrad quoted above is transformed into 0.0774π mm mrad after normalization with $\beta = 0.01032$, corresponding to 50 keV beam energy. The resulting disagreement of about 25% between these two totally independent methods does not appear to be too serious, all the more so because the simulated pattern has a rectangular shape while the rms formula applies to elliptical shapes. Either method clearly indicates the practical emittance limit that cannot be further reduced by varying the mechanical extraction system contours.

Understanding the emittance concept, one can again return to the general considerations made in Section 2. Assuming that truly aberration-free ion optical lenses exist, one does not have to create a parallel beam as Pierce attempted, but rather a laminar one. Convergent and divergent beams can have emittance values as low as those of parallel beams, with just different ellipse orientations and eccentricities. Therefore, one may as well allow the beam to expand immediately after leaving the emitting meniscus, which in this case is curved in the downstream direction [25]. The benefit of this method is twofold: The current density that causes this meniscus curvature is higher than one that leads to a concave meniscus, allowing for higher ion current densities under otherwise comparable conditions; and second, the beam is not compressed in the first extraction gap where its full space charge is acting. This helps to avoid any nonlinear effects that would lead to stronger aberrations. The question is, however, is the resulting emittance pattern straight enough or too distorted? Even with a uniformly curved meniscus, the equipotential surfaces near the screening electrode can cause undesired effects. Only detailed emittance measurements or simulations can give an answer for any particular system. Another problem may arise due to the fact that divergent beams need larger lens apertures and usually lead to stronger aberrations, just shifting the problem from the extraction to the transport section. The precise choice of the orientation of the starting emittance pattern therefore ultimately depends on the specific beam quality requirements and external factors such as availability of required beam transport elements.

5 VOLTAGE SCALING LAWS FOR BEAM CURRENT AND BRIGHTNESS

The availability of simulation codes allows the solution of virtually any extraction system problem, but it is still interesting to derive some analytical scaling rules for cases where no sharp boundary conditions are imposed on the system to be designed, and high brightness with reasonably high beam current are being looked for. To limit this task, an empirical case study is presented in the following that applies to round, single-aperture systems [26]. In this study, rather than a beam line acceptance, only an acceptance half-angle, amounting to 20 mrad, is arbitrarily assumed as one stringent experimental condition.

In formal accordance with the Child-Langmuir law (1), it has often been experimentally verified that the transported fractions of the extracted ion beam exhibits the dependencies

$$I_{tr} \propto (A/\zeta)^{-1/2} \tag{15}$$

and

$$I_{tr} \propto U^{3/2} \tag{16}$$

even though Eq. (1) is not related to any acceptance limit.

As a demonstration of the validity of Eq. (16), data obtained with an extraction system similar to the one shown in Figure 3.7, with contours represented by the broken lines, are plotted in Figure 3.9. Relation (15) is implicitly contained in the following discussion, in that we consider normalized transported currents $I_{tr,n}$ from here on; compare Eq. (4) above.

After combining the relations (15) and (16) into a scaling formula similar to Eq. (1), but with a different proportionality factor because now transported currents are under discussion, one would choose the largest extraction system aperture area

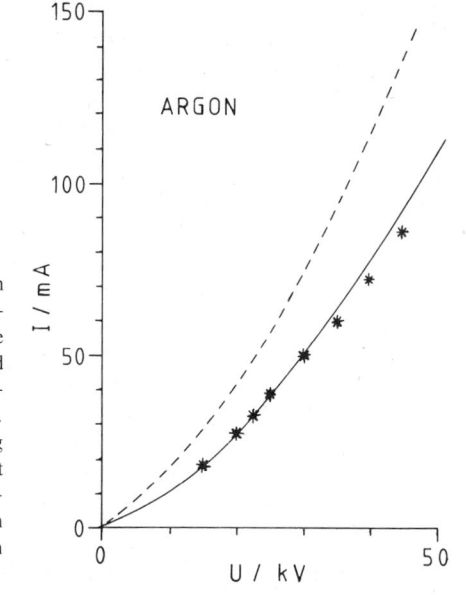

Figure 3.9 Dependence of transported ion beam currents I_{tr} on the extraction voltage U, for a 7-aperture, 7-mm-diameter (each hole) triode extraction system. The continuous curve is fitted to current values obtained at up to 30 kV extraction voltage, assuming a $U^{3/2}$ dependence. Broken curve: empirical current limit according to the formula derived in the text below. The limit is not reached here because the system was optimized for brightness and not for maximum current. From Ref. 19, reprinted with permission by Pergamon Journals, Ltd.

that can be uniformly filled with plasma of the matching density by the generator, because the resulting beam current increases with this area. Unfortunately, the proportionality between area and transported beam current holds only for apertures that are small compared to the extraction gap width, whereas for very large apertures a saturation current value will be reached. In a thorough experimental study using a duopigatron ion source [11] the following formula was derived:

$$I_{tr,n} = P^* U^{3/2} S^2 / (1 + aS^2) \qquad (17)$$

where $S = r/d$, aspect ratio of the extraction gap, r is the outlet aperture radius, and d is the gap width. The two factors a (aberration factor) and P^* (low-S perveance) depend upon the divergence half-angle α accepted by the transport system, see Figure 3.10a and also upon the chosen extraction electrode geometry. The quoted study states $a \cong 3$. For a more refined, two-gap extraction system [27], see Figure 3.11, the values $a = 1.7$ and $P^* = 6 \times 10^{-5}$ [mA/V$^{3/2}$] give the best fit to the transported current values measured within the acceptance half-angle of 20 mrad, see Figure 3.10b.

The existence of a saturation limit for the transportable currents with increasing aspect ratio S strongly suggests maintaining $S \cong 0.5$ when designing extraction systems, because the spilled beam extracted from larger apertures would cause quite unfavorable effects whereas the actual gain in useful current is very low. Only sophisticated systems like the one in Ref. 22, or two-gap systems as shown

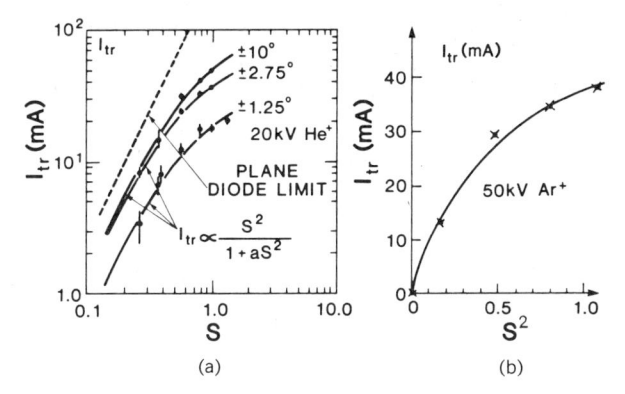

5 Voltage Scaling Laws for Beam Current and Brightness   41

Figure 3.10 Transported ion currents I_{tr} as a function of the aspect ratio S (a) or its square value (b). (a) Triode after Ref. 11, log–log plot. © 1973 AIP. (b) Pentode after Ref. 27, linear plot. © 1983 Springer-Verlag.

in Figure 3.11, permit the application of values $S \cong 1$ without losing too much in functional reliability.

Assuming now that the quoted values of P^* and a, found for the two-gap system [27] represent the actual performance limit of any single-aperture extraction system

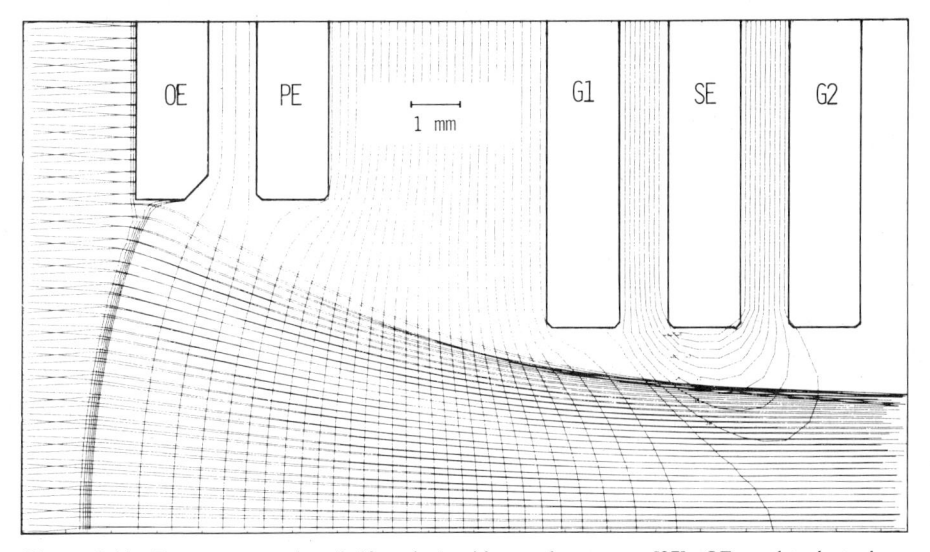

Figure 3.11 Two-gap extraction (half-section) with round apertures [27]. OE, outlet electrode, 50 kV; PE, puller electrode, 45 kV; G1, first ground electrode; SE, screening electrode, −5 kV; G2, second ground electrode. The screening electrode is here enclosed between two ground electrodes to improve the system reliability, according to Ref. 12. Trajectory plot by AXCEL-GSI [20]. © 1983 Springer-Verlag.

and putting $S = 1$, one can conclude that the maximum normalized current that can be obtained from a single round aperture and transported within a 20-mrad divergence half-angle is a function of the extraction voltage alone and, in convenient units, amounts to

$$I_{tr,n} = 0.703 U^{3/2} \quad [(\text{mA})/\text{kV}^{3/2}] \tag{18}$$

A graphical representation of Eq. (18) is given in Figure 3.12. Many published beam data, reduced to one emitting outlet aperture, support this current-limit formula in the sense that no current value significantly exceeds the line in Figure 3.12. This fact a posteriori justifies the assumption that Eq. (18) indeed describes the actual performance limit for plasma ion sources and their extraction systems. The data in Figure 3.12 were taken from Ref. 1, 4, 11, 12, 19, 22, and 27–42.

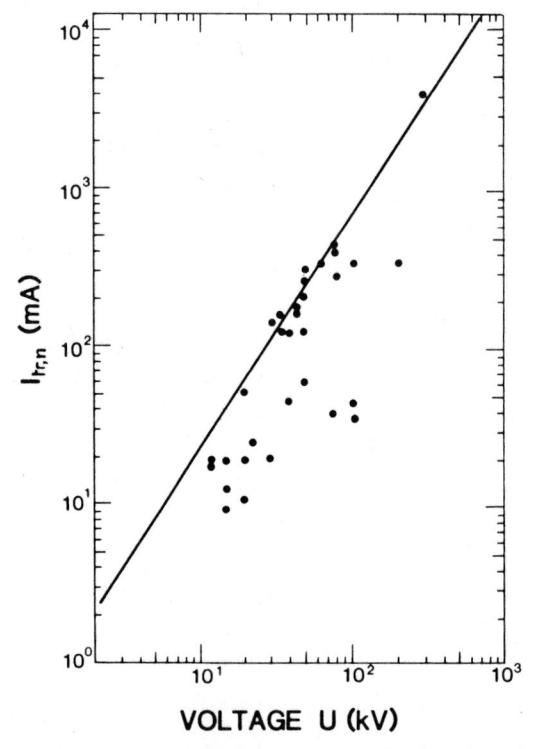

Figure 3.12 Voltage dependence of the normalized, transported current $I_{tr,n}$ according to Eq. (18). The points mark the $I_{tr,n}$ values of different actual sources, calculated for one emitting aperture only. Current values that lie significantly below the marked limit mostly belong to neutral injection experiments where other conditions than the maximum transported current are dominant. From Ref. 26, © 1986 AIP.

5 Voltage Scaling Laws for Beam Current and Brightness

It should be kept in mind that Eq. (18) was derived under the assumption that the outlet aperture diameter is enlarged proportionally as the extraction gap is increased, according to the condition $S = $ const. This is the reason the mechanical dimensions of the extraction gap do not explicitly appear in the formula. They are very important, however, as soon, as the effective absolute emittance of the beam is to be minimized. In the plane where the extracted beam has a waist, this emittance can easily be determined if the pattern can be encompassed by an elliptical border line:

$$\epsilon = \alpha_0 r_0 \quad [\pi] \tag{19}$$

where α_0 is the maximum divergence half-angle of the trajectories crossing the beam symmetry axis in the plane of observation, and r_0 is the beam waist radius. Assuming as before that $\alpha_0 = 20$ mrad, and also that the beam waist is half as wide as the outlet aperture, which is well enough justified by an inspection of simulated trajectory plots (see Figure 3.7 and 3.11 or plots in Refs. 19, 21, and 22), one can immediately show that

$$\epsilon = 20 \times 0.5r = 10d \quad [\pi \text{ mrad}] \tag{20}$$

where r is the outlet aperture radius and d is the mechanical extraction gap width (since $S = r/d = 1$ here).

The question of minimizing the emittance is then reduced to minimizing the gap width d, and its limit depends upon the high-voltage breakdown law for the given conditions. Collected data of many existing ion sources (see list of references below Eq. (18)), show that an implicit formula derived by Kilpatrick [43] for the dc case indeed marks the actual breakdown limits through two decades of voltages (see Figure 3.13), and that a reasonably conservative empirical limit fitted to the displayed data is given by the formula

$$d = 0.01414 U^{3/2} \quad [\text{mm}/\text{kV}^{3/2}] \tag{21}$$

permitting, for example, 50 kV to be applied to a 5-mm gap.

With the practical limit of Eq. (21), the minimum absolute emittance, Eq. (20), can now be written in the form

$$\epsilon = 0.1414 U^{3/2} \quad [\pi \text{ mm mrad kV}^{-3/2}] \tag{22}$$

The minimum normalized emittances then scale according to

$$\epsilon_n = 2.06 \times 10^{-4} (A/\zeta)^{-1/2} U^2 \quad [\pi \text{ mm mrad}/\text{kV}^2] \tag{23}$$

The conclusion to be drawn for the dependence of the brightness values on the extraction voltages looks quite unfamiliar: the emittance-normalized brightness,

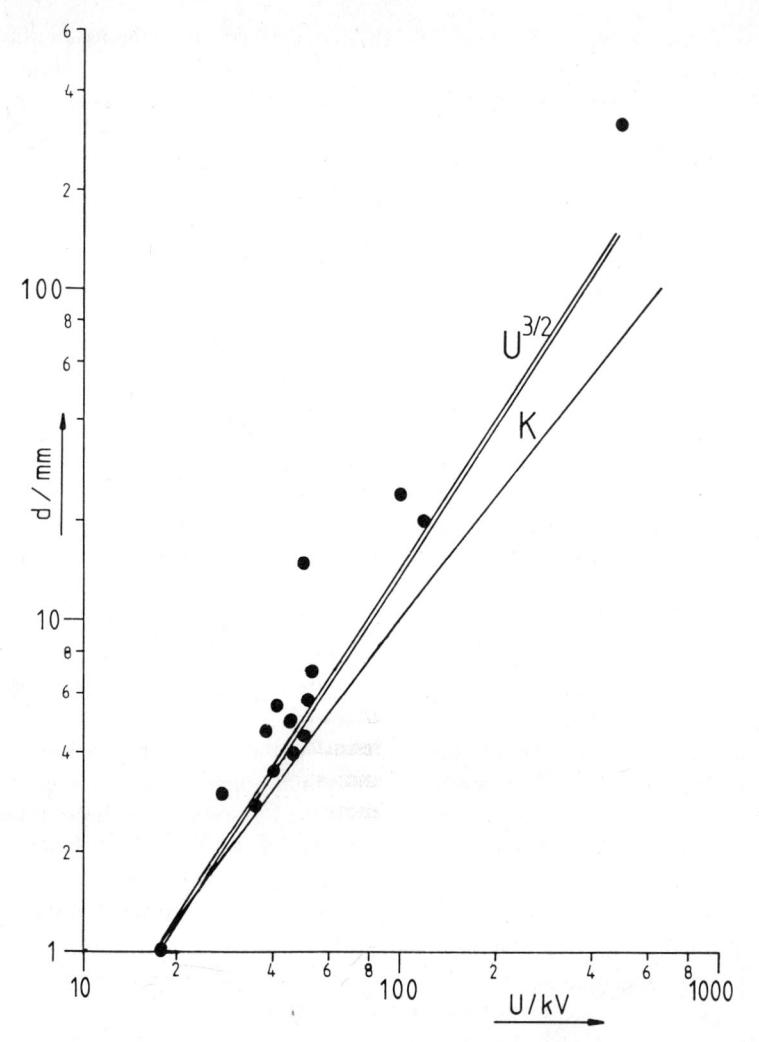

Figure 3.13 Breakdown limits for extraction gaps. K, Kilpatrick law [43]; $U^{3/2}$, empirical limit. After Ref. 42, reprinted with permission by Pergamon Journals, Ltd.

according to Eq. (11), as well as the current-normalized brightness, according to Eq. (12), both strongly decrease with increasing extraction voltage:

$$B_{en} = 1.65 \times 10^7 \, (A/\zeta)^{1/2} U^{-5/2} \quad [\text{mA kV}^{5/2}/(\pi \text{ mm mrad}^2)] \quad (24)$$

and

$$B_{cn} = 35.3 U^{-3/2} \quad [(\text{mA}) \, \text{kV}^{3/2}/(\pi \text{ mm mrad})^2] \quad (25)$$

Being derived from a single case study under well-specified conditions, this brightness scheme is in no way meant to show absolute physical constants in the numerical expressions given. Much brighter beams than described by the last formulae can be produced, for example, by sacrificing beam current and using only the core of the beam, as was explained at the beginning of the preceding section, or by allowing for the minimum divergence angle determined by the thermal ion motion, according to Eq. (13), only. But the tendencies expressed by Eqs. (18), (24), and (25) seem to be well confirmed by the available experimental data, and at least indicate the voltage scaling rules for current and brightness, starting from a well-known set of extraction parameters.

Assuming the presented scheme to be valid, maybe with different numerical factors for different optimization goals, one can divide Eq. (18) by (22) and finally arrive at a new invariant quantity that might be termed *emittance-specific current* [44]:

$$I_{tr,n}/\epsilon = 5 \ [(mA)/(\pi \text{ mm mrad})] \tag{26}$$

This ultimate expression appears to be a well-founded, constant limit for high-brightness ion beams generated from a high-current source. It does not directly depend on other quantities, but is still subject to the assumptions outlined within the present section.

6 SPECIAL EXTRACTION SYSTEMS

The current limitation expressed by Eq. (18) cannot be accepted in many cases, especially with neutral beam injection experiments requiring very high currents at relatively low energy. This problem can easily be solved by using multiaperture extraction systems with many holes in every electrode. As long as the generator yields a plasma of sufficiently large cross section to offer a constant current density to all outlet parameters, the delivered ion current will simply scale proportionally to the number of these apertures. The emittance of the entire beam, however, will suffer overproportionally from this introduction of more emitting apertures, because now the distances between the holes contribute to the effective beam waist size, too. This effect leads to a brightness loss of about a factor of 10 when changing from a single-aperture to seven aperture of equal total size, as illustrated in Figure 3.14 and 3.15.

For current requirements not much greater than the stated limit (18), it is worthwhile to examine if two, three, or four apertures are sufficient instead of seven, because the emittance increase would then be somewhat lower. For more than seven apertures, arranged in a hexagonal pattern, no further brightness loss is to be expected. Systems with more than 1000 apertures [35] have been successfully operated, yielding hydrogen ion currents in the 10- to 100-A range. The main problem associated with increasing the number of extraction apertures is the efficient cooling of the central part of the grid structures.

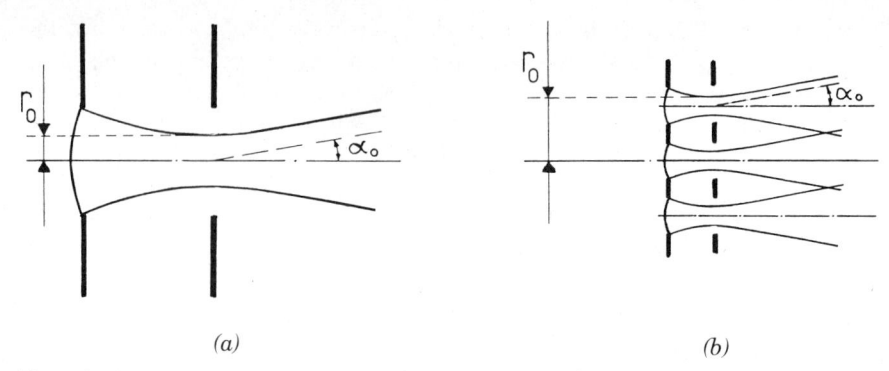

Figure 3.14 Comparison between single-aperture (*a*) and multiaperture (*b*) extraction. Both systems have the same emitting area, assuming seven apertures in case *b*. The increase of the effective beam waist radius r_0 leads to a brightness loss of a factor of 6.6 for case *b*. After Ref. 27, © 1983 Springer-Verlag.

One way to avoid the brightness loss of multiaperture systems is to make use of a single extraction slit, instead. With a slit, the two transverse emittance patterns are quite different from each other, but at least there are no spaces in between emitting areas along the slit direction. Thus, the intensity gain obtained by increasing the outlet area does not lead to any brightness loss. There are two major disadvantages associated with slits, however. First, in terms of ion optics, the slit

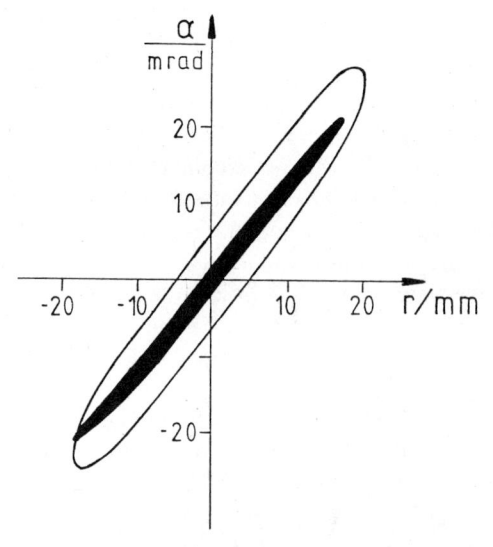

Figure 3.15 Comparison of the measured emittance patterns of two drifting argon beams, taken 1 m distant from the source. Full black, 26-mA beam obtained from a single-aperture system with B_{en} = 12.5 A/(π mm mrad)^2. Solid line, effective emittance contour for a 33-mA beam from a seven-aperture system, with B_{en} = 1.1 A/(π mm mrad)^2. After Ref. 27, © 1983 Springer-Verlag.

6 Special Extraction Systems

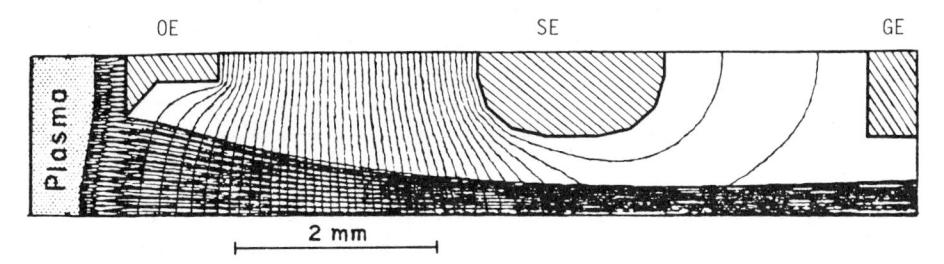

Figure 3.16 Slit extraction system (triode), shown as half-section. OE, outlet electrode, 20 kV; SE, screening electrode, −1.2 kV; GE, ground electrode. The system is designed to yield a proton beam of 650 mA/cm^2 current density in the outlet plane with an rms divergence half-angle of 23.8 mrad. After Ref. 23.

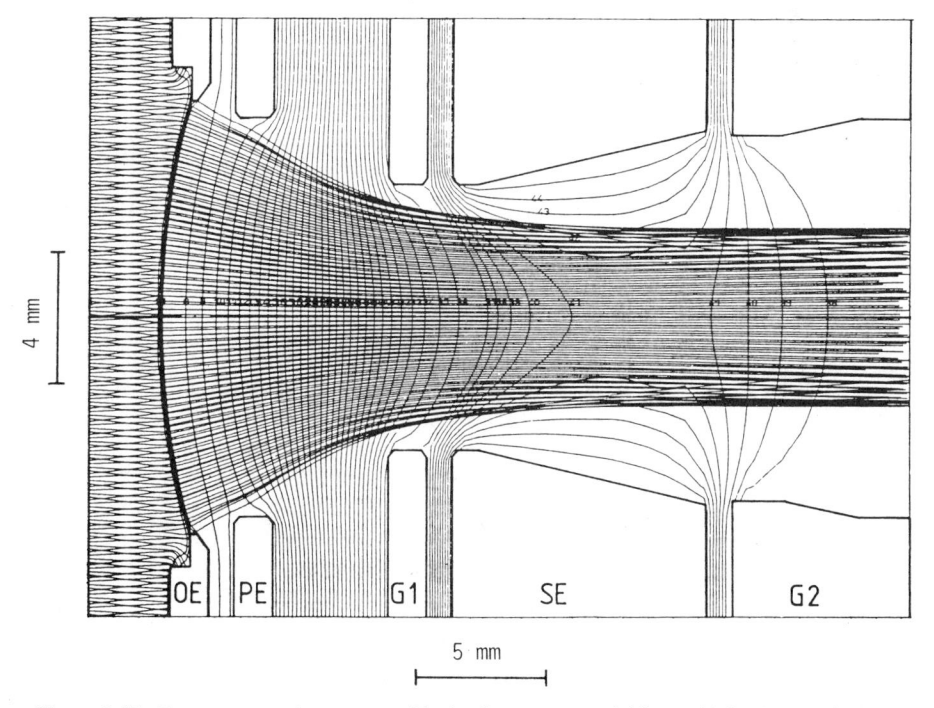

Figure 3.17 Two-gap extraction system with circular apertures, yielding a 41.5-mA argon beam at 50 keV energy. From Ref. 42, reprinted with permission by Pergamon Journals, Ltd. Note that the radial and axial distances are drawn in slightly different scales. OE, outlet electrode, 50 kV; PE, puller electrode, 45 kV; G1 and G2, first and second ground electrodes; SE, screening electrode, −4 kV, enclosed between two ground electrodes to improve the system reliability, following Ref. 12. Tapering of the screening electrode to reduce the backflow of secondary particles into the source was described in Ref. 48. Reprinted with permission by Pergamon Journals, Ltd.

48   *Ion Extraction*

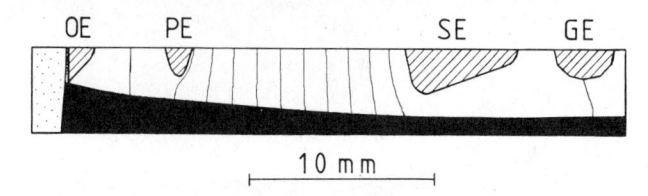

Figure 3.18 Two-gap slit extraction system (half-section), yielding a deuterium ion beam of 250 mA/cm^2 current density (for a realistic mixture of atomic and molecular ions) at the outlet plane within a divergence half-angle of 9.3 mrad. OE, outlet electrode, 120 kV; PE, puller electrode, 101 kV; SE, screening electrode, −2.3 kV; GE, ground electrode. After Ref. 45.

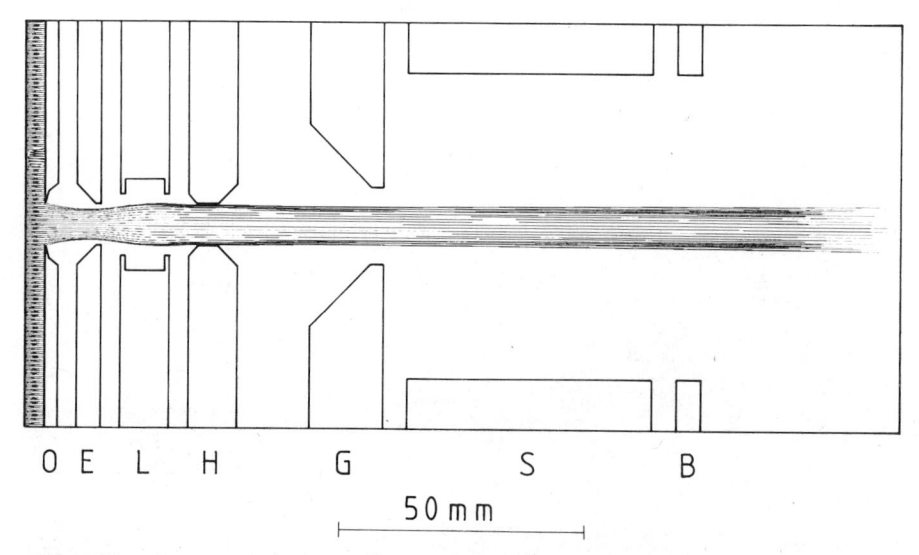

Figure 3.19 Compound extraction system after Ref. 47, as applied to the high-current source MEVVA [Ref. 49], described in a separate chapter of this book. O, outlet electrode, 159 kV; E, extractor electrode, 125 kV; L, einzel lens, 154 kV; H, high-potential gap electrode, 125 kV; G, grounded gap electrode, 0 kV; S, screening electrode, −1.5 kV; B, beam potential electrode, 0 kV. Courtesy U.S. Department of Energy.

Figure 3.20 Emittance pattern and density profile of a 15-mA, 159-keV uranium beam extracted from MEVVA [49] by the compound system, shown in Figure 3.19. The intensity in the dotted wings of the emittance pattern is less than 1% of the total intensity. The emittance was measured using a radial pepper pot technique. The core of the emittance pattern as determined by a photometrical density measurement of the entire beam (profile below) is marked in full black. The entire beam, without the emittance tails, has an emittance-normalized brightness of $B_{en} = 21$ A/(π mm mrad)^2; the value for the core alone is $B_{en} = 78$ A/(π mm mrad)^2 for 2.8 mA current as calculated after integrating the beam profile. Previously, with the same source and a conventional seven-aperture extraction system, 40 mA beam current with a brightness $B_{en} = 1.4$ A/(π mm mrad)^2 had been obtained. After Ref. 47, courtesy of U.S. Department of Energy.

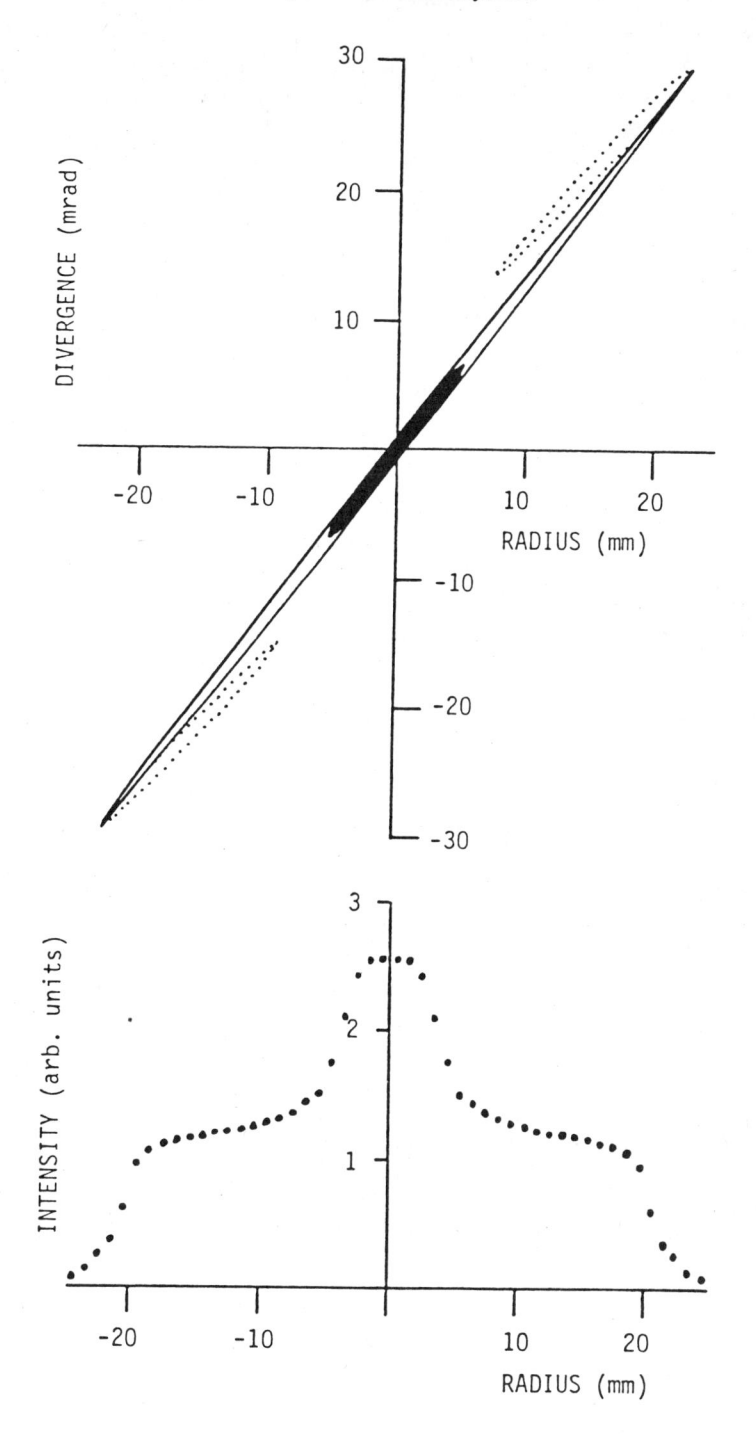

6 *Special Extraction Systems* 49

end zones are rather similar to two round apertures and would require a different value for the matching plasma density than needed by the main part of the slits. The beam parts stemming from these zones are therefore lost for any further transport and may even cause difficulties by striking the screening electrode. Thus, a slit should be at least ten or more times longer than wide, to reduce the consequences due to this effect. Second, it is much more difficult to fabricate and, during operation, to maintain perfectly aligned straight linear patterns like the slit edges, than circular contours. The mechanical accuracy required for the extraction electrodes is about a few thousandths of the aperture width because the extraction gap represents an ion optical lens of very short focal length, typically some mm, to be compared to the order of 1 m for an average transport length. But these problems can be solved with not too advanced technology, and the emittance anisotropy may even be desired, for example, for mass separation purposes.

An example of a high-current slit extraction system is shown in Fig. 3.16. With multislit systems the mentioned, overproportional emittance increase occurs in the transverse direction only, and therefore the total brightness is reduced by just a factor of about 3, instead of about 10.

Another way to overcome the brightness limits of Eqs. (24) and (25) is to use multigap structures. By dividing the extraction gap into sections one can make use of the nonlinear voltage scaling of the breakdown law of Ref. 43, or the approximation of Eq. (21), thus obtaining a shorter system than would be possible with a single gap structure [1]. In a practical case study [27] (see Figure 3.17), it turned out that 50 keV beam energy is about the lower limit for which such a design begins to have advantages over a single-gap arrangement. The action of the new "puller" electrode introduced into the main extraction gap is to correct the imperfections of the equipotential shapes, rather than to increase the extraction field strength to some greater extent. A two-gap slit extraction system (tetrode), according to Ref. 45, is shown in Figure 3.18.

Going a step further, one can finally incorporate an entire electrostatic einzel lens within the main extraction gap, thereby decoupling the optical action of the system from the acceleration process [46]. An example of such a *compound system* [47] is shown in Figure 3.19; the measured emittance pattern of a beam generated with this system is shown in Figure 3.20.

A compound system allows focusing the beam into the entrance plane of an RF accelerator with perfectly matched emittance orientation, no matter what the actually extracted ion current amounts to. Its length is short enough to avoid too high a rate of electron generation by ion impact ionization of the residual gas, an effect normally precluding the application of the electrostatic focusing principle to long-pulse or dc high-current ion beams.

REFERENCES

1. A. J. T. Holmes and M. Inman, Proceedings, 1979 Linear Accelerator Conf., BNL-51134, Brookhaven Nat. Lab., 1979, p. 424.
2. K. W. Ehlers and K. N. Leung, Report LBL-9107, Lawrence Berkeley Laboratory, Berkeley, CA, 1979.

References

3. S. A. Self, *Phys. Fluids* **6**, 1762 (1963).
4. J. Ishikawa, F. Sano, and T. Takagi, *J. Appl. Phys.* **53**, 6018 (1982).
5. R. G. Wilson and G. R. Brewer, *Ion Beams with Application to Ion Implantation*, Krieger, Malabar, 1973, p. 140.
6. C. D. Child, *Phys. Rev.* (Ser. 1) **32**, 492 (1911).
7. I. Langmuir and K. T. Compton, *Rev. Mod. Phys.* **3**, 251 (1931).
8. J. R. Pierce, *Theory and Design of Electron Beams*, Van Nostrand, Toronto, 1954, pp. 177 and 181.
9. C. R. Emigh and D. W. Mueller, Proceedings 2d Int. Conf. on Ion Sources, Vienna, Austria, Oesterreichische Studiengesellschaft fuer Atomenergie, Vienna, 1972, p. 226.
10. C. J. Davisson and C. J. Calbick, *Phys. Rev.* **38**, 585 (1931).
11. J. R. Coupland, T. S. Green, D. P. Hammond, and A. C. Riviere, *Rev. Sci. Instrum.* **44**, 1258 (1973).
12. J. D. Schneider and D.D. Armstrong, *IEEE Trans. Nucl. Sci.* **NS-30**, 2844 (1983).
13. C. Lejeune and J. Albert, *Applied Charged Particle Optics*, A. Septier (Ed.), Part 13A, Academic, 1980, New York, pp. 159–259.
14. F. J. Sacherer, *IEEE Trans. Nucl. Sci.* **NS-18**, 1105 (1971).
15. P. Lapostolle, Proceedings, 2d Int. Conf. on Ion Sources, Vienna, Austria, Oesterreichische Studiengesellschaft fuer Atomenergie, Vienna, 1972, p. 226.
16. J. Guyard and M. Weiss, CERN/PS/LIN 76-3, CERN Geneva, Switzerland, 1976, p. 4.
17. R. Keller, J. D. Sherman, and P. Allison, *IEEE Trans. Nucl. Sci.* **NS-32**, 2579 (1985).
18. A. Septier, Proceedings Symp. on Ion Sources and Formation of Ion Beams, BNL-503109, Brookhaven Nat. Lab., 1971, p. 9.
19. R. Keller, P. Spädtke, and H. Emig, *Vacuum* **36**, 833 (1986).
20. P. Spädtke, Report GSI-83-9, GSI Darmstadt (1983) (In German).
21. M. R. Shubaly, R. A. Judd, and R. W. Hamm, *IEEE Trans. Nucl. Sci.* **NS-28**, 2655 (1981).
22. B. Piosczyk, Proceedings, 1981 Linac Conf., LA-9234-C, Los Alamos Nat. Lab., 1982.
23. W. S. Cooper, K. Halbach, and S. B. Magyary, Proceedings, 2d Symp. on Ion Sources and Formation of Ion Beams, LBL-3399, Lawrence Berkeley Laboratory, Berkeley, CA, 1974, p. II-1-1.
24. P. Allison, J. D. Sherman, and H. V. Smith, Report LA-8808-MS, Los Alamos Nat. Lab., 1981.
25. I. Chavet and R. Bernas, *Nucl. Instrum. Methods* **47**, 77 (1967).
26. R. Keller, *Am. Inst. Phys. Conf. Proc.* **139**, 1 (1986).
27. R. Keller, P. Spädtke, and K. Hofmann, *Springer Ser. Electrophys.* **11**, 69 (1983).
28. D. Keefe and the HIF Staff, Report LBL-12594, Lawrence Berkeley Laboratory, Berkeley, CA, 1981, p. 3.
29. M. R. Shubaly, Inst. Phys. Conf. Ser., No. 54, Adam Hilger, Bristol, UK, 1980, p. 333.
30. R. P. Vahrenkamp and R. L. Seliger, *IEEE Trans. Nucl. Sci.* **NS-26**, 3101 (1979).
31. M. Olivo, Proceedings, 4th Linac Conf., GSI-84-11, GSI Darmstadt, 1984, p. 380.
32. R. Keller, *Nucl. Instrum. Methods* **189**, 97 (1981).
33. S. Ishida, J. Morikawa, N. Nihei, K. Ota, N. Inoue, and T. Uchida, Proceedings, Int. Ion Engineering Congr., Kyoto, Inst. Electr. Engineers of Japan, Tokyo, 1983, p. 31.
34. J. Ishikawa, Y. Takeiri, and T. Takagi, loc. cit. Ref. 33, p. 379.
35. Y. Ohara, loc. cit. Ref. 33, p. 447.
36. T. Kuroda, O. Kaneko, Y. Oka, K. Sakurai, S. K. Guharay, loc. cit. Ref. 33, p. 459.
37. J. F. Bonnal, J. Druaux, M. Fois, and R. Oberson, loc. cit. Ref. 33, p. 465.
38. T. Sato and H. Nishimura, loc. cit. Ref. 33, p. 493.
39. R. Keller, loc. cit. Ref. 33, p. 25.

40. J. P. Brainard and J. B. O'Hagan, *Rev. Sci. Instrum.* **54,** 1497 (1983).
41. R. Keller, *Springer Ser Electrophys.* **11,** 63 (1983).
42. R. Keller, F. Nöhmayer, P. Spädtke, and M.-H. Schönenberg, *Vacuum* **34,** 31 (1984).
43. W. D. Kilpatrick, *Rev. Sci. Instrum.* **28,** 824 (1957).
44. R. Keller, Proceedings, NATO Advanced Study Inst., Pitlochry, UK (1986), NATO ASI Series vol. 178, Plenum Press, New York, 1988, p. 395.
45. K. H. Berkner, W. S. Cooper, K. W. Ehlers, R. V. Pyle, and E. G. Hooper, Proceedings, Workshop Plasma Heating Requirements, Gaithersburg, MD, UCID-3987, US DOE CONF-771241, 1977.
46. M. R. Shubaly and R. W. Hamm, *IEEE Trans. Nucl. Sci.* **NS-28,** 1316 (1981).
47. R. Keller, Proceedings, 1986 Linac Conf., SLAC-303, SLAC Stanford, 1986, p. 232.
48. A. J. T. Holmes and T. S. Green, Inst. Phys. Conf. Ser., No. 54, Adam Hilger, Bristol, UK, 1980, p. 163.
49. I. G. Brown, J. E. Galvin, and R. A. McGill, *Appl. Phys. Lett.* **47,** 358 (1985).

4

Beam Transport

A. J. T. Holmes
UKAEA Culham Laboratory
Culham, England

The subject of beam transport covers a very wide area including ion implantation, particle accelerator design, and the heating of fusion plasmas by neutral beam injection. In all these areas the motion of ions is considered under the influence of electric and magnetic fields; hence, the full description of the beam behavior can be obtained, in principle, from the laws of motion and knowledge of the fields that exist. However, such a general approach runs into severe mathematical problems; therefore, in common with almost all treatments of ion beams, we limit the discussion to ion beams that have trajectories that are close to parallel with the beam axis. This paraxial approximation reduces the problems considerably and holds for almost all ion beams. Part of the following discussion is used to derive the paraxial ray equation that can be used to describe the beam motion under such circumstances.

A further simplification that is frequently made is the assumption of linear or Gaussian charged particle optics. This holds if the external forces acting on the ions are linearly proportional to distance from the axis. Apart from aberrations, this is true for many beam focusing/defocusing elements and results in a very close analog being developed between charged particle optics and light optics. This assumption is also true for the fields created by the beam itself which arise from the beam space-charge density and beam current, providing the beam has a uniform density. We use this approximation to derive the beam envelope equation under both external and internal fields. We also examine some of the lenses that have been designed to control the beam profile.

This picture must be extended to include the effects of the phase space volume occupied by the ensemble of particles that comprise the beam. In this instance the axes of phase space are the three usual spatial ones and the corresponding momentum axes. Liouville's theorem shows that this volume is conserved irrespective of the external forces acting on the beam, providing they are time reversible (i.e., nonstochastic). If, in addition, these forces act independently in the three spatial planes then the cross-sectional area of phase space in each plane is also a conserved quantity and this is known as beam emittance. The beam emittance can be incorporated into the beam envelope equation to describe the shape of the beam although under these circumstances the concept of individual ion motion is no longer a useful concept.

The beam space charge can become the strongest factor influencing the beam profile for high-current beams. The space-charge density is easily reduced, however, by the effects of adding slow particles of the opposite charge via gas ionization or other means. This process and the effect is has on the propagation of a drifting ion beam are discussed in the following sections. This particular effect is of great importance in the development of neutral beam heating for fusion plasmas.

In the following sections we begin by developing the paraxial ray equation which is then used to examine the effects of various lens designs on the beam. The paraxial ray equation is extended to the matrix notation for beam transport with Gaussian optics. The concept of beam envelope is introduced and the effects of both self fields and neutralized self fields on it are discussed for both positive and negative ion beams. These effects are incorporated into a modified paraxial ray equation which describes the beam envelope shape. Lastly, the effects of beam emittance on the beam envelope are included and the combined emittance and neutralized self field effects are examined.

In the following sections many symbols are used. Table 4.1 gives the frequently used terms, together with their SI units.

1 BEAM OPTICS IN THE ABSENCE OF COLLECTIVE EFFECTS

When all collective effects are absent, the beam can be considered as an ensemble of rays or trajectories, each of which would be followed by an individual ion moving under the influence of the applied fields. Hence, the form of the ion trajectory can be derived from the equations of motion which we present here in their full relativistic form. The most common form of this equation of motion in ion optics is known as the paraxial ray equation. The beam has a defined axis so that the ion motion is derived in terms of motion relative to this axis and the applied fields are expressed as first-order expansions of the field on the beam axis. The major assumption in the derivation is that the angle between the ion trajectory and the axis is always small, which is a good approximation in virtually all real beams even if the axis becomes curved (for example, in high-energy particle accelerations).

Since most beams possess a large degree of cylindrical symmetry, we will base all the following derivations on this assumption and derive the paraxial ray equation

1 Beam Optics in the Absence of Collective Effects 55

TABLE 4.1 Frequently Used Terms, Their Symbols, and SI Units

Symbol	Quantity	SI Unit
V	Potential	volts (V)
E	Electric Field	volts/meter (V/m)
B	Magnetic field	tesla (T)
ϕ_0	Magnetic flux	weber (Wb)
e	Electric charge	coulomb (C)
Z	Charge number	—
v or v_b	Beam ion velocity	meter/second (m/s)
c	Velocity of light	meter/second
m_e	Electron mass	kilogram (kg)
m or m_b	Beam ion mass	kilogram
m_i	Slow plasma ion mass	kilogram
r_0	Beam envelope radius	meter (m)
a	Beam waist radius	meter
z	Axial distance	meter
f	Focal length	meter
ρ_L	Larmor radius	meter
n_b	Beam ion density	meter^{-3} (m^{-3})
n_e	Electron density	meter^{-3}
n_i	Slow ion density	meter^{-3}
β	v_b/c	—
γ	$(1 - \beta^2)^{-1/2}$	—
ϵ_0	Permittivity of free space	Farad/meter
μ_0	Permeability of free space	Henry/meter
K	Generalized perveance	—
h	Fraction of beam space charge that is neutralized	—

in this form. The alternative form for a Cartesian (slab) beam is similar. In all the following analyses, SI units are used.

1.1 Busch's Theorem and the Paraxial Ray Equation

Busch [1] derived an expression describing the motion of a particle moving in an axially symmetric magnetic field which relates the angular velocity of this particle to the encircled magnetic flux. The situation is shown in Figure 4.1 where the Lorentz force acting on the ion can be equated to the rate of change of angular momentum following Newton's second law. Hence,

$$-Ze(\dot{r}B_z - \dot{z}B_r) = \frac{1}{r}\frac{d}{dt}(\gamma m r^2 \dot{\theta}) \qquad (1)$$

where Z is charge number, γ equals $(1 - v^2/c^2)^{-1/2}$, m is the ionic rest mass,

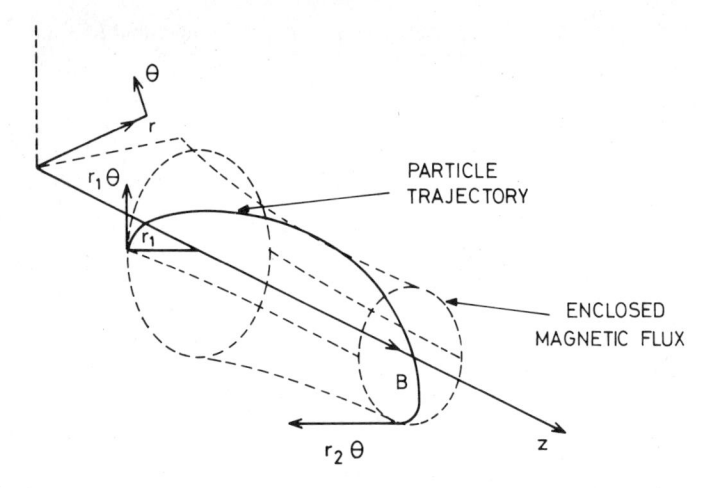

Figure 4.1 Busch's theorem, which relates the canonical angular momentum of particle to the enclosed magnetic flux.

and the dot above the symbol indicates the time derivative. The magnetic fields B_r and B_z refer to an axisymmetric system.

Motion in the axial and radial direction causes a change in the magnetic flux ϕ enclosed by the orbit motion of the particle in the field. This change, $\Delta\phi$, is expressed by

$$\Delta\phi = B_z 2\pi r \, \Delta r + \pi r^2 \left(\frac{\partial B_z}{\partial z}\right) \Delta z$$

$$= 2\pi r (B_z \Delta r - B_r \Delta z)$$

$$\therefore \quad \dot{\phi} = 2\pi r (B_z \dot{r} - B_r \dot{z}) \tag{2}$$

Substitution of Eq. (2) into Eq. (1), when integrated, yields an expression of Busch's theorem given by

$$\gamma m r \dot{\Theta} = \frac{Ze}{2\pi}(\phi_0 - \phi) \tag{3}$$

where ϕ_0 is the value of ϕ when $\dot{\Theta}$ is zero. Equation (3) can be expressed more formally as the conservation of canonical angular momentum P_θ such that

$$P_\Theta = Zer A_\Theta + \gamma m r^2 \dot{\Theta}^2 \tag{4}$$

where A_Θ is the azimuthal magnetic vector potential; the only potential that can exist in an axisymmetric beam.

1 Beam Optics in the Absence of Collective Effects

The paraxial ray equation is derived by combining the properties of paraxial fields, the conservation of canonical angular momentum and the conservation of energy. We begin by equating the radial acceleration to the Lorentz force acting on the beam ion, which yields

$$\frac{d}{dt}(\gamma m\dot{r}) - \gamma mr\dot{\Theta}^2 = Ze(E_r + r\dot{\Theta}B_z)$$

Using Busch's theorem, which is Eq. (3), and the condition that B_z is independent of r allows the elimination of $\dot{\Theta}$ to yield

$$\ddot{r} + \frac{\beta ZeE_z}{\gamma mc}\dot{r} + \frac{Z^2e^2B_z^2}{4\gamma^2m^2}r - \frac{Z^2e^2\phi_0^2}{4\pi^2\gamma^2m^2r^3} - \frac{ZeE_r}{\gamma m} = 0 \qquad (5)$$

This is the paraxial equation with time as the independent variable where $\dot{\gamma}$ has been replaced by $\dot{\gamma} \approx \beta ZeE_z/mc$.

However, a more useful form of Eq. (5) is with z as the independent variable. Providing $\dot{z} \gg \dot{r}$ we can transform Eq. (5) into the following expression:

$$r'' + \frac{\gamma'r'}{\beta^2\gamma} + \left[\frac{\gamma''}{2\beta^2\gamma} + \left(\frac{ZeB_z}{2\beta\gamma mc}\right)^2\right]r - \left(\frac{Ze\phi_0}{2\pi\beta\gamma mc}\right)^2\frac{1}{r^3} = 0 \qquad \boxed{6}$$

where the primes express derivatives with respect to z and the fact that div $\mathbf{E} = 0$ has enabled the radial electric field to be expressed in the relativistic limit as

$$E_r = -\tfrac{1}{2}rE'_z \approx -\tfrac{1}{2}r\gamma''mc^2/Ze \qquad (7)$$

A more convenient form of Eq. (6) makes use of the substitution

$$(\gamma - 1)mc^2 = \int_{z_0}^z ZeE_z = ZeV$$

where V is defined as the potential traversed by the ions from a point z_0, which is the position where the ions have zero kinetic energy. Hence, if V_0 is also defined by

$$V_0 = mc^2/Ze$$

then

$$\gamma = 1 + V/V_0$$

and

$$\beta^2 = V(2V_0 + V)/(V + V_0)^2$$

which gives

$$\frac{V(2V_0 + V)}{(V_0 + V)} r'' + V'r' + \left[\frac{V''}{2} - \left(\frac{ZeB_z}{2mc}\right)^2 \frac{V_0^2}{(V_0 + V)}\right] r$$
$$+ \left(\frac{Ze\phi_0}{2\pi mc}\right)^2 - \frac{V_0 r^3}{(V + V_0)} = 0 \qquad \boxed{8}$$

If there are only electric fields present and the particles are nonrelativistic (where $V_0 \gg V$) then Eq. (8) reduces to the more familiar form,

$$r'' + (V'/2V) r' + (V''/4V)r = 0 \qquad \boxed{8a}$$

1.2 Solutions of the Paraxial Ray Equation and the Matrix Method

It is convenient to modify the form of Eq. (8) by a change of dependent variable to achieve a more convenient form of the differential equation. Lawson [2] has used the transformation

$$R = r[(V/V_0)^2 + 2V/V_0]^{1/4} \qquad (9)$$

and if we assume the canonical angular momentum to be zero, then

$$R'' + \frac{1}{4}\left[V'^2 \frac{(V^2 + 2VV_0 + 3V_0^2)}{(V^2 + 2VV_0)^2} + \frac{V_0^2 Z^2 e^2 B_z^2}{V(2V_0 + V)m^2c^2}\right] R = 0$$

or

$$R'' + k(z)R = 0 \qquad (10)$$

Equation (10) has the solution,

$$R(z) = C(z)R_0 + S(z)R_0' \qquad (11)$$

where R_0 and R_0' are the initial values of R and R' at $z = z_0$, and $C(z)$ and $S(z)$ are two independent solutions of the homogeneous equation:

$$C'' + k(z)C = 0 \qquad S'' + k(z)S = 0$$

subject to the initial conditions

$$C(z_0) = 1 \quad \text{and} \quad \left.\frac{dC}{dz}\right|_{z=z_0} = 0$$

$$S(z_0) = 0 \quad \text{and} \quad \left.\frac{dS}{dz}\right|_{z=z_0} = 1$$

which can be derived from Eq. (11) and its first derivative with respect to z. The curves of C and S as a function of z are referred to as the *cosine* and *sine* trajectories.

Hence we can relate $R(z)$ and $R'(z)$ to the initial values of R and R' by a linear transformation of the form

$$\begin{bmatrix} R \\ R' \end{bmatrix} = \begin{bmatrix} C & S \\ C' & S' \end{bmatrix} \begin{bmatrix} R_0 \\ R_0' \end{bmatrix} \qquad (12)$$

This transformation matrix has a unit determinant, since

$$\frac{d}{dz}(CS' - SC') = CS'' - SC'' = -k(CS - SC) = 0 \qquad (13)$$

Some simple examples of the usefulness of this method of describing ion trajectories are given below. Drift space can be represented by the matrix

$$\mathbf{D} = \begin{bmatrix} 1 & d \\ 0 & 1 \end{bmatrix} \qquad (14)$$

where the determinant is unity. Electrostatic and magnetic thin lenses can be expressed by

$$\mathbf{L} = \begin{bmatrix} 1 & 0 \\ -1/f & 1 \end{bmatrix} \qquad (15)$$

where f is positive for converging lenses and negative for diverging lenses. Thick lenses can be represented by a combination of drift spaces and thin lenses using the concepts of principal planes as in light optics.

Accelerating gaps can also be described using the matrix approach. Pierce [3] has expressed this using the matrix

$$\mathbf{A} = \begin{bmatrix} 2(H-1)U_1/E & 1 \\ 1/H & 0 \end{bmatrix} \qquad (16)$$

where $H = (1 + U_2/U_1)^{1/2}$, U_1 and U_2 are the initial and final kinetic energies of the beam ion, and E is the uniform axial electric field. This latter expression has been used by Holmes and Thompson [4], and Okumura et al. [5] to describe the extraction of ions from the plasma source.

1.3 Focusing of Ion Beams

Apart from any beam collimation created by the accelerator at the ion source, it is frequently necessary to have other lenses along the beam transport channel to keep

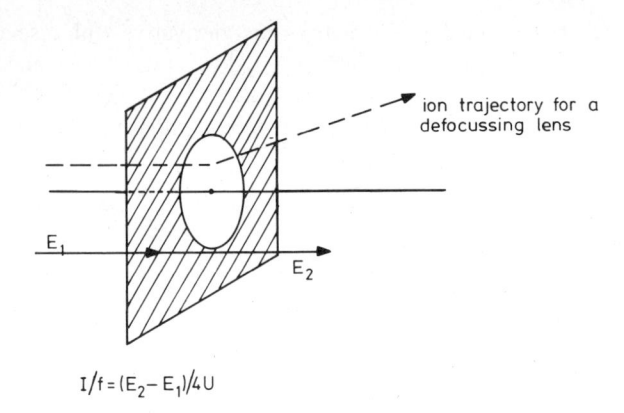

Figure 4.2 A simple aperture lens where the focal strength $1/f$ is proportional to the difference in electric fields.

the beam diameter approximately constant over a long distance. This is particularly true in particle accelerator applications where the axial distance travelled by the ion can be very large indeed. These lenses can be electrostatic or magnetic in nature and their properties can be derived from solutions of Eq. (8) or (10).

1.3.1 Electrostatic Lenses

Probably the most simple lens is the electrostatic aperture lens which is a thin disk with a hole through which the beam passes. When this disk separates two regions of differing field, as seen in Figure 4.2, a radial electric field is generated which creates a focusing or defocusing force that increases linearly with radius. The lens can hence be either diverging or converging according to the sign of the fields on either side. The focal length can be derived from Eq. (8) where the r' term is neglected in comparison with the other terms. Hence,

$$\frac{-r}{r_2'} = f = \frac{2V(2V_0 + V)}{(V + V_0)(V_2' - V_1')} \tag{17}$$

where $V_2' = E_2$ and $V_1' = E_1$ are the electric fields on the exit and entry sides of the lens. This expression reduces in the nonrelativistic limit to the formula derived by Davisson and Calbrick [6]:

$$f = 4V/(V_2' - V_1') \tag{18}$$

The next order of complexity in the structure of an electrostatic lens is to remove the long-range fields existing in the lens described above by placing the aperture between two planes at the beam space potential so that it can be used in a beam

Einzel lens
(always focusing)

Cylindrical lens
(always focusing)

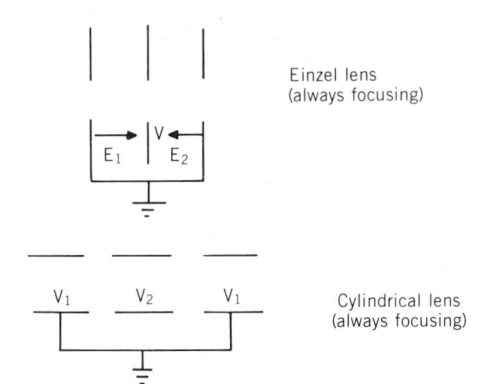

Figure 4.3 Two types of einzel lenses, which focus the beam without changing the beam energy (unlike the simple aperture lens).

transport channel without beam acceleration. Figure 4.3 shows a set of such lens designs, formed by a combination of tubes or disk apertures and usually termed *einzel lenses*.

Pierce [3] has shown that a simple einzel lens of three thin disks, separated by equal distances L and with beam energies V_1 at the outer disks and V_2 at the inner one, has a focal length f_e of

$$\frac{L}{f_e} = \frac{3}{8}\left(\frac{V_2}{V_1} - 1\right)\left[4 - \left(\frac{V_2}{V_1}\right)^{1/2} - 3\left(\frac{V_1}{V_2}\right)^{1/2}\right] \tag{19}$$

and a principal plane at a distance d_e from the center of the lens given by

$$\frac{d_e}{L} = \frac{4}{[3 - (V_2/V_1)^{1/2}][(V_2/V_1)^{1/2} + 1]} - 1 \tag{20}$$

The values of L/f_e for various values of V_2/V_1 are shown in Figure 4.4, where it can be seen that it is almost invariably focusing. The cause of this effective focusing derives from the small difference between focal strength of the lenses located at the outer disks relative to the stronger lens located at the center disk. As a result the effective focal length tends to be large relative to the lens dimensions. This can be seen more clearly if we expand Eq. (19) in the limit where $(V_2 - V_1) = \delta$ which is small compared with V_1. Then the dimensionless lens strength L/f_e is

$$\frac{L}{f_e} \approx \frac{3}{8}\frac{\delta^2}{V_1^2} \tag{21}$$

which is always positive (i.e., a focusing lens). A very large value of V_2/V_1 can give a defocusing lens but this is usually impractical because the beam energy at

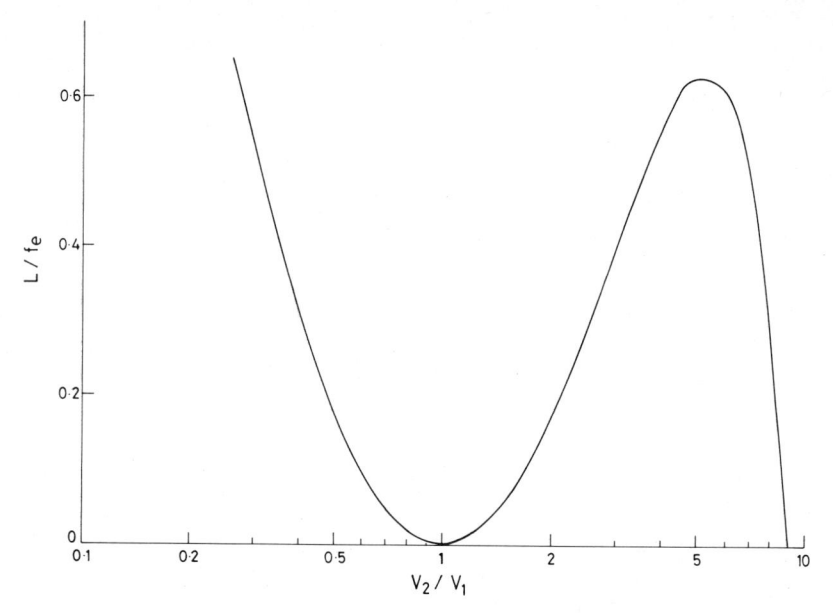

Figure 4.4 The effective strength of an einzel lens (L is the gap length) versus the ratio of beam energies at the inner and outer electrodes. The lens is virtually always focusing.

the central disk is then many times the injection energy. A more detailed analysis of realistic einzel lenses with dissimilar aperture diameters with finite thickness electrodes is discussed by El-Kareh and El-Kareh [7].

1.3.2 Magnetic Lenses

1.3.2.1 The Solenoid. The magnetic analog of the einzel lens is the solenoid, which creates cylindrically symmetric radial and axial magnetic fields using axicentered coils. The focal length can again be derived from Eq. (8) where V' is assumed to be zero. Hence,

$$\frac{1}{f} = \frac{R'_2}{R_1} = \frac{r'_2}{r_1} = \int_{-\infty}^{\infty} \frac{V_0^2 Z^2 e^2 B_z^2 dz}{4V(2V_0 + V)m^2 c^2} \tag{22}$$

In the nonrelativistic limit, the expression for the focal length f becomes

$$\frac{1}{f} = \frac{LB_z^2 Z^2 e^2}{(2mv)^2} \quad [\text{m}^{-1}] \qquad \boxed{23}$$

where B is the magnetic strength in tesla, mv is the beam momentum, and we have assumed the field to have a trapezoidal form so that the axial length L is the distance between the half-field intensity points.

As in the case of the einzel lens, the inverse focal length is proportional to the square of the field, hence making the lens always of the focusing type. The focal length is also proportional to the beam energy, and it is this factor that makes solenoids unsuitable for high-energy beams.

1.3.2.2 The Quadrupole. A more suitable lens for high-energy (or high-mass) beams is the quadrupole lens shown in Figure 4.5. In this case the motions in the x and y planes are decoupled so we cannot use the equations developed in Section 1.2. Now the fields are proportional to transverse displacement in the orthogonal quadrupole, giving

$$B_y = B_0 y/a$$
$$B_x = -B_0 x/a \qquad (24)$$

where a is distance from the axis to the pole face (see Figure 4.5). The orbit equations can be derived directly:

$$y'' = \frac{ZeB_0}{\gamma m v a} y$$

$$x'' = -\frac{ZeB_0}{\gamma m v a} x$$

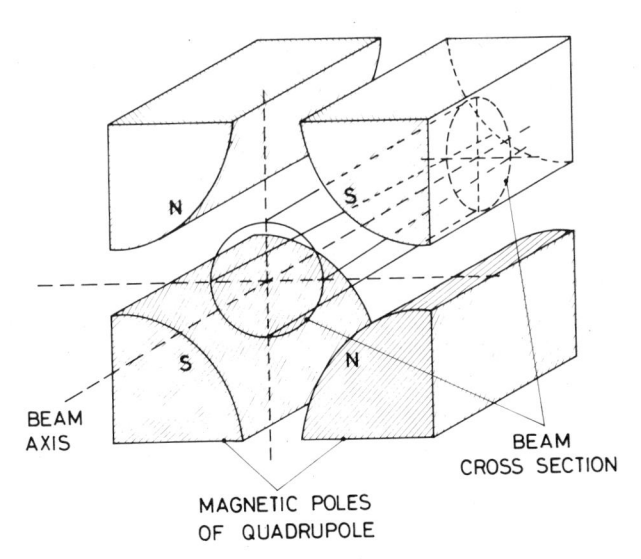

Figure 4.5 A magnetic quadrupole lens. This lens is focusing in one plane and defocusing in the other.

The solutions are

$$x = x_1 \cos(\sqrt{k}z) + x_1' \sin(\sqrt{k}z)/\sqrt{k}$$
$$y = y_1 \cosh(\sqrt{k}z) + y_1' \sinh(\sqrt{k}z)/\sqrt{k} \tag{25}$$

where x_1, y_1, x_1', and y_1' are the initial starting positions and angles and $k = ZeB_0/\gamma m v a$, which yields the following transfer matrices:

$$\begin{bmatrix} x \\ x' \end{bmatrix} = \begin{bmatrix} \cos(\sqrt{k}s) & \sin(\sqrt{k}z)/\sqrt{k} \\ -\sqrt{k}\sin(\sqrt{k}z) & \cos(\sqrt{k}z) \end{bmatrix} \begin{bmatrix} x_1 \\ x_1' \end{bmatrix}$$

$$\begin{bmatrix} y \\ y' \end{bmatrix} = \begin{bmatrix} \cosh(\sqrt{k}z) & \sinh(\sqrt{k}z)/\sqrt{k} \\ \sqrt{k}\sinh(\sqrt{k}z) & \cosh(\sqrt{k}z) \end{bmatrix} \begin{bmatrix} y_1 \\ y_1' \end{bmatrix} \tag{26}$$

Again the determinant of the transfer matrix is unity. A similar expression exists for electrostatic quadrupoles where

$$k_e = ZeE_0/\gamma m v^2 a \tag{27}$$

The quadrupole lens focuses in one plane and defocuses in the other, hence breaking rotational symmetry. However, the focal length is much shorter than a solenoid with equal field, and is given by

$$\left|\frac{1}{f}\right| = \sqrt{k}\sin(\sqrt{k}L) \quad [\text{m}^{-1}] \tag{28}$$

where L is the quadrupole length. The focal length is a linear function of magnetic field and length for short magnets.

Most quadrupoles are of the general design shown in Figure 4.5 (if of the magnetic type), where the pole tips should have a hyperbolic shape. However, it is frequently impractical to have a truly hyperbolic shape and instead the pole tips are machined as an arc of a circle. Dayton et al. [8] and Grivet and Septier [9] have investigated the fields produced by cylindrical pole tips and have found that for a completely symmetric magnet the first nonquadrupole field component is of the 12-pole type. For a pole tip to opposite pole tip separation of $2a$, the best radius for the pole tip is $1.15a$ with a sector angle of $90°$. The field created by this type of pole tip is pure quadrupole for radii less than $0.9a$. It is, however, always sensible to minimize the beam radius within the magnet. The ampere-turns required to produce a pole tip field B_0 are given by

$$\mu_0 NI = 2\int_0^a B_y\, dy$$

Hence

$$NI = B_0 a/\mu_0 \quad [\text{ampere-turns}] \qquad (29)$$

where NI is the ampere-turns in one of the "valleys" of the quadrupole. Iron losses have been ignored.

The effective magnetic length of the quadrupole is defined as $\int B_y dz/B_{y\,max}$ along an axis of constant y. This length is virtually $2a$ longer than the mechanical length of the magnet. A major cause of aberrations in these lenses is the change in magnetic length between the axial value and the value at the surface of the pole tip. The effective length is slightly larger at the pole face, being $2.15a$ longer than the mechanical length, which can lead to aberrations of typically 1 to 2% at full aperture. Grivet and Septier [9] have given details of cylindrical shims for the end faces of the quadrupole and Steffen [10] describes pole end shapes which give a quadrupole form to the end fringing field.

The beam when it passes through a quadrupole assumes an elliptical cross section because of the diverging action in one plane and the focusing action in the orthogonal plane. This problem is increased if a series of such lenses is used, and requires that the acceptance of the lenses be greater in one plane than the other. Hand and Panofsky [11] have designed such a magnet, which consists of an iron window frame yoke with four excitation coils located on the four inner faces of the iron yoke. Providing the current density and ampere-turns are the same in all the windings, the field is purely quadrupole over the entire aperture of the rectangle.

The field gradient B_0/a is

$$B_0/a = \mu_0 NI/(bc - S/2) \quad [\text{T/m}]$$

where $2b$ and $2c$ are the lengths of the open lens aperture (full width) and S is the area of one excitation winding with NI ampere-turns. Comparing this expression with the one for a conventional quadrupole we see that the ampere-turns required for an equal acceptance area and field gradient is greater by a factor of $\pi/2$.

1.3.2.3 Combination of Quadrupole Lenses. A common combination of quadrupoles is two (usually equal) lenses in series, one rotated through 90° so that each plane contains a focusing and defocusing element. This doublet focuses the beam in both planes, but the focal length for each plane cannot be made equal, as can be seen from the transfer matrices for the defocusing–focusing (DF) combination relative to the focusing–defocusing (FD) combination that exists in the orthogonal plane. For equal lens strengths k and lengths L we have in the limit of small $\sqrt{k}L$

$$\mathbf{C}_{DF} = \begin{bmatrix} 1 + kL^2 & 2L \\ -2k^2L^3/3 & 1 - kL^2 \end{bmatrix}$$

$$\mathbf{C}_{FD} = \begin{bmatrix} 1 - kL^2 & 2L \\ -2k^2L^3/3 & 1 + kL^2 \end{bmatrix} \qquad (30)$$

In whichever order these lenses are encountered, the result is always a focusing element, because the ion trajectories are always are always farther from the axis in the focusing part of the overall system than in the defocusing region.

It is possible to combine three quadrupoles to create a focusing lens that has an equal focal length in both planes, hence approximately restoring rotational symmetry. Such a unit is usually termed a triplet. The transfer matrix for a combination, where the central lens has twice the length, 2L, of the outer pair, is for both planes

$$\mathbf{C} = \begin{bmatrix} 1 & 2L \\ -3k^2L^3 & 1 \end{bmatrix} \quad (31)$$

1.3.2.4 The Magnetic Sector Lens.
In its simplest form, the sector lens consists of a gap with a uniform magnetic field extending over the poleface area. A typical magnet is shown in Figure 4.6. Each particle trajectory forms part of the arc of a circle in the uniform field B_0 of the gap and the radius of this trajectory is given by the Larmor radius, ρ_L,

$$\rho_L = \gamma m v / Z e B_0 \quad [\text{m}] \quad (32)$$

where the magnetic field is parallel to the y axis. It is this bending action on the particle trajectories which forms the most frequent use of this type of magnet, particularly in particle accelerators. The mass and velocity dependence of ρ_L also allows these magnets to be used as mass or momentum selection devices.

These magnets can also act as thick lenses when the total bend angle is significantly less than 180°. The method by which the focusing occurs is shown in

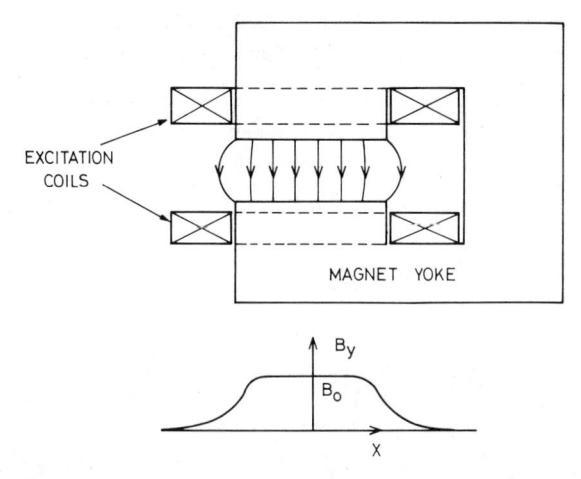

Figure 4.6 A bending magnet showing the best position for the excitation coils for minimum stray field.

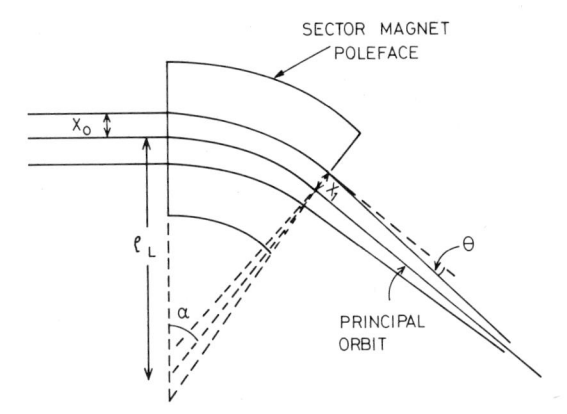

1 Beam Optics in the Absence of Collective Effects

Figure 4.7 The bend plane focusing action of a sector bending magnet.

Figure 4.7; the focusing effect is caused by the reduction in path length through the magnet for the inner trajectory. For a simple sector magnet (as shown in Figure 4.6) where the pole face edges are normal to the central ray of the beam, the reduction in separation between the central ray and a parallel ray at an initial separation x_0 is

$$x_1 = x_0 \cos \alpha$$

where α is the verticle angle of the magnetic sector. Using the geometry shown in Figure 4.7, we can see that

$$\Theta = -x_1 \sin \alpha / \rho_L$$

hence,

$$f = \rho_L / \tan \alpha \quad [\text{m}] \qquad \boxed{33}$$

It is often convenient to express these relations in matrix form, which for a sector magnet has the form

$$\mathbf{M}_m = \begin{bmatrix} \cos \alpha & \rho_L \sin \alpha \\ -\sin \alpha / \rho_L & \cos \alpha \end{bmatrix} \qquad \boxed{34}$$

As for the other matrices discussed above, the determinant is unity.

Focusing of the beam also occurs when the ion beam enters the gap at an angle to the edges of the dipole magnet, and arises from the component of the main field at the edge of the magnet that is orthogonal to the beam direction. This is seen in Figure 4.8. When the beam enters and leaves the magnet normal to the pole face, B_x is zero. Hence, no focusing would then be observed.

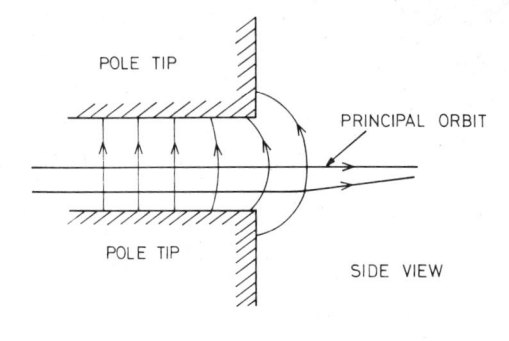

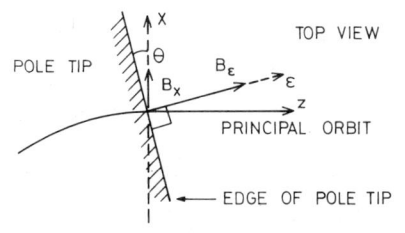

Figure 4.8 The pole edge focusing in the nonbend plane when the beam enters or exits nonorthogonally to the edge of the magnet.

The focal length of this effect can be derived by calculating the change in transverse momentum Δp_y gained by the particle in traversing an edge which is inclined at an angle Θ to the beam axis. A simple method of analysis, derived by Humphries [12], shows that

$$\Delta p_y = \int B_x Z e v_z dt$$

but $B_y = B_\epsilon \sin \Theta$ where B_ϵ is the field referenced to the edge of the magnet and

$$d\epsilon = dz \cos \Theta = v_z \cos \Theta \, dt$$

Hence,

$$\Delta v_y = \frac{Ze}{\gamma m} \int B_\epsilon \tan \Theta \, d\epsilon \tag{35}$$

Since the divergence of B is zero, it is possible to show that

$$\int B_\epsilon \, d\epsilon = B_0 y$$

where B_0 is the main gap field. Hence, since the transverse velocity is proportional to y, the concept of focal length is valid and it is possible to show that

$$f_e = \rho_L/\tan\Theta \quad [\text{m}] \qquad (36)$$

If Θ is greater than zero, then the magnet edges provide vertical focusing but the horizontal focusing is decreased. In this situation, the sector magnet can provide focusing in both planes and this concept has been used in the design of the zero-gradient synchrotron [13].

1.4 Beam Transport Channels

The matrix method of expressing the effect of a lens or drift space makes the solution of a series of such elements straightforward. If the ion trajectory encounters a series of elements expressed by the matrices **A**, **B**, **C**, $\cdots$, in that order, then the final trajectory r_f, r'_f can be written in the form

$$\begin{bmatrix} r_f \\ r'_f \end{bmatrix} = \mathbf{C} \times \mathbf{B} \times \mathbf{A} \times \begin{bmatrix} r_i \\ r'_i \end{bmatrix} \qquad (37)$$

It should be noted that the mathematical order of the matrices **A**, **B**, and **C** must replicate the geometric order in which the components are encountered, because matrix multiplication is not commutative.

As an example we consider a drift space of length $\ell/2$, a lens of focal length f, and another drift space. The combined matrix **C** is the product of

$$\mathbf{C} = \begin{bmatrix} 1 & \ell/2 \\ 0 & 1 \end{bmatrix} \begin{bmatrix} 1 & 0 \\ -1/f & 1 \end{bmatrix} \begin{bmatrix} 1 & \ell/2 \\ 0 & 1 \end{bmatrix}$$

$$= \begin{bmatrix} 1 - \ell/2f & 1 - \ell^2/4f \\ -1/f & 1 - \ell/2f \end{bmatrix} \qquad (38)$$

As with all transfer matrices, the determinant is again unity.

When the number of lenses and drift spaces becomes large, following an individual trajectory through the series of transfer matrices can become tedious, although it is straightforward using numerical techniques. If the matrices can be grouped into a number of identical cells, each with a transfer matrix **A**, then

$$\begin{bmatrix} r_f \\ r'_f \end{bmatrix} = \mathbf{A}^n \times \begin{bmatrix} r_i \\ r'_i \end{bmatrix} \qquad (39)$$

However, the ion trajectory motion in such an assembly is stable only if the modulus of the trace of the matrix is

$$\text{Tr}(\mathbf{A}/2) \leq 1 \qquad (40)$$

When this is the case, the orbit is periodic with an angular phase advance per call μ, given by

$$2 \cos \mu = \text{Tr } \mathbf{A} \tag{41}$$

If the absolute value of $\text{Tr}(\mathbf{A}/2)$ exceeds unity, the ion motion is unstable and the ion leaves the periodic channel.

2 THE TRANSPORT OF LAMINAR BEAMS WITH SELF FIELDS

In the previous section the motion of the beam particles was discussed under the assumption that the effects of any fields created by the beam itself had a negligible effect on the ion trajectories. This is an adequate description for low-current beams but is increasingly untrue for higher-current ion beams. We now turn to the question of the magnitude of these self fields and their effect on the beam, under the assumption that the ion trajectories do not cross (i.e., laminar flow) and the beam ion velocity distribution is monoenergetic.

Two fields are created by the beam: a radial electric field due to the beam space charge, and an azimuthal magnetic field arising from the beam current. If the beam space-charge density is a function of only r and Θ then the electric field is expressed by

$$\mathbf{E}(r, \Theta) = \int_A \frac{Zen_b(a, \Theta)}{2\pi\epsilon_0} \frac{(\mathbf{r} - \mathbf{a})}{|(r-a)|^2} 2\pi a \, da \, d\Theta \tag{42}$$

where $\mathbf{r}$ represents the vector to the point of interest, $\mathbf{a}$ is the vectorial position of the volume element, and A is the beam cross section.

For a uniform beam Eq. (42) reduces to

$$\mathbf{E}(r) = \frac{Zen_b \mathbf{r}}{2\epsilon_0} \quad (r < r_0) \tag{43}$$

where r_0 is the beam edge radius.

The magnetic field can be derived from the Biot-Savart law, and is

$$\mathbf{B}(r, \Theta) = \mu_0 \int_A \frac{Zen_b(a, \Theta)}{2\pi} \frac{(\mathbf{r} - \mathbf{a})}{|(r-a)|^2} \times \mathbf{v}_b \cdot 2\pi a \, da \, d\Theta \tag{44}$$

which again can be simplified to

$$\mathbf{B}(r) = \mu_0 Zen_b \mathbf{r} \times v_b \quad (r < r_0) \tag{45}$$

for a uniform beam.

2 The Transport of Laminar Beams with Self Fields

The force acting on an individual beam ion is given by the Lorentz force, which is expressed by

$$\mathbf{F}(r, \Theta) = Ze(\mathbf{E} + \mathbf{v}_b \times \mathbf{B})$$

$$= \int_A \frac{Ze^2 n_b(a, \Theta)}{2\pi} \frac{(\mathbf{r} - \mathbf{a})}{|(\mathbf{r} - \mathbf{a})|^2} 2\pi a \, da \, d\Theta \left(\frac{1}{\epsilon_0} - \mu_0 v_b^2\right)$$

$$= Ze\mathbf{E}(r, \Theta)(1 - \beta^2) \quad [\text{Newtons}] \qquad (46)$$

This relation is completely general and applies to all beams, when subject to transverse forces. It also applies to accelerated beams providing β is virtually constant over an axial distance equal to the beam radius.

2.1 Motion of an Ion Beam without External Fields

The motion of an ion subject to the Lorentz force derived in the previous section is given by

$$\ddot{r}\gamma m = Ze E(r)(1 - h - \beta^2) \qquad (47)$$

where h is the degree of space-charge neutralization (discussed in Section 3). Since the field E is a linear function of r for a uniform beam, the trajectories of all particles are identical apart from a factor r/r_0. Consequently, we can derive, without loss of generality, the outermost trajectory $r_0(z)$, which describes the beam envelope for laminar beams as

$$\ddot{r} = r_0'' \beta^2 c^2 = Ze E(r)/\gamma m$$

Then

$$r_0'' = \frac{Ze^2 n_b r_0}{2\gamma \epsilon_0 m \beta^2 c^2}\left(\frac{1}{\gamma^2} - h\right)$$

$$= \frac{I(1/\gamma^2 - h)}{4\pi \epsilon_0 U^{3/2}(2eZ/m)^{1/2} r_0}$$

$$= \frac{K}{r_0} \qquad (48)$$

where U and I are the beam energy and current and K is the generalized beam perveance. K can be expressed in the form

$$K = 6.49 \times 10^5 \frac{I}{U^{3/2}}(1/\gamma^2 - h)\left(\frac{A}{Z}\right)^{1/2} \quad [\text{A} \cdot \text{V}^{-3/2}] \qquad (49)$$

where A is the ion mass in amu.

This equation can be solved by integrating twice. Harrison [14] has derived a form of Eq. (48) that is valid when $r_0 - a$ is less than a, which is the value of r_0 when r_0' is zero. The solution is a hyperbola of the form

$$r_0^2 - Kz^2 = a^2 \tag{50}$$

with an asymptotic divergence angle of $\sqrt{K}$. This indicates that the paraxial form is valid only when K is much less than unity.

The exact solution of Eq. (48) is

$$\frac{z}{a} = \left(\frac{2}{K}\right)^{1/2} \int_0^{\ln(r_0/\rho)^{1/2}} \exp(u^2) \, du \qquad K > 0$$

$$\frac{z}{a} = \left(-\frac{2}{K}\right)^{1/2} \int_0^{\ln(\rho/r_0)^{1/2}} \exp(-u^2) \qquad K < 0$$

The integral can only be solved numerically, and a curve of r_0/a versus $z\sqrt{K}/a$ is shown in Figure 4.9. The curve for positive K is very close to being a hyperbola as can be seen in the figure. When K is negative, the beam converges rapidly and can propagate only a short distance along the beam axis before the trajectories

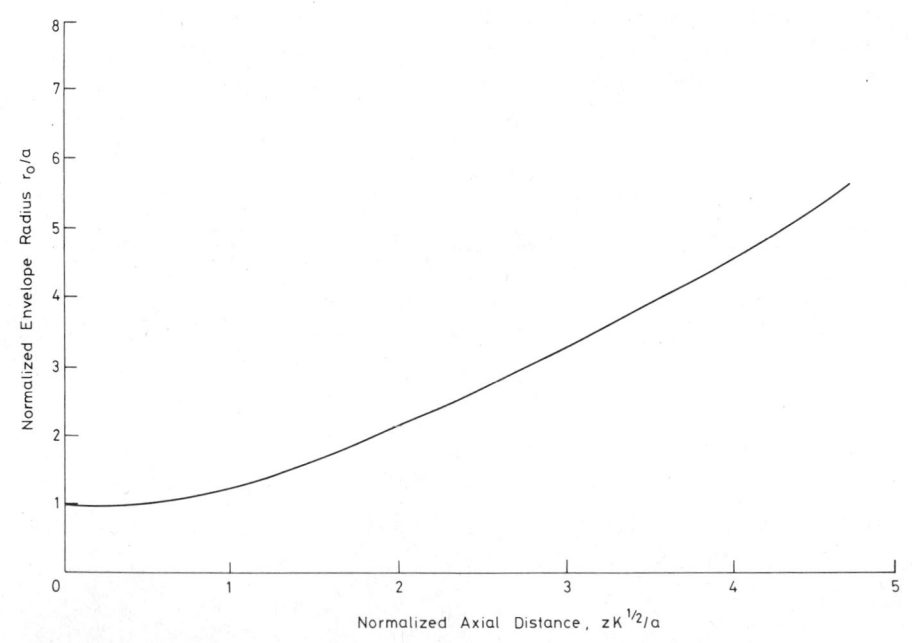

Figure 4.9 Beam expansion from a waist of a radius a and with a perveance K through the action of the unneutralized beam space charge.

cross the axis at the beam cusp. This latter situation is unusual and only occurs if h is very close to or in excess of unity.

An indication of the relative importance of the generalized perveance can be found by deriving a current density limit j_0 where the beam space charge would cause a 10% (as an example) increase in beam diameter. From Eq. (49) we see that

$$r_0^2 - a^2 = 0.21a^2 = Kz^2$$

Hence,

$$j_0 = \frac{0.84\pi\epsilon_0 U^{3/2}(2eZ/m)^{1/2}}{z^2(1/\gamma^2 - h)} \quad (51)$$

$$= \frac{3.24 \times 10^{-7} U^{3/2} Z^{1/2}}{A^{1/2} z^2 (1/\gamma^2 - h)} [A \cdot m^{-2}] \quad \boxed{52}$$

where A is the ion mass in proton units (U is expressed in volts and z in meters). Equation (52) shows that the value of j_0 we can transport over significant distances (i.e., many beam diameters) is very low when h is small compared to unity. If h is zero, then only relativistic beams can propagate over a significant distance with a high-current density. However, several methods of improving the beam transport exist and these are discussed in the following sections.

2.2 Self Fields and the Paraxial Ray Equation

If the self fields causing transverse motion of the ion beam generate ion velocities that are small in comparison with the axial velocity, then these fields can be incorporated in the paraxial ray equation derived in Section 1.1. Using Eq. (48) to derive the additional radial acceleration, we find that on adding Eq. (48) to Eq. (6) when $r = r_0$, we have

$$r_0'' + \frac{\gamma' r_0'}{\beta^2 \gamma} + \left[\frac{\gamma''}{2\beta^2\gamma} + \left(\frac{ZeB_z}{2\beta\gamma mc}\right)^2\right] r_0 - \left(\frac{Ze\phi_0}{2\pi\beta\gamma mc}\right)^2 r_0^{-3} - \frac{K}{r_0} = 0 \quad (53)$$

This equation indicates that a drifting beam can be propagated without divergence when ϕ_0 is zero if the generalized perveance K matches the axial field B_z through the relation

$$K = \left(\frac{ZeB_z r_0}{2\beta\gamma mc}\right)^2 \quad \boxed{54}$$

This condition is known as Brillouin flow and is discussed in greater detail in the next section.

2.3 Beam Propagation Using an Axial Magnetic Field

We have already described in Section 1.1 the motion of ions in an axial field (Busch's theorem) and the paraxial ray equation. In the simple case of a uniform nonrelativistic beam with space charge, where h is zero, we can write

$$mr\dot{\Theta}^2 = \frac{-n_b Z^2 e^2 r}{2\epsilon_0} - ZeB_z r\dot{\Theta}$$

where the beam is assumed to be in the state of a *rigid* rotator so that $\dot{\Theta}$ is a constant. The above equation can be reduced to

$$\dot{\Theta}^2 + \omega_p^2/2 - 2\dot{\Theta}\Omega_L = 0$$

where Ω_L is the ion Larmor frequency ($\Omega_L = ZeB_z/m$) and ω_p is the ion plasma frequency of the beam ($\omega_p = (n_b Z^2 e^2/\epsilon_0 m)^{1/2}$), which gives the solutions

$$\dot{\Theta} = \Omega_L \pm (\Omega_L^2 - \omega_p^2/2)^{1/2} \tag{55}$$

Providing Ω_L is greater than the plasma frequency, there are two real solutions for $\dot{\Theta}$, one on each side of the Larmor frequency. The condition that

$$\Omega_L = \omega_p/\sqrt{2} \tag{56}$$

is known as the Brillouin flow condition [15], described in greater detail below. First, however, Eq. (55) implies that the axial beam velocity is not constant, and it can be shown that

$$\beta^2 - \beta_0^2 = \frac{r^2}{2c^2}(\omega_p^2/2 - \dot{\Theta}^2) \tag{57}$$

where β_0 is the value of β on axis and c is the velocity of light. Because a transverse variation in beam axial velocity is difficult to achieve, it is preferable to ensure that $\dot{\Theta}$ is equal to $\omega_p/\sqrt{2}$, which again results in Brillouin flow.

For a nonrelativistic beam, Brillouin flow allows the beam to move as a rigid body without divergence. The individual beam ion trajectories are helixes around the beam axis, and for increasing beam ion density the pitch of the helix decreases. It is useful to derive the maximum beam current that can be transported for a given beam energy U while allowing the magnetic field to increase so as to maintain Brillouin flow. This is seen by the equalities

$$\frac{\omega_p^2}{2} = \Omega_L^2 = \dot{\Theta}^2$$

∴

$$\dot{\Theta}^2 = Z^2 n_b e^2/2\epsilon_0 m \tag{58a}$$

and
$$\frac{2eUZ}{m} = v^2 + r_0^2 \dot{\Theta}^2 \qquad (58b)$$

The beam current is given by

$$I = \pi r_0^2 n_b v_b Ze$$
$$= \pi r_0^2 \dot{\Theta}^2 2\epsilon_0 m v_b / eZ$$

Hence,

$$v = \left(\frac{2ZeU}{3m}\right)^{1/2}$$

and

$$I_{max} = \frac{16\pi\epsilon_0 (Ze)^{1/2} U^{3/2}}{3\sqrt{6} m^{1/2}} \quad [\text{A}] \qquad \boxed{59}$$

For protons, this corresponds to a perveance of 6.0×10^{-7} $\text{AV}^{-3/2}$, which is sufficiently large to permit the extraction and transport of high-current beams by this technique, although the axial field required is

$$B_z^2 = \frac{4}{3} \frac{mU}{Zer_0^2} \quad [\text{T}^2] \qquad \boxed{60}$$

This field can be many teslas and usually places a more stringent limit to the beam current than the perveance. If we express the current density in terms of the axial field we obtain

$$j_{max} = \frac{\sqrt{2}\pi\epsilon_0 Z^2 e^2 r_0 B_z^3}{m^2} \quad [\text{A} \cdot \text{m}^{-2}] \qquad \boxed{61}$$

For protons this current density is

$$j_{mp} = 3.65 \times 10^5 B_z^3 r_0 \quad [\text{A} \cdot \text{m}^{-2}] \qquad \boxed{62}$$

3 SPACE-CHARGE NEUTRALIZATION OF ION BEAMS

The effective charge density n of an ion beam can be reduced by adding a charge density of the opposite sign to that of the beam itself. This is achieved by allowing the beam to pass through a background gas where the beam ions have ionizing and charge exchange collisions with the gas molecules. As a result, slow ions and electrons are formed within the beam channel, and the value of n in this plasma of electrons, slow ions, and beam ions can be reduced to less than 1% of that of the unneutralized beam n_b (i.e., $h \geq 0.99$) when the pressure is in excess of 10^{-4} Torr. The value of n (or h) is a function of the gas pressure and increases at low pressures for positive ion beams. In the case of negative ion beams similar processes occur, but the role of the electrons in the beam is altered because of the reversal of the beam charge density. As a result, we treat the case of the negative ion beam separately; both cases are discussed in the following sections.

3.1 Neutralization of Positive Ion Beam Charge

A full treatment of the plasma formed in the channel of a positive ion beam consists of four relations: the continuity equations for the production of slow ions and electrons, which determine their densities; the energy balance of the plasma, which yields the electron temperature; and lastly Poisson's equation, which gives the radial electric field as a function of the other variables. Unfortunately, all these expressions are nonlinear, integral–differential equations that require very sophisticated numerical techniques for solution.

It is this mathematical difficulty that has been the cause for the differing approaches adopted by various authors. The early solutions did not use differential equations at all and simply created a zero-dimensional model. The earliest discussion of the problem [16] did not have the energy balance equation and simply assumed plasma neutrality to replace Poisson's equation. Gabovich et al. [17] ignored the electron continuity equation and also assumed plasma neutrality, while Hamilton [18] did not consider the effects of the slow ions in the beam plasma. Each of these zero-dimensional models ignores one of the four fundamental processes and hence cannot be considered as a good theory for describing the beam plasma.

Dunn and Self [19] described a model for the neutralization of electron beams to create a one-dimensional model for ion beams. This latter model uses the approximation of a uniform beam ion density and a sharp edge to the beam to simplify the one-dimensional model, and hence obtains analytic expressions for the potential ϕ in multiples of the electron temperature kT. No energy balance is included so the electron temperature remained a free parameter, hence forcing the solutions to be derived in dimensionless variables. Hooper et al. [20] extended the model of Dunn and Self to include Gaussian profile ion beams, but without the energy balance relation, and introduced the concept of ionic drift in the beam plasma rather than the ionic free-fall described by Dunn and Self. Green [21] has used the ionic free-fall integral to create a neutralization model but without including energy balance. Because the plasma normally has a low density, the latter model is more correct, although it causes considerably greater mathematical difficulties for a full solution.

Holmes [22] has used the four basic expressions, including ionic free-fall, to describe a model for the beam plasma. This model is, of necessity, mathematically

3.2 A Simple Model for a Positive Ion Beam Plasma

To avoid the mathematical difficulties of the one-dimensional models of Dunn and Self, Green, and Holmes, we present a simple one-dimensional model for the plasma that retains the free-fall integral for collisionless ion motion. This model contains all the continuity and energy balance equations as well as plasma neutrality, hence showing the physical basis of this approach. Since the expressions derived here have a much simpler form than those described above, yet retain similar scaling laws, we can go on to use these expressions in deriving the beam envelope in Section 5.

The beam plasma is formed by inelastic collisions between the beam ions and the background gas. For a positive ion beam these processes are described by the expressions

$$\underline{A^+} + X^0 \rightarrow \underline{A^0} + X^+ \quad \text{(charge exchange)}$$

$$\underline{A^+} + X^0 \rightarrow \underline{A^+} + X^+ + e \quad \text{(ionization)}$$

where A^+ is the original beam ion and X^0 is the gas molecule. The underline indicates the fast particle. It is clear that more slow ions are made than electrons as both processes contribute to ion production. The slow positive ions are expelled from the beam by the radial electric field and, providing the beam radius r_0 is less than the mean free path for ion–neutral collisions, $1/N\sigma_x$, where N is the neutral atom density, then the motion can be treated as free-fall. Hence

$$2\pi r \ell \, dn_i(r) v(\rho, r) = 2\pi \rho \, d\rho \, \ell \frac{\partial n_i}{\partial t} \tag{63}$$

where $dn_i(r)$ is the incremental slow ion density at r caused by slow ion production, $\partial n_i/\partial t$, at ρ. Integration and the substitution of

$$\frac{\partial n_i}{\partial t} = N n_b \sigma_i v_b \tag{64}$$

where $\sigma_i = \sigma_{\text{charge exchange}} + \sigma_{\text{ionization}}$, n_b is the uniform beam density, and v_b is the beam ion velocity, yields

$$n_i(r) = \frac{1}{r} \int_0^r \frac{n_b N \sigma_i v_b \rho \, d\rho}{(2e/m_i)^{1/2} [-\phi(r) + \phi(\rho)]^{1/2}} \tag{65}$$

where the ion temperature T_i is assumed to be small compared with the potential ϕ and is hence neglected, and m_i is the slow ion mass. The only finite solution for $n_i(r)$ occurs when

$$-\phi(r) = \phi_0 \frac{r^2}{r_0^2} + \phi_1 \frac{r^4}{r_0^4} + \cdots \tag{66}$$

which for small r gives the expression

$$n_i = n_b N\sigma_i v_b r_0 m_i^{1/2} (1 + 5\phi_1 r^2/6\phi_0 r_0^2)/(2e\phi_0) \tag{67}$$

Hence n_i is a constant (like the beam density) until the higher terms in the expansion of $\phi(r)$ become significant.

The thermal electron continuity equation is easily derived by equating production to loss out of the beam. Hence, we can write

$$\pi r^2 n_b N\sigma_e v_b = 2\pi r n_e(r) v_e/4 \tag{68}$$

where $\sigma_e = \sigma_{\text{ionization}}$ and v_e is the electron velocity. As the electrons are trapped in the beam by the space potential, they are fully thermalized and can only escape by gaining enough energy from the beam to escape. In this case the electron density is given by the Boltzmann distribution,

$$n_e(r) = n_{e0} \exp(e\phi/kT) \tag{69}$$

where T is the electron temperature. Using Eq. (66) and expanding Eq. (69) gives

$$n_e = n_{e0}(1 - \phi_0 e r^2/kT r_0^2) \tag{70}$$

To derive n_{e0}, we make the assumption that the electron velocity becomes equal to the thermal velocity outside the beam where the electrons are then essentially collisionless and can escape. Hence when $r = r_0$, substitution of Eq. (69) into Eq. (68) allows us to show that

$$n_{e0} = 2 r_0 n_b N\sigma_e v_b \left(\frac{\pi m_e}{8kT}\right)^{1/2} \exp(e\phi_w/kT) \tag{71}$$

where $\phi_w \approx -\phi_0$ is the potential at the beam edge.

The energy balance equation can be derived from the concept that each electron removes an energy of $2kT$ when it leaves the plasma [24]. This energy is the sum of the kinetic energy it possesses plus the potential energy gained in order to leave the plasma. If we reduce the energy transfer process from the beam ions to the electrons to a simple cross section ϕ_u then the volumetric energy input into the electron distribution is

$$H = n_b n_e \sigma_u v_b eU \tag{72}$$

where U is the beam energy (in eV). Integrating over the beam volume then gives

$$\int_0^{r_0} 2\pi r H \, dr = 2kT\pi r_0^2 n_b N\sigma_e v_b \tag{73}$$

After integration we obtain

$$\frac{U\sigma_u n_{e0}}{|\phi_w|}(1 - \exp e\phi_w/kT) = 2N\sigma_e \tag{74}$$

The final stage is to use the plasma neutrality condition to complete the description of the beam plasma. This assumption is based on the fact that the Debye length in the plasma is less than the beam radius r_0. On axis we can then write

$$n_b + n_{i0} = n_{e0} \tag{75}$$

where $n_{i0} = n_b N\sigma_i v_b r_0 m_i^{1/2}/(2e\phi_0)^{1/2}$

and is the axial slow ion density. On expansion we have

$$n_b + \frac{n_b N\sigma_i v_b r_0 m_i^{1/2}}{(2e\phi_0)^{1/2}} = 2r_0 n_b N\sigma_e v_b \left(\frac{\pi m_e}{8kT}\right)^{1/2} \exp(e\phi_w/kT)$$

This equation can be simplified using Eq. (74) to yield

$$\frac{n_b}{N} + \left(\frac{m_i}{m_b}\right)^{1/2}\frac{\sigma_i n_b r_0}{G^{1/2}} \approx 2G\frac{\sigma_e}{\sigma_u} \quad \boxed{76}$$

$$\text{where } G = \frac{|\phi_w|}{U} = \sigma_u n_b a \exp\left(-\frac{e\phi_w}{kT}\right)\left(\frac{m_e}{2kT}\right)^{1/2} \tag{77}$$

The term G is hence the ratio of the potential well depth to the beam energy (in eV). For simplicity it has been assumed that $e\phi_w/kT$ is significantly larger than unity in Eq. (74).

The plasma neutrality condition also holds away from the beam axis where

$$\frac{n_b N\sigma_i v_b r_0 m_i^{1/2}}{(2e\phi_0)^{1/2}}\frac{5\phi_1 r^2}{6\phi_0 r_0^2} = n_{e0}\frac{e\phi_0 r^2}{kT r_0^2}$$

Hence,

$$\phi_1 = \frac{6}{5} \phi_0^{5/2} \frac{e}{kT} \left(\frac{2e}{m_i}\right)^{1/2} \frac{1}{n_b \sigma_i r_0 v_b} \frac{2\sigma_e G}{\sigma_u} \tag{78}$$

Equations (76)–(78) completely describe the beam plasma as a function of the main beam parameters, n_b, r_0, U, and N, and the appropriate cross sections. Figure 4.10 shows the value of G as a function of n_b/N for several values of $n_b r_0$ with a 50-keV H$^+$ beam. This indicates that G decreases linearly with n_b/N from its limiting value of $29IA^{1/2}/U^{3/2}$ (where I is the beam current in mA, A is the ion mass in amu, and U is the beam energy in keV) at low values of N when the beam is fully unneutralized. At high gas densities, G tends to a constant value that increases with the product $n_b r_0$.

This behavior has been observed experimentally by Holmes [22], and ϕ_w measured by an emitting Langmuir probe is shown as a function of N for a 20-mA beam at 20 keV in Figure 4.11. More recently Klabunde and Schoenlein [23] have confirmed this dependence of ϕ_w with N using a 200-keV A$^+$ beam and measuring ϕ_w by the energy of slow ions emitted from the beam.

This saturation in the magnitude of the well depth G arises when the beam ion density becomes small relative to r_{b0} and n_{e0}. This is seen more explicitly in Figure 4.12 where the fractional densities n_{i0}/n_b and n_{e0}/n_b are plotted as a function of Nr_0 for two different values of $n_b r_0$ for a 50-keV proton beam. At high gas densities, the beam acts merely as a source of particles and the space potential then takes on a value so that electrons and ions have the same density. At low gas densities, the slow ion density is negligible although it does increase at approximately $N^2 r_0^2$, while the electron density is virtually equal to the beam ion density.

The variation in the reduced electron temperature $kT/e\phi_w$ as a function of Nr_0 is shown in Figure 4.13. This rises slowly with increasing pressure, while the value of ϕ_w decreases and then saturates. Hence, kT has a U-shaped dependence on Nr_0 and is virtually independent of beam density. Holmes [22] has observed this variation experimentally as seen in Figure 4.14. The electron temperature was measured by the energy distribution of the emitted electrons using an energy-analyzing probe.

3.3 Neutralization of Negative Ion Beams

It might be thought that the neutralization of negative ion beams would reduce effective beam space charge to a few percent of the value of an unneutralized beam. However, it has been found by Sherman et al. [25] and Goretskii and Naida [26] that the beam exhibits ion–ion instabilities at pressures below a critical value, when the beam is observed to have a negative plasma potential corresponding to a negative net space charge density. At higher pressures, the beam becomes stable and has a positive plasma potential which corresponds to a net positive space-charge density.

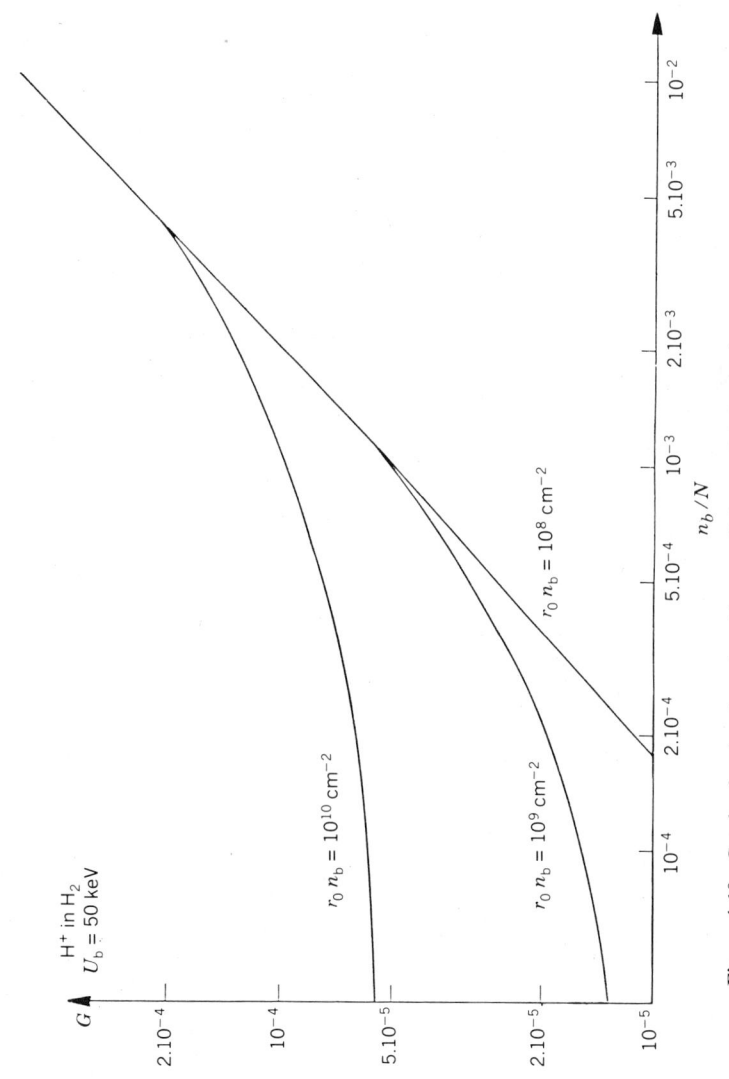

Figure 4.10 Graphs showing the reduced beam radial potential $G\,(=\phi/U_b)$ as a function of the ratio of beam density to gas density.

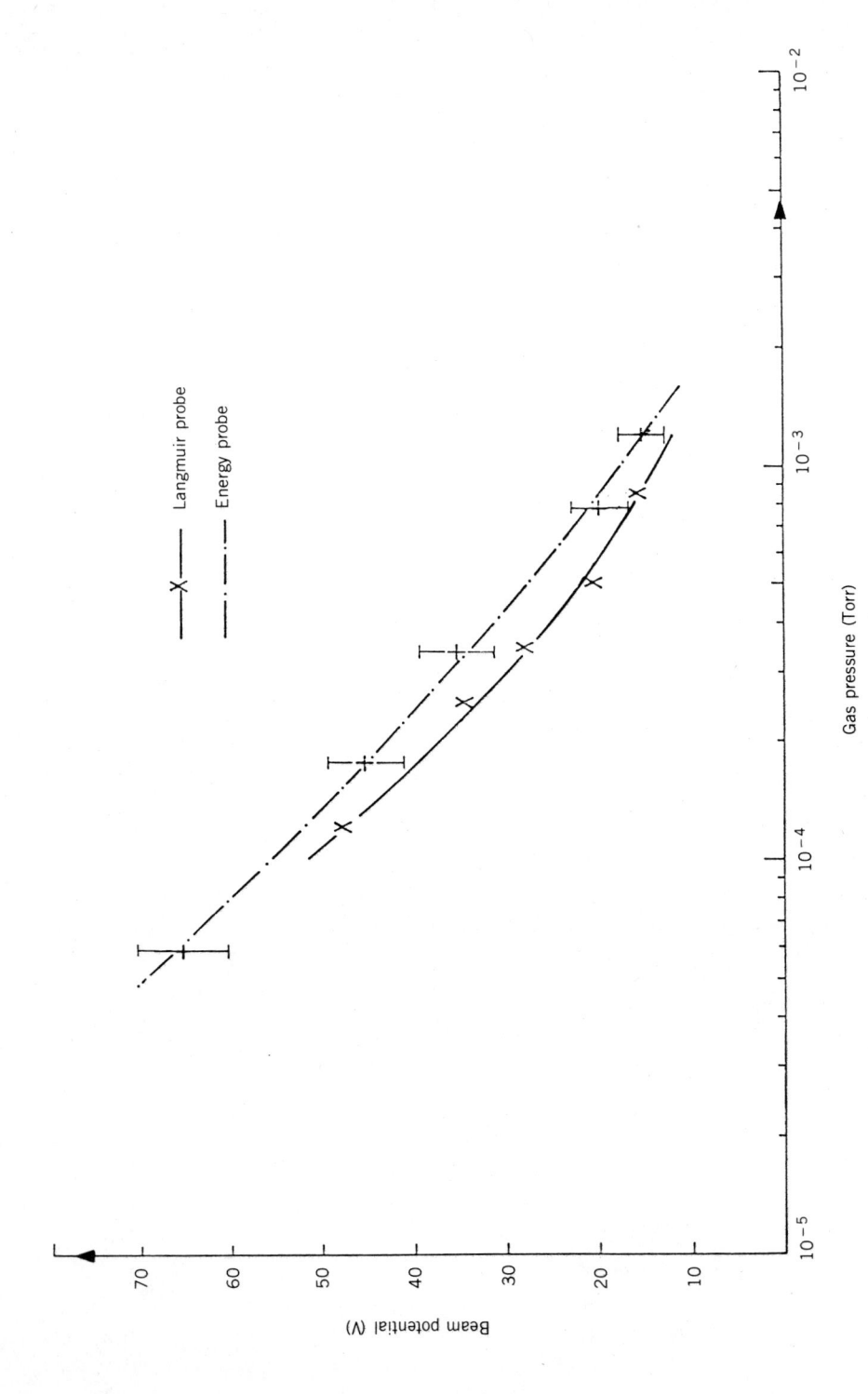

Figure 4.11 Experimental measurements of the radial potential of a 20-keV He^+ beam in He as a function of gas pressure [Ref. 22].

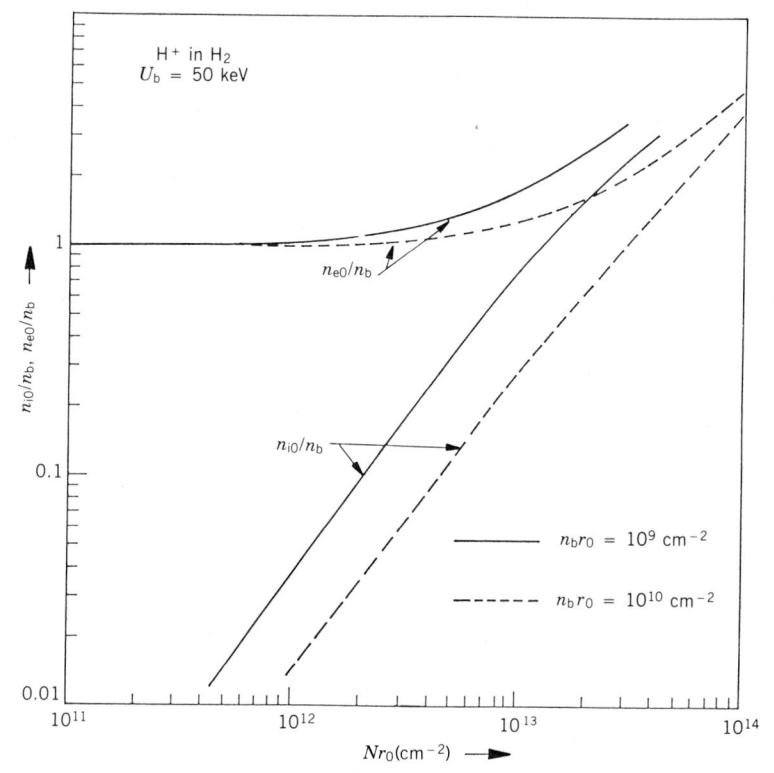

Figure 4.12 Theoretical calculation of the reduced densities of slow ions or electrons as a function of the product of gas density N_0 and beam radius r_0.

Consequently, the theoretical models that have been developed consider only the latter case, which is, of course, very similar to the description of the neutralization of the positive ion beam with exception of the reversed sign of the beam space charge. However, a recent model by Wright [27] does address the transition between the two neutralization states, although it is purely numerical and does not derive scaling laws. Lemons et al. [28] and Hooper et al. [20] have derived one-dimensional models based on plasma ion drift that describe a beam plasma with a positive potential well. However, none of these models includes the major energy input process to the electron distribution that comes via the negative ion stripping collisions with the background gas molecules, creating electrons with the same velocity as the beam ions. This latter process is of increasing importance as the beam energy rises.

The remaining processes are almost identical to those described for the positive ion beam. These are the balance between the production and loss of electrons as well as positive ions, and we again assume plasma neutrality. In the following section a simple one-dimensional model is described for the beam plasma of a negative ion beam, based on the above processes.

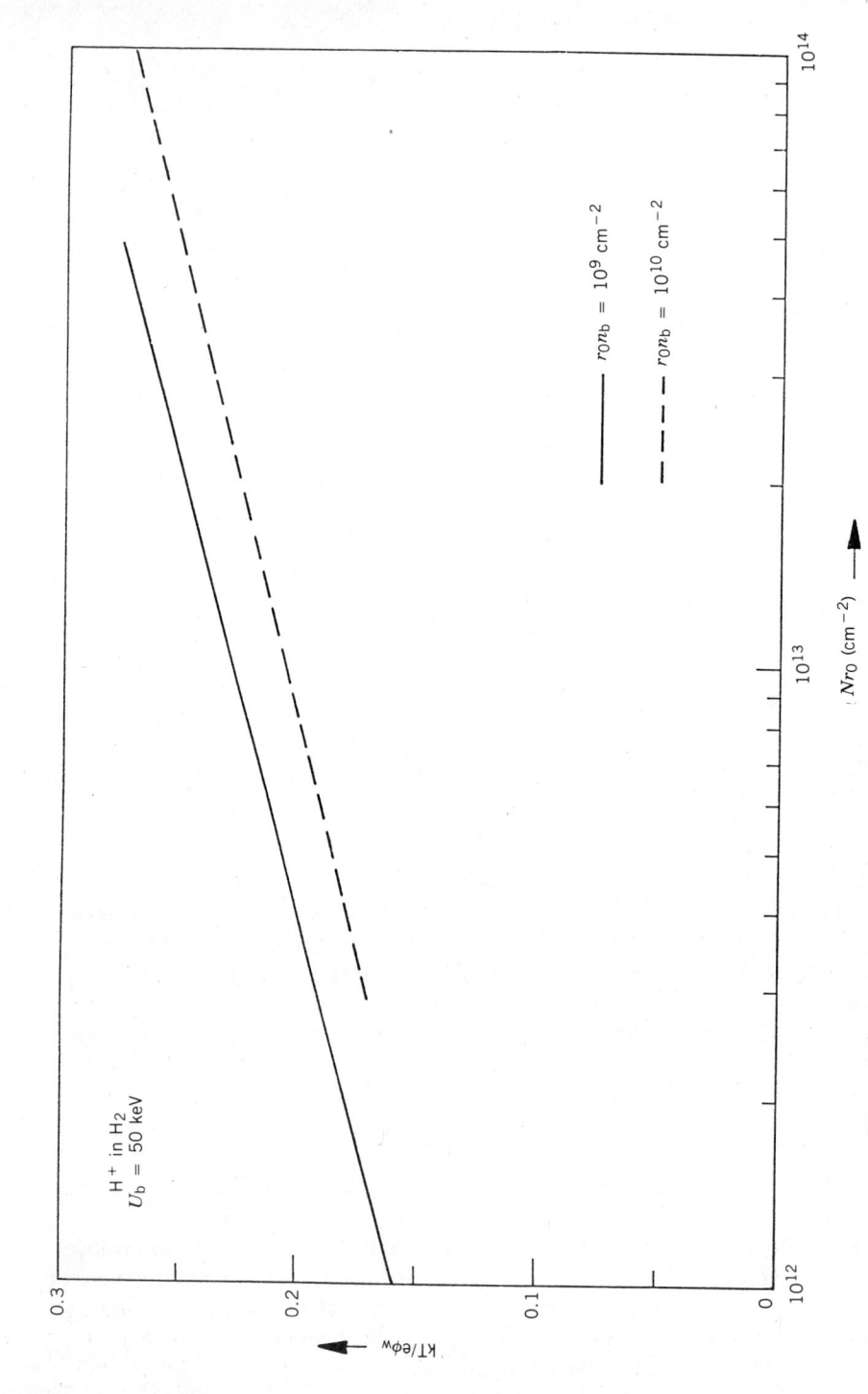

Figure 4.13 Theoretical calculation of the reduced electron temperature (relative to the beam radial potential) as a function of the product of gas density and beam radius.

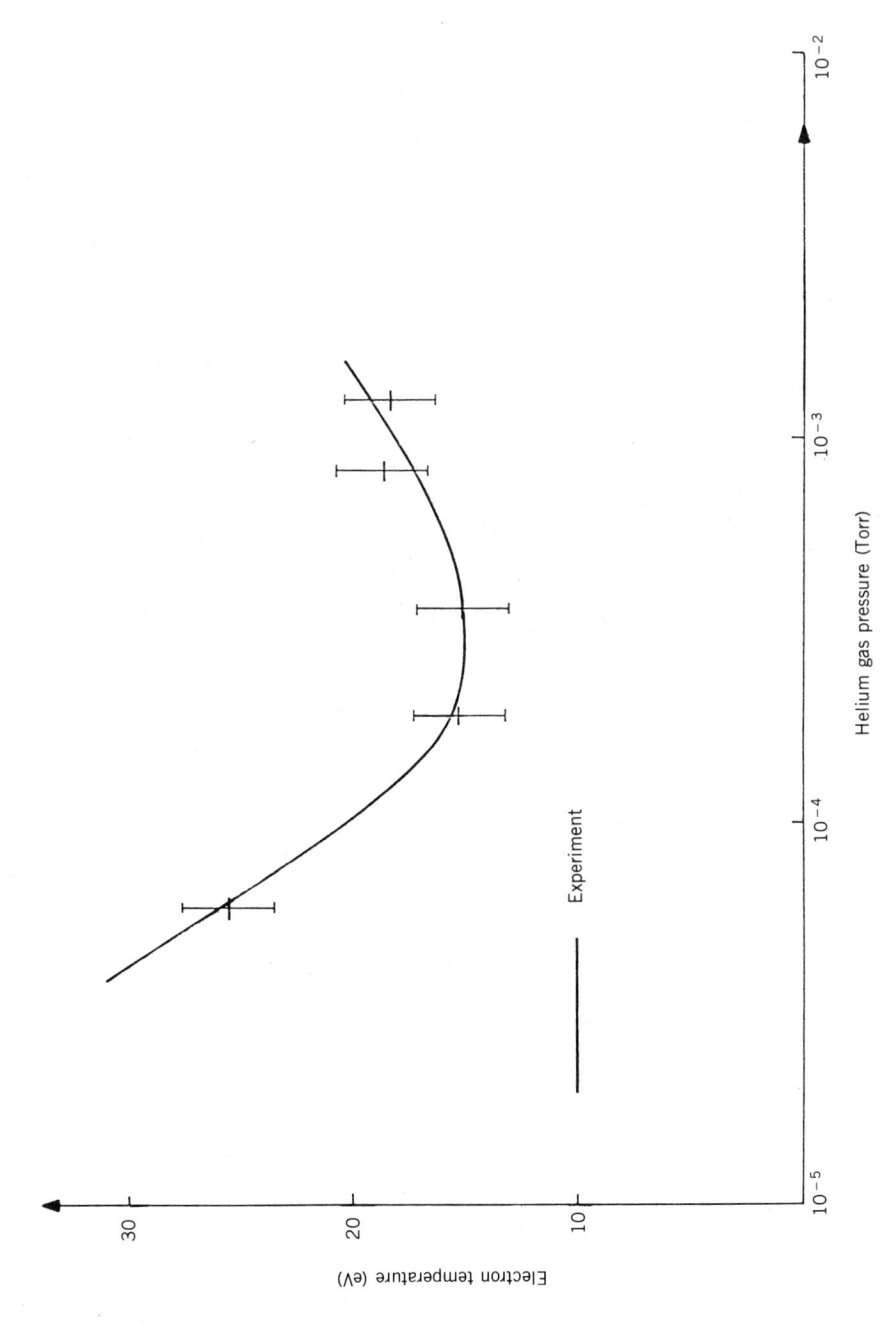

Figure 4.14 Experimental measurement of the electron temperature in a 20-keV He$^+$ beam in He as a function of gas pressure [Ref. 22].

3.3.1 A Simple Model of Negative Ion Beam Neutralization

Two basic interaction processes between the beam ions and gas molecules exist in a negative ion beam. These are

Ionization

$$\underline{A^-} + X^0 \rightarrow \underline{A^-} + X^+ + e$$

Electron detachment

$$\underline{A^-} + X^0 \rightarrow \underline{A^0} + X^0 + \underline{e}$$

or

$$\rightarrow \underline{A^0} + X^+ + \underline{e} + e$$

where A^- represents the beam ion and X^0 is the gas molecule. The underlines indicate fast particles. Unlike positive ion beams these processes generate more electrons than positive ions.

The slow positive ion continuity equation generates an expression that would be identical with Eq. (67) except that the space potential is likely to be so small at low gas pressures that the effect of the small, but finite, slow ion temperature has to be included. In the limit that the beam potential well does not exist, so that ϕ_0 is zero, the slow ion density is given by

$$n_i(r) = \frac{N v_b \sigma_i n_b r_0}{2(2kT_i/m_i)^{1/2}} \tag{79}$$

Hence, we can effectively combine Eq. (79) with Eq. (67) by linear addition of the effective ion energies in the denominator in the event that both a finite ion temperature and a beam potential coexist.

The resulting expression for the slow ion density is then

$$n_i(r) = \frac{N v_b \sigma_i n_b m_i^{1/2} r_0}{(2e)^{1/2} (\phi_0 + kT_i/4e)^{1/2}} (1 + 5\phi_1 r^2 / 6\phi_0 r_0^2) \tag{80}$$

where ϕ_0 and ϕ_1 have the values described in Eq. (66) and kT_i is the ion temperature.

In the earlier case for positive ions, T_i was very much smaller than ϕ_0 and so could be neglected. The plasma electrons obey the Boltzmann distribution so that

$$n_e = n_{e0} \exp(e\phi/kT)$$

$$\approx n_{e0}(1 - e\phi_0 r^2 / kT r_0^2), \tag{81}$$

3 Space-Charge Neutralization of Ion Beams

while the electron production and loss balance over the beam cross section gives

$$\pi r_0^2 n_b \sigma_e N v_b = 2\pi r_0 n_{e0} \exp(e\phi_w/kT) \frac{v_e}{4} \tag{82}$$

where ϕ_w is the potential at the beam edge and we again make the assumption that v_e equals the thermal speed at the beam edge. This yields an identical expression to that developed for positive ion beams.

However, the energy balance equation is significantly different from that for positive ion beams because of the presence of the energetic electrons created by electron loss from the negative beam ions. These electrons have an initial directed velocity along the beam axis which is then randomized by electron–electron and electron–molecule collisions. If we assume that this process dominates over all others at high beam energies, then the average energy gained per electron created ϵ_e is

$$\epsilon_e = \frac{m_e}{m} eU \frac{(\sigma_e - \sigma_i)}{\sigma_e} \tag{83}$$

where m is the mass of the negative beam ions, and the factor $(\sigma_e - \sigma_i)/\sigma_e$ arises from the fact that electrons produced by ionization do not have this directed velocity and are assumed to be formed with virtually no energy. We also neglect beam energy transfer to the electrons. The escaping electrons remove an energy of $2kT$ per electron; hence the energy balance equation is

$$\pi r_0^2 n_b N \sigma_e v_b \epsilon_e = 2\pi r \frac{v_e}{4} 2kT n_{e0} \exp(e\phi_w/kT)$$

which, when combined with Eq. (82), yields

$$\epsilon_e = 2kT \tag{84}$$

The final step is to use plasma neutrality to derive the relative potential at the beam edge G, which gives an expansion of the neutrality condition on axis:

$$n_b + n_{e0} = n_{i0} \tag{85}$$

Hence,

$$1 + 2Nr_0\sigma_e\left(\beta\frac{m_e}{m}\right)^{1/2} \exp(\beta G) = \left(\frac{m_i}{m}\right)\frac{Nr_0\sigma_i}{(G+G_0)^{1/2}} \quad\boxed{86}$$

where

$$\beta = \frac{2m}{m_e} \frac{\sigma_e}{(\sigma_e - \sigma_i)}$$

and

$$G = |\phi_w|/U \qquad G_0 = kT_i/4eU$$

G has an identical meaning to the analogous term for positive ions and is the ratio of the potential well depth to the beam energy.

Away from the beam axis we find that

$$\phi_1 = \frac{6}{5} \phi_0 \frac{n_{e0}}{n_{i0}} \frac{e\phi_0}{kT} \tag{87}$$

Equation (86) has a very different form than the analogous expression for positive ion beams. The reduced beam edge potential $|\phi_w|/U$ is not a function of n_b and depends solely on the product of Nr_0. An illustration of the potential for a 50-keV H$^-$ beam is shown in Figure 4.15 where it can be seen that it rises from zero at low pressures and attains a saturation value. The pressure corresponding to a zero value for ϕ_w/U depends on the slow ion temperature (presumably 300 K). This positive space potential may be sufficiently large to provide a significant amount of gas focusing for the beam ions.

The normalized slow ion and electron densities are plotted in Figure 4.16 for the case described above. These normalized values are again independent of the beam ion density, but otherwise show a similar dependence to that observed for positive ion beams except that the roles of slow ions and electrons are reversed.

3.4 The Effect of Magnetic Fields

The models described above assume that the beam is uniform in the axial direction, but in reality this is rarely the case. In most beam transport systems, the beam is usually divergent and the gas density varies axially due to the location of gas feeds and pumps. Fortunately, these effects have only a modest effect on the degree of beam neutralization because the electrons in the beam potential well are highly mobile in the axial direction and are only confined by the ground plane at the accelerator and beam dump. As a result, this electron mobility has the effect of averaging out local variations in gas density, hence preserving charge neutrality even in regions where it would otherwise be weak due to a low gas density.

The addition of magnetic fields causes a large reduction in the electron mobility. If the field is transverse to the beam axis, as in a bending magnet, then the downstream beam is totally decoupled from the upstream part and would require

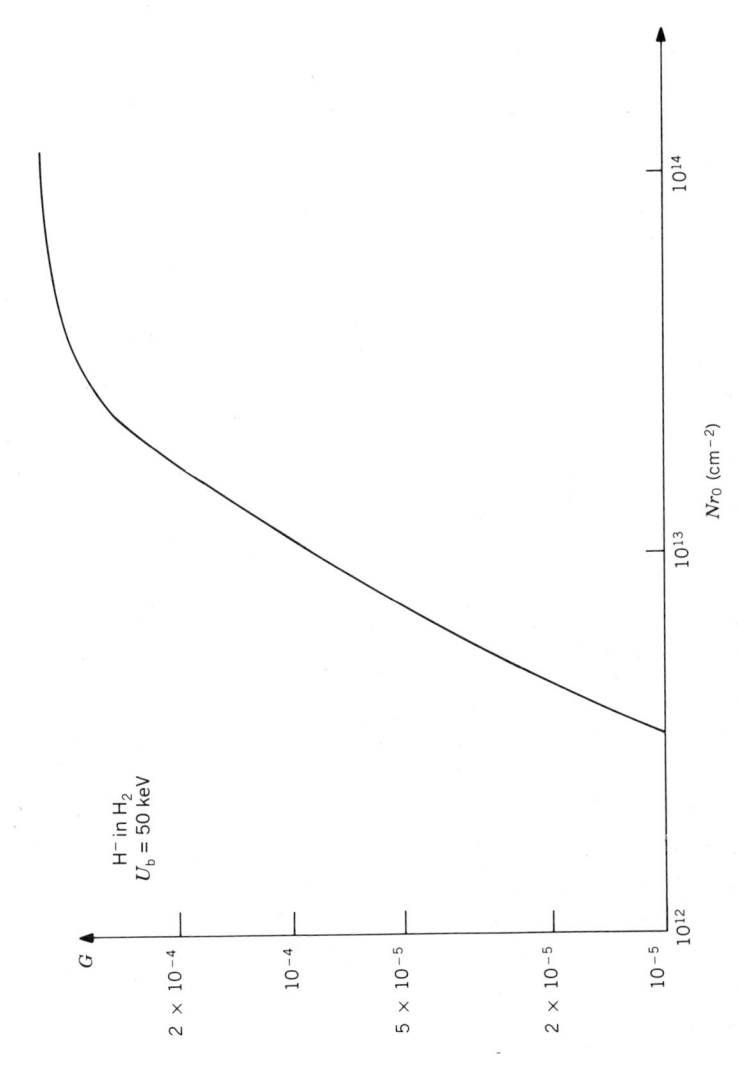

Figure 4.15 Theoretical calculation of the reduced radial potential for a negative H^- beam in H_2 as a function of the gas density and beam radius.

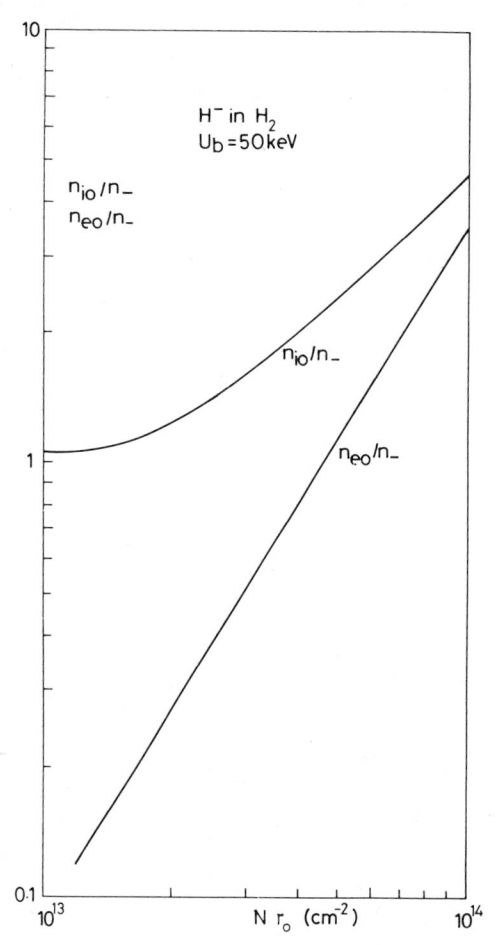

Figure 4.16 Calculation of the slow ion and electron densities relative to that of an H⁻ beam as a function of the gas density and beam radius product.

a separate gas supply to reestablish beam charge neutralization. Magnetic quadrupoles have essentially the same effect.

The rapidly varying degree of neutralization caused by these magnets makes the beam envelope difficult to calculate and indicates that a strong focusing beam channel of this type is not easily compatible with space-charge neutralization.

Longitudinal fields such as those generated in Brillouin flow or a solenoid magnet do allow axial motion of the electrons and space-charge neutralization does not appear to be appreciably modified. This field would have the effect of reducing the transverse diffusion of the electrons without retarding the slow ions; hence, the space potential is likely to be reduced by the application of an axial field, which reduces the space-charge expansion of the beam.

4 NONLAMINAR BEAMS

4.1 Liouville's Theorem

Hitherto we have considered the flow of ions as laminar so that in principle beams can be brought to a point focus. Unfortunately, real beams are not that simple and at any point in space, there will be a range of trajectory angles relative to the principal axis, resulting in nonlaminar flow. We treat this behavior using the concept of phase space where the major axes are x, y, p_x, and p_y for the transverse dimensions and z and p_z for the axial motion, where $p_{x,y,z}$ are the momenta of the ion in these directions.

Motion in phase space is subject to Liouville's theorem, which we derive below as the first step in understanding nonlaminar beam transport. This theorem is proved here by using Hamiltonian methods but a more general method is described by Goldstein [29]. We begin by writing the Lagrangian L for a collection of particles subject to external potentials V and $\mathbf{A}$ arising from electrostatic and magnetic fields.

$$L = -mc^2/\gamma - Ze(V - \mathbf{v} \cdot \mathbf{A}) \tag{88}$$

The Lagrangian does not include effects arising from nonconservative forces such as scattering. The canonical momentum p_i is defined as

$$p_i = \partial L/\partial \dot{x}_i$$

where x_i is the *position* of the ith particle. The Hamiltonian is expressed as

$$H = \sum_i p_i \dot{x}_i - L \tag{89}$$

Hence,

$$-\dot{p}_i = \frac{\partial H}{\partial x_i} \quad \dot{x}_i = \frac{\partial H}{\partial p_i} \quad \frac{\partial H}{\partial t} = -\frac{\partial L}{\partial t} \tag{90}$$

The equations above are known as Hamilton's canonical equations and there is one set of these equations for each degree of freedom, that is, one for each axis in the phase space of the beam. In many instances where the forces in each degree of freedom are decoupled, we can simplify the concept of phase space considerably, but for the moment we retain complete generality.

Liouville's theorem derives from this concept of phase space and states that the density of noninteracting particles in six-dimensional phase space, subject to conservative forces, is invariant when measured along a particle trajectory. The proof of this theorem is straightforward: the density function $f(p_i, x_i, t)$ obeys the continuity equation so that

$$\frac{\partial f}{\partial t} + \frac{\partial}{\partial x_i}(f\dot{x}_i) + \frac{\partial}{\partial p_i}(f\dot{p}_i) = 0 \tag{91}$$

where summation of $i = 1$ to 3 is assumed. Expansion and substitution of the expression

$$\frac{\partial \dot{p}_i}{\partial p_i} + \frac{\partial \dot{x}_i}{\partial x} = -\frac{\partial^2 H}{\partial x_i \partial p_i} + \frac{\partial^2 H}{\partial x_i \partial p_i} = 0$$

yields

$$\frac{\partial f}{\partial t} + \dot{x}_i \frac{\partial f}{\partial x_i} + \dot{p}_i \frac{\partial f}{\partial p_i} = 0$$

or

$$\frac{df}{dt} = 0 \tag{92}$$

following a particle trajectory. Equation (92) is Liouville's theorem. This is analogous to an incompressible gas which indicates that the volume in phase space occupied by an ensemble of particles such as a beam can change its shape but not its volume.

4.2 Emittance and Brightness

For most ion beams, where the magnetic vector potential is zero (ignoring the self field of the beam), we can consider the motion in the two transverse planes as mutually independent and decoupled from axial beam motion. If the ion trajectories are also closely paraxial, such that $p_x = \gamma mc\beta x'$ and $p_y = \gamma mc\beta y'$, then the concept of conservation of phase space volume reduces to the conservation of transverse phase space area $A(x, x')$ and $A(y, y')$ in the two transverse planes. This area is known as the beam emittance multiplied by π. The concept of emittance in the z direction is not meaningful unless the beam is modulated in time, such as could occur in radiofrequency accelerators, and is not discussed here.

Very frequently, the beam has a nonuniform density and it is useful to introduce the concept of the fraction of all the beam particles that lie within the area of the emittance contour along a line of equal density. An illustration of this is shown in Figure 4.17. The case of a Gaussian beam profile derived from a thermal ion distribution (which is the most common) has been intensively studied, and in this instance the beam emittance area ϵ, which contains a fraction f of all beam ions,

4 Nonlaminar Beams

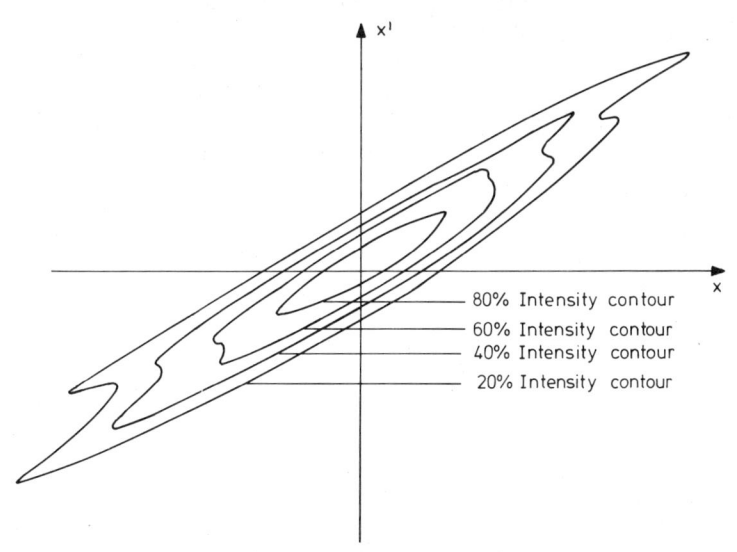

Figure 4.17 A transverse emittance diagram showing the contours of equal intensity. This behavior is characteristic of a quasithermal ion velocity distribution in the direction orthogonal to the beam axis.

is related to f by the expression

$$f = 1 - \exp(-\epsilon/2\epsilon_{rms}) \tag{93}$$

where ϵ_{rms} is the root mean square emittance and has been defined by Lapostolle [30] to be

$$\epsilon_{xrms} = [(\overline{x^2}\,\overline{x'^2}) - (\overline{xx'})^2]^{1/2} \tag{94}$$

where the averages for x and x' are weighted by the beam intensity. This definition differs from the original one by dividing the value of ϵ_{xrms} by a factor of 4. A similar expression exists for ϵ_{yrms}.

One cause of a finite beam emittance is the nonzero ion temperature in the ion source where the ions that form the beam are created by ionization. It is possible to relate the rms emittance to this ion temperature directly. We make direct use of Eq. (94), where the ions are assumed to leave the plasma parallel to the beam axis (apart from the transverse motion imparted by the ion temperature) and with velocity v_b. If the beam ions leave the source through a circular aperture of radius a, then for a uniform emitter we find that

$$\overline{x^2} = \int_0^a 2\pi x^2 y\, dx \bigg/ \int_0^a 2\pi y\, dx$$

$$= a^2/4 \tag{95}$$

For a Maxwellian velocity distribution characterized by an ion temperature T_i we have

$$\overline{x'^2} = \frac{\int_{-\infty}^{\infty} \left(\frac{v_\perp}{v_b}\right)^2 \exp(-mv_\perp^2/2kT_i)\, dv_\perp}{\int_{-\infty}^{\infty} \exp-(mv_\perp^2/2kT_i)\, dv_\perp}$$

$$= kT_i/mv_b^2 \qquad (96)$$

The x and x' terms are uncorrelated so their product average is zero, which yields a value for the rms emittance of

$$\epsilon_{x\mathrm{rms}} = \frac{a}{2}\left(\frac{kT_i}{mv_b^2}\right)^{1/2} = \epsilon_{y\mathrm{rms}} \qquad (97)$$

If the beam energy is not constant then the concept of normalized beam emittance is useful; this is defined as

$$\epsilon_n = \beta\gamma\epsilon \qquad (98)$$

Thus,

$$\epsilon_{nx\,\mathrm{rms}} = \frac{a\gamma}{2}\left(\frac{kT_i}{mc^2}\right)^{1/2} \qquad (99)$$

The emittance is closely related to the concept of beam brightness, which is usually defined as

$$B = \eta I/(\pi^2 \epsilon_{x\mathrm{rms}} \epsilon_{y\mathrm{rms}}) \qquad (100)$$

where η is of order unity (see Section 4.5). The normalized brightness is more useful and is defined as

$$B_n = \eta I/(\pi^2 \epsilon_{nx\,\mathrm{rms}} \epsilon_{ny\,\mathrm{rms}}) \qquad (101)$$

If we substitute Eq. (97) to eliminate the emittance terms we find that

$$B_n = \frac{4\eta j_i}{\pi}\frac{mc^2}{\gamma^2 kT_i} \qquad (102)$$

where j_i is the ion current density at the extraction aperture adjacent to the ion source.

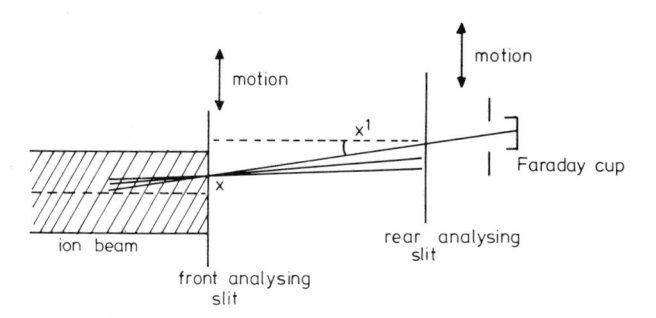

Figure 4.18 The concept of emittance measurement where the position and angle are measured separately.

Hence, the beam brightness in this instance depends solely on the parameters determined by the ion source and is maximized by using sources that produce cold ions at high-current densities. Ion beams with a high ion mass have an improved brightness since this reduces the transverse velocity for a given temperature.

4.3 Measurement of Beam Emittance

To measure the angle of a trajectory precisely, a two-slit approach is used in virtually all diagnostics; this is shown conceptually in Figure 4.18. The spatial position of the front slit determines the position of the diagnostic, while the rear slit measures the angle and angular spread of the trajectories passing through the front slit. Many diagnostics have been developed using this approach and a review of them is presented by van Steenbergen [31]. A more recent and elegant version of this technique, which is capable of very precise measurements of the emittance, has been devised by Allison [32] where electrostatic deflection plates allow the rear slit to remain fixed relative to the front slit, hence reducing the independent motion of both slits to motion of the entire unit as shown in Figure 4.19.

The data from such diagnostics can be analyzed using Eq. (93); Figure 4.20 shows results derived from a low-current (8-mA) H^- beam as described by Holmes [33]. A straight line is obtained, indicating that the beam can be described by a Gaussian or Maxwellian distribution. The transverse ion temperature in this case is 0.25 eV.

4.4 Emittance and Beam Propagation

The paraxial ray equation for one of the transverse planes for a drifting beam can be written in the form

$$X'' + P(z)X = 0 \qquad (103)$$

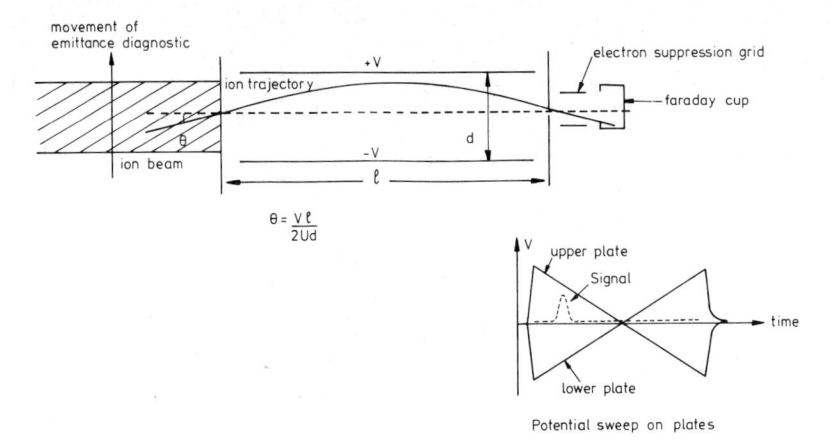

Figure 4.19 The emittance diagnostic, designed by P. Allison, which removes the need for two separate mechanical movements.

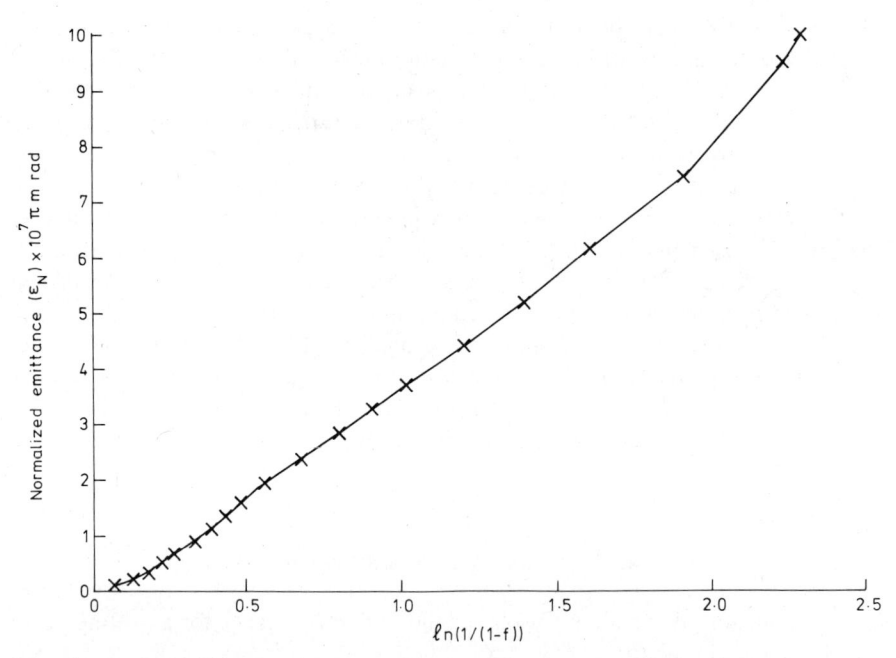

Figure 4.20 For thermal ion distributions the emittance within a contour containing a fraction f of the beam is a linear function of $\ln[1/(1-f)]$. This figure illustrates typical data from an ion beam extracted from a multipole source.

4 Nonlaminar Beams

where

$$X = (\beta\gamma)^{1/2} x$$

and $P(z)$ represents the sum of external focusing forces and linear self-focusing fields in a beam of uniform density. This expression is frequently referred to as *Hill's equation*. A solution to this equation can be expressed by

$$X(z) = aw(z) \cos[\phi(z) - \delta] \qquad (104)$$

and

$$X'(z) = a\left(w' \cos[\phi(z) - \delta] - \frac{1}{w} \sin[\phi(z) - \delta]\right) \qquad (105)$$

where $\phi' = 1/w^2$.

Substitution back into (103) shows that w must satisfy the equation

$$w'' + P(z)w - \frac{1}{w^3} = 0 \qquad (106)$$

When ϕ is eliminated between Eq. (104) and (105) the following expression is obtained:

$$a^2 = \gamma_0 X^2 + 2\alpha_0 XX' + \beta_0 X'^2 \qquad (107)$$

where

$$\alpha_0 = -ww'$$
$$\beta_0 = w^2$$

and

$$\gamma_0 \beta_0 - \alpha_0^2 = 1 \qquad (108)$$

Equation (107) describes an ellipse in the $X'X$ plane and an ensemble of points with different values of δ will follow the ellipse as they move along the axis z subject to the forces described in Hill's equation. Particles that have different values of a will follow similar ellipses, which will be larger or smaller but will have the same orientation relative to the X and X' axes. The area of the ellipse described by Eq. (107) is given by

$$S = \pi a^2 (\beta_0 \gamma_0 - \alpha_0^2)^{1/2} \qquad (109)$$
$$= \pi a^2$$

If the beam consists of an ensemble of trajectories with different values of a, they all move on a set of nested ellipses of constant shape; consequently, the density of particles is constant as expected from Liouville's theorem. Hence, the area S is identical with the normalized emittance multiplied by π. The beam envelope, which is the maximum value $X(z)$ for a given value of S, is

$$X_m(z) = \sqrt{\epsilon_n \beta_0} \qquad (110)$$

and the divergence $\Omega(z)$ is

$$\Omega(z) = \sqrt{\epsilon_n \gamma_0} \qquad (111)$$

The equation of the ellipse enclosing the beam is, hence,

$$\gamma_0 X^2 + 2\alpha_0 XX' + \beta_0 X'^2 = \epsilon_n \qquad \boxed{112}$$

An illustration of the ellipse is shown in Figure 4.21.

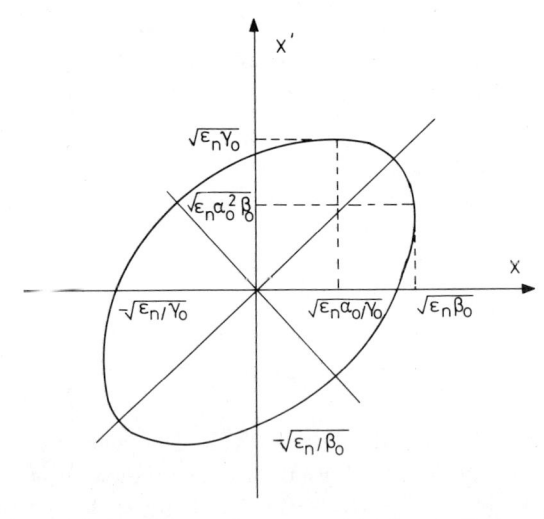

Figure 4.21 The beam emittance diagram of area $\pi\epsilon_n$ and the Courant-Snyder parameters α_0, β_0, and γ_0 are related to the beam divergence and radius as shown in the above illustration.

4 Nonlaminar Beams

To derive the shape of the beam envelope we use Eqs. (106) and (110) to obtain

$$X_m'' + PX_m - \epsilon_n^2 X_m^{-3} = 0 \tag{113}$$

where X_m is the normalized beam envelope radius. A simple example is a uniformly focusing channel where P is a constant. Under these circumstances the ellipse remains upright and the beam is said to be "matched" to the channel. The value of X_m is then

$$X_m = \epsilon_n^{1/2} P^{-1/4} \tag{114}$$

If the beam is to be transferred from a channel of strength P_1 and rematched to a new channel of strength P_2, it has to pass through a region of the intermediate focusing strength P_m where

$$P_m^2 = P_1 P_2 \tag{115}$$

The length of this region should be such that the emittance ellipse rotates by 90 degrees.

4.5 The Kapchinskij–Vladimirskij Distribution

This concept was first proposed by Kapchinskij and Vladimirskij [34] and is referred to hereafter as *the K-V distribution*. It describes a uniformly filled hyperellipsoidal three-dimensional shell in four-dimensional phase space, x, x', y, y'. In a uniform matched focusing channel, the hyperellipsoid axes are parallel to the coordinate axes and the distribution is then

$$f(x, y, x', y') = \delta\left(\frac{x^2}{a^2} + \frac{y^2}{b^2} + \frac{a^2 x'^2}{\epsilon_x^2} + \frac{b^2 y'^2}{\epsilon_y^2} - 1\right) \tag{116}$$

where δ represents the delta function. This distribution has the geometrical property that all two-dimensional projections are uniform so that the charge density in the beam is constant and the self fields are linear with radius. If $a/b = \epsilon_x/\epsilon_y$, then the transverse energy of all particles is the same. Walsh [35] has shown that the form factor η used in connection with beam brightness is 2 for the K-V distribution.

Unfortunately, the K-V distribution is physically not very realistic because beams tend to have a density in four-dimensional phase space that decreases with radius rather than being hollow. However, the K-V distribution provides a convenient model that can be used to discuss self effects and yields results that are in good agreement with experiment.

Equation (113) is the envelope equation with an additional term relating to beam emittance, but with axial fields excluded. If the beam is axisymmetric then the envelope radius r_0 replaces X_m in Eq. (113), and we then have

$$r_0'' + \frac{\gamma' r_0'}{\beta^2 \gamma} + \left[\frac{\gamma''}{2\beta^2 \gamma} + \left(\frac{ZeB_z}{2\beta \gamma mc}\right)^2\right] r_0 - \frac{K}{r_0} - \frac{\epsilon_n^2}{\beta^2 \gamma^2 r_0^3} = 0 \qquad (117)$$

where the effects of the external fields have been written out in their full form and K is the generalized perveance.

If both r_0'' and r_0' are zero, so that the beam is matched and we rely solely on the beam self fields to oppose the expansion caused by the emittance, then

$$K = -\epsilon_n^2 / \beta^2 \gamma^2 r_0^2 \qquad (118)$$

which requires either a magnetic pinch or an overneutralized ion beam. If the beam is noncircularly symmetric, Kapchinskij and Vladimirskij [34] showed that for drifting beams the envelope equations are

$$a'' + k_x a - \frac{2K}{(a+b)} - \frac{\epsilon_x^2}{a^3} = 0 \qquad (119)$$

$$b'' + k_y b - \frac{2K}{(a+b)} - \frac{\epsilon_y^2}{b^3} = 0 \qquad (120)$$

where a and b are the envelope dimensions in the x and y planes, k_x and k_y are the external focusing forces, and K represents the effect of the self fields. The analogous trajectory equations are

$$x'' + \left[k_x - \frac{2K}{a(a+b)}\right] x = 0 \qquad (121)$$

$$y'' + \left[k_y - \frac{2K}{b(a+b)}\right] y = 0 \qquad (122)$$

4.6 Emittance Growth and Nonlinear Optics

So far only linear focusing fields have been considered. However, most lenses or self fields contain terms where the applied field varies as the cube or higher power of the beam radius. These aberrations distort the phase space ellipses while retaining the original area of the ellipse. For the purposes of beam transport it is not the ellipse area that is of importance but rather the area of the new ellipse, which can enclose the distorted phase space ellipse and is, of course, larger than the original ellipse. An illustration of this is seen in Figure 4.22 where the progressive filamentation of the original ellipse causes a continuous growth of the effective beam emittance along the beam axis.

This problem is of particular importance in periodic focusing channels where the filamentation can develop to cause very significant emittance growth. The

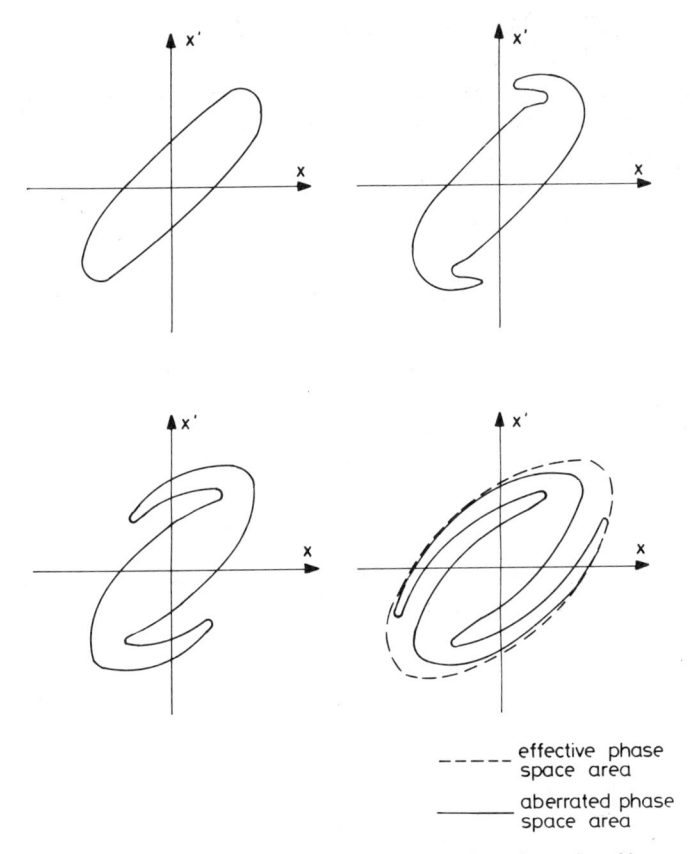

Figure 4.22 The effective emittance growth caused by passage through a series of lenses. The actual true emittance is, however, unchanged.

problem is increased by the fact that beams tend to have nonuniform radial profiles, with the result that the space-charge field is nonlinear while the lens fields are linear, leading to a mismatch between the lens and space-charge fields. This mismatch has been discussed by Hofmann [36], who has shown that the emittance growth is proportional to the beam current in the channel and to the mismatch between the ideal uniform profile and the observed beam profile. Experimental measurements made by Klabunde et al. [37] have confirmed these theoretical models.

5 THE BEAM ENVELOPE OF A DRIFTING BEAM

In the previous sections we have seen the development of the paraxial ray equation from a simple ion trajectory equation to a form that includes the contribution of the self fields and the beam emittance. In this last section we incorporate the effects of beam space-charge neutralization and derive the envelope shape of a drifting beam, a solution that is of importance for many types of ion sources, in particular the neutral beam injectors used in fusion applications. We make the fundamental assumptions that there is no axial magnetic guide field and that the canonical angular momentum is zero, which reduce Eq. (117) to the form

$$r_0'' - \frac{K}{r_0} - \frac{\epsilon_n^2}{\beta^2 \gamma^2 r_0^3} = 0 \tag{123}$$

where the terms in γ' and γ'' are also zero because of the absence of accelerating electric fields.

The term K is the generalized beam perveance (Eq. (48)) which can also be expressed in terms of the beam density:

$$K = 2r_0^2 n_b Z^2 e^2 (1/\gamma^2 - h)/4\epsilon_0 \beta^2 \gamma mc^2 \tag{124}$$

However, in Sections 3.2 and 3.3 the effects of space-charge neutralization were introduced in the form of $|\phi_w|/U$ where ϕ_w is the space potential at the beam edge. The perveance and space potential can be reconciled by the definition of h, which is

$$1 - h = -\frac{\phi_w}{\phi_i} \tag{125}$$

where $\phi_i = \dfrac{Zn_b e r_0^2}{4\epsilon_0}$

Hence,

$$h = 1 + \frac{4\phi_w \epsilon_0}{r_0^2 n_b Ze} \tag{126}$$

and

$$K = -\frac{r_0^2 n_b Z^2 e^2}{2\epsilon_0 \beta^2 \gamma mc^2} \left(\frac{4\phi_w \epsilon_0}{r_0^2 n_b Ze} + \beta^2 \right)$$
$$= -2Ze\phi_w/(\beta^2 \gamma mc^2) - r_0^2 n_b Z^2 e^2/(2\epsilon_0 \gamma mc^2) \tag{127}$$

It should be noted that ϕ_w is a negative quantity for both positive and negative ion beams; hence, K is positive but small for a positive ion beam except at relativistic velocities, and is always negative for a negative ion beam. In the nonrelativistic limit K becomes

$$K_{nr} = -\frac{2\phi_w Ze}{mv_b^2 \epsilon_0} \tag{128}$$

5 The Beam Envelope of a Drifting Beam

5.1 Solutions of the K-V Equation

In Section 4 it was seen that ϕ_w is a function of N, n_b, and r_0 for a given ion species and beam energy; there is no simple expression that can be inserted into Eqs. (123) and (128) to derive a general analytic solution. Numerical solutions can be obtained readily, however, and indicate that the beam envelope for positive ion beams is close to having a hyperbolic shape. However, it is useful to look for an analytic solution to gain an insight into the scaling of the beam divergence with major beam parameters.

Equations (123) and (128) can be written in the form

$$r_0'' + a_1 r_0 + a_2/r_0 + a_3/r_0^3 = 0 \qquad (129)$$

where $a_1 = n_b Z^2 e^2 / 2\epsilon_0 \gamma m c^2$

$a_2 = 2Ze\phi_w / \beta^2 \gamma m c^2$

$a_3 = \epsilon_n^2 / \beta^2 \gamma^3$

However, if the beam current is conserved, then

$$I_b = \pi r_0^2 n_b Z e v_b \qquad (130)$$

The beam space potential of a neutralized positive ion beam at low gas densities can be derived from Eq. (76) and is

$$\phi_w \approx -\frac{Zn_b}{N} \frac{\sigma_u U}{2\sigma_e} \qquad (131)$$

Combining Eqs. (129)–(131) yields

$$r_0'' + b_1/r_0 - b_2/r_0^3 = 0 \qquad (132)$$

where $b_1 = \dfrac{ZeI_b}{2\pi\epsilon_0 v_b \gamma m c^2}$ and $b_2 = \dfrac{ZI_b \sigma_u}{2\pi N \sigma_e v_b e} + \dfrac{\epsilon_n^2}{\beta^2 \gamma^3}$

In the nonrelativistic limit, the b_1 term becomes small and the solution to Eq. (132) approaches a hyperbola with an envelope equation of the form

$$r_0^2 = A^2 + b_2 z^2 / A^2 \qquad (133)$$

where A is the envelope radius at the beam waist, where dr_0/dz is zero.

The hyperbolic waist can be expressed in terms of the initial envelope radius and convergence angle R and R' so that

$$A^2 = R^2 \left(1 + (RR')^2 / b_2\right)^{-1} \qquad (134)$$

and

$$z_w = -\frac{R^3 R'}{b_2\left(1 + (RR')^2/b_2\right)} \quad (135)$$

where z_w is the axial position of this waist from the start position. The beam target radius r_t is now expressed by

$$\begin{aligned}r_t^2 &= A^2 + (L - z_w)^2 b_2/A^2 \\ &= \frac{R^2}{\alpha} + \frac{L^2 b_2 \alpha}{R^2} + 2LRR' + \frac{R^4 R'^2}{b_2 \alpha}\end{aligned}$$

where L is the distance to the target and $\alpha = 1 + (RR')^2/b_2$. Hence,

$$r_t^2 = R^2 + \frac{L^2 b_2}{R^2} + L^2 R'^2 + 2LRR' \quad (136)$$

The value of r_t is minimized when R' is equal to $-R/L$ so that

$$r_{tmin} = Lb_2^{1/2}/R \quad \boxed{137}$$

Returning to Eq. (132) we can express b_2 in the form

$$b_2 = \left(\frac{Zj_+ \sigma_u}{2N\sigma_e v_b e} + \frac{T_i}{8U\gamma}\right) R^2 \quad (138)$$

where j_+ is the initial current density in the beam. Hence,

$$r_t = L \left(\frac{Zj_+ \sigma_u}{2N\sigma_e v_b e} + \frac{T_i}{8U\gamma}\right)^{1/2} \quad (139)$$

Holmes [22] has exploited this effect, for a fixed value of L and j_+, by increasing the value of R so that the effective beam divergence angle Ω, which is defined as $(r_t - R)/L$, decreases to a very low value while the total beam current increased as R^2. This result is shown in Figure 4.23. At high beam energies, the space-charge term eventually dominates.

In the case of negative ion beams, an analytic form for the beam envelope shape becomes difficult to derive as the space-charge and emittance effects oppose each other. At high gas pressures the space potential becomes independent of pressure and beam current. Hence, we would expect the envelope to take on an oscillatory

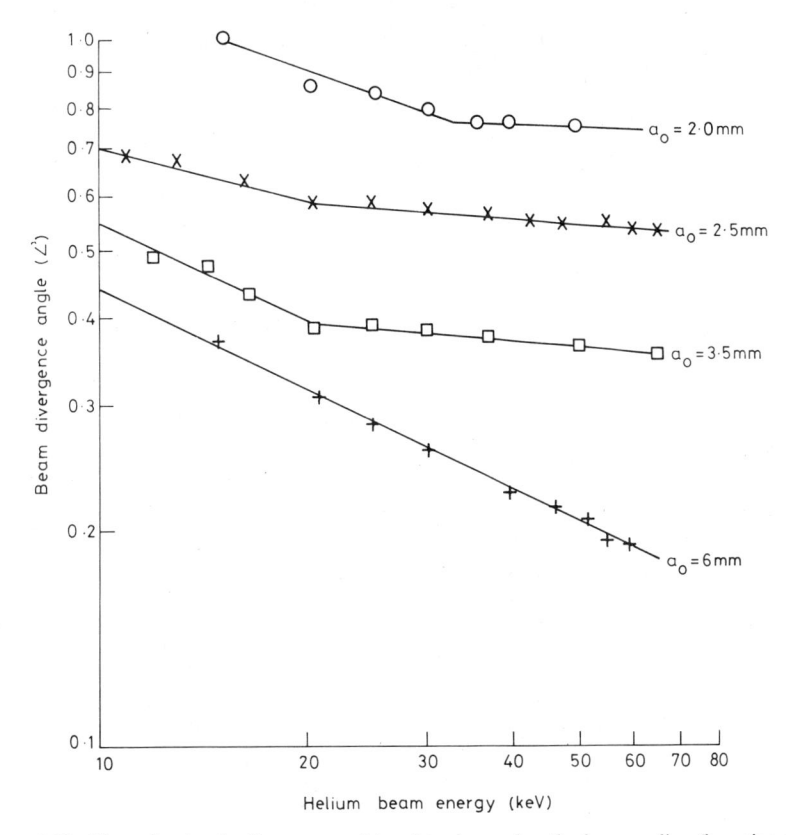

Figure 4.23 The reduction in divergence achieved by increasing the beam radius (by using larger extraction apertures) at constant beam perveance density for an He$^+$ beam in He [Ref. 22].

form, as the emittance term scales as the inverse cube of the beam radius while the space-charge term varies only inversely with radius. The magnitude of these oscillation is critically dependent on the initial starting conditions.

REFERENCES

1. H. Busch, Z. Phys. **81,** 974 (1926).
2. J. D. Lawson, *The Physics of Charged Particle Beams*, Clarendon, Oxford, 1977.
3. J. R. Pierce, *Theory and Design of Electron Beams*, 2d ed., Van Nostrand, Princeton, NJ, 1954.
4. A. J. T. Holmes and E. Thompson, Rev. Sci. Instrum. **52,** 172 (1981).
5. Y. Okumura, Y. Mizutani, and Y. Ohara, Rev. Sci. Instrum. **51,** 471 (1980).
6. C. J. Davisson and C. J. Calbick, Phys. Rev. **28,** 525 (1931).
7. A. B. El-Kareh and J. C. B. El-Kareh, *Electron Beams, Lenses and Optics*, vol. 1, Academic, New York, 1970.
8. I. E. Dayton, F. C. Shoemaker, and R. F. Mozley, Rev. Sci. Instrum. **25,** 485 (1954).

9. P. Grivet and A. Septier, *Nucl. Instrum. Methods* **6,** 126, 243 (1960).
10. K. G. Steffen, BNL Conference, 1961, p. 437.
11. L. N. Hand and W. K. H. Panofsky, *Rev. Sci. Instrum.* **30,** 927 (1959).
12. S. Humphries, Jr., *Principles of Charge Particle Accelerators*, Wiley, New York, 1986.
13. A. C. Crewe, *Proceedings, Int. Conf. High Energy Accelerators*, CERN, Geneva, 1959, p. 359.
14. E. R. Harrison, *J. Electron. Control* **4,** 193 (1958).
15. L. Brillouin, *Phys. Rev.* **67,** 260 (1945).
16. V. S. Anastasevitch, *Sov. Phys. Tech. Phys.* **2,** 1448 (1955).
17. M. D. Gabovitch, L. P. Katsubo, and I. A. Solochenko, *Sov. J. Plasma Phys.* **1,** 162 (1975).
18. G. W. Hamilton, Proceedings, Int. Symp. Ion Sources, Brookhaven Natl. Lab. Report, BNL 50310, 1971, p. 171.
19. D. A. Dunn and S. A. Self, *J. Appl. Phys.* **35,** 113 (1964).
20. E. B. Hooper, O. A. Anderson, and P. A. Willmann, *Phys. Fluids* **22,** 2334 (1979).
21. T. S. Green, *Appl. At. Collision Phys.* **2,** 339 (1984).
22. A. J. T. Holmes, *Phys. Rev. A* **19,** 389, (1979).
23. J. Klabunde and A. Schoenlein, Proceedings, Linear Accelerator Conference, Stanford, SLAC Report 303, Palo Alto, CA, 1986, p. 296.
24. P. Harbour, Culham Laboratory Report CLM-P. 535, 1978.
25. J. D. Sherman, P. Allison, and H. V. Smith, Jr., 1985 Particle Accelerator Conf., Vancouver, BC, May 13–16, 1985; *IEEE Trans. Nucl. Sci.* **NS-32,** 1973 (1985).
26. V. P. Goretskii and A. P. Naida, *Sov. J. Plasma Phys.* **11,** 227 (1985).
27. L. Wright, 4th Symp. on Negative Ion Beams, p. 520, Brookhaven National Laboratory, 1986.
28. D. S. Lemons, M. E. Jones, A. Kadish, H. Lee, and B. S. Neuberger, *J. Appl. Phys.* **57,** 4962 (1985).
29. H. Goldstein, *Classical Mechanics*, Addison-Wesley, Reading, MA, 1950.
30. P. Lapostolle, *IEEE Trans. Nucl. Sci.* **NS-18,** 1101 (1971).
31. A. van Steenbergen, *IEEE Trans. Nucl. Sci.* **NS-12,** 746 (1965).
32. P. W. Allison, J. D. Sherman, and D. B. Holtkamp, *IEEE Trans. Nucl. Sci.* **NS-30,** 2204 (1983).
33. A. J. T. Holmes, Proceedings, Particle Accelerator Conference, **1,** 259, Washington, DC, March 1987.
34. I. M. Kapchinskij and V. V. Vladimirskij, *Proceedings, Int. Conf. on High Energy Accelerators*, CERN, Geneva, 1959, p. 274.
35. T. R. Walsh, *J. Nucl. Energy* **5,** 17 (1963).
36. I. Hofmann, *Nucl. Instrum. Methods* **187,** 281 (1981).
37. J. Klabunde, A. Schonlein, R. Keller, T. Kroll, P. Spaedtke, and J. Struckmeier, Proceedings, Linear Accelerator Conference, Seeheim, Germany, GSI Report 84-11, 1984, p. 315.

5

Computer Modeling

P. Spädtke
GSI, Gesellschaft fur Schwerionenforschung
Darmstadt, Federal Republic of Germany

In this chapter we consider the application of computer programs to the design of charged particle systems. These systems are electron or ion sources, accelerating or retarding systems, and transport sections.

Ion beams have to be produced by ion sources. This causes a lot of trouble, and the optical or accelerating systems following the ion source suffer from the imperfections (divergence, aberrations) of the beam generation process. Because the quality of the overall system is limited by its worst component, it is important to optimize the beam generation system.

Some design considerations can be expressed in closed analytical form. These are helpful in the general layout and include deciding whether to use round or slit apertures, whether a single- or multiaperture extraction system is best, and determining the number of extraction holes or slits to use and the aspect ratio. Furthermore, the general layout of a transport section can be determined, for example, whether electrostatic or magnetostatic lenses should be used, as described elsewhere in this book or in the literature [1, 2].

In these simplified approaches technical aspects are neglected. For example, infinitely thin electrodes with the extraction system are not possible in the real world of sputtering. Misalignments, even when small, are the norm, and should be considered.

Computer simulation codes provide a powerful tool for the optimization of charged particle systems. However, the right program has to be selected to take full advantage of the computer. This chapter is intended to provide the user with the knowledge to make the right choice.

There are three classes of computer programs for the simulation of charged particle beams: (1) envelope models; (2) particle tracking programs; and (3) Ray tracing programs.

In the first class, the envelope models, the beam is described by ellipses in different projections of phase space. These ellipses are transformed by a matrix, called the transfer matrix, for each optical element of the beam line. Knowledge of the optical behavior of each element is therefore necessary. An example of the output of such a code [3] is shown in Figure 5-1.

The advantage of envelope models is that they are very fast (only a few parameters need to be transformed). Therefore, these codes can be made interactive very easily. For example, it would be possible to simulate a beam line with a few lenses, and to increase the focusing strength by "turning a knob." The envelope of the beam can be displayed on a graphic display and the effect of the lenses on the beam can be seen on line. The disadvantage of this kind of code is that aberrations

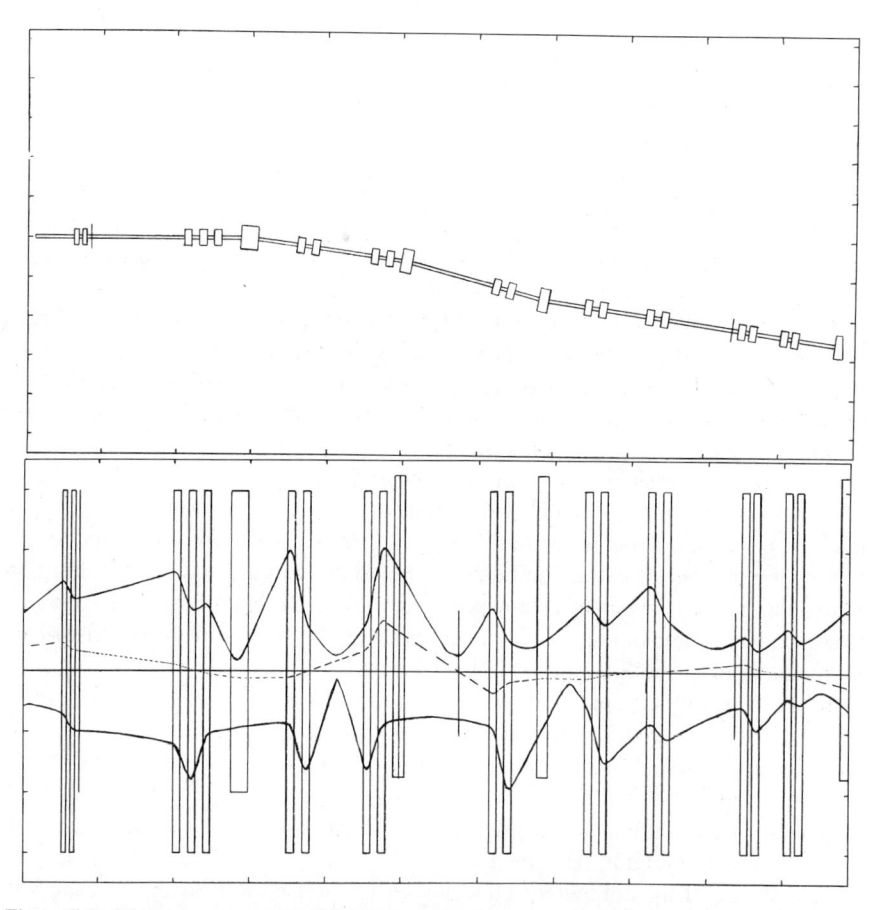

Figure 5.1 Output from program MIRKO [3]. This program can be run on an ATARI as well as on a large IBM 3090. In the survey plot at the top, a beam line is displayed (note that the length of the beam line is more than 110 m). This beam line consists of magnetic quadrupole lenses (doublets and triplets) and dipole magnets. In the lower part of the picture the envelope (horizontal and vertical) is displayed. The dashed line represents the dispersion function.

normally are not considered (the calculations are made in first order; therefore, the size of the ellipse remains constant, and only the ellipse parameters are changed).

Whereas this method can be used to analyze and optimize static beam transport sections (such as drift, electrostatic lenses, solenoids, quadrupoles, or bending magnets), it is not applicable to the extraction of a beam from an ion source.

There is a second way to use transfer matrices: by tracking single particles. Now, instead of three ellipse parameters α, β, γ (for the notation see [4]) for each 2D plane, the two phase space coordinates x, x' and y, y' need to be transformed; however, a few hundred or a few thousand particles need to be transformed. The advantage of this method is that higher order calculations are possible. This gives a good picture of the aberrations the beam suffers in the different lens systems.

The disadvantage of this approach is that it is more time-consuming than the simple envelope model, and the transfer matrix of the optical elements must be known precisely even in higher order. Ion beam extraction systems cannot be simulated by this method.

The ray tracing method is perhaps the only kind of program that can simulate the behavior of a charged particle beam within the extraction system and for low energy, if aberrations are not negligible. Ray tracing in this sense means the following procedure: The equation of motion for a ray is solved exactly. Each ray represents a tube in which charged particles flow. To execute the integration to solve the equation of motion, the potential distribution (or better, the field distribution) must be known exactly. This distribution can be found by some numerical method. Normally the boundary conditions of the electrostatic and/or magnetostatic problem must be known (either the potential or its derivative on a closed boundary).

If time-dependent fields are present, the beam becomes bunched and simple ray tracing programs are not applicable any more, because they assume dc beams. Particle-in-cell (PIC) codes must be used; for example, the programs (PARMILA, PARMTEK, or PARMTRA [5]. Here single particles are traced either with matrix formalism or by integration through a system. Because the Coulomb interaction between different particles has to be simulated individually, a large number of particles is necessary. This type of program is obviously very time-consuming and therefore restricted to mainframe machines. A good description of PIC codes can be found in [6].

In the user manual for the program TRANSPORT [7] I have found the following warning, which I would like to extend to all the programs mentioned here: "The program is superb at doing the numerical calculations for the problem but not the physics. The user must provide a reasonable physics input if he (or she) expects complete satisfaction from the program."

All these computer programs need a lot of computer CPU time, usually measured in minutes. The results, however, may save a lot of experimental time.

In the following I will restrict myself to the ray tracing programs, describing a little bit of the history, as well as the present situation of this kind of program. Examples of a 2D and a 3D code are given; these examples demonstrate how trajectory codes are used.

1 HISTORY

When the first trajectory codes were developed, computer storage space available was restricted to 64 kB[1] on most machines. This is why only 2D codes could be treated by the existing computers. The reason is that the discretization of a physical problem is limited by the number of mesh points. This number, in turn, is restricted by the storage available on the computer. With improvements in computer hardware as well as in the operating system (the program that controls input and output, and running of the application program), more storage has become available. It became possible to develop 3D codes. One 3D code was developed at GSI, which is named KOBRA 3 [8]. Other 3D codes have been developed in Japan by Ose et al. [9], and in Oak Ridge by Whealton [10]. The KOBRA3 code is described in more detail in a later section.

Some of the well-known programs of the early 1970s are the SLAC [11] code, written by W. B. Herrmansfeldt, and the AXCEL [12] program, which was originated in Oak Ridge by Jaeger and Whitson. The original Poisson solver of AXCEL can be traced back to 1965. The SLAC code in its different versions is now spread all over the world. The main application of this program is electron gun design. The AXCEL code, on the other hand, is used mostly for the design of ion source extraction systems. The BEAM [15] code, developed by M.R. Shubaly, is an improved version of the original version of the AXCEL code. There are other codes, developed in other laboratories, such as the WOLF [16] code, program SNOW [17] written by J. E. Boers, and ION written by D. Dimirkis.

There are no major differences between these programs and there are a lot of publications dealing with the applications (see for example, [18–22]). Individual advantages are due to the special applications for which the programs are written. This is why some codes are well suited for fixed cathode, and others are better adapted to plasma ion sources. The principal way of solving the physical problem is similar for all these programs and is described in a later section. There is a compendium of useful computer programs for accelerator application available from the Los Alamos computation group [23], which might be very useful for the selection of a program.

Today, some of the codes have become interactive; this improves the feasibility of optimization. The philosophy is that the user should be flexible during the program execution about how to proceed with the program.

Input errors (wrong geometry input, for example) are detected before execution of the time-consuming main parts of the program (solving the potential equation or the ray tracing) and can be corrected easily. The influence of small changes in the problem parameters, such as potential, geometry, and current density, can be displayed after carrying out a few iterations. Feedback from the computer to the user is very fast. There is no question that productivity can be improved greatly with an interactive program. On the other hand, it must be pointed out that on a

---

[1] kB is a measure of computer memory size.

1 History 111

computer with many users there might be problems (especially time problems) in running a lot of interactive programs at the same time.

A well-known interactive 2D code is the AXCEL-GSI [28] code, which runs on several different computers (IBM, CRAY, VAX, ATARI). Like AXCEL-GSI, the three-dimensional program KOBRA3 is an interactive code that also runs on these computers.

After this brief look into history, it is interesting to project into the future. Parallel to the increasing power of the mainframes, small computer systems are getting faster and storage problems diminished. So it is no surprise that today the above-mentioned simulation codes are available on quite small computers. A new type of computer architecture, the transputer, will have more CPU power than the largest mainframes of today; these computers are arranged in clusters, to solve the numerical problem in parallel.

With the increasing power of very small computers (16-bit PC are available in every laboratory and the 32-bit generation is yet to come) and the increasing costs of CPU-time on the mainframe, it makes sense to use the trajectory codes even on tiny computers (as an example of the efficiency of these small computers, see Figure 5.2). It is common today to have a few Mbytes of fast storage medium available, nearly in the size of a pocket calculator. Of course, the CPU-speed differs from the cycle time of a mainframe, but on the other hand, in a large computer center the user has to share the computer with a few hundred other users.

To compare different computer types, benchmark tests are used. For meaningful comparisons, a special benchmark program for each application should be used. For the problems described here, the Laplace equation will serve as a good benchmark. A FORTRAN program to do this is listed in the appendix to this chapter as a proposed standard; the source code is easy to translate into other languages. Table 5.1 shows the CPU time necessary to solve this benchmark program on different computers using different languages. In this benchmark program the only boundary conditions given on a mesh of 100×100 mesh points are that the lower and left sides are set to zero potential, and the upper and the right sides are set to the unity potential. The simple five-point rule (for definition see the appendix) is used to perform the discretization. With the help of the successive overrelaxation method (with a fixed relaxation parameter of 1.8) the set of equations is solved. The iteration is carried out 300 times. This numerical method is described in a later section.

It should be pointed out that the rate of convergence differs on different machines. This depends on whether the calculation is carried out in single or double precision, and on the quality of the arithmetic subroutines (the firmware) on each computer.

The solution times of this particular problem give some feeling for the performance on different computers. At first glance it is surprising that the vectorizing machines, such as the CRAY, do not show larger time advantages. It turns out that if the program structure is not adapted to the vector facility, there is no, or only poor, advantage of that feature. This adaptation to the vector facility, however, might be difficult for a special algorithm.

112   *Computer Modeling*

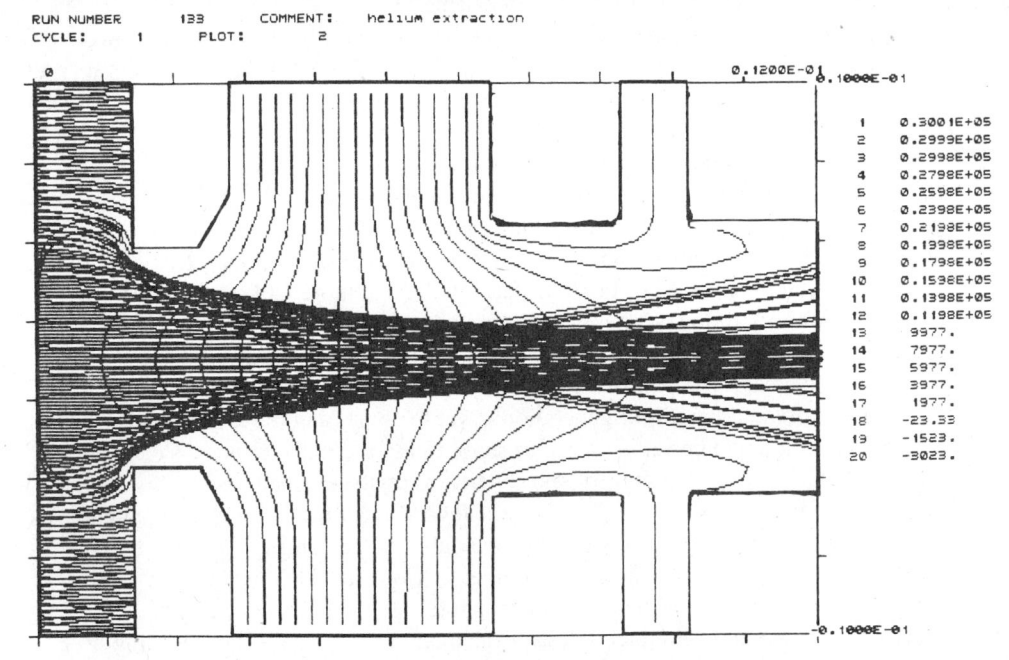

Figure 5.2 Output of a trajectory code, produced on an ATARI. In fact, the program is identical to the AXCEL-GSI code, which runs on the IBM 3090. The cylindrically symmetric geometry of an extraction system, trajectories, and potential lines are plotted for the first three iterations.

The small PCs are included in this table to demonstrate that the time differs by only about a factor of 1000 or less. On the other hand, machine cost is much cheaper than this factor, and most small systems are single-user systems, which decreases the advantage of the mainframes still further.

2 TECHNIQUES OF RAY TRACING PROGRAMS

The equations to be solved are known as the Poisson problem:

$$\delta^2 U/\delta x^2 + \delta^2 U/\delta y^2 + \delta^2 U/\delta z^2 - \rho/\epsilon = 0 \qquad (1)$$

In the 3D case, or in the 2D rotationally symmetric case,

$$\delta^2 U/\delta z^2 + \delta^2 U/\delta r^2 + 1/r \; \delta U/\delta r - \rho/\epsilon = 0 \qquad (2)$$

The space-charge term can be calculated by the Lorentz equation,

$$d(p)/dt = q(\mathbf{E} + \boldsymbol{v} \times \mathbf{B}) \qquad (3)$$

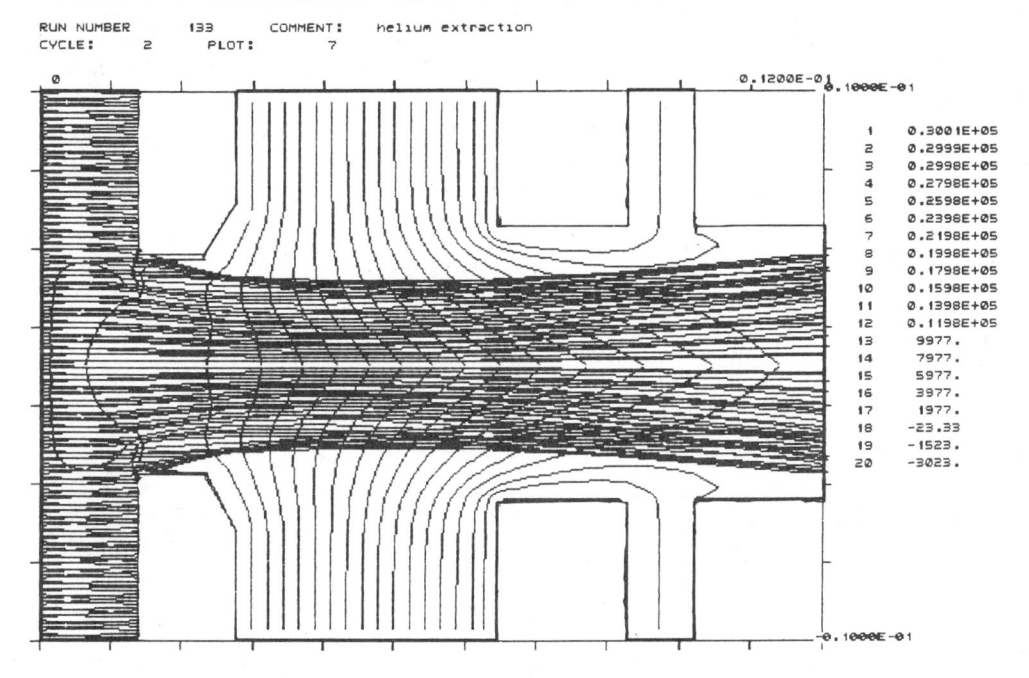

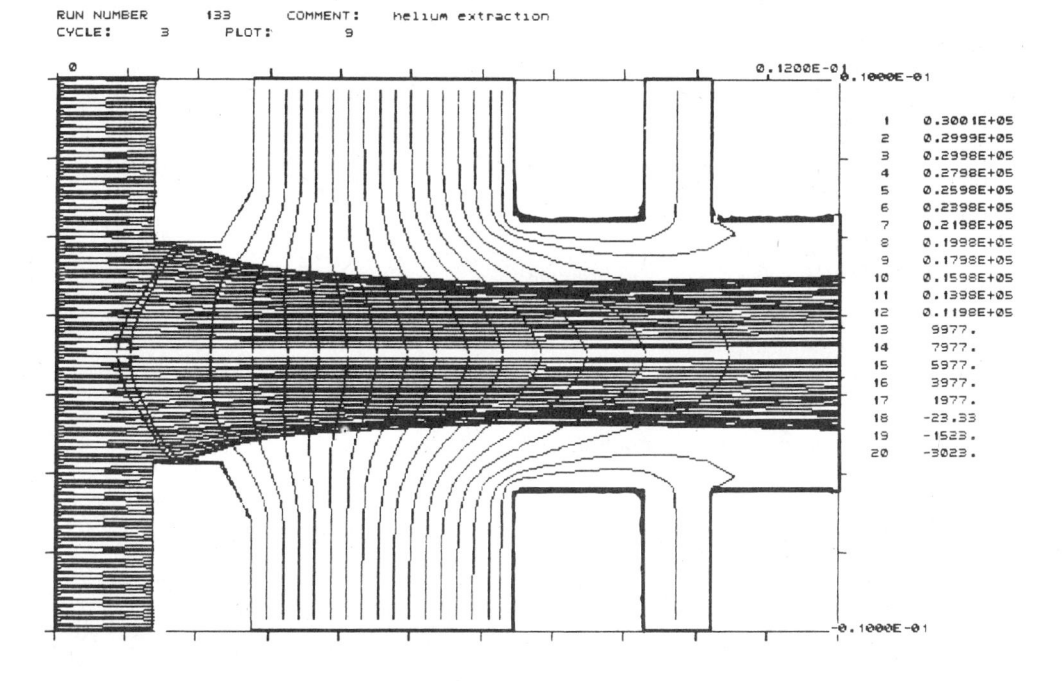

Figure 5.2 (*Continued*)

TABLE 5.1 CPU Time (s) for the Benchmark Program Described in the Appendix[a]

Computer	Language	Single Precision	Double Precision
McIntosh	Pascal*	60000.00	
McIntosh	Basic*	54000.00	
Commodore C64	Assembler	18000.00	
ATARI ST	GFA Basic*	7720.00	
HP 9816	Basic*	7710.00	
IBM PC	Fortran	7055.00	
McIntosh	Fortran	2700.00	
ATARI ST	GFA Basic	2670.00	
ATARI ST	Fortran	2520.00	
COMPUPRO	Fortran	1975.00	
ATARI ST	Assembler	1800.00	
IBM XT (with 8087)	Fortran	1723.00	
IBM AT-02	Fortran	1280.00	
μVax I	Fortran	356.00	
HP 9000/320	Fortran	263.00	
PROTEUS (Inmos T414)	Occam	257.00	700.00
Apollo 3000	Fortran	187.00	204.00
SUN 3/110	Fortran	159.00	175.00
Apollo DN580	Fortran	158.00	174.00
μVax II	Fortran	80.00	114.00
SUN	Fortran	49.00	72.00
Apollo 9000	Fortran	43.00	
PROTEUS (Inmos T800)	Occam	45.10	56.00
SUN 3/300	Fortran	39.00	
VAX II/785	Fortran	30.00	
μ VAX II FPA	Fortran	29.00	
VAX 8600	Fortran	11.10	21.10
VAX 8700	Fortran	10.50	15.80
SPERRY II/91	Fortran	6.70	
CRAY II	Fortran	2.89	
IBM 3090-200	Fortran	2.61	3.70
CRAY XM-P	Fortran	2.20	

[a] This table has been compiled with the help of R. Becker [29].
*The language is interpretive.

and the law of charge conservation,

$$\text{div } \mathbf{j} = 0 \qquad (4)$$

In these equations U is the potential, $\mathbf{E}$ the electric field, $\mathbf{B}$ the magnetic induction, $\mathbf{v}$ the velocity of the particles, ρ the charge, and ϵ the permittivity. p is the momentum, t is the time, and x, y, z, and r are the space coordinates.

2 Techniques of Ray Tracing Programs

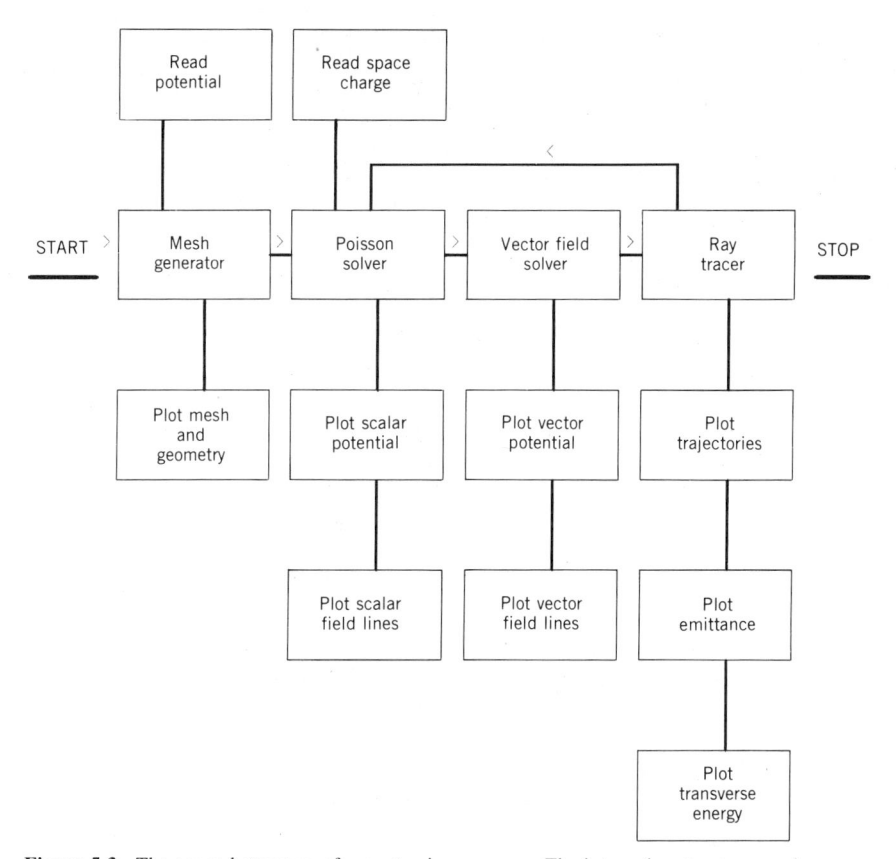

Figure 5.3 The general structure of a ray tracing program. The interactive structure can be seen.

Clearly, the equations are coupled. In most cases, such a system of coupled equations cannot be solved analytically, and an iterative procedure has to be used. The structure of the iteration scheme to be used here is shown in Figure 5.3. This major iteration is repeated until convergence is achieved. As a criterion for convergence the deviation in the potential calculation or the result of the ray tracing can be used.

2.1 Poisson Equation

The question of whether the FEM (Finite Element Method) or FDM (Finite Difference Method) should be used to solve Poisson's equation has not been answered definitively, but FDM is used in all the programs described here.

The advantage of the FDM method is that it is more straightforward. The mesh generation is normally made by the computer, because it is easy. Attention must be given to suitable interpolation methods for geometric shapes. In most of the

FDM programs the Poisson equation is discretized by the five-point rule (sometimes the nine-point rule is used). For the definition of these rules see the appendix. The distances are normally the mesh size, but if a boundary is closer to the midpoint, it should be replaced by the correct distance.

The FEM, on the other hand, has advantages in fitting the mesh to the problem geometry, but the disadvantage is that dividing up the regions into finite elements usually has to be done by the user, which is time-consuming.

In both cases a large number of linear equations is to be solved. Numerically this problem can be solved either by an iterative solver or by a direct solver. For the iterative solver, a successive overrelaxation method is usually used (Gauss Seidel method [24]). The iteration function can be written as

$$U^{n+1} = \omega U + (\omega - 1) U^n \tag{5}$$

with

$$0 \leq \omega \leq 2 \tag{6}$$

where U is the result of the five-point rule equation, U^n is the result of the previous iteration, U^{n+1} is the new result, and ω the relaxation factor.

For ω larger than one the procedure is called overrelaxation, for ω less than one it is underrelaxation. To improve the rate of convergence an optimum value for ω must be chosen. The rate of convergence can increase greatly, as can be seen in Figure 5.4.

In this example the Laplace equation has been solved on a mesh of 100×100 mesh points with given boundary conditions. The same test was carried out for the 3D case, and similar results were obtained. Here the benchmark test program was extended simply by inserting another loop. With increasing number of mesh points, the overrelaxation parameter should be closer to two. This procedure is damped even for a small mesh size and large ω.

2.2 Cathode Simulation

If the emitting surface is fixed, as in surface ionization sources or electron guns, the current can be calculated by the Child-Langmuir law,

$$\mathbf{j} = \frac{4\epsilon_0}{9} \sqrt{\frac{2q}{m}} \frac{\phi^{3/2}}{d^2} \tag{7}$$

where $\mathbf{j}$ is the current density, q the particle charge, m the particle mass, ϕ the applied voltage, and d the extraction gap. This option is included in most of the trajectory codes. In this case no space-charge compensation is considered and the results are quite reliable so long as the source is operated within the space-charge-limited regime.

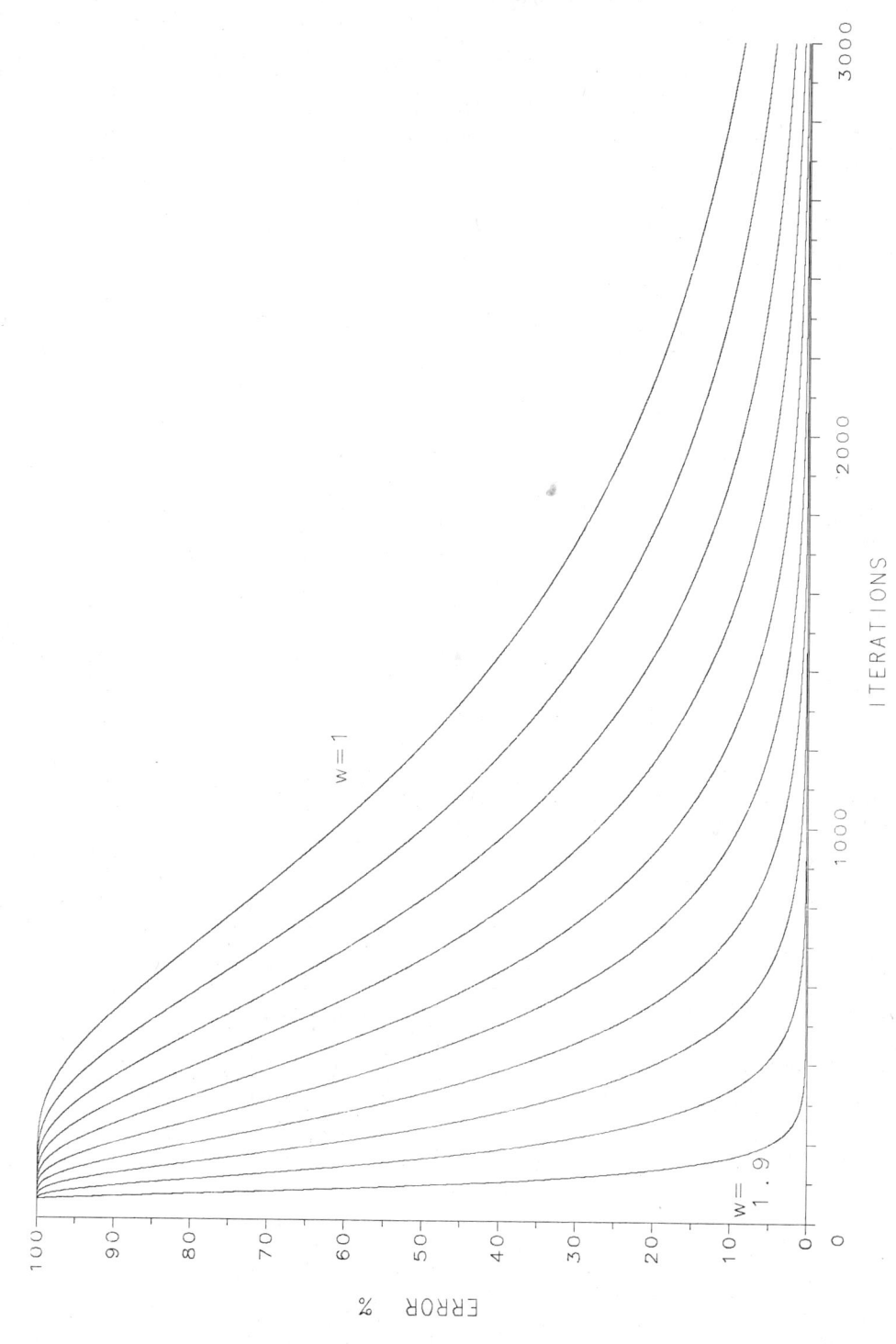

Figure 5.4 Rate of convergence for the SOR iteration, using different relaxation parameters. The benchmark program was used to demonstrate the effect of different ω.

2.3 Plasma Simulation

To simulate the extraction of an ion beam accurately, a self-consistent plasma sheath algorithm must be included within the program. This ensures correct starting conditions for the ions.

So far we have taken into account the space charge of one charge species. But in the case of ion beam extraction systems, for example, there are at least two species, positive ions and negative electrons. More difficulties will arise in the case of negative ion sources (volume production of negative ions, for example). In that case there are three populations to be taken into account.

The space charge (ρ) is composed of two or even three populations, positive (n_{i+}) and negative ions (n_{i-}) and electrons (n_{e-}). For the case of a plasma where negative ions can be neglected we can write

$$(n_{i+} + n_{e-})q = \rho \tag{8}$$

and where negative ions cannot be neglected,

$$(n_{i+} + n_{i-} + n_{e-})q = \rho \tag{9}$$

For the case of positive ions it can be shown that the electron density generated within the plasma depends on the potential: Within the plasma,

$$n_{e0} = n_{i0} \tag{10}$$

close to the plasma boundary,

$$n_e = n_{e0} e^{-(\Delta U/kT)} \tag{11}$$

and far away from the plasma,

$$n_e = 0 \tag{12}$$

where ΔU is the deviation of the potential at a given place from the plasma potential and kT is the electron temperature. The number density of ions n_{i+} is assumed to be equal to n_{e-} within the plasma. Furthermore, a Boltzmann distribution of the electrons is assumed within the plasma [25]. It should be noted that this density distribution is an analytical expression, and it must be demonstrated that this formula is applicable to the given problem. That the density of the neutralizing particles depends on the potential is the reason the potential equation (Eqs. (1), (2)) now becomes nonlinear.

One possible way to solve such a nonlinear equation is to iterate. For each mesh point a self-consistent potential and electron density has to be found. The iteration for each point can be done by a simple Newton-Raphson iteration. Now it is clear why a direct Poisson solver is not a good choice in solving this special plasma problem: There is no way to carry out such an inner iteration loop when the set of equations is solved with a direct Poisson solver.

For negative ion sources the space-charge-compensating electrons are replaced by positive ions. Then two negatively charge components will be extracted—negative ions and electrons. A self-consistent model can be used to determinate the correct space charge. Within the plasma the charge neutrality condition is

$$n_{i+0} = n_{i-0} + n_{e0} \tag{13}$$

close to the plasma boundary the positive ion density decreases according to

$$n_{i+} = n_{i+0} e^{(\Delta U/kT)} \tag{14}$$

and far away from the plasma,

$$n_{i+} = 0 \tag{15}$$

The space charge of the negative charges decreases with increasing potential.

Preliminary attempts have been made using KOBRA3 [26] to solve this multi-species problem without any analytical formula to simulate the plasma, but three different species have been included in the plasma. Space-charge neutrality must be ensured by a proper ratio of positive to negative particles (rays) and their starting velocities. Furthermore, it must be ensured that particles are not trapped within a potential dip, because the resulting very high space charge would lead to a divergent behavior.

In summary of this special topic, there are two different ways of representing the space charge close to the plasma boundary: by an analytic description and by a multispecies trajectory code. For the second method, first computer experiments have been carried out.

2.4 Ray Tracing

The space charge that is used to solve the Poisson equation is created during ray tracing. Therefore, in the first iteration there is normally no space charge. To distribute the space charge onto the mesh smooth enough, a minimum number of traces are calculated by integration of the equation of motion.

Normally this initial value problem of second order is replaced by a system of first-order differential equations. This system can be solved by any integration method. If the position and the velocity of each ray is known, the charge distribution can be calculated easily.

An exact calculation of the fields is important for the accuracy of the ray tracing. Normally first-order interpolation is sufficient if mesh density is high enough.

Higher order interpolation becomes very time-consuming especially if 3D ray tracing is required.

3 REPRESENTATION OF RESULTS

One of the most important parts of a computer program is how to output the results. There is no advantage to providing, for example, printed tables of potential maps; nobody would look at them. All the results should be provided graphically, if possible. In the case of potential, this should be field lines or lines of constant potential. Sometimes it might be useful to plot the potential along a given trace. Therefore, the output data should be easy to access. Users should have the option to write their own diagnostic programs. As well as potential plots, trajectory plots and emittance diagrams should be available. Sometimes current density profiles are necessary. More examples are given below.

The quality of a beam can be specified by the emittance. However, it must be clear which definition of emittance is used in any particular case. The following definitions are commonly in use:

2D emittance $x, x' \mid _{y=y_0, y'=y_0'}$

2D emittance integrated in the perpendicular plane $\iint x, x' \, dy \, dy'$

The resulting area of each emittance figure, independent of its definition, can be described in different ways:

The area of the emittance, calculated using the rms quantity; or

The total (minimized) area of an ellipse which encloses all points or a given fraction [27].

These quantities are either multiplied by π or not, depending on the author. Sometimes the absolute emittance is used, sometimes the normalized one. To complete the confusion, the normalization can be made with respect to the velocity or to the mass. See Chapter 3, Ion Extraction, for further discussion. Independent of the special definition of the emittance, the emittance together with the current information is a very good measure of beam quality.

Another quantity that describes the beam quality (especially in use for a beam in a magnetic field) is the transverse energy. Principally, this gives the same information as the emittance picture, but along the path of the trajectory, rather than at a single location. Sometimes this is more useful than the emittance.

Other diagnostics one might consider are the current density profile of the beam and its development. A more sophisticated diagnostic is, for example, the integration of electric and magnetic field components along a particle path to investigate the influence of field errors (this might be a good application for a 3D code).

4 A 2D EXAMPLE

The following is an example of how to optimize the design of a high-current ion beam generating system used for an accelerator. It should be pointed out that very special conditions are chosen in this example, but the technique of the optimization

4 A 2D Example

is of importance here, not the example itself. Special care has been taken to demonstrate the advantages of an iterative program [28].

4.1 Ion Beam Extraction

We consider a plasma ion source from which a positive ion beam (in this case argon) is extracted. It is very important to match the current density or plasma pressure close to the extraction region to the extracting field strength so as to create the correct plasma boundary. This effect is demonstrated in Figure 5.5, where the current density is increased from the top to the bottom.

The correct plasma boundary, however, depends on the special design goals of the extraction system, which could be, for example, that the extracted current should be as high as possible within a given acceptance, or the beam should have a minimum of abberations or minimum divergence.

Typical values for normal source operation for accelerator injection might be a divergence half-angle of 20 mrad and a spatial size of 20 mm radius at 1 m from the source. Of course, each optimization of an extraction system depends on the application. It should be pointed out that the optimization does not depend on the specific ion source so long as the required ion species can be produced and the source is capable of operation over a wide range of current density.

The physics to be solved is the following: The source plasma is space-charge neutralized, as well as the beam in the drift section behind a screening electrode. Inside the extraction system, full space-charge forces act on the beam. Therefore, Self's law (as described in Section 2.3) can be applied. After a few iterations we will find a self-consistent solution for the current density inside the source for the given geometry and given potentials on these electrodes. Then the current density has to be changed until the maximum current is inside the acceptance of the beam line. Which part of the extracted beam is inside the acceptance can be determined in the emittance picture, which can be displayed at any position of the calculated problem. Such emittance pictures are shown in Figure 5.6. The electrode shaping as well as optimization of the spacing of the electrodes can start at this point. After each modification of the geometry, the plasma density has to be rematched to the changed extraction field, to find the optimum current within the acceptance. When an optimum design has been found, it might be of interest to postaccelerate the beam. Such a possibility is shown in the next section.

4.2 Drift Space

After a drift distance of 0.5 m, and under the assumption of 95% space-charge neutralization, the emittance will change its shape, as shown in Fig. 5.7. The input emittance is shown in Figure 5.7a. The emittance at the end of the drift sections without space charge is shown in Fig. 5.7b, and that with space charge and the assumption of 95% space-charge compensation is shown in Fig. 5.7c. The space-charge distribution in this example is, of course, not homogeneous, due to aberrations of the beam.

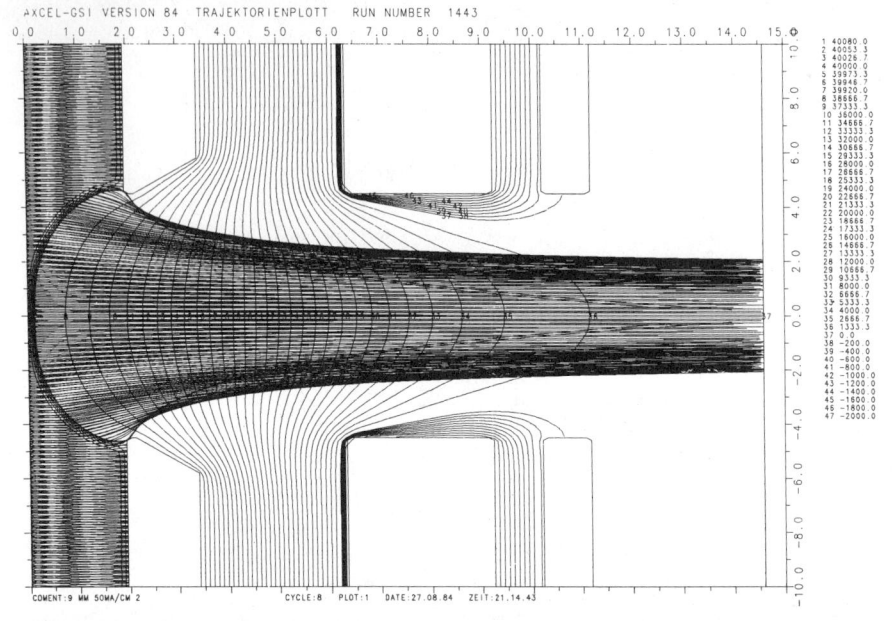

Figure 5.5 Influence of current density (particle density) close to the extraction system for constant extraction field. (Current density increases from top to bottom.)

4.3 Acceleration

The output coordinates of the beam from the drift section are used to accelerate the beam to the energy of 256 keV. For a correct simulation it is necessary to simulate the space-charge compensation on the high-potential side of the acceleration gap. The beam is injected into an acceleration column until a gap length is found that produces a parallel beam. The accelerating gap with the beam is shown in Figure 5.8.

5 A 3D EXAMPLE

There are a lot of problems that cannot be solved by 2D codes, for example, extraction from the ion source using slits or a multiaperture extraction system, beam steering caused by electrode displacement, or beam optics in the presence of general magnetic fields. Two examples are shown in Figures 5.9 and 5.10.

For a 3D simulation the problem of geometry input becomes a much more severe problem compared to the 2D case. This is true for the FDM as well as for FEM programs. Some preprocessors for different programs are available, but it should be pointed out that the use of present 3D codes is not an easy task. Good knowledge of the special program is necessary to obtain correct results. Automatic control of the input data for errors becomes much more important.

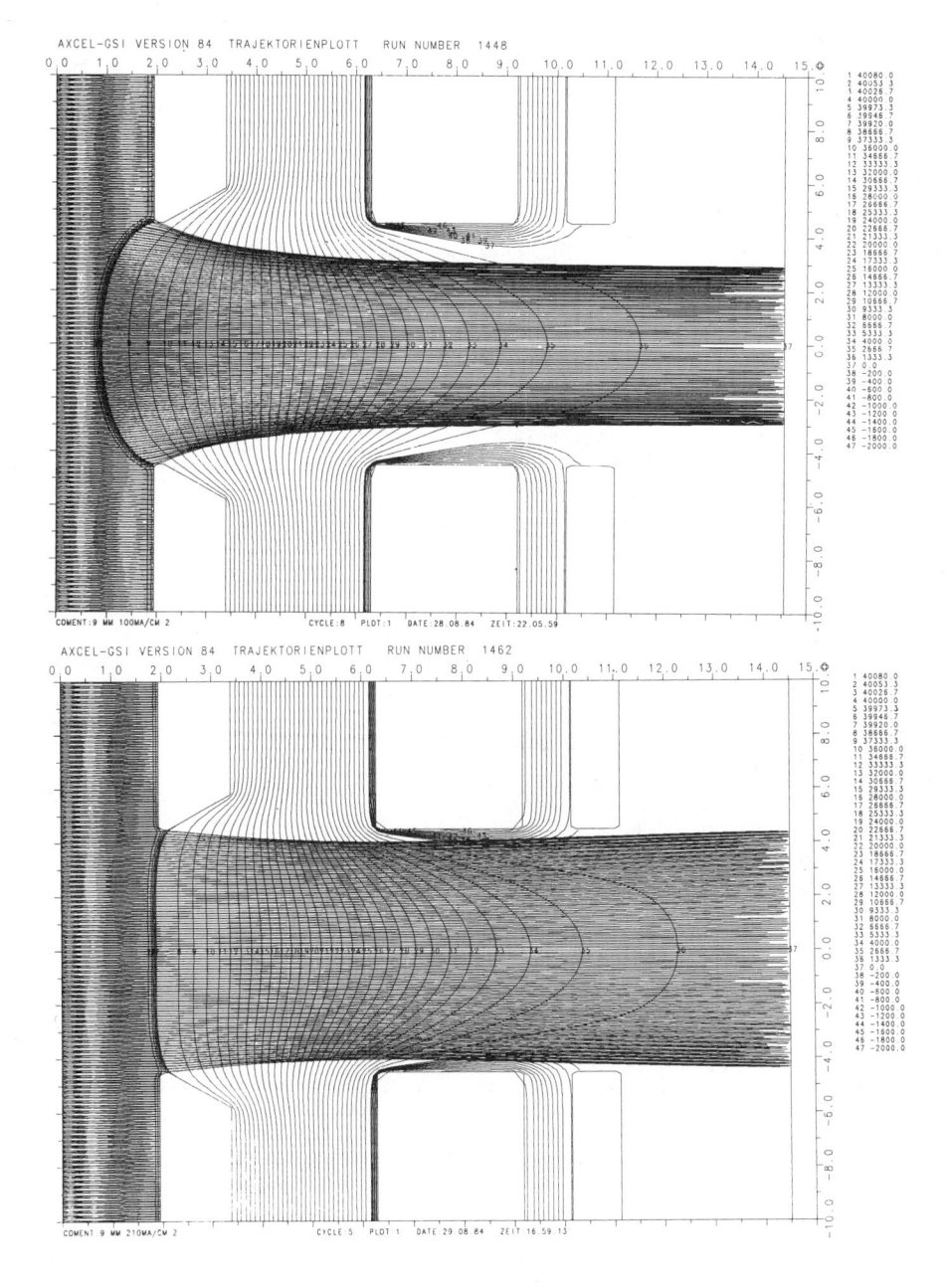

Figure 5.5 (*Continued*)

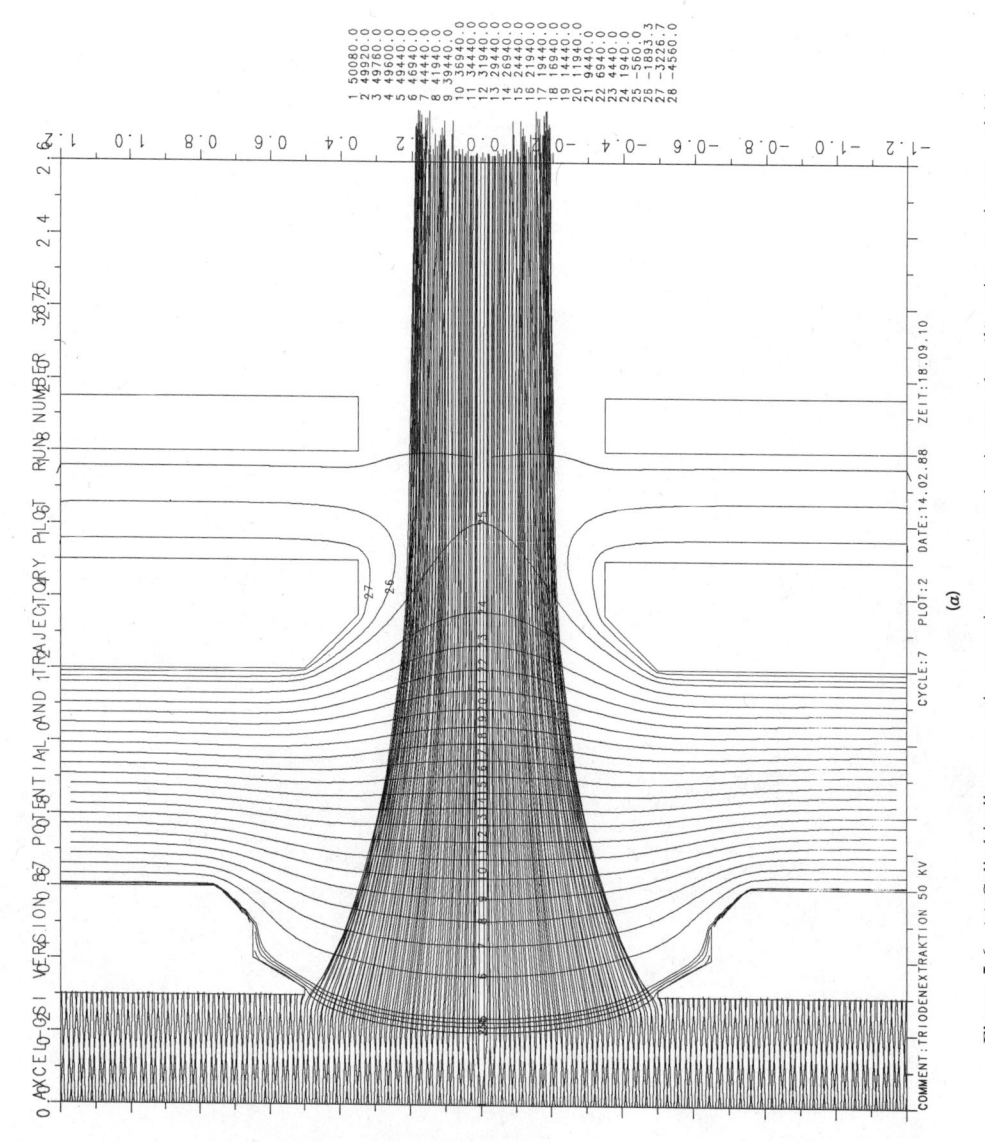

Figure 5.6 (*a*) Cylindrically symmetric extraction system and trajectory plot; (*b*) emittance picture; and (*c*) transported current as a function of the acceptance angle for given source operation parameters.

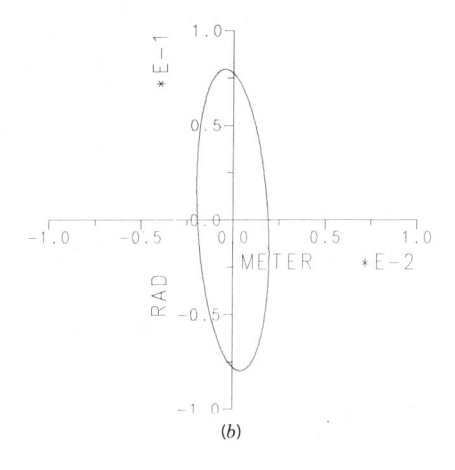

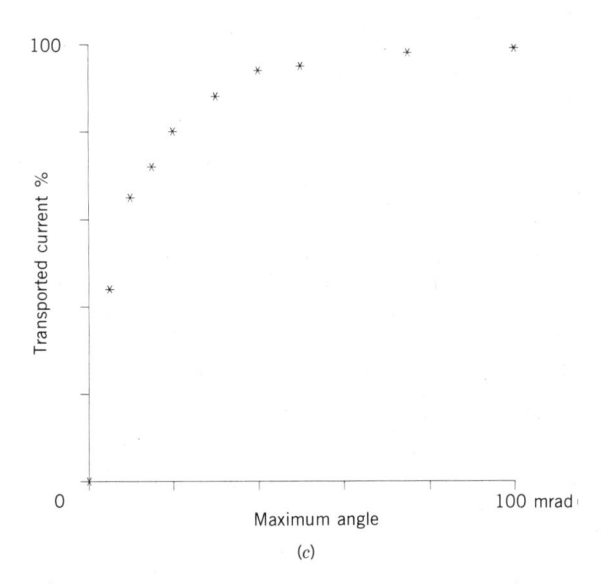

Figure 5.6 (*Continued*)

The mathematical iteration procedure itself, however, remains the same as in the 2D case, even if the number of mesh points will increase drastically. This means that running times on the computer are, of course, much longer than with 2D codes. Therefore, any possibility to reduce CPU time should be used. Optimizing the code is one way to do this; another way is clever use of previously obtained results.

In Figure 5.9 the steering of a 50-keV proton beam extracted from the source using a tetrode system with displaced electrodes is shown. The steering is caused by the third (negative) electrode being not perfectly aligned to the other electrodes.

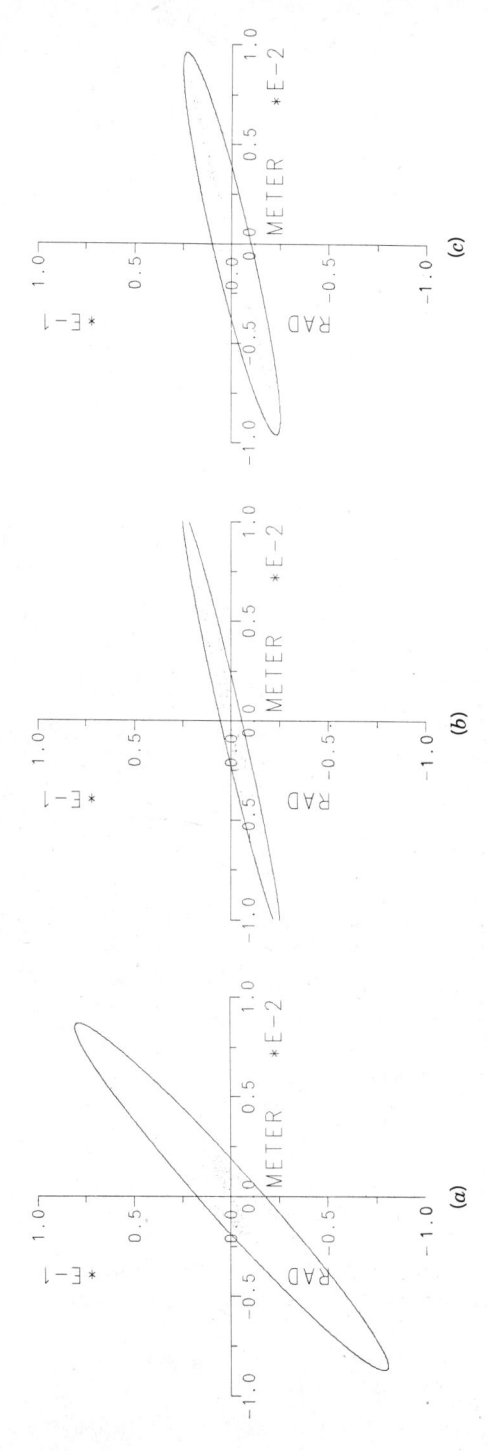

Figure 5.7 Development of the emittance in a drift section of length 0.5 m. (*a*) Input emittance; (*b*) output emittance with 100% space-charge neutralization; (*c*) output emittance with 95% space-charge neutralization.

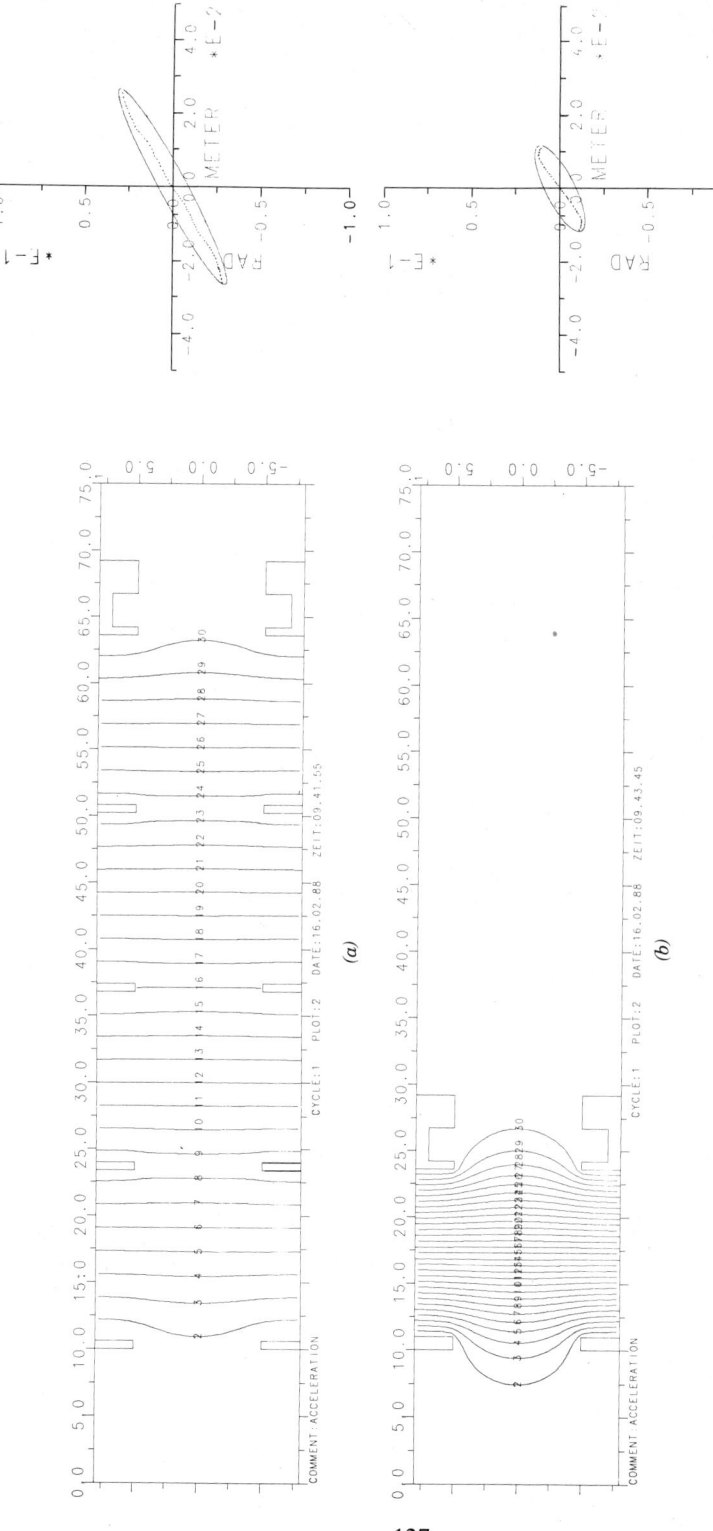

Figure 5.8 Acceleration of the ion beam shown in Figure 5.7 by two different acceleration columns: (*a*) a low-gradient multigap acceleration system, and (*b*) a high-gradient single-gap acceleration system. An emittance picture of the accelerated beam as well as a potential plot for each case is shown.

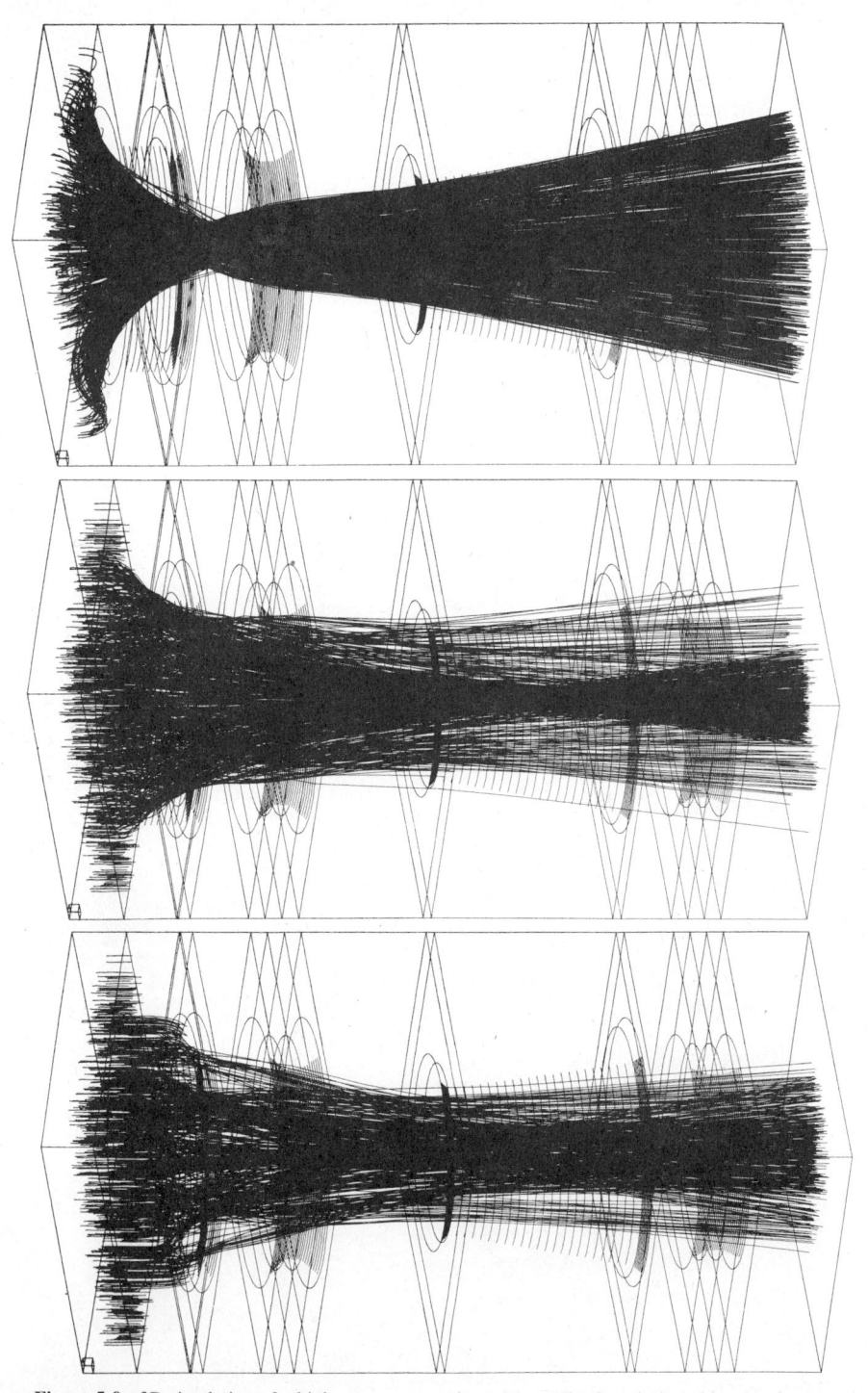

Figure 5.9 3D simulation of a high-current extraction system. The first six iterations are shown.

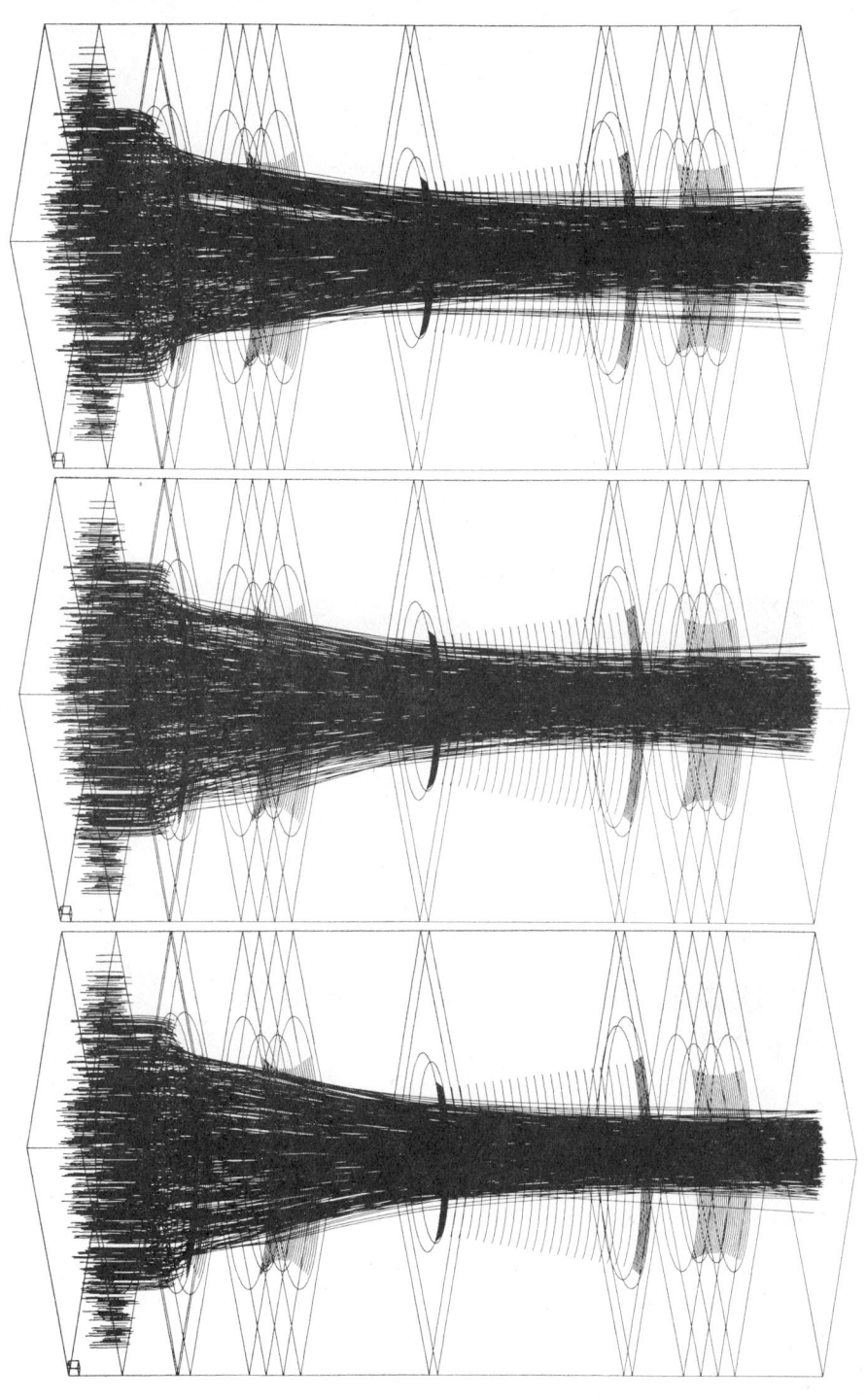

Figure 5.9 (*Continued*)

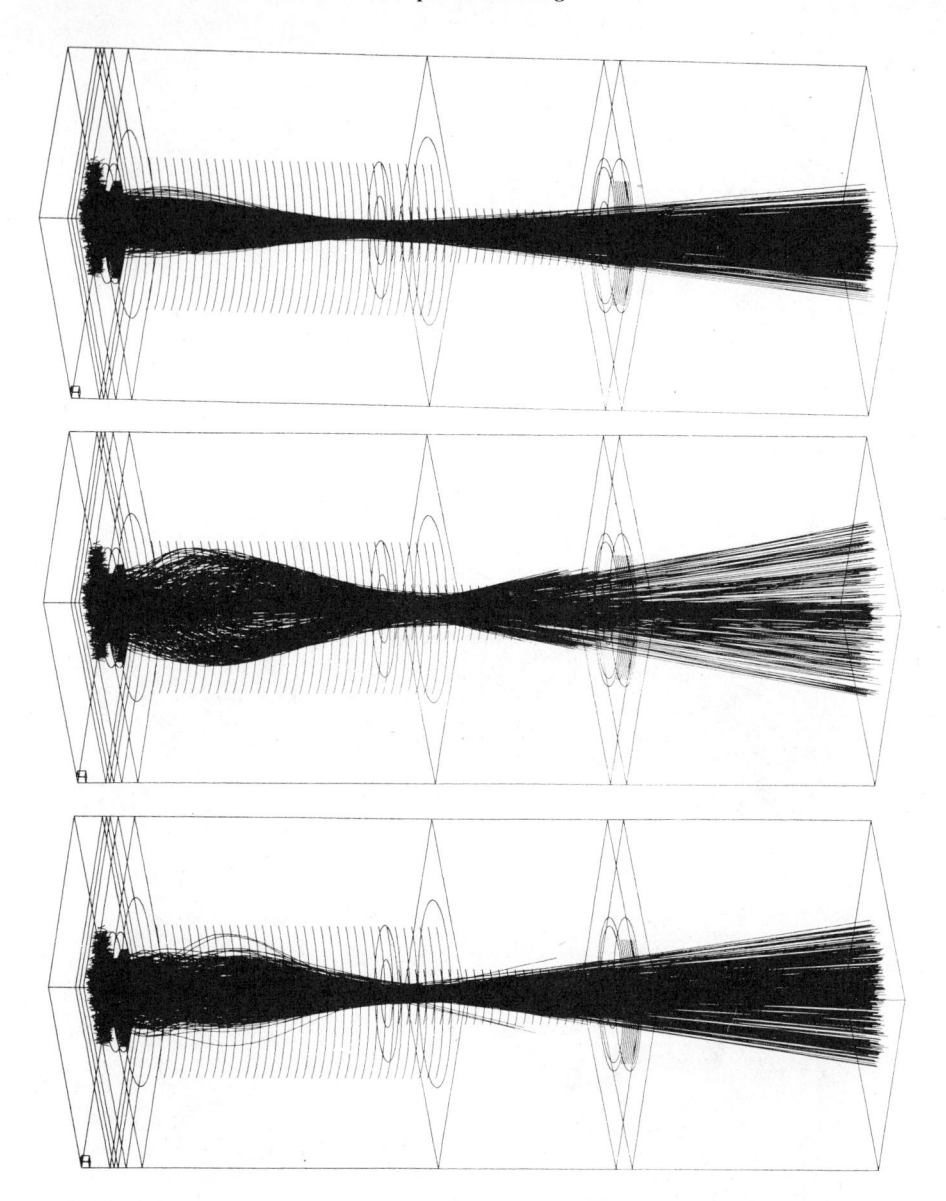

Figure 5.10 3D simulation of an ECR extraction system. The first six iterations are shown.

These calculations include very time-consuming plasma boundary determination. To save computer time in that example, solutions of previous calculations can be used if the only change has been displacement of the electrode.

The first six iterations are shown in the figure. In these iterations the development of the plasma boundary and the resulting effect on the ion beam is demon-

5 A 3D Example

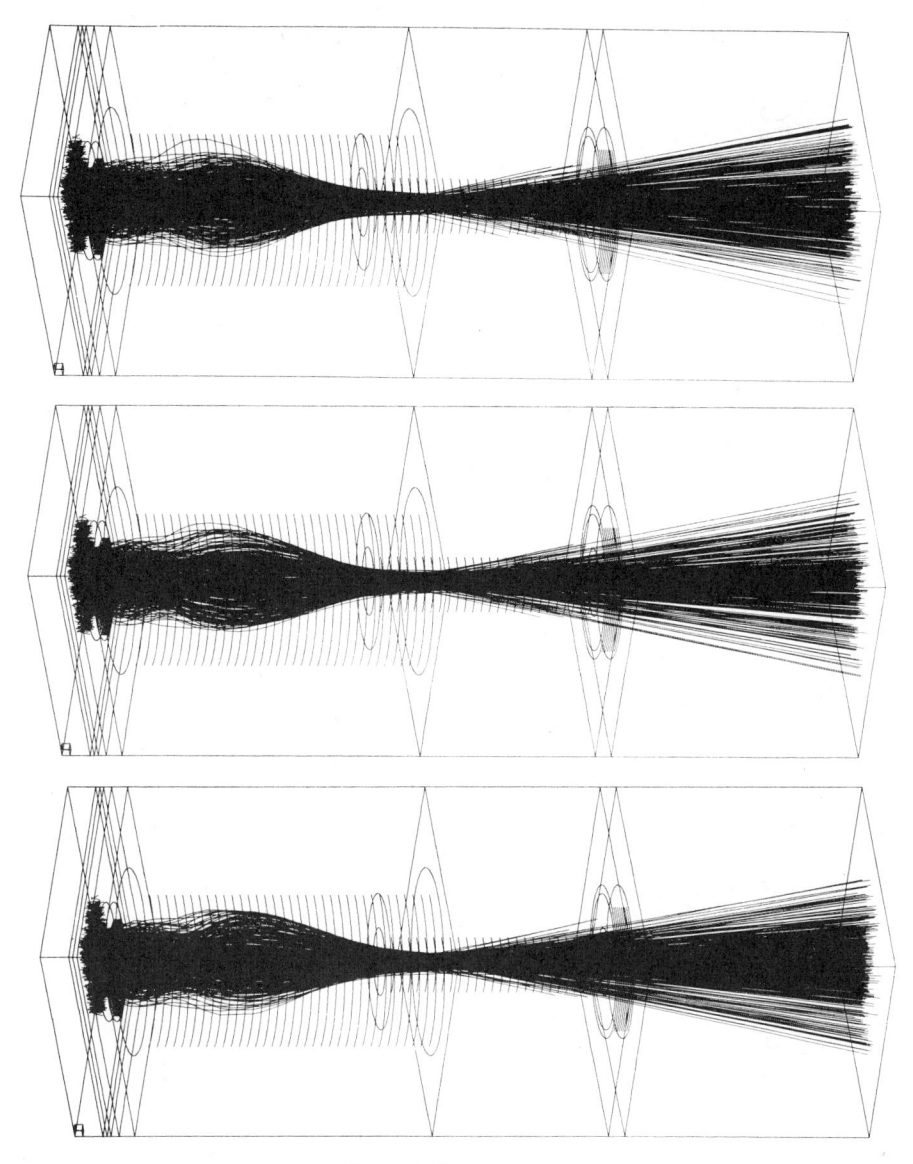

Figure 5.10 (*Continued*)

strated. The first iteration is without space charge and therefore without plasma boundary. The applied extraction voltage penetrates deep into the source region. The extracted beam is focused very close to the outlet aperture. Due to this strong focus the space charge for the second iteration will be overestimated at that point. The plasma boundary of the second iteration is therefore wrong, but by proceeding with the iterations the change in location and form of the plasma boundary small and a self-consistent plasma boundary is found.

Space charge is important in cases other than high-current extraction systems. Even in sources with relatively low current densities, space charge is not negligible, as can be seen in Figure 5.10. Here a simple diode extraction system for an ECR (Electron Resonance Source) source is shown. Due to the special design of the extraction system, the electrostatic field close to the outlet aperture of the source is very low. This causes the plasma to boil out of the source. The plasma behaves like in an extraction system with expansion cup. Normally such a problem could be solved by a 2D program; here a 3D program was used to investigate the effect of the magnetic field on the beam.

APPENDIX

Benchmark program used (written in Fortran77) to compare different computers with each other for the special case of the Poisson equation. The times necessary are listed in Table 5.1.

```
C    ...CHOOSE PARAMETER NDX AND NDY = EVEN
C    ...
C        NDX AND NDY ARE MESH LINE NUMBERS, MAXIT IS THE
         NUMBER OF
C        ITERATIONS TO BE CARRIED OUT.
C
         PARAMETER (NDX=100, NDY=100, MAXIT=300)
         DIMENSION U(0:NDX,0:NDY)
C
C    ...SET BOUNDARY CONDITIONS
C
         DO 100 I = 1,NDX
         U(I,NDY) = 1
         U(I-1,0) = 1
         U(0,I-1) = 1
         U(NDX,I) = 1
100      CONTINUE
C
C    ...CHOOSE OVER RELAXATION PARAMETER
C
         OMEGA = 1.8
C
C    ...SOLVE LAPLACE
```

```
      C
            DO 200 I = 1,MAXIT
      C
            DO 300 IY = 1,NDY-1
            DO 300 IX = 1,NDX-1
            DELTAU = (U(IX,IY + 1) + U(IX-1,IY) + U(IX+1,IY) +
            U(IX,IY-1))
            */4 - U(IX,IY)
            U(IX,IY) = U(IX,IY) + OMEGA*DELTAU
      300   CONTINUE
      C
            ERROR = U(NDX/2,NDY/2) - 0.5
            IF (MOD(I,10).EQ.0) PRINT*,I,ERROR
      200   CONTINUE
            END
```

Discretization

There are different methods of doing the discretization of Poisson's equation to the mesh. The five-point rule for the Cartesian 2D case in its simpliest version is defined as

$$\delta^2 U/\delta x^2 + \delta^2 U/\delta y^2 = \frac{U_{i-1,j} - 2U_{i,j} + U_{i+1,j}}{\Delta x^2} + \frac{U_{i,j-1} - 2U_{i,j} + U_{i,j+1}}{\Delta y^2}$$

where $U_{i,j}$ denote the potential at the logical coordinates of the mesh.

A more sophisticated five-point rule takes into account the exact distance of the mesh point U_0 to the closest neighbors U_i, even if this mesh point is within an electrode.

If more accuracy is needed the nine-point rule can be used. In the following equation $\Delta x = \Delta y$ is assumed.

$$\delta^2 U/\delta x^2 + \delta^2 U/\delta y^2$$
$$= \frac{U_{i-1,j+1} + 4U_{i+1,j+1} + U_{i+1,j+1}}{\Delta x^2} + \frac{4U_{i-1,j} - 20U_{i,j} + 4U_{i+1,j}}{\Delta x^2}$$
$$+ \frac{U_{i-1,j-1} + 4U_{i,j-1} + U_{i+1,j-1}}{\Delta x^2}$$

REFERENCES

1. Septier, *Focusing of Charged Particles*, vols. 1 and 2, Academic, New York, 1967.
2. R. G. Wilson and G. R. Brewer, *Ion Beams with Application to Ion Implantation*, Wiley, Malabar, FL, 1973.

3. B. Franczak, MIRKO—*An Interactive Program for Beam Lines and Synchrotrons*, Lecture Notes in Physics, Springer-Verlag, Berlin, 1984.
4. W. Joho, Representation of Beam Ellipses for Transport Calculations, SIN-Report TM-11-14, Schweizer Institut für Nuklearforschung, Switzerland, 1980.
5. J. Struckmeier, User manual for PARMTRA program, TN GSI-6/87 GSI, West Germany, 1981.
6. R W. Hockney and J. W. Eastwood, *Computer Simulation Using Particles*, McGraw-Hill, New York, 1981.
7. K. L. Brown, D. C. Carey, Ch. Iselin, and F. Rothacker, TRANSPORT, A Computer Program for Designing Charged Particle Beam Transport Systems, SLAC 91, 1974.
8. N. Schmitt, KOBRAS—Entwickeln und Austesten eines Programmes zur Bestimmung der elektronen- und ionenoptischen Eigenschaften elektrostatischer und magnetostatischer Anordnungen, Thesis, FH Wiesbaden, West Germany, 1983.
9. Y. Ose, T. Takagi, and K. Miki, Numerical Simulation of 3D Ion Beam Optics by Boundary-Fitted Coordinate Transformation Method, Proceedings, 10th Symposium on ISIAT, Tokyo, 1986.
10. J. H. Whealton, R. W. McGaffey, and P. S. Meszaros, A Finite Difference 3-D Poisson-Vlasov Algorithm for Ions Extracted from a Plasma, *J. Comput. Phys.* **63,** 20 (1986).
11. W. B. Herrmannsfeldt, SLAC Electron Trajectory Program; SLAC 266, 1979; original SLAC 166, 1973.
12. E. F. Jaeger and I. C. Whitson, Numerical Simulation for Axially Symmetric Beamlets in the Duopigatron Ion Sources, ORNL/TM-4990, Oak Ridge, TN, 1975.
13. J. S. Hornsby, A Fortran Program for the Analysis of Electrostatic Lenses, CERN Computer Centre, Program Library, W126, 1965.
14. J. S. Hornsby, A Computer Program for the Solution of Elliptic Partial Differential Equations, CERN Computer Centre, Program Library, D300, 1977.
15. M. R. Shubaly, BEAM, an Improved Beam Extraction and Accelerating Modelling Code; *IEEE Trans. Nucl. Sci.*, **NS-28,** 2655 (1981).
16. W. S. Cooper, K. Halbach, and S. B. Magyary, Computer Aided Extractor Design, Proceedings of the Second Symposium on Ion Sources and Formation of Ion Beams, Berkeley, LBL-3399, 1974.
17. J. E. Boers, SNOW, a Digital Computer Program for the Simulation of Ion Beam Devices, Sandia National Laboratory Report SAND 79-1127, 1979.
18. C. Weber, The Use of Computers in Electron- and Ion-Gun Design, *Comput. Phys. Commun.*, 5 (1973).
19. J. C. Whitson, E. F. Jaeger, and J. H. Whealton, Optics of Ion Beams of Arbitrary Perveance Extracted from a Plasma, *J. Comput. Phys.* **25,** (1978).
20. J. C. Whitson, J. Smith, and J. H. Whealton, Calculations Involving an Ion Beam Source, *J. Comput. Phys.* **28,** 408 (1978).
21. J. H. Whealton and J. C. Whitson, Space Charge Ion Optics Including Extraction from a Plasma, *Particle Accelerators* **10,** 235 (1979).
22. J. H. Whealton, Ion Extraction and Optics Arithmetic, *Nucl. Instrum. Methods* **189,** 55 (1981).
23. Los Alamos Code group AT6, Computer Codes Used in Particle Accelerator Design, ATN 86-26, LANL, Los Alamos, NM, 1986.
24. J. Stoer, R. Bulirsch, *Introduction to Numerical Mathematics*, Springer-Verlag, Berlin, 1973.
25. S. A. Self, Exact Solution of the Collisionless Plasma-Sheath Equation, *Phys. Fluids* **6,** 1762 (1963).
26. P. Spädtke, Computer Simulation of High Current DC Ion Beams; Proceedings, 1984 LINAC Conference, Seeheim, 1984.

27. R. Keller, J. D. Sherman, and P. Allison; Use of a Minimum Ellipse Criterion in the Study of Ion Beam Extraction Systems, *IEEE Trans. Nucl. Sci.* **NS-32,** 2579 (1985).
28. P. Spädtke, AXCEL-GSI Interaktives Simulationsprogramm zur Berechnung von zweidimensionalen Potentialverteilungen elektrostatischer Anordnungen sowie von Ionenbahnen in elektrostatischen Feldern unter Berücksichtigung der Raumladung, GSI-Report 9, GSI, West Germany, 1983.
29. R. Becker, private communication; IAP J. W. Goethe University, Frankfurt, 1987.

6

Some Basic Ion Sources

I. G. Brown
Lawrence Berkeley Laboratory
University of California
Berkeley, California

The need occurs in the laboratory, from time to time, for an ion source whose beam characteristics are not too important, possibly for a quick check of another diagnostic or instrument, or for development of a technique or detector. One then would like to be able to assemble a source simply and quickly, using not much more than inexpensive laboratory stock and the modern equivalent of string and sealing wax. What is simple and inexpensive to one experimenter, however, may be out of the question to another, depending on experimental background and the kind of laboratory facilities available. The concept of a "simple" source is certainly an ambiguous one.

In essence, an ion source is no more than a plasma source from which some of the ions are formed into a beam. Roughly speaking, the kind of plasma determines the kind of ions and the way the extraction is done determines the beam parameters. For a simple ion source, therefore, one needs to combine a modest plasma device with a rudimentary ion extractor in the simplest possible way. The details of how this is done depend on the experimenter's requirements, and govern the ion beam characteristics.

In this chapter several easily constructed and inexpensive sources that have been described in the literature are presented. This chapter is not intended to present a complete survey of all of the many kinds of sources that have been developed and described in the literature. Rather, it is meant to provide food for thought and an initial guide to some of the literature.

1 SOURCES MADE FROM BAYARD-ALPERT GAUGES

The structure of the commercially available ion gauge of the Bayard-Alpert type is well suited for conversion into a simple ion source. This kind of vacuum gauge is, of course, already an ion source of a kind—the pressure is determined by measuring the ion current resulting from ionization of the ambient gas by the electrons emitted from a heated filament. In normal operation, the ions so created are collected by a negatively biased electrode and the ion current thus measured is proportional to the gas pressure. The gauge can be converted into a usable ion source by appropriate modification of the electrode structure and applied voltages, and extraction of ions from the ionization region in the form of a beam.

Sources of this kind have been made that produce beams of gaseous ions with currents up to about a microampere or so at energies of up to several kilovolts. The energy spread of the extracted ion beam is a few to a few tens of electron volts. The ions are predominantly singly ionized. Typical operating pressure may be from below 10^{-5} Torr up to about 10^{-3} Torr. In the simplest setup, the electrical system to drive the ion source can be the ion gauge power supply provided by the gauge manufacturer.

The configuration utilized by Leffel [1] is shown in Figure 6.1. The ion beam current was measured at 1.6×10^{-13} A into a solid angle of 2.2×10^{-4} st for a source pressure of 2.7×10^{-5} Torr, an ion energy of 100 eV, and an energy spread of 4 eV, using an electron current in the ion gauge of 1 mA. The source was operated at a beam extraction voltage of up to 3 kV.

Another variant of this concept has been described by Shoji and Hanawa [2]. They have constructed a source that is mounted completely on the flange of a

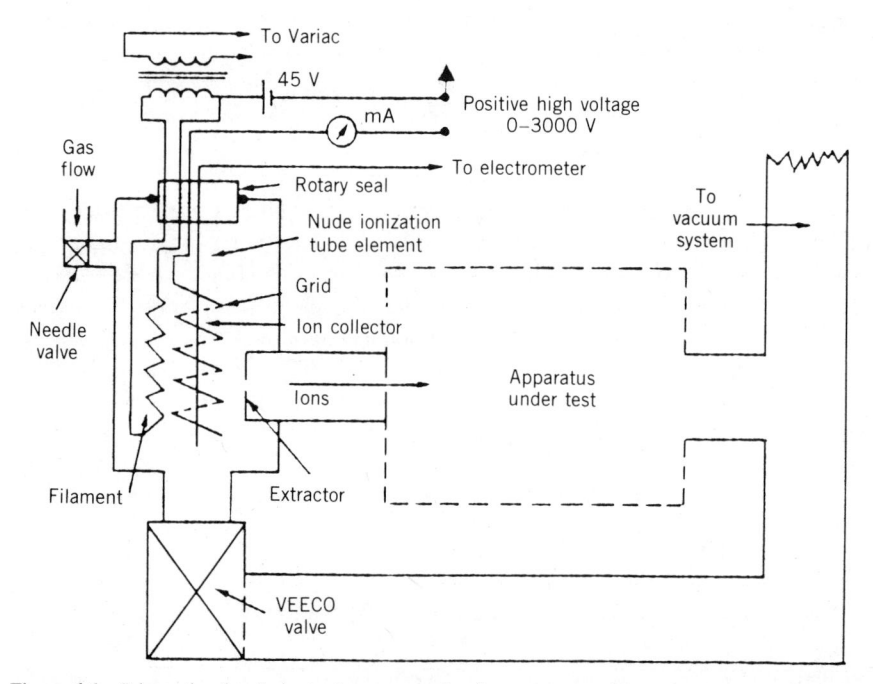

Figure 6.1 Schematic of nude ionization gauge tube element mounted as an ion source. After Leffel [1].

2 Other Electron-Bombardment Sources

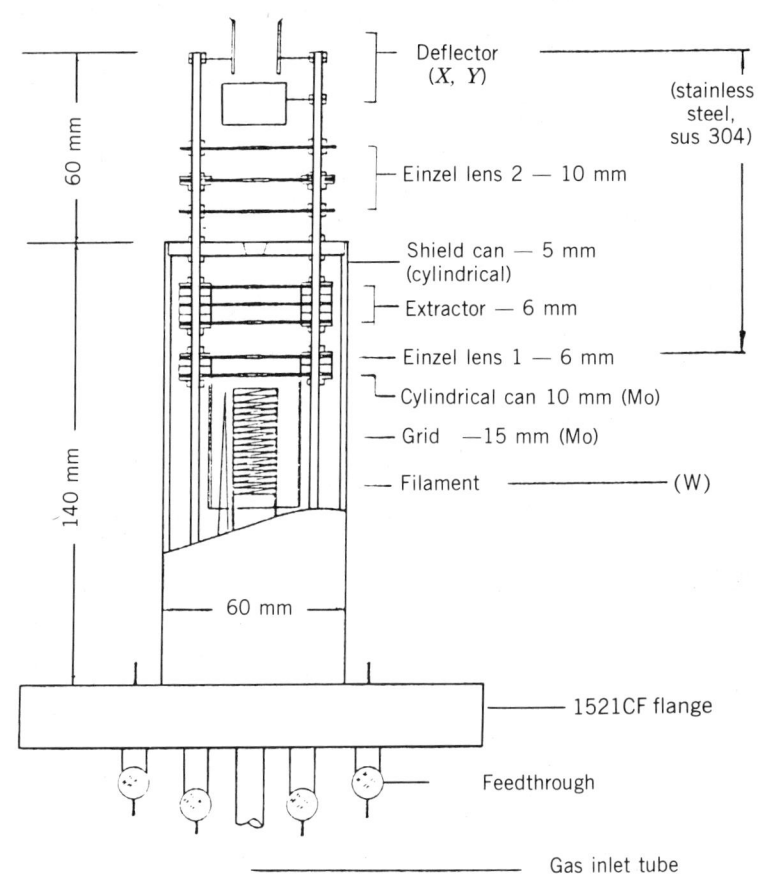

Figure 6.2 Ion source constructed from an ionization gauge, and incorporating focusing lens and beam steering plates. After Shoji and Hanawa [2].

Bayard-Alpert gauge, and that incorporates two 3-grid einzel lenses for beam focusing and two pairs of deflection plates for beam steering. An outline of this source is shown in Figure 6.2. Beam extraction voltage was up to 1 kV and the energy spread for helium was estimated at about 1 eV. The ion beam current was up to a few tenths of a microampere.

Kirschner [3] has described a source that produces an Ar^+ ion beam current of several microamperes at an energy of 1 keV. This source is shown in Figure 6.3. The source is bakeable up to 300°C, and is provided with beam focusing electrodes.

2 OTHER ELECTRON-BOMBARDMENT SOURCES

One can extend the concept typified by the sources described above to source types in which the electron-emitting filament and the plasma containment region are

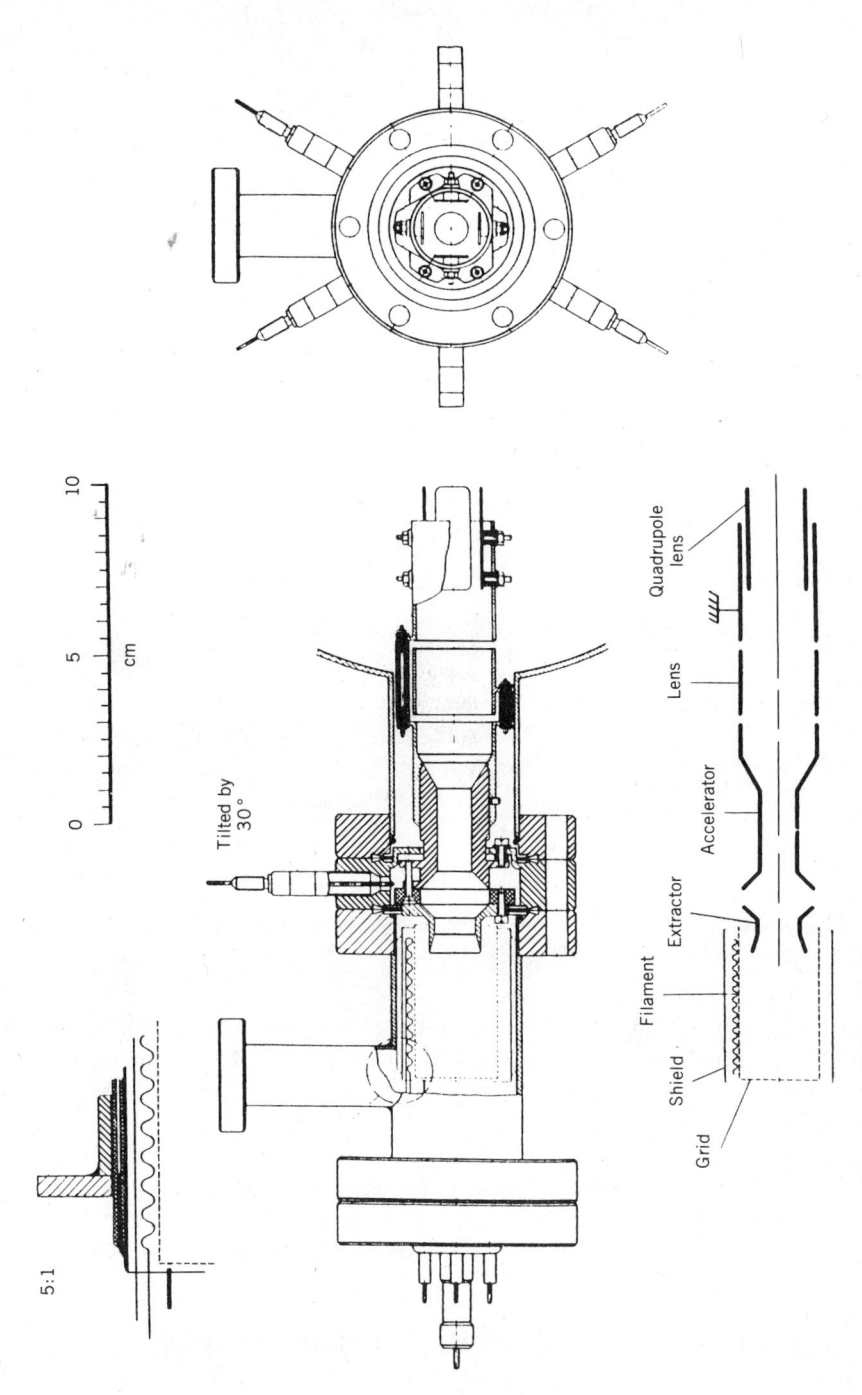

Figure 6.3 Side and front view of an ion source constructed from an ionization gauge, mounted on a 35-mm flange. After Kirschner [3].

designer-made, rather than being of the configuration determined by preexisting commercially available hardware. The ion source can still be quite simple and cheap, but it might now incorporate one or two features that better suit a specific requirement.

A simple electron-bombardment ion source has been developed by Dworetsky et al. [4], and this source is shown in Figure 6.4. The source of electrons in this case is an oxide-coated cathode. Beams of noble gas ions as well as nitrogen and hydrogen were produced at an energy up to 5 kV and a beam current of about 1 μA. Another similar source is that described by Khan and Schroeer [5]; this version produced a beam current of about 1 μA of Ar^+ at an argon pressure of 5×10^{-6} Torr.

The beam current can be greatly increased by providing some containment of the plasma produced by electron bombardment, and by increasing the size of the aperture through which ions are extracted. Then the plasma reservoir from which the ions are to be extracted is of higher density than it would otherwise be, for the same input power, and a larger area beam can be extracted. These two separate improvements have been incorporated into the ion source described by Stenzel and Ripin [6], shown schematically in Figure 6.5. In this source the plasma is confined by using a multipole array of permanent magnets that surround the plasma chamber, and the plasma is established by an arc that is drawn between the emitting filaments and the vessel wall. Thus, in this case there is also another "performance-enhancing feature" that has been added—namely, increasing the ionization by making use of a dc arc as opposed to simply an emitting filament. This source produced beams of 10 mA of hydrogen ions at 5 kV extraction voltage. It is interesting to note that this ion source concept is that which has been developed to a fine pitch for the

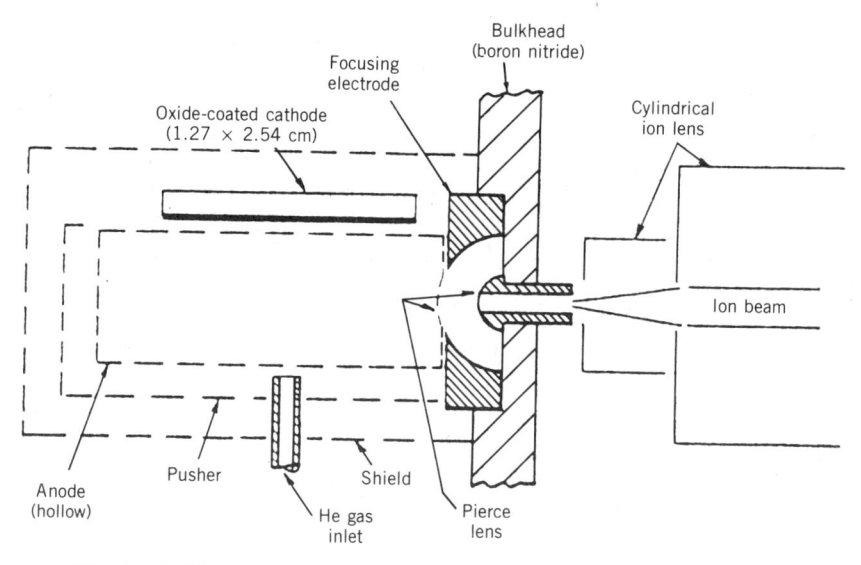

Figure 6.4 Electron-bombardment ion source developed by Dworetsky et al. [4].

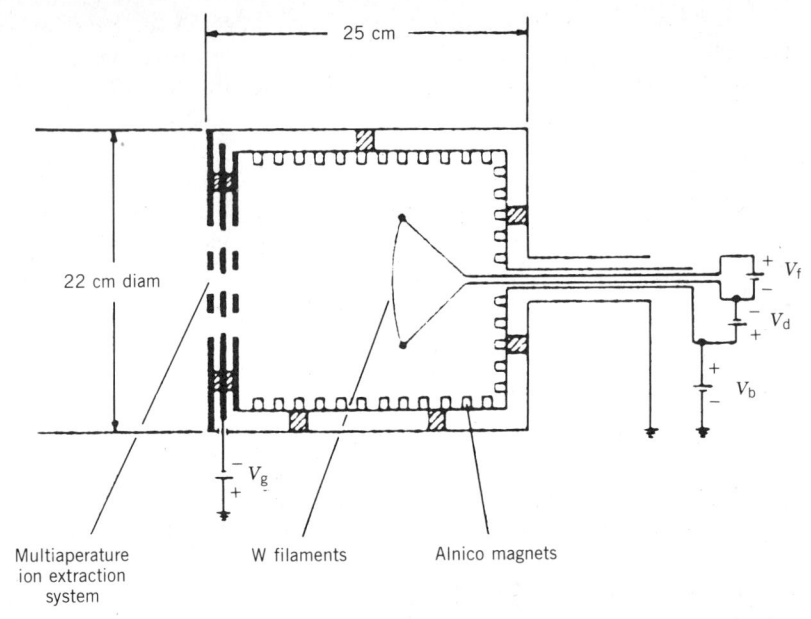

Figure 6.5 Ion source using permanent magnet multipole plasma confinement and multiaperture extraction system. After Stenzel and Ripin [6].

neutral beam development program; beams of current up to about 100 A (equivalent) at particle energies of up to 100 keV or more have been produced [7].

3 METAL ION SOURCES

Metal ion sources fall naturally into several varieties, according to the boiling point (vapor pressure) of the metal species used and the ionization potential of the neutral atom. Thus, if the metal has a particularly low boiling point, or more specifically a high vapor pressure at a low temperature, then the metal vapor can be produced with only moderate heating and fed into an ion source of more-or-less standard design. A concern that needs to be kept in mind in the design of this or any other source in which the working substance is a condensable vapor is the buildup of solid deposits and their effect on source operation. A source of this kind has been developed by Lejeune and Gautherin [8], and is shown in Figure 6.6. In this source arsenic is heated in an oven to a temperature of about 300°C and fed into the hot discharge cavity. A beam of arsenic ions (polyatomic ions) with current up to 200 μA was produced at an extraction voltage of several kilovolts. In the source described by Sugiura [9] (Figure 6.7), a discharge in antimony vapor was used to produce a beam of up to 200 μA at 1 kV. Another version of source of this general kind is that described by Hasan et al. [10]; they obtained In^+ beams with current up to about 200 μA at an extraction voltage of several hundred volts.

3 *Metal Ion Sources* 143

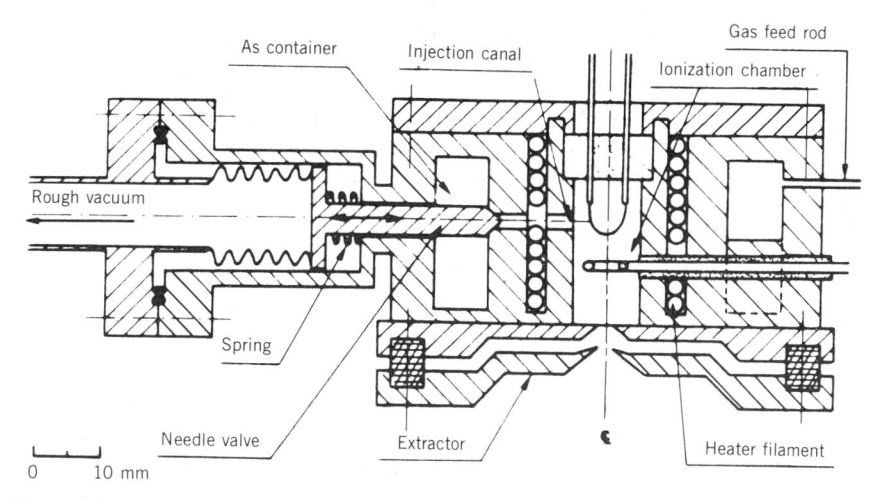

Figure 6.6 Arsenic ion source developed by Lejeune and Gautherin [8]. Reprinted with permission from *Vacuum* **34**, C. Lejeune and G. Gautherin, "Small, Simple and High Efficiency Arsenic Ion Source," c1984.

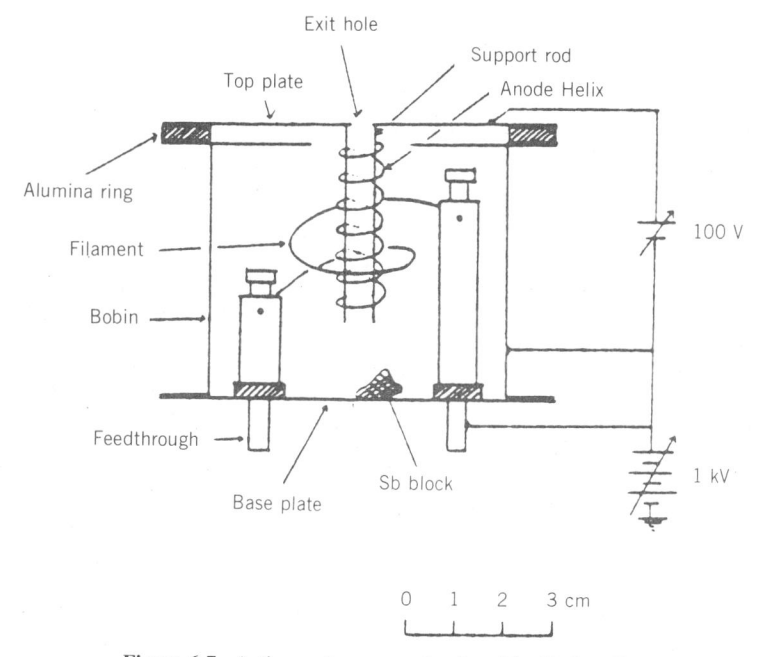

Figure 6.7 Antimony ion source developed by Sugiura [9].

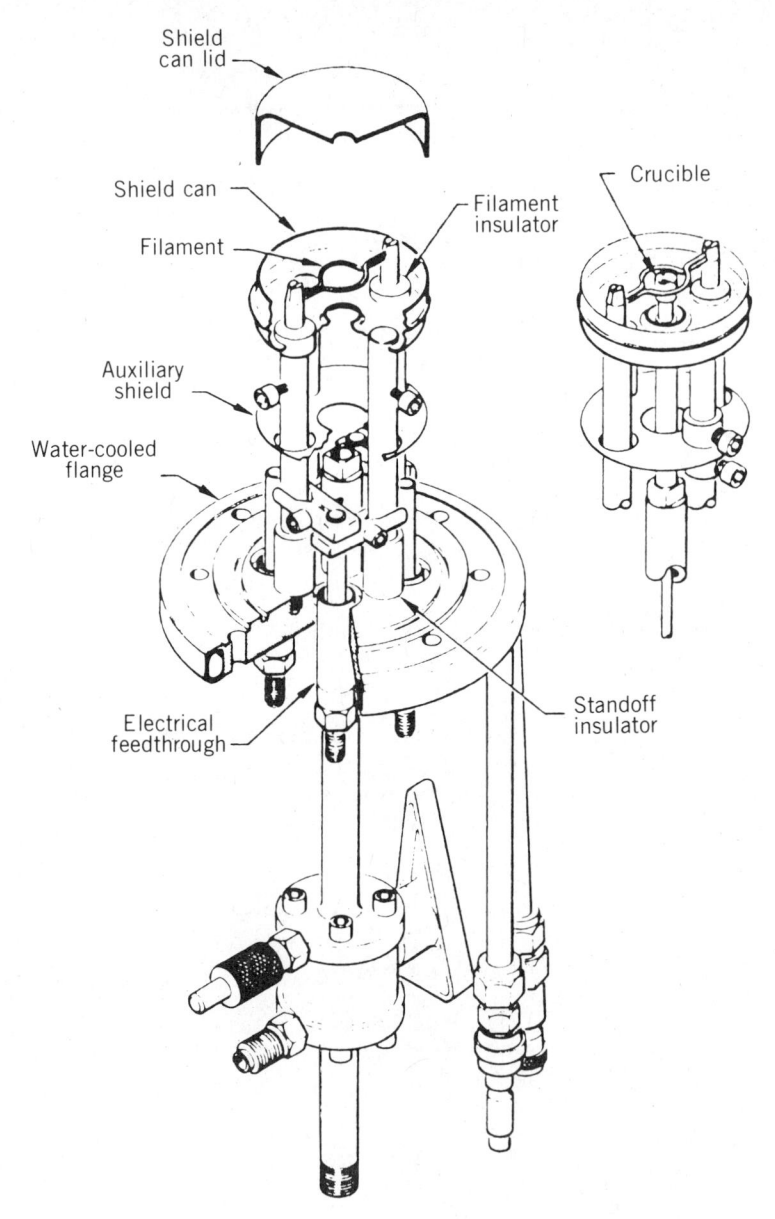

Figure 6.8 Surface ionization source for rare earth metal ions, as developed by Johnson et al. [13]. Courtesy of North-Holland Physics Publishing Co., and the University of California, LLNL, and US DOE.

For metals having relatively low ionization potential, such as the alkali metals and rare earths, surface ionization can be used for the generation of ions directly from a hot metal surface with which the neutral metal atoms are caused to come into contact. The generation of ions in this way is a technique that has been developed extensively by the plasma physics community for the generation of synthesized alkali metal plasmas [11, 12]. Thus, a metal ion source can be made utilizing this phenomenon. The principal experimental difficulty is the high temperatures required for efficient surface ionization, in the range 2500–3300°C. A source of this kind has been described by Johnson et al. [13], and their source is shown in Figure 6.8. Beams of a wide range of rare earth metals of current of order 10 μA were produced.

A particularly high emission of metal ions from a hot surface is obtained in the case when the hot surface, for example, tungsten, is coated with an aluminosilicate of the metal [14–16]. The composition $Li_2O \cdot Al_2O_3 \cdot 2SiO_2$ is the mineral β-eucryptite, and the related mineral spodumene has composition $Li_2O \cdot Al_2O_3 \cdot 4SiO_2$; zeolite-A is a synthetic alkali aluminosilicate of the form $6X_2O \cdot 6Al_2O_3 \cdot 12SiO_2$, where X is the alkali metal. Thus, sources of this kind have been referred to as aluminosilicate-type sources, β-eucryptite sources, spodumene sources, zeolite sources, or Blewett-Jones-type sources after the early pioneers of this kind of emission [16]; the basic phenomenon utilized in all cases is that of the enhanced surface ionization and emission from these kinds of hot surfaces. An early source of this kind was described by Allison and Kamegai [17], who produced 100-μA beams of Li^+ from a source of lifetime about 100 h. Another version, still relatively uncomplicated in design, is that of Moeller and Kamke [18]; this source yielded Li^+ beams with current of up to about 400 μA for a lifetime of many hours and up to 1.5 mA in pulses of several minutes duration. Feeney et al. [19] have reported on their filament sources for producing ion beams from Li, Na, K, Rb, Cs, and Tl. A particularly simple source, using a dispenser-type source with a sodium silicate matrix, has been described by Hirsch and Varga [20], and the construction of their emitting surface is shown in Figure 6.9. The emission characteristics of sources of this kind have been reported by Heinz and Reaves [21], Haq [22], and Satoh et al. [23]. Fabricated emitters for a wide range of elements can be obtained commercially also [24].

For metals of high melting point, the methods described above produce only minute beam currents. The MEVVA ion source is particularly suitable in this case. This kind of source uses a metal vapor vacuum arc as the plasma formation mechanism, and is described in detail in a separate chapter. We note here, however, that the MicroMEVVA source is of unusually simple design, and can be used to produce beams of a wide range of metal ion species at current levels over 10 mA at extraction voltages of up to 20 kV in pulses of up to 100 μs or so duration.

4 MICROWAVE ION SOURCES

The ECR (electron cyclotron resonance) ion source for producing dc beams of very highly stripped ions, and the microwave ion source for the production of very

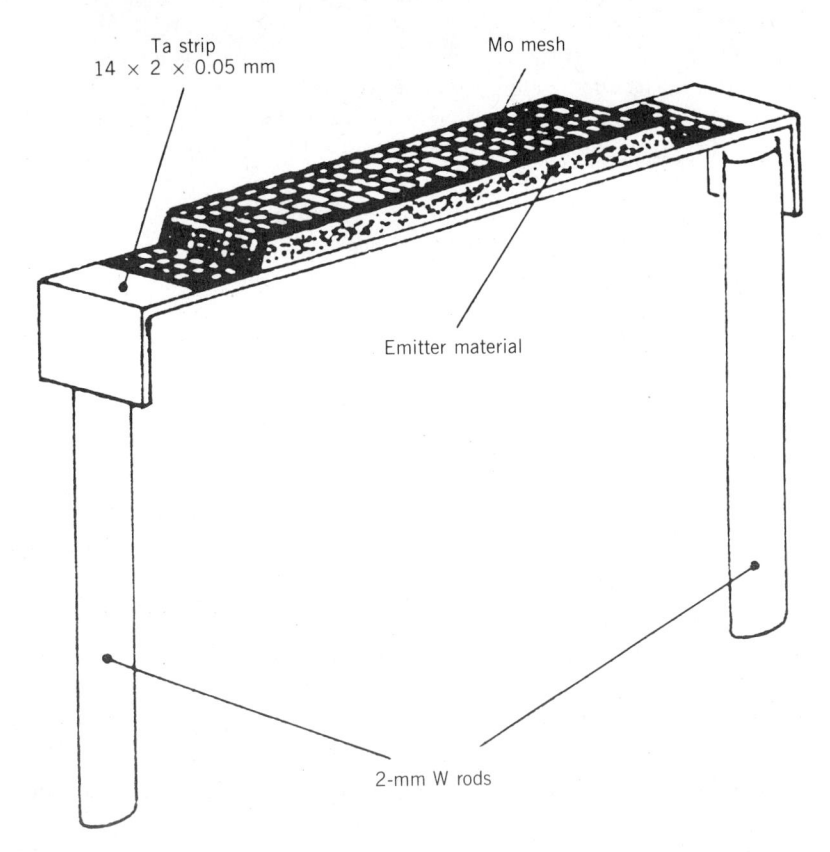

Figure 6.9 Construction of the emitter of a dispenser ion source with sodium silicate matrix. After Hirsch and Varga [20].

high current beams of singly ionized ions are two kinds of ion sources that make use of plasmas established by microwave breakdown of the working gas. These kinds of sources are described in detail elsewhere in this book. Here we consider briefly several other sources that employ microwave-produced plasmas and that are simply constructed.

Asmussen and Root have described a "microwave plasma disk ion source" [25, 26] in which 2.45-GHz microwaves at a power of about 100 W are coupled into a disklike region of plasma, located at one end of a tuned cylindrical cavity, as shown in Figure 6.10. Using argon gas, a beam current of up to nearly 300 mA was produced through the 6-cm-diameter, multiaperture, two-grid extractor. An interesting feature of this source is the hot tungsten filament located just on the beam-formation side of the extractor that is used to supply neutralizing electrons to the ion beam so as to suppress space-charge blowup and consequent loss of beam. This novel ion source has been developed further by the addition of a multicusp magnetic field established by rare earth permanent magnets [27, 28], and has been used in several applications [29–31].

4 *Microwave Ion Sources*   147

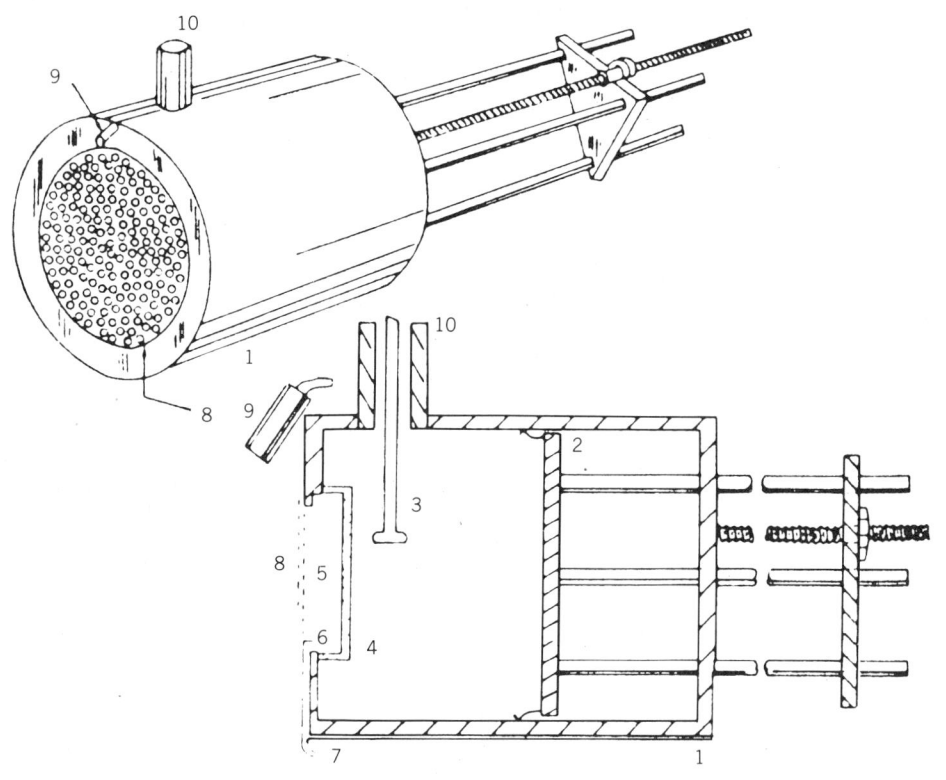

Figure 6.10 Microwave plasma disk ion source developed by Asmussen and Root [25]. 1, microwave cavity; 2, sliding short; 3, excitation probe; 4, quartz cup; 5, plasma region; 6, annular gas feed; 7, gas inlet line; 8, extraction grids; 9, hot tungsten filament.

Another version of a simple microwave ion source has been developed by Leung and coworkers [32, 33]. In this compact design, a quartz tube passes through a tuned microwave cavity. Gas is introduced into one end of the tube and a simple two-grid extractor at the other end extracts beam from the plasma that is formed inside the quartz tube. Using 2.45-GHz microwaves at a power of up to 400 W, a beam current of up to about 5 mA was produced. One embodiment of this concept is shown in Figure 6.11.

A significant part of the cost of any microwave ion source is the cost of the basic high-power microwave generator. The wide availability of low-cost microwave ovens operating at a frequency of 2.45 Ghz and powers of several hundred watts provides a possible avenue for obtaining an inexpensive source of microwave power. This has been done by Meiners and Alford [34]. They have described in detail their conversion of a small portable microwave oven into a 0- to 600-W cw microwave power source and the coupling of this power into a cylindrical microwave cavity to produce a plasma inside a quartz tube passing through the cavity. Although no ion extraction was done, this presumably could be readily added.

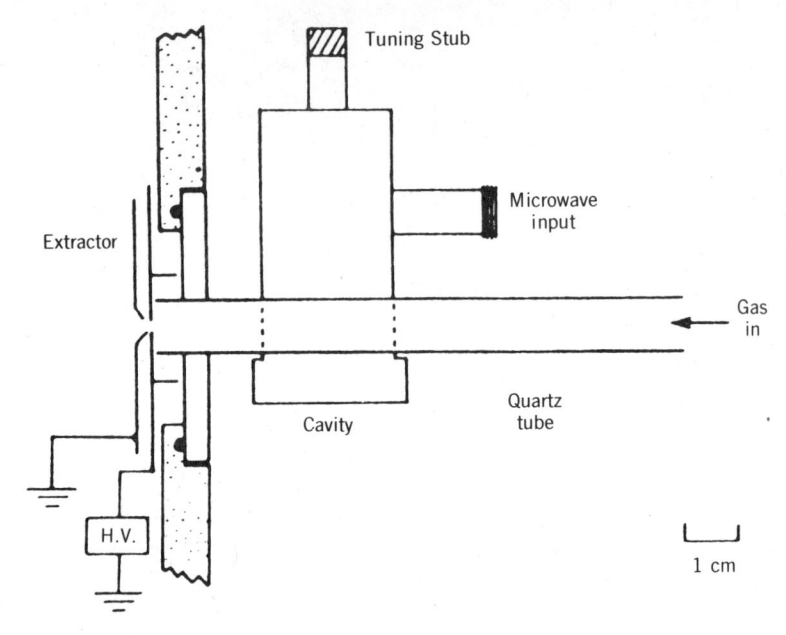

Figure 6.11 Compact microwave ion source developed by Leung et al. [32].

Alternatively, the microwave supply could be used to power either of the microwave ion source embodiments described above.

5 MISCELLANEOUS SOURCES

The breakdown of a gas by rf fields provides a simple and often convenient method of plasma formation that can be utilized in an ion source. The rf is usually in the range from around 1 MHz to a few tens of megahertz. This kind of source can be extremely simple, consisting of little more than a glass container for the gas/plasma, some kind of extraction aperture(s), and a coupling loop surrounding the vessel to couple inductively into the plasma. A power of a few tens of watts can be adequate. Sources of this kind have been described by several authors [35–37]; an inductively coupled rf plasma source, from which ions were not extracted but which provides a good example of this kind of plasma formation mechanism, has been described by Petty and Smith [38].

Many kinds of ion sources can be purchased commercially. A few examples of the kinds of sources that can be obtained in this way follow. As mentioned above, aluminosilicate ion emitters for low currents of alkali metals, the rare earths, and other elements also, can be obtained from Spectra-Mat, Inc. [24]. Broad beam sources (Kaufman type), with ion energies up to about 1.5 keV and total extracted ion current (typically argon) up to several amperes and with beam size up to as high as 50 cm, can be obtained from a number of manufacturers, including Anatech

Ltd. [39], CVC Products Inc. [40], Commonwealth Scientific Corp. [41], Oxford Instruments Ltd. [42], and Veeco Instruments Inc. [43]. Liquid metal sources can be obtained from Kratos Analytical Inc. [44], Leybold-Heraeus GmbH [45], and VG Instruments Inc. [46]. Low-energy, 100-mA hollow cathode ion sources are offered by Ion Tech, Inc. [47], and both rf sources and positive and negative ion duoplasmatron sources are available from National Electrostatics Corp. [48]. Kimball Physics, Inc. [49] offers both alkali metal sources and duoplasmatrons. Several low-current ion sources as well as an ECR plasma device are offered by Microscience, Inc. [50]. A complete ion beam kit, including ion source, accelerating system, deflection system, velocity filter, and decelerator, is available from Colutron Research Corp. [51]. An extensive range of ion beam sources and beam handling equipment of all kinds is offered by Danfysik [52] and by General Ionex Corp. [53]. There are many more ion source manufacturers than these; the potential customer would be well advised to research the market, for example, via the annual Buyers Guides put out by several organizations [54, 55], and to contact a wide sampling of manufacturers and distributors.

The sources outlined in this chapter are a very few examples of the many kinds of simply constructed ion sources that are available to the experimenter. The job of putting together a rudimentary ion source facility need not necessarily be expensive, lengthy, complicated, or overwhelming in any way. It really depends on what the required ion beam parameters are: the less demanding the parameters, the simpler the chore. In the limit, an ion source can be created easily by adding a simple extractor to whatever kind of plasma formation hardware is most readily available. The examples presented here will hopefully provide some inspiration and guidance.

REFERENCES

1. C. S. Leffel, Jr., *Rev. Sci. Instrum.* **41**, 285 (1970).
2. F. Shoji and T. Hanawa, *J. Phys. E: Sci. Instrum.* **14**, 90 (1981).
3. J. Kirschner, *Rev. Sci. Instrum.* **57**, 2640 (1986).
4. S. Dworetsky, R. Novick, W. W. Smith, and N. Tolk, *Rev. Sci. Instrum.* **39**, 1721 (1968).
5. M. Khan and J. M. Schroeer, *Rev. Sci. Instrum.* **42**, 1348 (1971).
6. R. L. Stenzel and B. H. Ripin, *Rev. Sci. Instrum.* **44**, 617 (1973).
7. W. B. Kunkel, in *Fusion*, Vol. 1, Pt. B, E. Teller (Ed.), Academic, New York, 1981.
8. C. Lejeune and G. Gautherin, *Vacuum* **34**, 251 (1984).
9. H. Sugiura, *Rev. Sci. Instrum.* **50**, 84 (1979).
10. M.-A. Hasan, J. Knall, S. A. Barnett, A. Rockett, J.-E. Sundgren, and J. E. Greene, *J. Vac. Sci. Technol. B* **5**, 1332 (1987).
11. N. Rynn and N. D'Angelo, *Rev. Sci. Instrum.* **31**, 1326 (1960).
12. R. W. Motley, *Q-Machines*, Academic, New York, 1975.
13. P. G. Johnson, A. Bolson, and C. M. Henderson, *Nucl. Instrum. Methods* **106**, 83 (1973).
14. J. L. Hundley, *Phys. Rev.* **30**, 864 (1927).
15. K. T. Bainbridge, *J. Franklin Inst.* **212**, 317 (1931).

16. J. P. Blewett and E. W. Jones, *Phys. Rev.* **50,** 465 (1936).
17. S. K. Allison and M. Kamegai, *Rev. Sci. Instrum.* **32,** 1090 (1961).
18. W. Moeller and D. Kamke, *Nucl. Instrum. Methods* **105,** 173 (1972).
19. R. K. Feeney, W. E. Sayle II, and J. W. Hooper, *Rev. Sci. Instrum.* **47,** 964 (1976).
20. E. H. Hirsch and I. K. Varga, *Rev. Sci. Instrum.* **46,** 338 (1975).
21. O. Heinz and R. T. Reaves, *Rev. Sci. Instrum.* **39,** 1229 (1968).
22. F. U. Haq, *J. Phys. E: Sci. Instrum.* **19,** 275 (1986).
23. Y. Satoh, M. Takebe, and K. Iinuma, *Rev. Sci. Instrum.* **58,** 138 (1987).
24. Spectra-Mat, Inc., 1240 Highway 1, Watsonville, CA 95076.
25. J. Asmussen and J. Root, *Appl. Phys. Lett.* **44,** 396 (1984).
26. J. Root and J. Asmussen, *Rev. Sci. Instrum.* **56,** 1511 (1985).
27. M. Dahimene and J. Asmussen, *J. Vac. Sci. Technol. B* **4,** 126 (1986).
28. L. Mahoney, M. Dahimene, and J. Asmussen, *Rev. Sci. Instrum.* **59,** 448 (1988).
29. J. Asmussen and M. Dahimene, *J. Vac. Sci. Technol. B* **5,** 328 (1987).
30. T. Roppel, D. K. Reinhard, and J. Asmussen, *J. Vac. Sci. Technol. B* **4,** 295 (1986).
31. J. Hopwood, M. Dahimene, D. K. Reinhard, and J. Asmussen, *J. Vac. Sci. Technol.* **B6,** 268 (1988).
32. K. N. Leung, S. Walther, and H. W. Owren, *IEEE Trans. Nucl. Sci.* **NS-32,** 1803 (1985).
33. S. R. Walther, K. N. Leung, and W. B. Kunkel, *Rev. Sci. Instrum.* **57,** 1531 (1986).
34. L. G. Meiners and D. B. Alford, *Rev. Sci. Instrum.* **57,** 164 (1986).
35. P. C. Thoneman, *Prog. Nucl. Phys.* **3,** 219 (1953).
36. C. Lejeune, J. P. Grandchamp, O. Kessi, and J. P. Gilles, *Vacuum* **36,** 837 (1986).
37. R. Lossy and J. Engemann, *Vacuum* **36,** 973 (1986).
38. C. C. Petty and D. K. Smith, *Rev. Sci. Instrum.* **57,** 2409 (1986).
39. Anatech Ltd., 5510 Vine St., Alexandria, VA 22310.
40. CVC Products Inc., 525 Lee Road, Rochester, NY 14603.
41. Commonwealth Scientific Corp., 500 Pendleton St., Alexandria, VA 22314.
42. Oxford Instruments Ltd., Osney Mead, Oxford OX2 0DX, England.
43. Veeco Instruments Inc., Terminal Drive, Plainview, NY 11803.
44. Kratos Analytical Inc., 535 East Crescent Ave., Ramsey, NJ 07446.
45. Leybold-Heraeus GmbH, Bonner Strasse 498, D-5000 Koln 51, West Germany.
46. VG Instruments, Inc., 32 Commerce Center, Cherry Hill Drive, Danvers, MA 01923.
47. Ion Tech, Inc., 2330 East Prospect, Fort Collins, CO 80525.
48. National Electrostatics Corp., Graber Road, Middleton, WI 53562.
49. Kimball Physics, Inc., Kimball Hill Road, Wilton, NH 03086.
50. Microscience, Inc., 41 Accord Park Drive, Norwell, MA 02061.
51. Colutron Research Corp., 2321 Yarmouth Ave., Boulder, CO 80301.
52. Danfysik, DK 4040 Jyllinge, Denmark.
53. General Ionex Corp., 19 Graf Road, Newburyport, MA 01950.
54. Annual Telephone Directory (Buyers Guide), Research and Development, (Research and Development, Barrington, IL).
55. Annual Buyers Guide, *Physics Today*, (August issue annually), American Institute of Physics, New York.

7

High-Current Gaseous Ion Sources

R. Keller*

GSI, Gesellschaft für Schwerionenforschung
Darmstadt, Federal Republic of Germany

Both terms marking this group of high-current ion sources need some specific explanations. *High current* strictly speaking means that for the beams delivered by these sources space-charge phenomena are a major concern. In absolute numbers, the ion beam currents may vary between 10 mA and 100 A, but still the essential features of the sources are similar enough to include them all in one group. *Gaseous* here takes on a rather broad meaning, including vapors of pure or compound materials. In reviewing such sources, one should distinguish between the plasma generator or source proper, and the extraction system; both together may then be called *ion gun* for complete clarity. The distinction is needed for the simple reason that different plasma generators can be used with identical extraction systems, and different extraction systems can well be used on a given plasma generator.

The present chapter is concerned primarily with plasma generators, leaving a thorough discussion of extraction systems to Chapter 3. There is one stringent condition that both of these two units must fulfill to enable a complete ion gun to work: The plasma generator must provide a plasma of the correct density to match a given extraction system; or, seen from the other side, the extraction system must be designed in such a way that it requires a plasma density value to be created by the generator that is within the working range of the latter.

To facilitate a comparison between the beam intensities that will be reported from various ion sources in the following, it is best to use normalized beam current values. This normalization takes into account that, according to the Child-Langmuir law [1, 2], heavier ions are harder to extract than lighter ones, see the chapter on ion extraction, Eqs. (1) and (2). The conversion between absolute current I and

---

*Current affiliation: Lawrence Berkeley Laboratory, University of California, Berkeley, California.

normalized current I_n is done according to

$$I_n = I \, (A/\zeta)^{1/2}$$

where A is the atomic mass number and ζ is the charge state; and to clearly mark normalized currents their units (mA) will be put in parentheses throughout the present chapter.

Historically, high-current ion sources have been developed for several completely different purposes: neutral injection to heat magnetically confined fusion plasmas [3], space propulsion [4], and materials treatment including ion implantation [5] at first, and in the last decade supplying particle accelerators with ions for fundamental research and inertial confinement fusion projects [6]. All of these aims have imposed quite different principal requirements on the ion sources, such as power efficiency, fuel efficiency, reliability, ample choice of the ion species, and beam quality. The dominance of one of these criteria over others, together with the common fact that a design philosophy is rarely changed if the device once had some degree of success, has led to fairly differentiated source types, but still there are many similarities between them.

The main features that are common to all high-current sources treated here are the following: The ions are created by electron-impact ionization of a gaseous medium, forming a plasma; this plasma has a considerable width, several centimeters at least and up to tens of centimeters; the electron density is about 1×10^{13} cm^{-3} and is homogeneous over that part of the plasma from which ions are to be extracted; the ion temperature usually lies well below 1 eV [7]. Quite frequently discharges are employed that are sustained by a thermionic cathode (often erroneously called *arc discharges*), but also rf discharges are well suited to create a plasma and preferred if electrode corrosion would be of serious concern.

All modern high-current ion sources employ magnetic fields to confine the plasma and enhance the ionization rate per electron, and here the so-called minimum-B configuration [8] is quite advantageous: Stable plasma containment over a large density range can be achieved only if the magnetic field lines are curved away from the region of higher plasma density, thus leaving a zone of low-field values at the center of the discharge chamber. Such configurations, see Figure 7.1, can be realized in the form of a linear multipole (usually created by strong permanent magnets), an axial cusp arrangement (generated by permanent ring magnets or coils with antiparallel current flows), a yin/yang or baseball current loop, or simply as a diverging axial field produced by permanent magnets or a single coil. In the latter case, the plasma volume is mechanically confined to a region on one side of the coil center plane only.

One family of sources makes use of a two-stage discharge where the plasma of the first discharge stage acts as a plasma cathode and yields primary electrons for sustaining the main stage [11-13]. Such an arrangement leads to very good gas and power efficiencies but is more likely to suffer from plasma oscillations, in turn.

In the following, several sources are presented which are typical or outstanding specimens of their group. It should be kept in mind that the performance data

1 Sources with Single-Stage Discharge

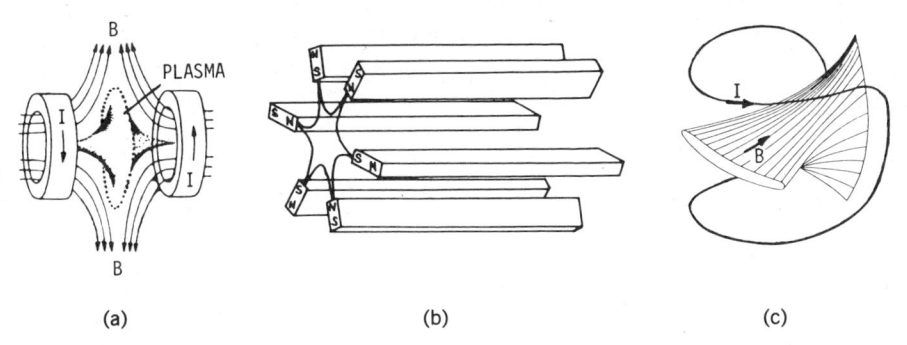

Figure 7.1 Minimum-B magnet configurations. (*a*) Monocusp. After Ref. 8, © 1976 Academic Press. (*b*) Multicusp, here a linear sextupole created by permanent magnets. From Ref. 9, © 1982 VDI Bildungswerk. (*c*) Yin/yang or baseball device. After Ref. 10, © 1970 Vieweg Verlag.

reported here are not always the real limits for the individual sources but in many cases are imposed by other conditions such as available power supplies or simply the given project goals. True beam current and brightness limits for all ion source types are determined by the physics of the extraction process.

1 SOURCES WITH SINGLE-STAGE DISCHARGE

An example of a very simple but reliable source is the so-called *Penning* source developed for a heavy-ion fusion project [14], see Figure 7.2. This high-current source should not be mistaken for the PIG (Penning Ionization Gauge) source for

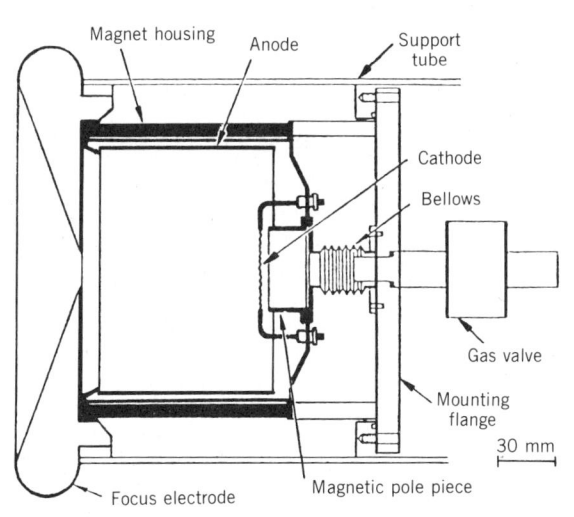

Figure 7.2 Penning high-current ion source. The beam is extracted to the left; the "focus electrode" is usually called the outlet electrode in this chapter. After Ref. 14, © 1979 IEEE.

multiply charged ions, which is treated in a separate chapter. The high-current Penning source is equipped with permanent magnet rods that are disposed around the anode cylinder and create an axial magnetic field diverging from the pole piece near the cathode toward the outer source border on the extraction side. Xenon and mercury have been used as feed materials, and xenon ion currents of 30 mA, or 344 (mA) normalized, were obtained from a 6.7-cm^2 single-aperture extraction system at 100 kV and with a discharge power as low as 150 W. As required by the intended application, in the quoted work the discharge was pulsed with duty cycles of about 0.001. Other sources of this type are operated under dc conditions as well [15].

The monocusp ion source, see Figure 7.3, was developed for a neutron generator, optimizing the discharge for high deuteron yield as well as for high gas efficiency [16]. The limitation of the magnetic field to one ring cusp and the careful positioning of the anode ring, just touching those field lines that pass through the edges of the cathode filaments, effectively minimize plasma losses. Further, there exist two electron populations in the discharge volume, with quite different mean energy values. By choosing a suitable discharge voltage one of these values can be adjusted to the maximum cross section for ionization of H_2 molecules, while the other one is much lower, favoring the dissociation of H_2^+ molecular ions. The disadvantage found with this source is the pronounced radial variation of the plasma density which severely limits the plasma size useful for extraction. Thus, only

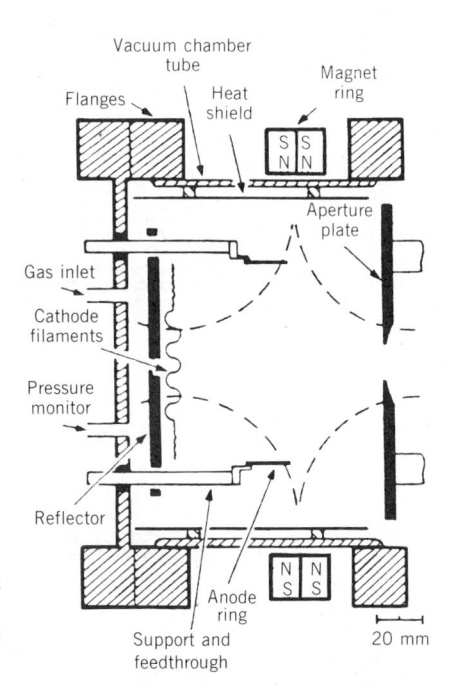

Figure 7.3 Monocusp ion source [Ref. 16]. Broken lines: magnetic field lines. The beam is extracted to the right; the "aperture plate" is usually called the outlet electrode in this chapter. © 1983 AIP.

1 Sources with Single-Stage Discharge

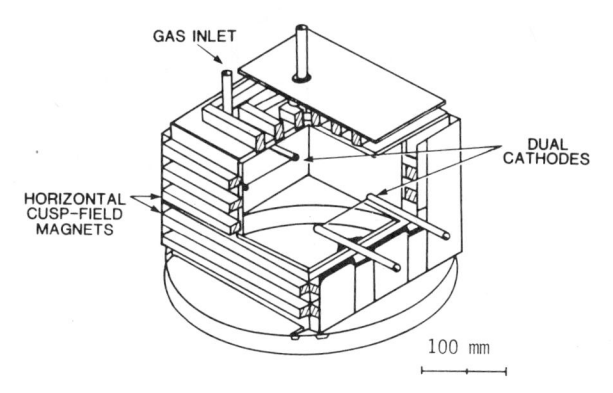

Figure 7.4 Multicusp ion source [Ref. 18]. The beam is extracted in the downward direction.

single-aperture extraction systems can be applied to this plasma generator. Total deuterium ion currents up to 800 mA dc with about 60% deuterons, or 1385 (mA) normalized, are reported for 200 kV extraction voltage and 6 kW discharge power.

Multicusp sources are the ones most frequently used in the class with a single discharge stage, presently under discussion, for one reason: Their detailed mechanical layout is far less critical than that of the two preceding source types. Any discharge vessel lined with permanent magnets will work satisfactorily in that it can yield a plasma with the required properties. This simplicity has led to the nickname "bucket source" for such plasma generators [17]. With the simplest version, see Figure 7.4, nearly the entire discharge chamber is lined by magnets and acts as anode; primary electrons are created by thermionic emission from filaments forming the cathode, and only the outlet electrode is biased close to cathode potential. The effective anode area is determined by the electron loss zones and amounts to the entire length of all the cusp lines together, times twice the electron gyro-diameter, typically a few millimeters [19]. The source according to Ref. 18 delivers 100 mA dc total hydrogen ion current—about 122 (mA) normalized—at 100 kV from a single-aperture extraction system with 1-cm^2 outlet area.

Other multicusp sources are more refined, to obtain special plasma conditions desired for the intended application. As an example, by introducing a magnetic dipole field inside the discharge chamber, across the plasma column, electron populations of different energies can be separated, favoring the production of protons in the case of hydrogen feeding gas [20]. This configuration is sometimes called a "picket-fence" [21].

Minimizing the power requirements for a given ion current output led to a source design, the *Kaufman source*, where the anode is composed of narrow metal strips positioned between two adjacent magnet pole pieces [15], see Figure 7.5. Such a configuration forces the discharge electrons to transversely cross the magnetic field lines and at higher densities invariably causes plasma oscillations to be excited, precluding the formation of ion beams with low divergence. In fact, quoted diver-

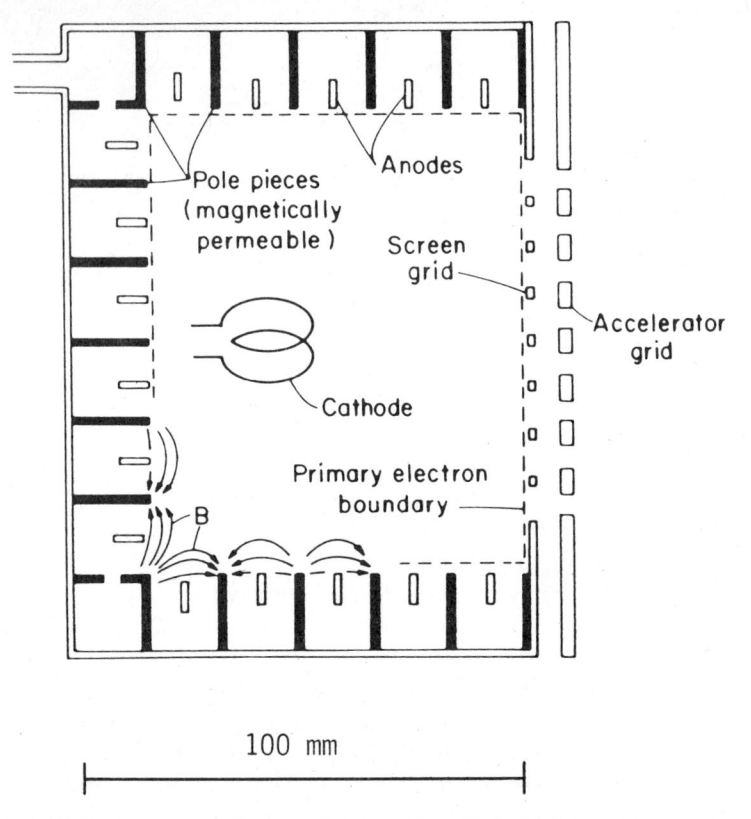

Figure 7.5 Multicusp ion source for industrial applications [Ref. 15]. As stated in the reference, the actual source dimensions may be varied widely; the scale shown gives an idea of a possible size. The beam is extracted to the right; the "screen grid" is usually called the outlet electrode in this chapter. © H. R. Kaufman.

gence values for this source are in the order of 10°, rather than 1° as customary for accelerator sources, but for applications such as ion beam milling through masks or sputter deposition this is not harmful at all. A typical argon ion current per extraction hole is 1 mA dc—6.3 (mA) normalized—at 2 kV, and such sources may have 1000 and more outlet apertures; only 540 W discharge power is then needed to sustain a plasma of the matching density.

Another specimen is the reflex discharge multicusp source CHORDIS [22], featuring the cylindrical anode alone lined with magnets and both end plates of the discharge chamber on or close to cathode potential, see Figure 7.6. While the electrons are well contained in this configuration, the ions will be accelerated towards both end plates. This may even favor the extraction process, but certainly leads to more ion losses on the other side of the source, to be paid for by higher discharge power than with other multicusp sources. The technical benefit lies in

1 Sources with Single-Stage Discharge

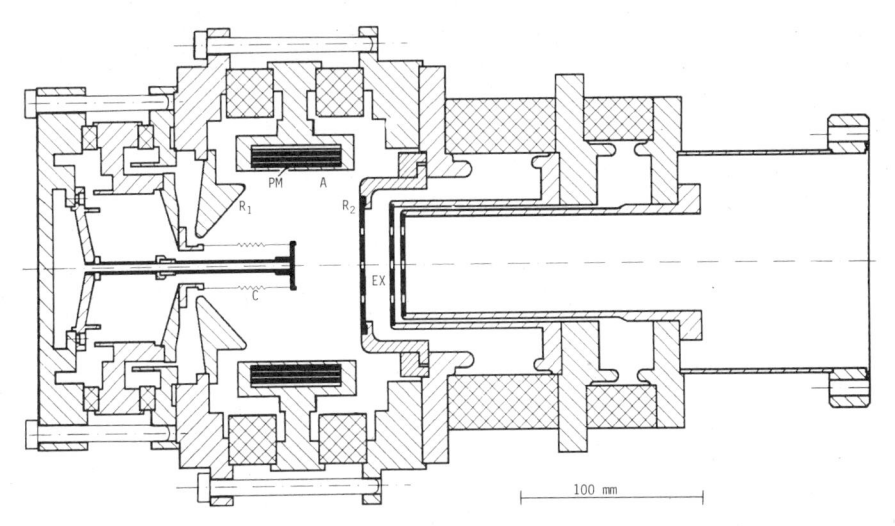

Figure 7.6 Cold version of CHORDIS, for gases [22]. A, anode; C, cathode filament; EX, accel-decel extraction system; PM, permanent magnet; R_1, R_2, reflector electrodes; R_2 acts as outlet electrode as well. The beam is extracted to the right.

the possibility to install more complex structures on the rear side of the source, in order to process materials other than gases. One example of an ion beam obtained from the cold source version is 71 mA dc of xenon ions, 814 (mA) normalized, at 50 kV, out of 2 cm^2 outlet area at 1.8 kW discharge power.

Other multicusp sources are presented in Ref. 23, describing the ion gun for a spallation neutron source project, and Refs. 7 and 24, regarding neutral injection sources for magnetic confinement fusion projects that can deliver several tens of amperes beam current during 10 second pulses at typically 80 kV from about 150 cm^2 total outlet area.

As previously mentioned, the discharge can also be maintained by rf power with typically 2 MHz frequency [25], rather than by a dc voltage between a thermionic cathode and an anode. The problem with rf discharges lies in the varying plasma impedance, making a careful matching to the rf coupling antenna imperative. One example of a successful source model uses inductive coupling by a water-cooled antenna disposed around the cylindrical quartz discharge vessel [26], see Figure 7.7. A 10-cm-diameter source of this type is able to generate hydrogen ion current densities of 200 mA/cm^2 with about 85% protons, at 3.2 kW discharge power. The outcoming beam then has about 2° divergence. The antenna may also be inserted into the discharge chamber, allowing the walls to be lined with permanent magnets to create a multicusp array [27]. With microwaves of 2.45 GHz frequency, capacitative coupling into a resonator of adjustable dimensions appears to be another suitable technique to create a plasma of the needed density [28]. Microwave sources in general are separately treated in another chapter of this book.

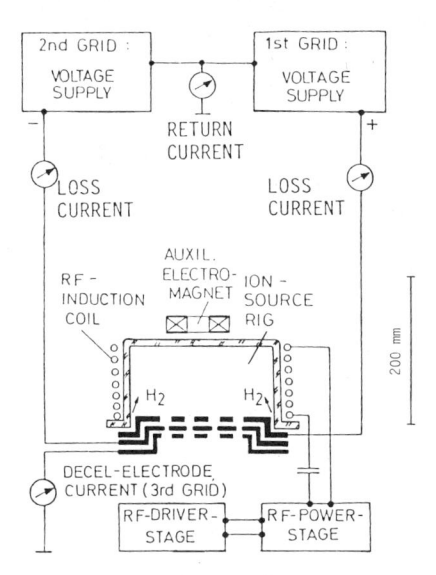

Figure 7.7 RIG 20 ion source with rf discharge [26]. The beam is extracted in the downward direction. The "first grid" is usually called the outlet electrode in this chapter.

2 SOURCES WITH TWO-STAGE DISCHARGES

The most important pioneer work in the development of high-current ion sources was performed with duoplasmatrons [11, 12] and duopigatrons [13], that is, two-stage discharges. With both, a discharge is maintained at relatively high pressure (about 3×10^{-2} Torr) and low voltage (typically 10 V) between a thermionic cathode and an intermediate electrode, acting as primary anode. The plasma is then guided by a strong axial magnetic field through an aperture within the intermediate electrode into the second discharge chamber, where the discharge runs at much lower pressure (about 2×10^{-3} Torr) and higher voltage (typically 80 V), between the intermediate electrode, now acting as cathode for this stage, and the main anode. In the case of a duoplasmatron, see Figure 7.8, the plasma created in the second stage flows out through a small aperture in the anode and expands into a third chamber, the so-called expansion cup [30]. Duopigatrons are modified duoplasmatrons, with largely increased anode aperture and the end wall of the expansion cup, functioning as source outlet electrode for the extraction system, biased negatively with respect to the anode potential. This arrangement causes the electrons of the second discharge stage to be reflected between outlet and intermediate electrode, thus leading to a reflex discharge with much better power and gas efficiencies than found with the duoplasmatron. The magnetic field that guides the plasma from the first into the second chamber expands toward the outlet electrode and thus the second stage of the duopigatron closely resembles the Penning high-current ion source mentioned above. An advanced duopigatron is shown in Figure 7.9. It yields, for example, 98 mA dc of 35 keV xenon ions—

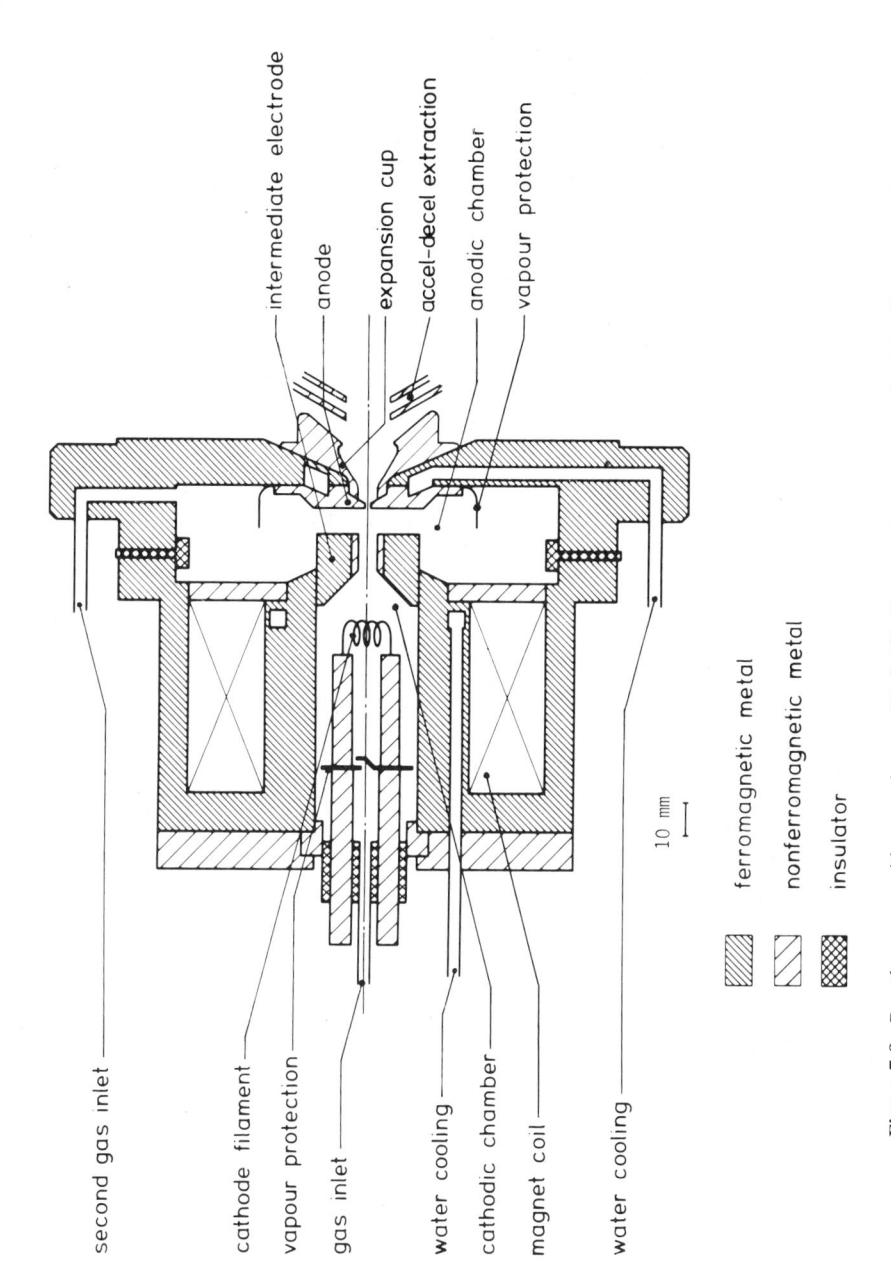

Figure 7.8 Duoplasmatron with expansion cup [Ref. 29]. The beam is extracted to the right. The anode and expansion cup together act as the outlet electrode. Reprinted by permission of the author and publishers. © 1979 Gordon and Breach Science Publishers, Inc.

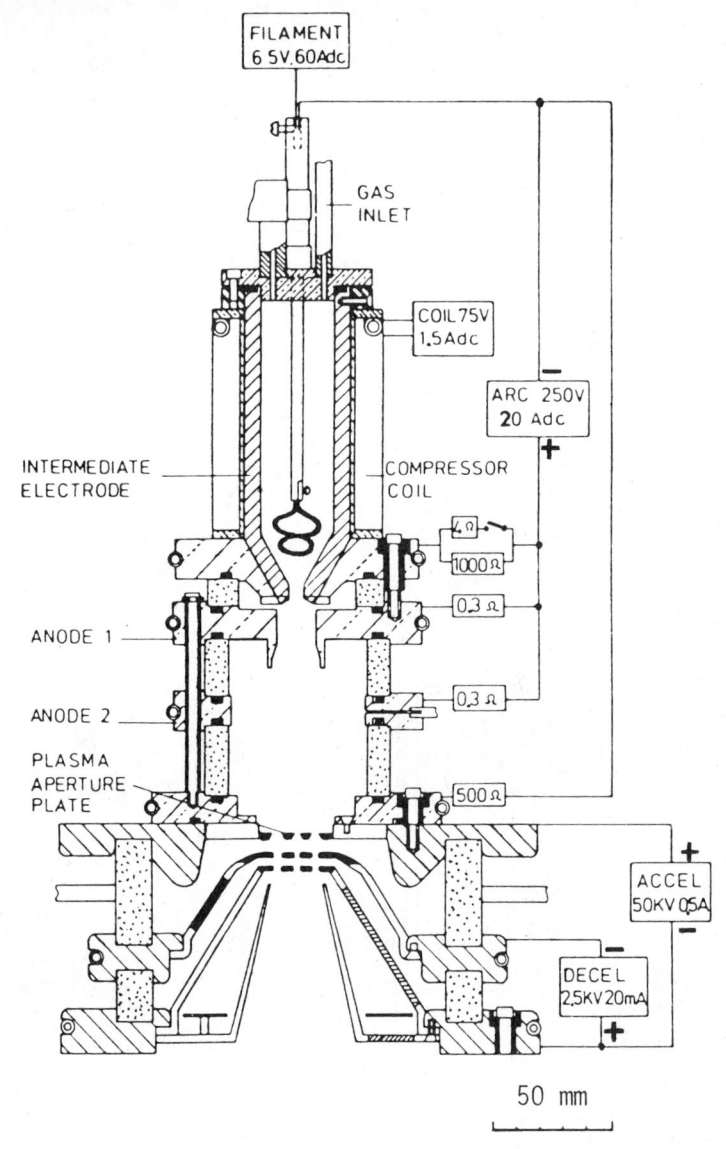

Figure 7.9 Duopigatron [Ref. 31]. The beam is extracted in the downward direction. The "plasma aperture" is usually called the outlet electrode in this chapter. © 1980 IOP Publishing Ltd.

1123 (mA) normalized—from 1.4 cm^2 aperture area at 520 W discharge power [31]. Another outstanding example is the MATS III source, yielding 1500 mA of 20 keV hydrogen beam (70% protons, 25% double, and 5% triple ion molecules) out of 8 cm^2 aperture area at 4 kW discharge power. 900 mA of this beam are then delivered to a remote target within 20 mrad divergence half-angle [32].

As a general remark, it appears to be more difficult to design a properly working two-stage discharge ion source than, for example, a multicusp source, but once having succeeded with this task, there is the benefit of the higher efficiency as compared to other sources. Moreover, a special operating mode becomes possible where a protective gas is fed into the first chamber, impeding fast consumption of the filaments, and a reactive gas into the second chamber, where no more hot metal electrodes are present. This two-gas technique [29] allows the stable production of many interesting ion species such as from high-melting-point elements. One has to take into account, however, that the desired ions have to share the total beam current with other species, and their fraction typically lies in the order of 15%. For oxygen, using argon and pure oxygen gas, 138 mA dc O_1^+ ions—552 (mA) normalized—out of 250 mA total current were obtained from 1.4 cm^2 outlet aperture at 52 kV, running a duopigatron at 1.25 kW discharge power [33].

In pursuing the idea of a two-stage discharge one could also go a step further and combine an rf discharge as first stage with a reflex chamber, lined with permanent magnets or equipped with a diverging axial field, to completely avoid any cathode erosion but still have the good efficiencies, wide range of operation parameters, and homogeneity found with stationary discharges.

3 MULTIPLY CHARGED IONS

The plasma generators presented in this chapter do not generally lend themselves to a copious production of multiply charged ions. This is partially due to the requirements listed earlier: In optimizing a discharge toward generating a quiet, cold plasma, one automatically ends up at conditions where singly charged ions are being produced preferentially. Further, when looking for maximum beam brightness values, one would hardly like to share the beam current among different ion charge states, but this is a necessary condition associated with steady state formation of multiply charged ions.

However, in cases where the highest brightness values are not the predominant requirement for a source, multiply charged ions might still be of interest, and there are some investigations of this topic being carried on presently. The most prominent results have been obtained with multicusp sources: In one experiment using an argon plasma and 250 V discharge voltage traces of Ar^{7+} were found [34], and in another experiment the following beam currents for multiply charged ions have actually been transported [35]: 11.6, 10.2, 5.0, 2.2, 0.8, and 0.11 mA for Xe^+ through Xe^{6+}, and 15.8, 9.9, 3.4, 0.44, 0.16, and 0.06 mA for Kr^+ through Kr^{6+} (both electrical, not particle currents). To obtain these values, the discharge had to be run at 400 V and 58 A, and such power loads imply pulsed operation with duty factors below 25%.

It is interesting to note that with the same source and extraction system that yielded the multiply charged ions listed above the discharge current was lower by a factor of three when a maximum share of singly charged ions was being produced. Considering that the plasma density must be equal in both cases to match the identical extraction field strength, one can conclude that at least the electron temperature, and most probably the ion temperature as well, is higher by about a factor of 10, compared to the normal values, when the discharge is tuned to favor the generation of multiply charged ions. Such a trend would further inhibit the formation of maximum-brightness beams, in addition to the facts mentioned above.

4 IONS FROM ELEMENTS WITH LOW VAPOR PRESSURE

General hints regarding the processing of nonvolatile elements in ion sources were published some time ago [36]. The most universal approach is to use volatile compounds, such as chlorides, fluorides, or in some cases oxides and sulfides. On-line chlorination techniques, guiding a flow of Cl_2 or CCl_4 over the solid material of interest, can also give good results in some cases. But there are three major drawbacks associated with these techniques. First of all, the total beam current has to be shared among different ion species, that is, the main components of the compound and usually a series of different molecular ions. The exact distribution depends on the chemical equilibrium in the discharge plasma and cannot be known in advance. Second, all the mentioned compounds have reactive constituents, and their use will shorten the filament lifetime considerably for all sources with steady discharges and thermionic cathodes. Employing a two-gas technique, as described in the section on two-stage discharges, will only partially solve this problem. And third, one must take care to prevent condensation of the nonvolatile constituent on cold source walls, which will occur whenever a compound preferentially dissociates before being ionized.

Thus, hot walls inside a source are very useful in any case, and the two well-established ways to realize them are either by using a discharge chamber made from graphite [37] or by installing metal liners inside a cold chamber [38]. In both cases, the radiation from the cathode filaments will keep the walls hot, independently of the discharge power that may not be distributed uniformly over the wall surfaces.

From a discharge chamber with hot walls it is only a small step to a source with incorporated oven, to produce vapors from pure elements. As an example of such a source, the hot version of the multicusp/reflex discharge CHORDIS [22], mentioned above, is shown in Figure 7.10. It can deliver the same ion current values as the cold basic version, and additionally permits processing any material with more than 2 Torr vapor pressure at 1000 °C. As a precaution, however, the extraction gap has to be made wider than the minimum value for operation with gases only, see the chapter on Ion Extraction, and this leads to a decrease of beam current and brightness. For bismuth, a dc ion current of 37 mA—535 (mA) normalized—was obtained from 2 cm^2 outlet aperture at 36 kV, with as little as 340 W discharge power, due to the low ionization energy of the metal. Adding an auxiliary gas such as argon to the vapor appears to be helpful in getting a stabler discharge, but still the ion beam contains as little as 5 to 10% gas ions, because their ionization energies are considerably higher than those of metals.

4 Ions from Elements with Low Vapor Pressure

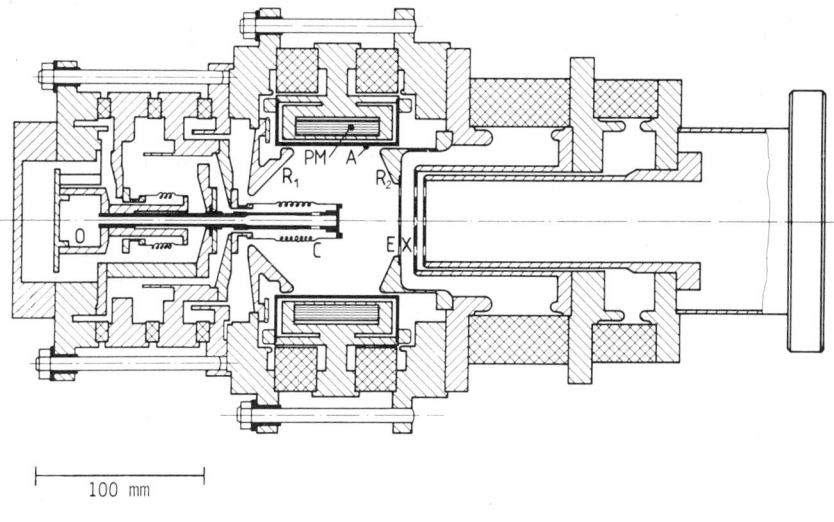

Oven/cathode chamber   Discharge chamber   Extraction system

Figure 7.10 Hot version of CHORDIS, with integrated oven [Ref. 22]. A, hot-running anode; C, cathode filament; EX, accel–decel extraction system; PM, permanent magnet; R_1, R_2, reflector electrodes; R_2 acts as outlet electrode as well; O, oven, to be charged from the back side of the source. The vapor is transported through a pipe from the oven to the cathode support tube and flows radially into the discharge chamber. The beam is extracted to the right.

The relatively low ionization energies of metallic elements are also advantageous when using a sputtering technique, well known from other ion source types such as the PIG [39] or Duoplasmatron [40]. In this case, an auxiliary discharge with one of the heavier noble gases is sustained in the source, and an electrode covered by the metal of interest is biased to some hundred volts negative with respect to the anode. This potential attracts ions from the discharge plasma, and they sputter atoms away from the surface; these will then be ionized by the discharge. The sputtering technique can, in principle, be applied to all metals, even those with low vapor pressures, and in the form of powder mixtures with metals pressed into shape, nonconducting materials can be processed, too. The particle production rate can easily be controlled through the current drawn by the sputter electrode, and this electrode should be well cooled, to avoid runaway conditions. As a demonstration of this technique, applied to a high-current ion source, an experiment has been performed using a modified CHORDIS where the outlet electrode was covered by an aluminum ring and electrically insulated against the other reflector electrode. A share of 20% aluminum ions was reached in the extracted beam at 250 V sputter voltage, using argon as auxiliary gas [41]. For elements with lower sputtering coefficients [42], shares in the 10% range can be expected.

Following experience with PIG sources for multiply charged ions [43], the internal parts of a sputtering high-current ion source should be kept hot to avoid recondensation of the material of interest, in the same way as when using volatile chemical compounds. The advantage of the sputtering technique lies in the fact that the use of reactive materials can be completely avoided.

5 OUTLOOK

A description has been presented of the basic physical working principles of gaseous, high-current sources, including some special topics such as multiply charged ions and low vapor pressure elements. This is intended as an introduction to the field, however, and is not meant to enable newcomers to start building the ion source of their choice right away without further research. But at least the fundamentals will now rest on a solid foundation.

One important aspect of ion source design has been almost completely omitted in this brief overview: the technological point of view. Besides the physical principles discussed, the problems of vacuum, insulation, high voltage, high temperature, and power dissipation problems have to be addressed at the same time. The reader should refer to the literature for in-depth accounts of how these problems have been solved in different situations. For this purpose, not only the publications concerned with individual sources—as quoted in the text—are recommended, but also more general ones [36, 45]. In addition, the reader will without doubt find much valuable information in the conference proceedings of Vienna [45], Kyoto [46], and a conference series in Great Britain [47, 48]. Basic material properties, without direct connection to ion sources, are collected in Refs. 49 and 50.

REFERENCES

1. C. D. Child, *Phys. Rev.* (Ser. 1) **32,** 492 (1911).
2. I. Langmuir and K. T. Compton, *Rev. Mod. Phys.* **3,** 251 (1931).
3. J. G. Cordey, *3rd Int. Meeting on Theoretical and Experimental Aspects of Heating of Toroidal Plasmas, Grenoble, France*, Vol. 2, Commissariat a l'Energie Atomique, Paris, 1976.
4. E. Stuhlinger, *Ion Propulsion for Space Flight*, McGraw-Hill, New York, 1964.
5. G. Dearnaley, J. H. Freeman, R. S. Nelson, and J. Stephen, *Ion Implantation*, North-Holland, Amsterdam, 1973.
6. R. Bock, *IEEE Trans Nucl. Sci.* **NS-30,** 3049 (1983).
7. A. J. T. Holmes and M. Inman, Proceedings, 1979 Linear Accelerator Conf., BNL-51134, Brookhaven Nat. Lab., 1979, p. 424.
8. F. F. Cap, *Handbook on Plasma Instabilities*, Vol. 1, Academic, New York, 1976, pp. 149–151.
9. R. Keller, VDI-Bildungswerk BW 41-18-02/BW 5244, Verein Deutscher Ingenieure, Düsseldorf, West Germany, 1982, p. 5 (in German); English translation: Ion Sources for High-Frequency Accelerators, LA-Tr-85-16, Los Alamos National Laboratory, 1985.
10. F. F. Cap, *Einführung in die Plasmaphysik*, Vol. 2, Vieweg, Braunschweig, 1970, p. 31 (in German).
11. M. von Ardenne, *Tabellen zur Angewandten Physik*, Vol. 1, VEB Verlag der Wissenschaften, Berlin, 1962, p. 653 (in German).
12. H. Fröhlich, *Nukleonik* **1,** 183 (1959).

References

13. R. A. Demirchanov, H. Fröhlich, U. V. Kursanov, and T. I. Gutkin, BNL-767 (C-36), Brookhaven Nat. Lab., 1962, p. 224.
14. R. P. Vahrenkamp and R. L. Seliger, *IEEE Trans. Nucl. Sci.* **NS-26,** 3101 (1979).
15. H. R. Kaufman and R. S. Robinson, *Am. Inst. Aeronaut. Astronaut. J.* **20,** 745 (1982).
16. J. P. Brainard and J. B. O'Hagan, *Rev. Sci. Instrum.* **54,** 1497 (1983).
17. R. Limpaecher and K. R. McKenzie, *Rev. Sci. Instrum.* **44,** 726 (1973).
18. J. D. Schneider, H. L. Rutkowski, E. A. Meyer, D. D. Armstrong, B. A. Sherwood, and L. L. Catlin, 1979 Linac Conf. BNL-51134, Brookhaven, 1979, p. 457.
19. A. Goede, T. S. Green, and B. Singh, 8th Europ. Conf. Contr. Fusion and Plasma Phys., Prague, 1977.
20. K. W. Ehlers and K. N. Leung, *Rev. Sci. Instrum.* **53,** 1423 (1982).
21. K. N. Leung, N. Hershkowitz, and K. R. MacKenzie, *Phys. Fluids* **19,** 1045 (1976).
22. R. Keller, P. Spädtke, and F. Nöhmayer, Proceedings, Int. Ion Engineering Congr., Kyoto, Inst. Electr. Engineers of Japan, Tokyo, 1983, p. 25.
23. B. Piosczyk, Proceedings, 1981 Linear Accelerator Conf. LA-9234-C, Los Alamos, 1982.
24. Y. Ohara, loc. cit. Ref. 22, p. 447.
25. H. Loeb, AIAA 7th Electric Propulsion Conf., Williamsburg, Am. Inst. Aeronautics and Astronautics, paper 69-285, 1969.
26. J. Freisinger, loc. cit. Ref. 22, p. 39.
27. J. R. Bayless, D. Arnush, W. F. DiVergilio, V. V. Fosnight, H. Goede, and P. W. Kidd, loc. cit. Ref. 22, p. 45.
28. J. Root and J. Asmussen, *Rev. Sci. Instrum.* **56,** 1511 (1985).
29. R. Keller, *Radiation Effects* **44,** 201 (1979).
30. O. B. Morgan, G. G. Kelley, and R. C. Davis, *Rev. Sci. Instrum.* **38,** 467 (1967).
31. M. R. Shubaly, Institute of Physics Conference Series, Vol. 54, Adam Hilger, Bristol, UK, 1980, p. 333.
32. J. E. Osher and G. W. Hamilton, Proceedings, 2d Symp. on Ion Sources and Formation of Ion Beams, LBL-3399, Lawrence Berkeley Laboratory, 1974, p. VI-7-1.
33. M. R. Shubaly, R. G. Maggs, and A. E. Weeden, *IEEE Trans. Nucl. Sci.* **NS-32,** 1751 (1985).
34. K. N. Leung, A. S. Schlachter, J. W. Stearns, R. E. Olson, and J. R. Mowat, Proceedings, 2d Neutralizer Workshop, Brookhaven Nat. Lab., 1986, p. 279.
35. R. Keller, Proceedings, 1987 Particle Accelerator Conf., Washington, DC, March 16–19, 1987.
36. J. H. Freeman and G. Sidenius, *Nucl. Instrum. Methods* **107,** 477 (1973).
37. J. H. Freeman, *Nucl. Instrum. Methods* **22,** 306 (1963).
38. G. D. Magnusson, C. F. Carlston, P. Mahadevan, and A. Comeaux, *Rev. Sci. Instrum.* **36,** 136 (1965).
39. B. F. Gavin, *Nucl. Instrum. Methods* **64,** 73 (1968).
40. R. H. V. M. Dawton, *Nucl. Instrum. Methods* **67,** 341 (1969).
41. R. Keller, P. Spädtke, and H. Emig, *Vacuum* **36,** 833 (1986).
42. H. H. Andersen and H. L. Bay, in *Topics in Applied Physics*, vol. 47, R. Behrisch (Ed.), Springer, Berlin, 1981, Chapter 4.
43. H. Schulte, W. Jacoby, and B. H. Wolf, *IEEE Trans. Nucl. Sci.* **NS-23,** 1042 (1976).
44. G. W. Hamilton, Proceedings, Symp. on Ion Sources and Formation of Ion Beams, BNL-503109, Brookhaven Nat. Lab., 1971, p. 171.
45. F. Viehböck, H. Winter, and M. Bruck (Eds.), *Proceedings 2d Ion Source Conf., Vienna 1972*, Vol. 2, Inst. of Experimental Physics, Vienna Technical University, Vienna, 1972.
46. T. Takagi (Ed.), loc. cit. Ref. 22.

47. Proceedings, Int. Conf. on Low Energy Ion Beams, 1977 and 1980, Institute of Physics Conference Series, Vol. 38, 1978, Vol. 54, 1980, The Institute of Physics, Bristol, UK.
48. Proceedings, Int. Conf. on Low Energy Ion Beams, 1983 and 1986, *Vacuum* **34** 1984, **36** (1986), Pergamon, Oxford, UK.
49. C. J. Smithells, *Metals Reference Book* Vols. 1–3, Butterworths, London, 1967.
50. I. E. Campbell and E. M. Sherwood (Ed.), *High-Temperature Materials and Technology*, Wiley, New York, 1967.

8

PIG Ion Sources

B. F. Gavin
Lawrence Berkeley Laboratory
University of California
Berkeley, California

The PIG ion source has been used as a source of multiply charged gaseous ions for over three decades, and more recently as a source of metallic ions. PIGs operate immersed in a magnetic field, which serves also to separate the charged ion species, with currents of multiply stripped ions up to milliamperes. This kind of source has been extensively used in injectors for particle accelerators—cyclotrons, synchrotrons, and linacs—and this is the focus of the scientific community that has provided most of the research and development of the PIG ion source.

1 BACKGROUND

The PIG source derives its name from the vacuum gauge invented by Penning [1]. This gauge was then manufactured by the Philips Company and the origin of the acronym is also given as the *Philips Ionization Gauge*. A simplified schematic of the PIG source is shown in Figure 8.1. Two cathodes are located at either end of a tubular anode within an axial magnetic field. Electrons emitted from either cathode are accelerated into the hollow anode as a beam, and a large fraction of them are trapped, axially by the electrostatic well and radially by the magnetic field. The primary beam electrons ionize the background gas to form the dense plasma from which the ion beam is to be extracted. The electrons eventually diffuse to the hollow anode, and the ions diffuse to the anode or accelerate to the cathodes.

A neutral gas background allows plasma and sheath formation, and gas must be fed in at a rate such as to provide equilibrium with the ion loss rate from the PIG plasma. The characteristics of the PIG plasma are determined in large part by

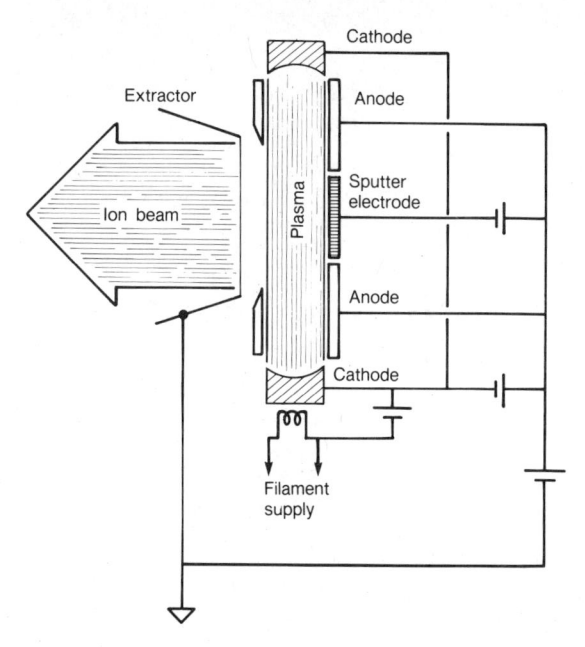

Figure 8.1 Schematic of the PIG source and its power supplies.

the pressure of the neutral gas within the discharge region. The low-pressure Penning mode occurs for pressures less than about 10^{-4} Torr [2–4]. This mode is not usually employed for ion source operation and will not be considered here. The high-pressure Penning mode—pressures above 10^{-3} Torr—is characterized by a plasma potential that is close to the potential of the most positive element of the source; it is usual for the plasma potential to be within 10 V or so of the anode voltage. Thus, the arc voltage drop occurs across the cathode sheath. The voltage applied across the arc (anode–cathode voltage) might typically be 700 V or more, and so the cathode sheath drop can be quite substantial; this is the voltage that accelerates the primary electrons. In this high-pressure mode of operation the discharge is largely unaffected by the magnitude of the axial magnetic field [5, 6]. Sizeable ion currents can be extracted from the plasma, either radially through an axial slit in the anode or axially through a small central hole in one of the cathodes. Eventually the cathodes are consumed by sputtering due to energetic ion bombardment. It is the cratering of the cathodes (as their material is sputtered away), and also the constriction of the hollow anode (as material is deposited there), that accounts for the slow decay in ion beam current that is typical of the PIG source.

The PIG source literature has been reviewed by a number of authors [7–13]. Here we discuss some aspects of PIG ion sources that are representative of source fundamentals.

2 CHARACTERISTICS OF COLD CATHODE PIGS

Emission of the primary electrons from the cathode is achieved either by secondary emission (cold cathode), thermionic emission (hot cathode), or a combination of both. Cold cathode sources require two well-cooled cathodes. The primary electron energy and the ion production rate are determined by the cathode material, neutral gas, and source dimensions. Typically about 75% of the primary electrons are trapped, and each trapped electron generates a half dozen or so ions from the background gas [5, 14].

A cold cathode discharge is characterized by a positive resistance current–voltage relation. Figure 8.2 shows the I–V relation in a PIG xenon discharge for a number of cathode materials [15]. With some gases, the formation of compounds on the cathodes complicates the consistency of the secondary electron emission coefficient—electron emission depends on the work function which in turn depends on the type of metal, surface purity, and surface crystal orientation.

The ion production rate in the plasma can be controlled via the arc current. For maximum yield of the higher ionic charge states, the arc current is driven as high as possible, consistent with a positive resistance arc characteristic. The cathodes may become at risk of melting, a circumstance that is evidenced by an abrupt change in the arc impedance, brought about by the asymmetric arc geometry as the cathode deforms. Thus, the duty factor at which the source is operated sets a limit to the maximum arc current that can be supplied. Usually the duty cycle is kept less than 25%; even at a low duty cycle, say 1%, the arc current can be

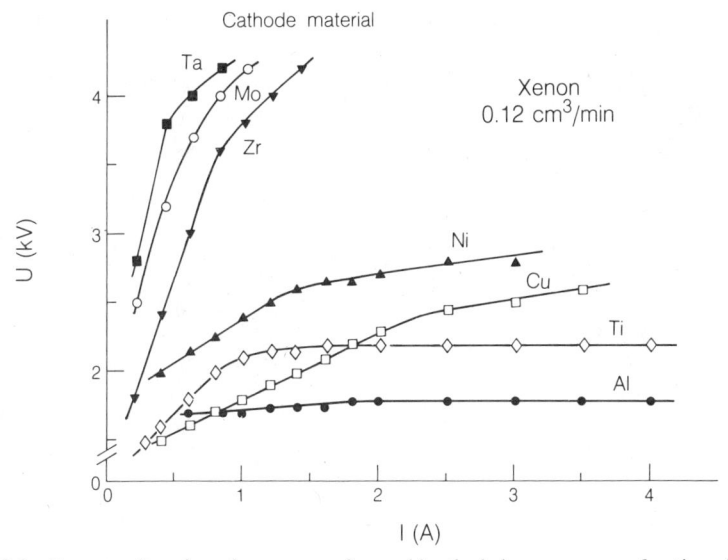

Figure 8.2 Current-voltage impedance curves for a cold cathode ion source, as a function of cathode material. Support gas xenon, magnetic field strength 5 kG. From Wolf [15].

increased only by a factor of about 2 over that typical of high duty cycle operation. Titanium, a popular cathode material, is not operated above 5 A due to thermal gradients across the cathode surface.

The energy of the primary electrons in cold cathode PIGs is usually above 1 keV and their density around 2×10^{12} cm^{-3}. The ion confinement time is of order several microseconds, and the important parameter $n_e t_i$ is thus approximately 10^7 cm^{-3} s.

One of the earliest attempts to extract beam from the cold cathode PIG was the work of Heil in the early 1940s; 1 to 2 mA of radially extracted mercury and hydrogen beams were reported [16]. Later, in 1954, Jones and Zucker produced 9-mA beams of N^{4+} [17]. Anderson and Ehlers made a similar source suitable for accelerator application [18]. Other sources were developed by Morosov [19] and by Wolf [15]. Some sources were big [17] and some were small [20], or specifically designed for high charge state, low duty factor pulsed operation [21]. A small, cold cathode source with radial extraction, developed for the SuperHILAC high-pressure capsulated injector at LBL [20], is seen in Figure 8.3, and is shown positioned directly above its normal location inside the magnet pole pieces.

Axial extraction was first investigated by the Manchester group [22]. Intense

Figure 8.3 LBL ADAM source, shown above support magnet structure as if in "exploded view." Cathode separation 4.1 cm, analyzing magnet gap 3.6 cm. Beam exit hole from the magnet assembly is on the left. Scale is in inches. From Gavin [57].

beams of multiply charged nitrogen were also reported by Mineev and Kovpik [23], who used an axial plasma expansion cup. A long-lived, low-power, axial extraction PIG source for low charge state ions was developed by Baumann [24]. A disadvantage of axial extraction is the erosion of the axial extraction aperture, even at modest power levels, and this has discouraged extensive application.

3 CHARACTERISTICS OF HOT CATHODE PIGS

The hot cathode PIG currently used in many accelerators utilizes an indirectly heated cathode of large enough mass to assure a long cathode lifetime. The source is generally large and suitable for high duty factor operation. Electron emission is predominantly thermionic. Cathode temperature is raised by ion bombardment from the plasma or by using a supplementary electron gun directed at one of the two cathodes. Figure 8.4 shows the typical lower impedance arc characteristic of a hot cathode PIG for three slightly different supplementary cathode heating powers [25], and allows the arc potential to be tuned independently of the arc current, a useful knob.

Early hot cathode PIGs used a directly heated filament as cathode; see, for example Jones and Zucker [17] and Livingston and Jones [26]. The filament was directly exposed to ion bombardment, and led to short lifetime and nonuniform emission. Mills and Barnett [27] used refractory metal cathode rods purposely not cooled. This kind of plasma heated cathode will give the arc a negative impedance characteristic when driven to high arc power levels [28].

An impressive advance in high-power, high charge state performance came with the work of Morozov et al. [19] and later by Pasyuk and Tretyakov [29, 30] at the Joint Institute of Nuclear Research in Dubna, USSR. One of the two cathodes was refractory and well cooled. Arc currents were as high as 50 A. The source was operated in a pulsed mode, with pulse width 2 ms and repetition rate 150 Hz. Pasyuk reports measuring 5 μA of Xe^{13+} and 15 μA of Kr^{11+}. A smaller version of the Dubna source has been described by Bieth [31], upgraded later by Bex for Ganil [32].

Another source, developed at the LBL, Berkeley, has a *dual head* (plasma cavity) configuration; the working head can be changed from one to the other without interruption of vacuum [33], and is seen in Figure 8.5. The source is shown detached from its mount and directly below its normal position. The source magnet is on the left. All modern hot cathode PIGs continue to use the Dubna design for cathode layout, as seen in cross section; see Figure 8.8.

4 PIG DISCHARGE MECHANISMS

An analysis of the energy flow within the hot cathode PIG was carried out by Basile and Lagrange [34–36]; their results compared favorably with experimental values for the energy needed to produce each ion. Axial and radial particle loss mecha-

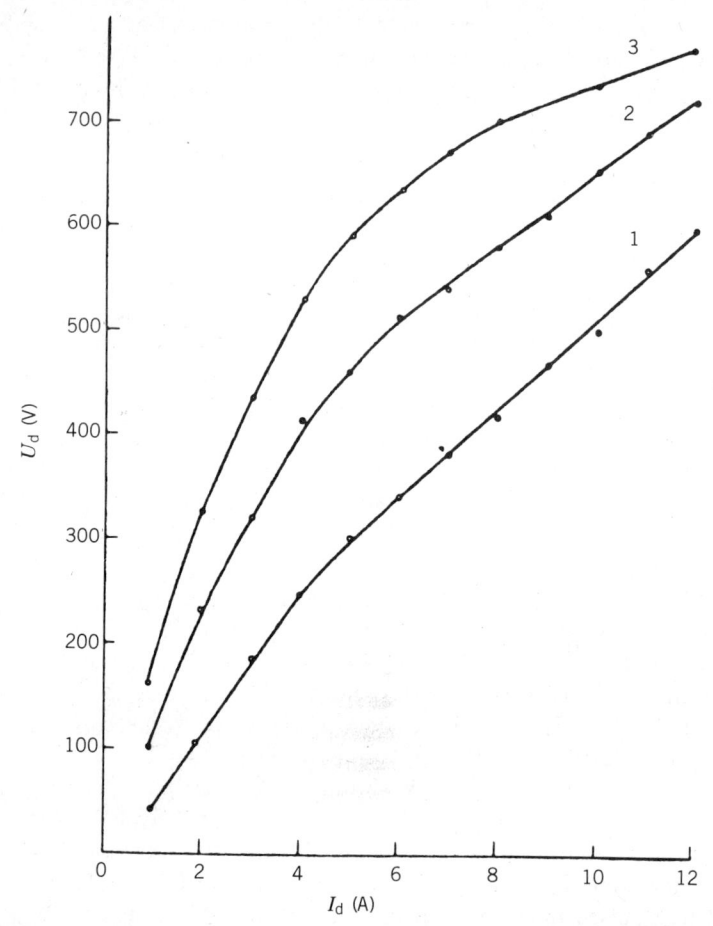

Figure 8.4 Current-voltage impedance curves for a hot cathode ion source of the Dubna type. Support gas xenon, flow rate 0.2 cm^3/min, 2-ms pulse width at 100 Hz. Cathode heating powers (external): (1) 1.19 kW, (2) 1.06 kW, (3) 1.00 kW. From Makov [25], © 1976 IEEE.

nisms were added into the theory by Green [37]. This kind of theoretical analysis was carried further by Schulte et al. [38], who included in the treatment the experimental observation that in the Dubna type source there is an electron flux of around 1 A/cm^2 to the cold cathode—against the cathode potential. This is a surprising result; the current to the cold cathode is not pure ion current but is a combination of ion current and an unusually large electron current. Somewhat less than 50% of the primary hot electron current is to the cold cathode (the ion particle currents to both cathodes are usually quite similar). Consequently, Schulte et al. proposed the generation of strong turbulence in the plasma due to the primary electron current (i.e., a beam–plasma instability), with the consequent heating of the beam electrons. The velocity distribution evolved by the primary electrons, as predicted by this theory, is indicated in Figure 8.6. The growth rate of this instability is

4 PIG Discharge Mechanisms

Figure 8.5 LBL SuperHILAC source doublet, shown detached from pivotable support structure, in a "swung-out" position. Axial separation between cathodes is 10.4 cm; transverse separation between the two sources is 7.3 cm; analyzing magnet gap (on left side) is 17.7 cm. From Gavin [57].

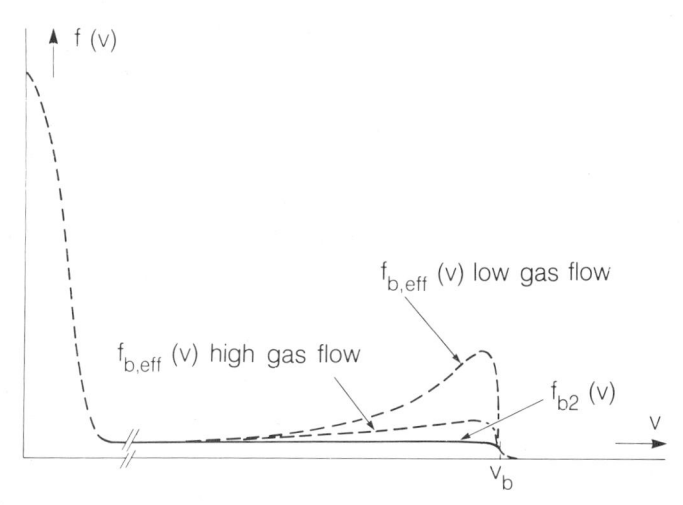

Figure 8.6 Postulated effective electron velocity distribution for a Dubna-type PIG source. The distributions f_b are shown as a function of plasma particle density. From Schulte [38], © 1976 IEEE.

large, and the turbulence is saturated in a time typically a few tens of nanoseconds, depending on the plasma parameters such as arc potential, discharge conditions, ionization cross sections, and neutral gas pressure. The main bulk of plasma electrons are relatively cold; for typical conditions this temperature has been measured by Pigarov and Morozov [39] at around 15 eV. The high-energy tail of the distribution is more important for ion stripping than is the low-temperature component. In this model the gas flow rate is important in that new neutrals are needed to refuel the hot core of the plasma; optimally, the mean free path for ionization will allow neutrals to reach the center, where the stripping is highest.

Makov [25] measured the perpendicular pressure (nkT) of the PIG plasma. He found that the pressure increases rapidly with decreasing gas flow (see Figure 8.7), in qualitative agreement with Schulte's effective velocity distribution. Makov estimated a mean perpendicular electron energy of approximately 150 eV, and n_e approximately 10^{13} cm^{-3} for an arc current of 11 A and arc voltage of 600 V.

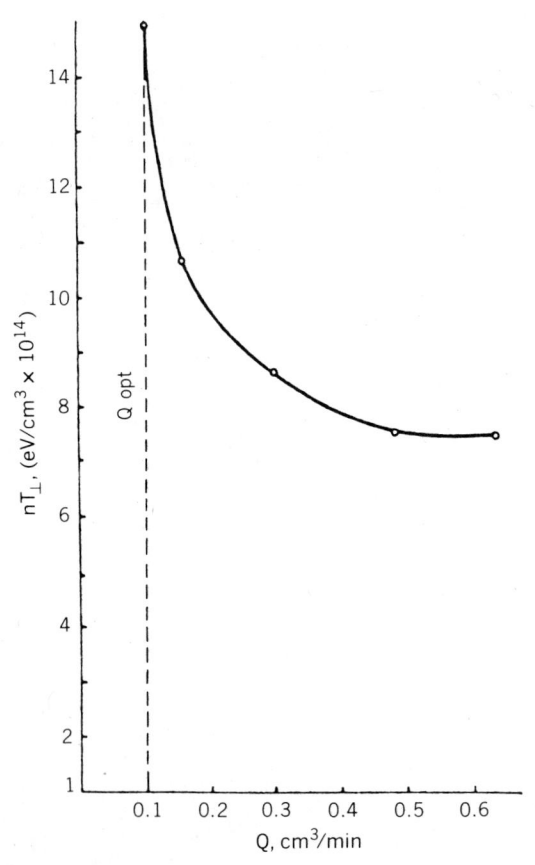

Figure 8.7 Measured transverse plasma pressure in a Dubna-type source. Xenon gas, flow rate 0.15 cm^3/min (STP). From Makov [25], © 1976 IEEE.

The radial potential gradient in the PIG discharge was measured by Gabovich [41]. He measured about 100 V for a large diameter anode, and showed that the potential increased linearly with magnetic field, and approximately as the square of the anode diameter. Jones [42] proposed that a fast neutral reflux of gas from elastic wall collision could lead to a delay in arc starvation and an accentuation of the primary electron temperature close to the axis, suppressing the plasma potential there and increasing the ion containment time. The fast neutrals are measured to have an energy roughly the same as the electron temperature. This fraction of recycled fast neutrals become significant, according to Jones, at arc starvation.

5 HIGH CHARGE STATE ION PRODUCTION

The high charge state ions created in the PIG ion source are produced via a stepwise ionization process. Winter and Wolf have investigated the evolution of the charge state distribution for both cold and hot PIGs [10]. The distribution is determined largely by the parameter $n_e t_i$, where n_e is the plasma electron density and t_i is the ion confinement time within the plasma. The electron temperature T_e is also an important parameter, and must be sufficiently high to provide electrons to strip the ions by electron–ion collisions. Thus, the ions reside within the plasma where the stripping is provided by the energetic electrons, and the final charge state depends on the time that the ions spend in the stripping region, the density of the electrons doing the stripping, and the energy of those electrons.

For maximum production of high charge state ions from the PIG, one wants a high primary electron energy, a high density of primary electrons, and a long ion confinement time. Note that as the charge state distribution is increased toward higher charge states, the fraction of low charge state ions is at the same time reduced.

The ion loss rate from the plasma due to recombination and charge exchange is not large [43], and the principal limitation to the ion confinement time is the ion transport across the boundary layers, that is, diffusion across the sheaths [44, 45].

For radial extraction of ions from the PIG plasma, as is the usual case, the ion confinement time is limited by transverse diffusion across the axial magnetic field, which occurs at the anomalously fast rate given by Bohm [44]. On this basis, Winter [43] has obtained an ion confinement time of

$$t_i = 10 r^2 B / T_e \qquad (1)$$

where r is the plasma radius in centimeters, B is the axial magnetic field in kilogauss, T_e is the electron temperature in electron volts, and t_i is in microseconds. For typical PIG ion source parameters—say, $r = 0.3$ cm, $B = 3$ kG, $T_e = 10$ eV—the ion confinement time is of order 10 μs. Attempts to increase the high charge state ion production by, for example, increasing the plasma radius r, have met with limited success. Plasma instabilities of various kinds are in fact seen in PIG discharges (Section 6), which limit the ion confinement time from high values that might be expected from Eq. (1).

While high charge state distributions are favored by cold cathode PIGs, the much larger currents of the lower charge states from the hot cathode PIG make the latter attractive. Some typical charge state distributions are given in Table 8.1.

6 OPERATING PARAMETERS

The charge state distribution of ions within the PIG plasma varies along the length of the hollow anode, as documented by Makov [25]. Gaseous ion yields are found to be lower near the cathodes. These effects may be due to the nonuniform axial electric field. The introduction of a nonuniform magnetic field near the cathodes is believed to alter the axial ion density and curb the influx of cathode ions into the extraction region. This may be helpful for small ion sources required to produce metallic ions [20]. It is conventionally assumed that the PIG magnetic field needs to be of high quality; that is, the field lines should be straight and parallel to the anode axis. Pasyuk [46] observed maximum ion yield with a homogeneous magnetic field, but noted that a small inhomogeneity is without drawback, provided the cathode positions are modified. He also found evidence clearly indicating that source axis should be parallel with the magnetic field lines; a deflection of one of two cathodes by 1 mm substantially reduced the Ar^{3+} and Ar^{8+} output. However, experimental work at LBL [47] showed clear improvement of N^{5+} and Ar^{7+} yields with a mirror field geometry in the vicinity of the cathodes. A decrease in minimum gas flow at starvation was observed in the pinched field configuration, probably due to an increased ion confinement time. Tauth [48] has presented a case for pinching the field lines in the center of the discharge. At GSI, Germany, a compact PIG source is presently being developed; this source has a large field inhomogeneity, about 0.2 kG/cm, and surprisingly the plasma stability and the ion beam charge state distribution are unchanged from "normal" operation [49] (see Figure 8.8). For further information regarding PIG geometry and its influence on performance, see Refs. 50–53.

Ion beams extracted from PIGs, either cold or hot, appear to be intensity modulated due to plasma oscillations at frequencies from 100 kHz to 1 MHz. In general, the closer the source is run to starvation (i.e., a high charge state tune) the more severe the modulation. Beam intensity oscillations aggravate neutralization of the beam in the downstream beamline, and may even be amplified by beam–plasma effects [54, 55]. The fundamental frequency seems to be independent of magnetic field. Amplitude may abruptly change with modal variations of arc voltage. Stronger intensity modulation is usually seen at the beam edge than in the center [56]. An oscillogram of typical beam current is shown in Figure 8.9 [57]. Beam noise has a fundamental frequency in the range 100–500 kHz and harmonics up to 3 MHz, identifiable as a plasma rotational type instability examined in one

TABLE 8.1 Tabulation of Typical PIG Source Ion Yields

Element	Discharge I (Amps)	Discharge V (Volts)	Sputter I (Amps)	Sputter V (Volts)	Gas Support	Slit Size (mm)	Emittance (Normalized) (cm-mr)	Extractor (kV)	Duty factor (%)	Ion charge state tuned	1	2	3	4	5	6	7	8	9	10	Measurement location, Comments	Date	Reference #
															Ion currents, eμA peak								
Ca	9.5	600			Xe						3000	23×10³	22×10³	14×10³	4500	1000	180	35	120		Cathode Separation ~24 cm	1972	82
W	9	360			Xe								20×10³	17×10³	13×10³	6800	3300	700					
Ca	7.5	–550	0.3	–650	N	20×0.8		20	25			–800	1000	500	200	800	320	60	600		Measured 80 cm from source magnet	1979	32
A	7.5	–800				20×0.8		20	25		1500	4500	9000	5200	2250	2000	2000	1600			"	"	"
Xe	7.5	–580				20×0.8		20	25				1250	1500	1750						"	"	"
Xe	2.6	1100				45×1.1		20	20	3		4700	5300	3600	3400	2400	1800	1000	300	130	Measured 50 cm from source magnet	1981	33
Au	2.7	1100			⁸⁶Kr			20	20	8			1500	1900	1700	1600	1700	650	≤400		"	"	"
Au	2.5	1350			Xe			20	20	4			≤5100	4600	4100	3100	2400	≤600	≤400		"	"	"
U	3.6	750			Kr			20	20	5				2800	3600	3300		1200		≤200	"	"	"
A	3.2	–800				45×1.1	X 0.05 π		20	1	4500										Percent area, emittance plot = 50%	1981	33
A	"	"					Z 0.05 π		20	1	6100										" = 68%	"	"
¹³⁵Xe	"	"					X 0.05 π		20	3			3300								" = 67%	"	"
¹³⁵Xe	"	"					Z 0.05 π		20	3			4000	4000							" = 85%	"	"
Au	"	"	(80v–800v)		Xe		X 0.05 π		20	4											" = 86%	"	"
Au	"	"	"		Xe		Z 0.05 π		20	8					3000			500			" = 80%	"	"
U	"	"	"		Kr		X 0.05 π		20	5						2300					" = 83%	"	"
U	"	"	"		Kr		Z 0.05 π		20	6											" = 76%	"	"
Ar	4.0	600			—	45×0.8	X 0.05 π	15	0.02	3			6700			4500					Stable 6 hour operation	1981	83
Xe	8.8	500			—	45×0.5	X 0.04 π	21	0.02	6						2400					"	"	"
Au	4.0	800			—	32×2.0	X 0.02 π	12	0.02	6											"	"	"
N	2.5	2600			CO₂	16×1.5		80 keV	0.02	5		500	180	32	10						Cathode separation = 8 cm	1972	21
C	2.5	2600			CO₂	16×1.5		80 keV	0.02	4		340	240	28							"	"	"
Ne	2.2					16×2.4	X,Z 0.06 π	17	0.02	3			1050								Measured 3 m from source	1985	84
Si	2.5					16×2.4	X,Z 0.06 π	17	0.02	4				400							"	"	"

177

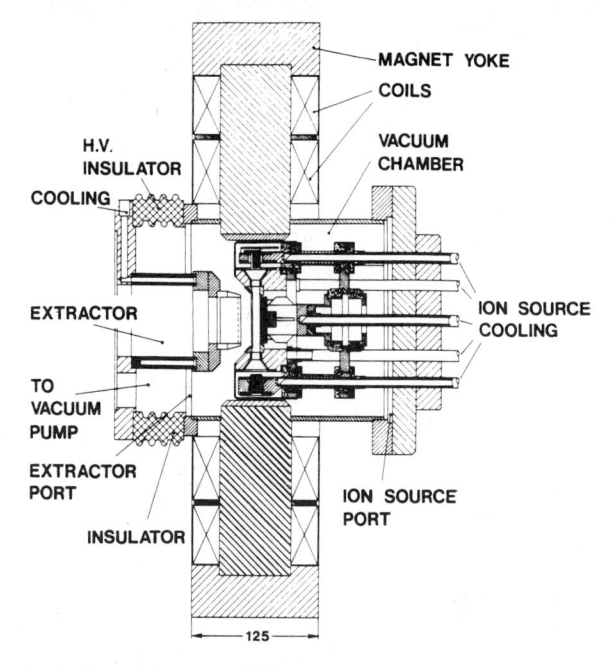

Figure 8.8 Sectional view of "compact PIG source" and extractor. Scale is in millimeters. From Mueller [49].

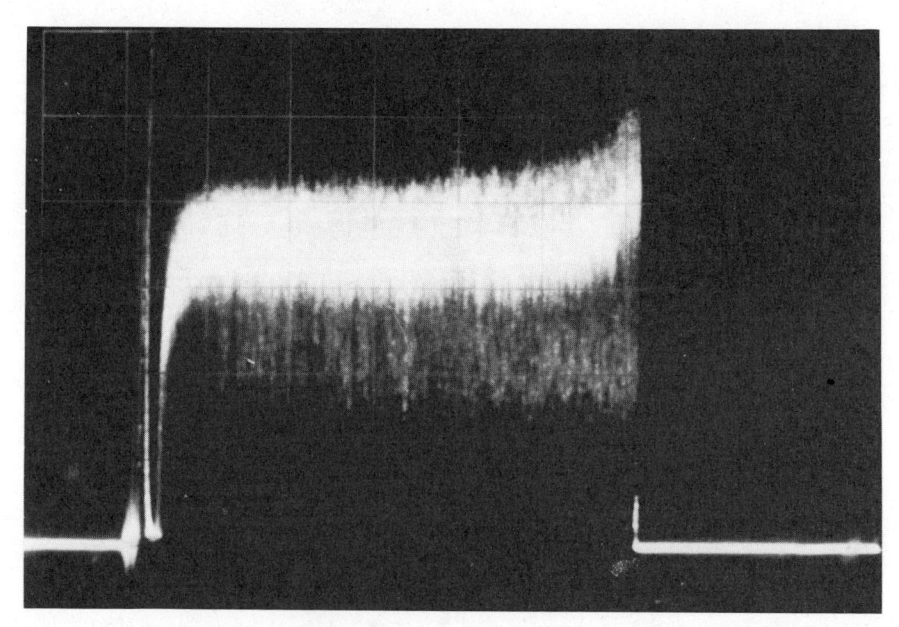

Figure 8.9 Beam current noise of a Au^{5+} ion beam. Fundamental frequency is 287 kHz, harmonics to several megahertz; beam current is 5.7 mA. From Gavin [57].

instance by Hoh [58, 59]. Little appears to be possible for reduction of this kind of beam noise other than increasing the neutral particle density, which of course greatly reduces the high charge state ion fraction; see Figure 8.7.

Source lifetime is shortened by large dc power input to the plasma. The tungsten cathode sputtering rates of hot PIGs are independent of most gases used, since the cathodes are bombarded by tungsten—so-called self-sputtering. Pasyuk [60] has compared theoretical prediction with empirical data; according to Pasyuk, the cathode use rate is proportional to arc current and the square of the arc voltage. For pulsed PIGs with refractory cathodes 0.4 g/h is typical; if the cathode axis is not well aligned, asymmetrical cratering can occur and the sputtering rate may increase substantially. Cold cathode PIG sputtering appears more dependent upon gas type, possibly a result of surface compound interaction. Additionally, the more deeply cupped a cathode, the lower will be the average extracted ion charge state.

7 METALLIC ION PRODUCTION

One method for ionizing low vapor pressure metallic materials is to vaporize the material in an oven and duct the vapor into the reflexing electron stream of the PIG. This requires the ion source to be maintained at a still higher temperature than the oven and vapor transport line. Dc (only) beams of metallic ions can be produced reliably with such a vapor feed. A heated cathode is necessary to avoid unfavorable cathode surface chemistry. Pasyuk [61] developed a PIG source of this kind for the production of calcium and zinc ions. With careful adjustment, Ca^{9+} and Zn^{10+} were discerned; consumption rates were high (about 100 mg/h for calcium). A Li^{3+} ion source has also been reported [62]. Vapor feed sources are by nature of higher efficiency than a sputtering-type feed where support gas is mandatory.

Alternatively, lithium ion beams were generated by allowing the plasma-heated "cold cathode" to sublimate lithium vapor through a hole in the cathode(s) [63]. Gold, mixed with the cathode material, has been extracted with an unusually high charge state distribution [64]. Similarly nickel, copper, and chromium were vaporized by judicious localized heating of the metal in the PIG [65].

Another technique, initially developed for an isotope separation source by Druaux [66] and later for accelerator application by Gavin [67], uses an electrode of the required metallic species, held at a negative potential with respect to the anode plasma column, as a source of sputtered material. Different electrode sputtering techniques were soon employed for accelerator-type sources [30, 68]. The sputtering feed is well suited to pulsed operation since the element to be ionized is conserved during off-time. Usually a substantial flow of an inert support gas is required to sustain the arc, diluting the metallic ion beam current. (For information on sputtering yields, angular dependence, etc., see Ref. 69.)

Most materials can be readily ionized in the PIG, using a sputter electrode. The usual technique is to clamp the electrode material in a cooled holder. The sample material can be screwed or soldered to, say, a copper block. Even a silver-loaded epoxy conductive adhesive is satisfactory (up to about 40 W/cm^2 can be transferred across the adhesive).

PIG sputter electrodes need to have a sufficiently high melting point to remain solid and electrically conductive while in use. Also, a large measure of austeniticity is required, so as not to perturb the magnetic field lines within the hollow anode too much. Special techniques have been developed for certain elements. For example, mercury is well handled as a 45% (atomic) amalgam with silver. Certain ferromagnetic metals (nickel, for example) are usually satisfactory in large sources but often cause failure in small sources unless special alloys are chosen; such alloys are frequently produced for specialized industrial applications. Cobalt is successfully extracted from a low permeability alloy, *elgiloy* [70]. Iron beams are extracted from stainless steel, or better still a manganese alloy called "submarine steel" [71]. A nonconductive compound such as CaF_2 can be mixed with silver powder and hot pressed to shape with adequate conductive properties [57]. Rare elements such as ^{64}Ni or ^{58}Fe are conveniently diluted with a noninterfering metal such as gold; this can be done in-house at the large accelerator facilities. Care must be exercised in mixing metals in the liquid state. Of relevance, convective currents are common in inductive furnaces, which helps the mixing process. For further discussion of raw materials and electrode preparation see Refs. 71 and 72.

It is found that either the metallic or the gaseous ion yields can be peaked by a careful optimization of both the input sputtered atomic flux and the support gas. The location of the support gas inlet may well be critical to the amount of support gas required, and the pumping effect of deposited metal no doubt tailors the gas cutoff point. For one source (see Figure 8.3) calcium flow rate amounted to about 50% of the support gas flow rate [20]. Approximately one out of six sputtered calcium atoms were ionized and extracted, and about 80% of the remaining atoms can be collected for reuse. These remaining atoms are collected on internal walls of the anode (or "sleeve") The sputter electrode of this source is shaped in ring fashion to collect material sputtered from itself. Mueller (GSI) has described [73] an arrangement of three sputter electrodes each occupying one quadrant of the anode bore, reasoning that a disabled electrode serves well as a collector. Each one of the three electrodes is activated in turn. See Figure 8.10. In this way nearly half of the surrounding surface is set for recycled use [74]. It is reported that the sputter electrode weight loss for a given beam of V^{3+} is reduced by a factor of three by using multielectrodes, (i.e., the consumption efficiency was increased by a factor of three) [75].

The sputtered metal flow rate and the charge state tuned for are interrelated. An optimized high charge state yield of, say, Pb^{11+}, calls for a very low sputter rate. Long sources are more suitable for high charge state ion extraction by virtue of the fact that the sputter electrode is not physically close to the cathodes, which otherwise are covered by the vaporized/sputtered cathode material. If it is necessary to tune the PIG to yield a maximum current of lower charge state metallic ions, very large sputtering rates are favored, in which case the radial exit slit

7 *Metallic Ion Production* 181

Figure 8.10 GSI PIG with three sputter electrodes; radial extraction slit removed. Filament for indirectly heating one cathode may be seen at the top. Compare with Figure 8.8. From Mueller [73].

slowly closes from buildup of the sputtered material as the source operates. Thus, the source output falls. (The author has biased the PIG slit to allow the slit buildup to be sputtered away; source lifetime was increased by 15% in this way. Associated extractor damage has been a setback to the use of this technique.)

A 50% reduction of the width of the sputter electrode opposite the radial exit slit (4 to 2 mm) produced a modest gain in sputter efficiency [75], but the advantages associated with multielectrodes may preclude this technique, given variable tuning conditions, and so on. Material consumption rates and efficiencies for a number of ion species have been given by Mueller et al. [75]. A few typical values are given below. Source efficiency is defined as the ratio of the total number of ions of one charge state species collected at the entrance of the prestripper accelerator (i.e., after injector acceleration and beam transport), to the total number of ions sputtered (electrode weight loss).

Species	Consumption Rate (mg/h)	Efficiency
Ca^{3+}	1.2	2×10^{-3}
Ni^{3+}	37	2.2×10^{-4}
Cr^{4+}	14.5	5×10^{-4}
Pb^{9+} (cold)	350	1.5×10^{-5}
Pb^{9+} (hot)	65	1.3×10^{-4}
U^{10+}	437	4×10^{-6}

The cold and hot Pb data refer to the temperature of the source itself. Evaporative conditions are dominant in the latter case.

8 PIG ION BEAM CHARACTERISTICS

The extraction of ions from a plasma is discussed in Chapter 3, and also by other authors [37, 45, 76-80]. In the usual PIG embodiment, ion extraction is through an axial slit in the anode cylinder, and is thus in the direction transverse to the static magnetic field. Slit width is customarily less than 2 mm, so as to prevent excessive extractor damage. The ion flow is improved with thin slits, often with a Pierce-type electrode configuration. "Accel-decel" extraction systems have not proven useful. Extraction transverse to a strong magnetic field and the multiplicity of ion charge states complicate theoretical understanding. One interesting manifestation of the complexity of the physics of beam extraction from the PIG is provided by the observation that it is usual to obtain the best beam when one edge of the extractor slit intercepts a portion of the extracted beam; perhaps bombardment of the extractor by the beam provides secondary electrons, necessary for space charge neutralization of the beam.

The slit length is determined mostly by the acceptance of the conjoining dipole magnet and beam line. In principle, a flooding of the beam line acceptance in the vertical plane is found to increase the beam brightness, possibly because beam lost in the vertical plane of the analyzing field will generate electrons that can provide neutralization for the beam passing across the field lines to which the electrons are tied. In one case [81], a doubling of the source slit length, while increasing the overall vertical emittance of a U^{6+} beam, nevertheless increased the beam intensity (measured at a normalized emittance of 0.05π cm. mrad.) by a factor of 1.7. In fact, since the beam core is brighter than the halo, the actual increase in brightness would be expected to be greater still.

Under typical operating conditions, ion current densities of argon beams immediately after extraction are typically several hundred mA/cm^2, with normalized emittance in the vicinity of 0.05π cm mrad in the plane transverse to the slit length. To be useful, transverse divergence is usually held below 5° by the acceptance of the analyzing dipole [33].

Some performance parameters for PIG source from Dubna, LBL, and GSI are given in Table 8.1.

9 CONCLUSION

The PIG source will continue to play a vital role in today's particle accelerators. Rapid change of atomic species is now practicable. Certain elements with low melting points, such as lead, continue to be produced primarily by PIG sources, and lost, unwanted charge state species in the postextractor region can be useful for beam neutralization.

Source lifetimes are limited to from 4 to 8 h under typical dc conditions, but with the advent of rapid source change and consistent source component assembly, the lifetime drawback is minimized. Multiple-source electrodes may not be useful for source life extension since cathode cupping is the dominant feature. However, should the need for expensive rare isotopes continue, small improvements in multi-electrode configurations could contribute decisively.

ACKNOWLEDGMENTS

The author thanks many of the researchers at GSI, Darmstadt, for their contributions, and applauds the cooperative spirit developed between their laboratory and LBL, Berkeley.

REFERENCES

1. F. M. Penning, *Physica* **4,** 71 (1937).
2. Yu. E. Kreindel, *Sov. Phys. Tech. Phys.* **3,** 883 (1963).
3. E. B. Hooper, Jr., *Adv. Electron. Electron Phys.* **27,** 295 (1969).
4. V. I. Farenik, V. V. Vlasov, A. M. Rozhkov, and A. A. Kozharin, *Sov. Phys. Tech. Phys.* **19** (9), 1258 (1975).
5. J. Backus, *J. Appl. Phys.* **30,** 1866 (1959).
6. B. H. Wolf, GSI-Report 73-13, GSI/Darmstadt, FRG 1973.
7. M. D. Gabovich, Plasma Ion Sources, U.S. Air Force Trans. FTD-MT-65-229, AD623-822, 1964 Pub. Foreign Technology Division, U.S. Air Force Systems Command.
8. E. D. Vorobev and A. S. Pasyuk, Multicharged Ion Sources, Argonne National Laboratory Translation S76, Conf. 700555-1 (1971), Joint Insititute for Nuclear Research, Dubna, USSR, Report P7-5177.
9. J. R. J. Bennett, *IEEE Trans. Nucl. Sci.* **NS-19** (2), 56 (1972).
10. H. Winter and B. H. Wolf, Proceedings, 2d Symposium on Ion Sources and Formation of Ion Beams, Berkeley, CA, 1974, Lawrence Berkeley Laboratory, Berkeley, CA.

11. J. Arianer, Proceedings, 7th International Conference on Cyclotrons and Their Applications, Zurich, Switzerland, 1975, Swiss Institute of Nuclear Research, Zurich.
12. L. Valyi, *Atom and Ion Sources*, Wiley, New York, 1977.
13. G. Gautherin and C. Lejeune, Proceedings, 8th International EMIS Conference on Low Energy Accelerators and Mass Separators, Skorde, Sweden, 1973, Goteburg University, Sweden.
14. J. Backus and N. E. Huston, *J. Appl. Phys.* **31,** 400 (1960).
15. B. H. Wolf, UNILAC-Report 73-13, GSI/Darmstadt, FRG 1973.
16. H. Heil, *Z. Phys.* **120,** 213 (1942).
17. R. J. Jones and Z. Zucker, *Rev. Sci. Instrum.* **25,** 562 (1954).
18. C. E. Anderson and K. W. Ehlers, *Rev. Sci. Instrum.* **27,** 809 (1956).
19. P. M. Morozov, B. N. Makov, and M. S. Ioffe, *At. Energ.* **3,** 272 (1957).
20. B. Gavin, *IEEE Trans. Nucl. Sci.* **NS-23** (2), 1008 (1976).
21. H. A. Grunder, R. M. Richter, M. M. Tekawa, and E. Zajec, *IEEE Trans. Nucl. Sci.* **NS-19** (2), 208 (1972).
22. G. N. Nassibian, J. R. J. Bennett, D. Broodbent, S. Devons, R. W. R. Hoisington, and V. E. Miller, *Rev. Sci. Instrum.* **32,** 1316 (1961).
23. F. I. Mineev and O. Kovpik, *Sov. Phys. Tech. Phys.* **8** (12), 1072 (1964).
24. H. Baumann and K. Bethge, *Nucl. Instrum. Methods* **122,** 517 (1974).
25. B. N. Makov, *IEEE Trans. Nucl. Sci.* **NS-23** (2), 1035 (1976).
26. R. S. Livingston and R. J. Jones, *Rev. Sci. Instrum.* **25,** 552 (1954).
27. C. B. Mills and C. F. Barnett, *Rev. Sci. Instrum.* **25,** 1200 (1954).
28. J. R. J. Bennett and B. Gavin, *Particle Accelerators* **3,** 85 (1972).
29. A. S. Pasyuk, Yu. P. Tretyakov, and S. K. Gorbachev, *At. Energ.* **24,** 21 (1968).
30. A. S. Pasyuk and Yu P. Tretyakov, Report JINR-P7-6668, Dubna, USSR, 1972.
31. C. Bieth, M. P. Bourgarel, A. Cabrespine, R. Gayraud, and P. Attal, *IEEE Trans. Nucl. Sci.* **NS-19,** 93 (1972).
32. L. Bex, P. Cardin, and L. Backouche, Ganil Report 79R/032/IS/01, March 1979, Ganil Laboratory, France.
33. B. Gavin, S. Abbott, R. MacGill, R. Sorensen, J. Staples, and R. Thatcher, *IEEE Trans. Nucl. Sci.* **NS-28,** 2684 (1981).
34. R. Basile and J. M. Lagrange, *Nucl. Instrum. Methods* **31,** 195 (1964).
35. R. Basile and J. M. Lagrange, *J. Phys. Radium Phys. Appl.* (Suppl. 3), **69A,** 24 (1963).
36. R. Basile and J. M. Lagrange, Proceedings, VIth International Conference on Ionization Phenomena in Gases, Paris, 1963, International Union of Pure and Applied Physics, Paris, 1963. p. 203.
37. T. S. Green and C. Goble, *Nucl. Instrum. Methods* **116,** 157 (1974).
38. H. Schulte, B. H. Wolf, and H. Winter, *IEEE Trans. Nucl. Sci.* **NS-23** (2) 1053 (1976).
39. Y. D. Pigarov and P. M. Morozov, *Sov. Phys. Tech. Phys.* **6,** 336, 342 (1961).
40. A. S. Pasyuk, Yu P. Tretyakov, and S. K. Gorbachev, *At. Energ.* **24** (1), 19 (1968).
41. M. D. Gabovich, *Sov. Phys. Tech. Phys.* **5** (3), 320 (1960).
42. R. Jones, *Phys. Lett.* **67A,** 194 (1978).
43. *Lecture Notes in Physics*, No. 83, *Experimental Methods in Heavy Ion Physics*, K. Bethge (Ed.), Springer-Verlag, Berlin, 1978, pp. 1–80.
44. D. Bohm, A. Guthrie, and R. K. Wakerling (Eds.), *The Characteristics of Electrical Discharges in Magnetic Fields*, McGraw Hill, New York, 1949, Chapter 3.
45. T. S. Green, *Rep. Prog. Phys.* **37,** 1257 (1974).
46. A. S. Pasyuk, I. P. Kuznetsova, and V. B. Kutner, *At. Energ.* **39** (2), 139 (1975).

References

47. L. Bex, D. J. Clark, C. E. Ellsworth, W. S. Flood, R. A. Gough, W. R. Holley, J. R. Meriwether, and D. Morris, *IEEE Trans. Nucl. Sci.* **NS-22,** 1702 (1975).
48. T. Tauth, G. Hadinger, G. Hadinger-Espi, and M. Bajard, Proceedings, 2d International Conference on Ion Sources, Vienna, Austria, Sep. 11-15, 1972, SGAE, Vienna, p. 620.
49. M. Mueller, GSI, Darmstadt, FRG, Annual Scientific Report, 1982.
50. Yu P. Grigorev and P. M. Morozov, *Sov. Phys. Tech. Phys.* **16** (11), 1926 (1972).
51. S. N. Popov, *Sov. Phys. Tech. Phys.* **6** (12), 1047 (1962).
52. M. D. Gabovich, *Sov. Phys. Tech. Phys.* **12** (5), 638 (1967).
53. I. Chavet, M. Kanter and M. Menat, *Nucl. Instrum. Methods* **139,** 47 (1976).
54. J. D. Lawson, *Nucl. Instum. Methods* **139,** 22 (1976).
55. T. S. Green, *IEEE Trans. Nucl. Sci.* **NS-23,** 926 (1976).
56. M. Mueller, H. Schulte and B. Wolf, GSI Darmstadt, FRG Annual Scientific Report, 1980.
57. Author's logbook.
58. F. Hoh, *Phys. Fluids* **5,** 22 (1962).
59. F. Hoh, *Phys. Fluids* **6,** 1184 (1963).
60. A. S. Pasyuk, Yu P. Tretyakov, and V. Stanku, *Sov. Phys. Tech. Phys.* **3,** 42 (1965).
61. A. S. Pasyuk, E. D. Vorobiev, R. I. Ivannikov, V. I. Kuznetsov, V. B. Kutner, and Yu. P. Tretyakov, JINR Report P7-4488, Dubna USSR, 1969.
62. G. Haushahn, Proceedings, 2d International Conference on Ion Sources, Vienna, Austria, Sep. 11-15, 1972, SGAE, Vienna, p. 825.
63. C. E. Anderson, Proceedings, Conference on Reactions between Complex Nuclei, Oak Ridge, TN, 1960, Oak Ridge National Laboratory, Oak Ridge, TN, p. 67.
64. B. Gavin, Proceedings, 2d International Conference on Ion Sources, Vienna, Austria, Sep. 11-15, 1972, SGAE, Vienna, p. 519.
65. E. J. Jones, *IEEE Trans. Nucl. Sci.* **NS-19,** 101 (1972).
66. Y. Druaux and R. Bernas, Proceedings of Conference on Electromagnetically Enriched Isotopes and their Spectrometry, Harwell UK, 1955, p. 30.
67. B. Gavin, *Nucl. Instrum. Methods* **64,** 73 (1968).
68. A. S. Pasyuk and Yu P. Tretyakov, Proceedings, 2d International Conference on Ion Sources, Vienna, Austria, Sep. 11-15, 1972, SGAE, Vienna, p. 512.
69. G. Carter and J. S. Colligon, *Ion Bombardment of Solids*, Elsevier, New York, 1968.
70. Elgiloy Co., Elgin, IL.
71. M. Mueller, K. Leible, B. H. Wolf, and N. Angert, GSI, FRG Darmstadt, Report 82-11, 1982.
72. K. Leible and B. H. Wolf, International Conference on Low Energy Ion Beams, University of Salford, Sep. 5-8, 1977, Conference Series No. 38, Institute of Physics, London, p. 96.
73. M. Mueller, GSI, Darmstadt, FRG Annual Scientific Report, 1983.
74. M. Mueller, GSI, Darmstadt, FRG Annual Scientific Report, 1984.
75. M. Mueller, GSI, Darmstadt, FRG Annual Scientific Report, 1986.
76. T. Consoli and R. B. Hill, *Nucl. Fusion* **3,** 237 (1960).
77. M. D. Gabovich, *Instrum. Exp. Techniques* **2,** 1985 (1963).
78. A. Septier, F. Prangere, H. Ismail, and G. Gautherin, *Nucl. Instrum. Methods* **38,** 41 (1965).105
79. T. S. Green, *Nucl. Instrum. Methods* **115,** 615 (1974).
80. T. S. Green, *IEEE Trans. Nucl. Sci.* **NS-23,** 918 (1976).
81. Author's measurements with LBL source, Ref. 33.
82. A. S. Pasyuk and Yu. P. Tretyakov, JINR Report P7-6668, Dubna USSR, 1972.
83. M. Mueller and Z. Weijiang, GSI, Darmstadt, FRG, Annual Scientific Report, 1981.
84. G. Stover and E. Zajek, *IEEE Trans. Nucl. Sci.* **NS-32,** 1806 (1985).

9

The Freeman Ion Source

D. Aitken*
Applied Materials Implant Division
Horsham, England

The Freeman source was developed in the early 1960s as an improved source for electromagnetic isotope separators [1]. The primary requirement was for a high beam current with the high resolving power necessary for the separation of the isotopes of heavy elements. A high extracted beam current from the source does not necessarily achieve this requirement if the quality of the beam after extraction and magnetic analysis is not sufficient to give the necessary high-quality focus at the resolving slit. These requirements lead to a ribbon beam system where the beam is extracted from a long slot. To achieve the necessary optical quality the output needs to be consistent along the length of the slot and, even more important, the plasma from which the beam is extracted must be free from high-frequency electrical oscillations, often referred to as "hash." These oscillations result in a modulation of the beam that destroys space-charge neutralization and results in the inability to achieve a good focus at the resolving slit.

The most important characteristic of the Freeman source is the ability to achieve consistent hash-free ion beams. Earlier high-current isotope separators had invariably been based on Calutron-type [2] ion sources or miniaturized versions of the Calutron source using independent source magnets and with the filament cathode outside the arc chamber [3] or inside the arc chamber [4]. A significant amount of tedious tuning was often necessary for these sources to achieve both high current and high resolving power in the isotope separator due to the tendency to run under hashy arc conditions. The strong crossed magnetic and electrical fields in the extraction region can lead to electrical breakdown and, in the case of separate source magnet, inconvenient deflection of the beam before entering the analyzing magnet. The interaction of the filament magnetic field with the source magnetic

---

*Current affiliation: Superion, 10 St. Leonards Rd., Surbiton, Surrey KT6 4DE, England.

field can lead to distortions in the collimation of the arc column which can lead to complex, unpredictable output characteristics and a high degree of nonuniformity of the output along the length of the extraction slot.

Probably the most important feature of the Freeman source is the long, straight filament placed close to and parallel to the extraction slot. This effectively controls the position and stability of the plasma, confining it to a region close to the slot from which the beam is to be extracted. The magnetic field is parallel to the slot as in the Calutron-type sources but is considerably lower in intensity, typically being about 150 G.

The Freeman source has been used extensively in high-current ion implanters [5–7] because of its consistent performance, simplicity, and high-current output.

1 THE EARLY FREEMAN SOURCE

This was first reported in 1962 [1] as a source primarily designed for the separation of the isotopes of plutonium. For this requirement a high resolving power is necessary if efficient isotopic separation is to be achieved. The major limitation with previous ion sources, in general based on the Calutron [2] design, was the tendency to produce a hashy discharge which leads to a reduction in the degree of space-charge neutralization of the beam and the consequent need for painstaking optimization of the discharge parameters to achieve the required quiet plasma conditions. A particular feature of the source is the relatively low (0–150 G) auxiliary magnetic field. The high magnetic fields used in Calutron-type sources are known to encourage hash, lead to electrical breakdown in the extraction region, and produce undesirable bending of the beam in the dispersion plane as the beam passes through the extraction electrodes.

The source is shown schematically in Figure 9.1. The filament, which is situated parallel to and about 3 mm away from the extraction slot, is biased up to 200 V negative with respect to the arc chamber. Filaments of various shapes, sizes, and materials were tested. The geometry chosen was a 4.5-mm-diameter tantalum rod machined to a rectangular section 4.5 × 1 mm at the required emitting region 45 mm long. The extraction slot was 42 mm long and 1–2.5 mm wide. Tungsten was tried but machining was more difficult, and graphite filaments behaved well but were found to be too fragile for reliable long-term operation. Other geometries were tried—flat foil, 2-mm-diameter circular rod and helical filaments—but were not found to offer any performance advantages although they would perhaps have been a more economic proposition compared with the machined tantalum design. The filament life was found to be surprisingly good considering that the filament is situated directly in the center of the arc. This is primarily because this source could operate effectively at significantly lower arc voltages than the Calutron types and the sputtering rate of the filament is very sensitive to ion bombardment energy.

The performance of the source was not found to be critically dependent on the filament position and the source could operate satisfactorily with the filament

1 The Early Freeman Source

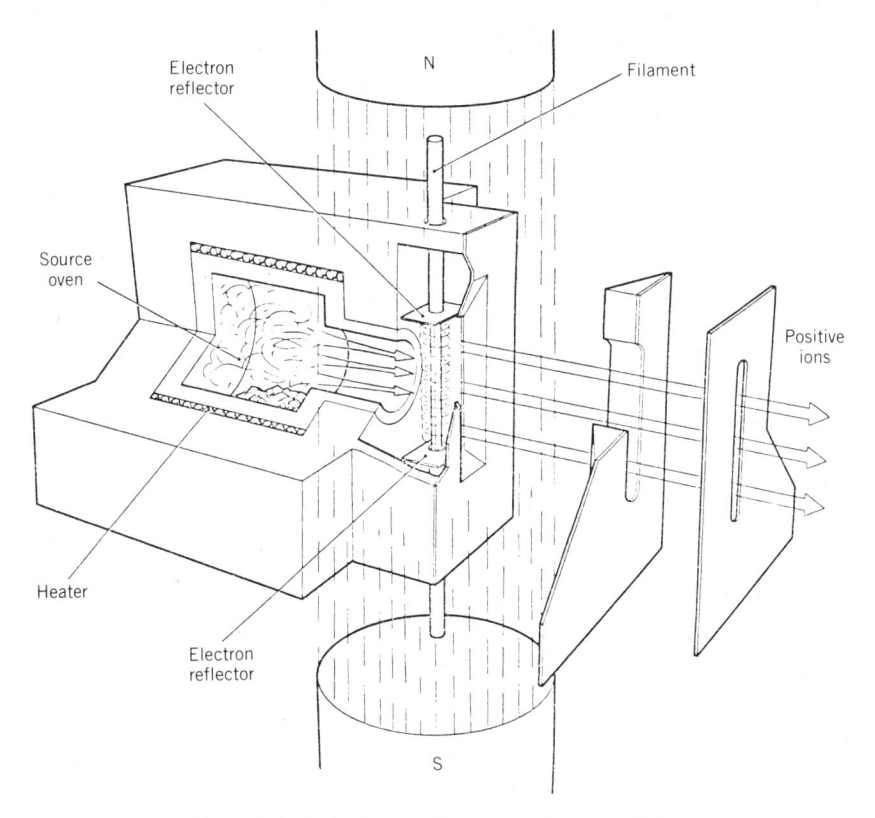

Figure 9.1 Early Freeman ion source. (Source: ref. 1.)

anywhere between 3 and 10 mm from the extraction slot. The original design had the option of fitting electron reflectors to the ends of the emitting section of the filament and this resulted in a 60% increase in source efficiency. For practical reasons these were not generally used, but were considered to be an attractive option for a more compatible design.

An auxiliary source magnet, external to the vacuum system, provided a 0- to 150-G magnetic field parallel to the axis of the filament. Unlike the Calutron-style sources the geometry of the filament and the interaction of the filament magnetic field with the externally applied field are both advantageous. The filament produces an intense arc in the most advantageous position, directly behind the extraction slot, which is fixed in space by the filament. The magnetic field does not have a spatial positioning role for the plasma column and this is the main reason that the intensity of the externally applied field can be so low compared with Calutron sources. The function of this external field is solely to produce a field that interacts with the filament field to produce a spiral geometry. This causes the primary electrons in the plasma to follow a spiral path which leads to efficient ionization at a low arc chamber pressure.

2 FREEMAN SOURCES IN LABORATORY SCALE SEPARATORS AND EARLY IMPLANTERS

2.1 The Original Harwell Design

A version of the Freeman source [8], designed for use in the laboratory scale Harwell Isotope Separator, has become the basis of the majority of sources used in commercial high-current ion implanters. A schematic of this source is shown in Figure 9.2. The early version had a tantalum arc chamber with heat shielding which together with a boron nitride or silicon nitride vaporizer allowed operation of the source with solid feed materials up to a temperature of 1100 °C. This was ideal for a flexible laboratory isotope separator where a wide range of ion species was required. For the more limited requirements of semiconductor ion implantation where boron, phosphorus, arsenic, and antimony represented virtually all the required species a simple, robust, and more economic graphite arc chamber was retrofittable into the standard source frame. The temperature of the arc chamber in this case, for typical arc conditions (say 70 V, 2 A) would be sufficiently high to prevent the condensation of elemental antimony, (i.e., 700–800 °C). Other likely feed materials are either more volatile solids such as elemental phosphorus and arsenic or gaseous feeds such as the halides and hydrides of boron, phosphorus, arsenic, and antimony. This graphite arc chamber had a significant advantage over the tantalum version, in which embrittlement or corrosion of tantalum was a serious problem. The major disadvantage of graphite is its porous nature, which leads to significant memory effects and limits the speed with which one can change species and obtain the maximum current of the second species. Despite this, commercial ion implanters use graphite arc chambers extensively.

2.2 Source Variants

A number of variants of this basic source design were used in the Harwell Isotope Separator and are illustrated in Figure 9.3a–g.

2.3 The Harwell Isotope Separator as an Implanter

The first commercial high-current ion implanter was produced by Lintott Engineering [5] and consisted of the Harwell Isotope Separator and a mechanically scanned wafer processor [9]. The isotope separator was at ground potential and an analyzed beam up to 40 keV was produced which could then be postaccelerated by floating the wafer processor at a negative potential.

The beam was extracted from a 42 × 1.5-mm extraction slot and beam alignment was achieved by using stationary extraction electrodes and tilting the entire source assembly on a vacuum bellows. As can be seen in Figure 9.2, the vacuum flange of the source has large high-voltage feedthroughs and is at ground potential, but the arc chamber assembly floats at the preanalysis acceleration potential of up

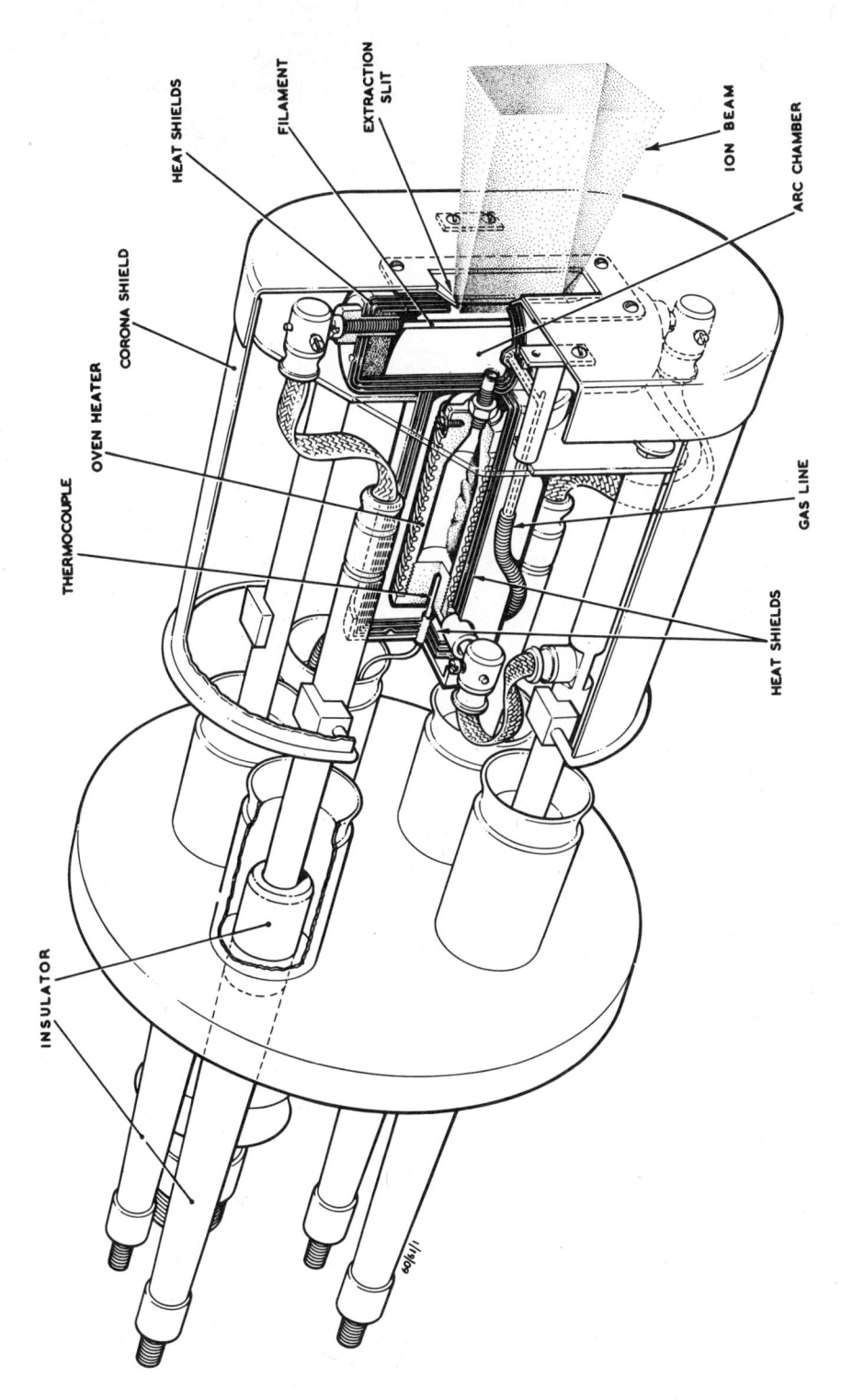

Figure 9.2 The Freeman ion source. (Source: ref. 8.)

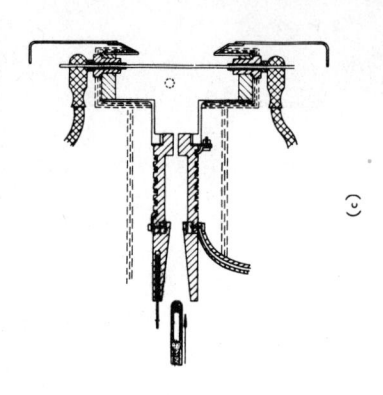

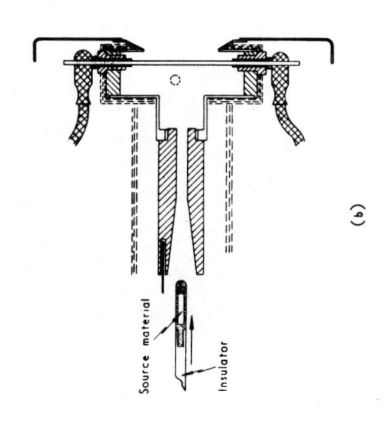

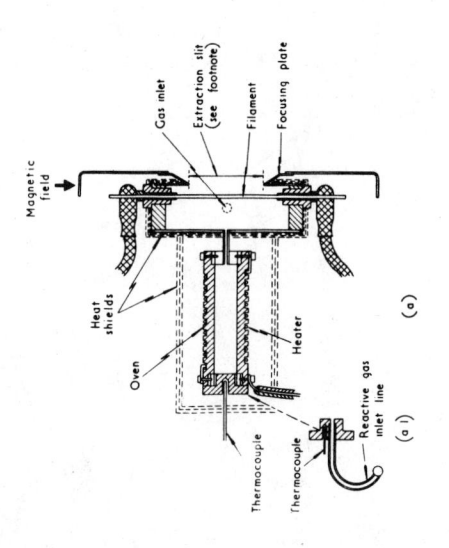

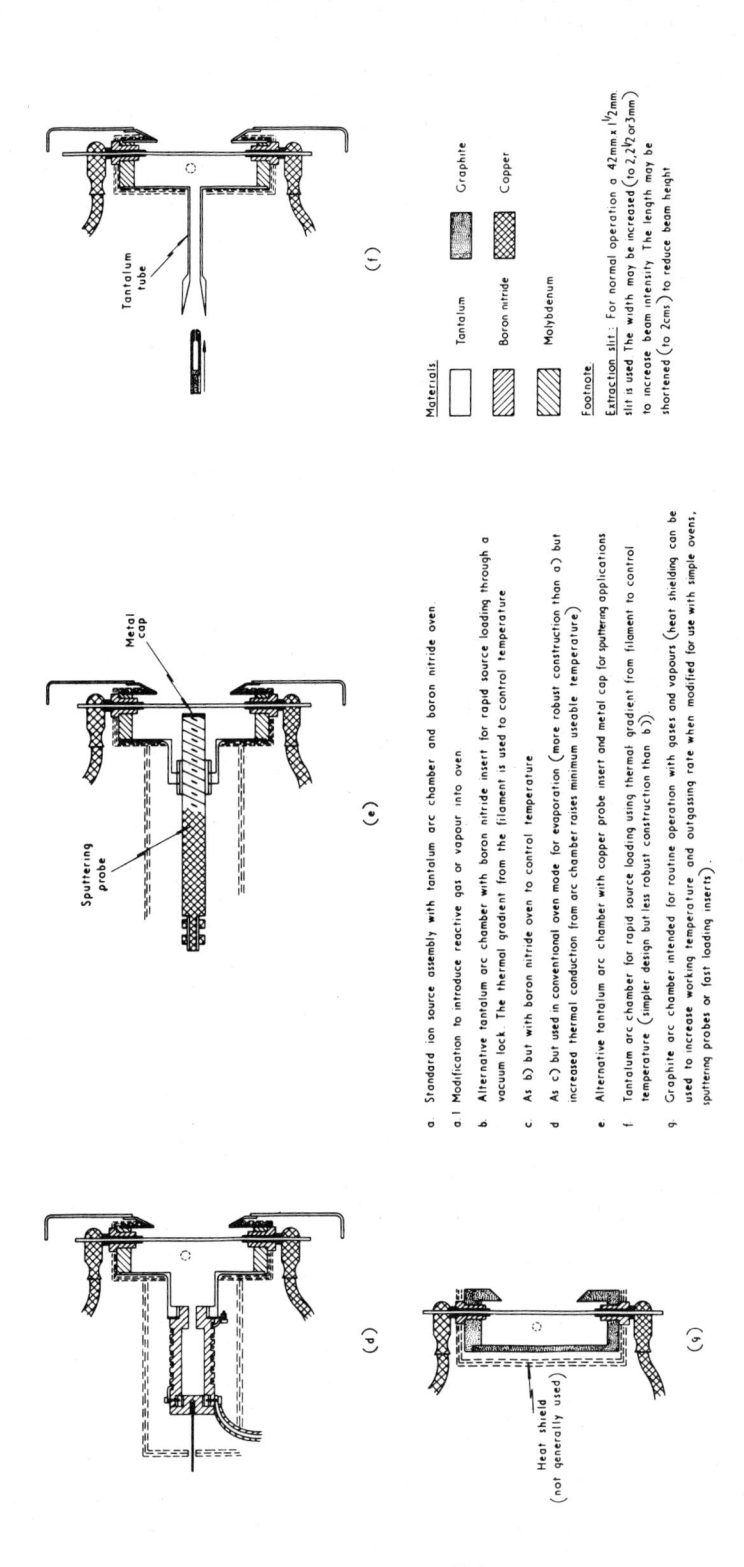

a Standard ion source assembly with tantalum arc chamber and boron nitride oven

a1 Modification to introduce reactive gas or vapour into oven

b Alternative tantalum arc chamber with boron nitride insert for rapid source loading through a vacuum lock. The thermal gradient from the filament is used to control temperature

c As b) but with boron nitride oven to control temperature

d As c) but used in conventional oven mode for evaporation (more robust construction than a) but increased thermal conduction from arc chamber raises minimum useable temperature)

e Alternative tantalum arc chamber with copper probe insert and metal cap for sputtering applications

f Tantalum arc chamber for rapid source loading using thermal gradient from filament to control temperature (simpler design but less robust construction than b))

g Graphite arc chamber intended for routine operation with gases and vapours (heat shielding can be used to increase working temperature and outgassing rate when modified for use with simple ovens, sputtering probes or fast loading inserts)

Materials:
- Tantalum
- Boron nitride
- Molybdenum
- Graphite
- Copper

Footnote

Extraction slit: For normal operation a 42mm x 1½mm slit is used. The width may be increased (to 2, 2½ or 3mm) to increase beam intensity. The length may be shortened (to 2cms) to reduce beam height

Figure 9.3 Arc chamber variants for the Harwell isotope separator. (Source: ref. 16.)

to 40 kV. The source magnet, which provides a magnetic field parallel to the filament of the source, is situated outside the source vacuum chamber at ground potential.

2.4 The Freeman Source in Commercial Ion Implanters

The Freeman source has been used extensively in commercial high-current ion implanters for semiconductor applications. Figure 9.4 shows a source used in the Lintott/Applied Materials [5] Series III machine. This incorporated twin vaporizers which were generally used to produce arsenic, phosphorus, or antimony ions from the elemental feed materials. In conjunction with gaseous BF_3 as the feed material for boron there were, therefore, three dopant species readily available without the necessity for a source change. Figure 9.5 shows a source produced by Nova [7] with the front plate removed. In both these sources the front plate was a removable component with a curved extraction geometry (to reduce the height of the beam at the analyzing magnet). Figure 9.6 shows the source used in the Applied Materials Precision Implant 9000 machine. This has a large extraction slot (90 × 5 mm) and is capable of phosphorus and arsenic currents in excess of 30 mA.

3 SOURCE CHARACTERISTICS

The source with a 2-mm-diameter tungsten filament operates well in the 40- to 70-V arc voltage range with arc currents in the range 0.1–3 A. At these modest arc voltages filament life at maximum output can be respectable, generally in the range 50–100 h. For ions that are relatively difficult to ionize, such as boron, the arc voltage may need to be in the 80- to 100-V range and this combined with some chemical erosion caused by the halide feed material (in the case of boron) results in a significant reduction in filament life, typically 20–30 h at maximum output capability and an arc current of 3–5 A.

For a source with a 40 × 2-mm extraction slot and an extraction voltage of 40 kV the ion beam capability is typically 1–2 mA B^+ from boron trifluoride and 3–7 mA of P^+, As^+ from elemental feed material. Under favorable circumstances significantly higher currents can be obtained.

A rigorous understanding of the characteristics of this source requires an appreciation of the effect of the long, relatively large-diameter filament on the arc characteristics.

3.1 Arc Current Control

The traditional arc current control technique is to control the arc by controlling the filament temperature, which in turn determines the electron emission from the filament. The filament temperature is primarily a function of the filament current, although the power dissipated in the arc and ion bombardment also make a significant contribution. The large filament in this source has significant thermal inertia,

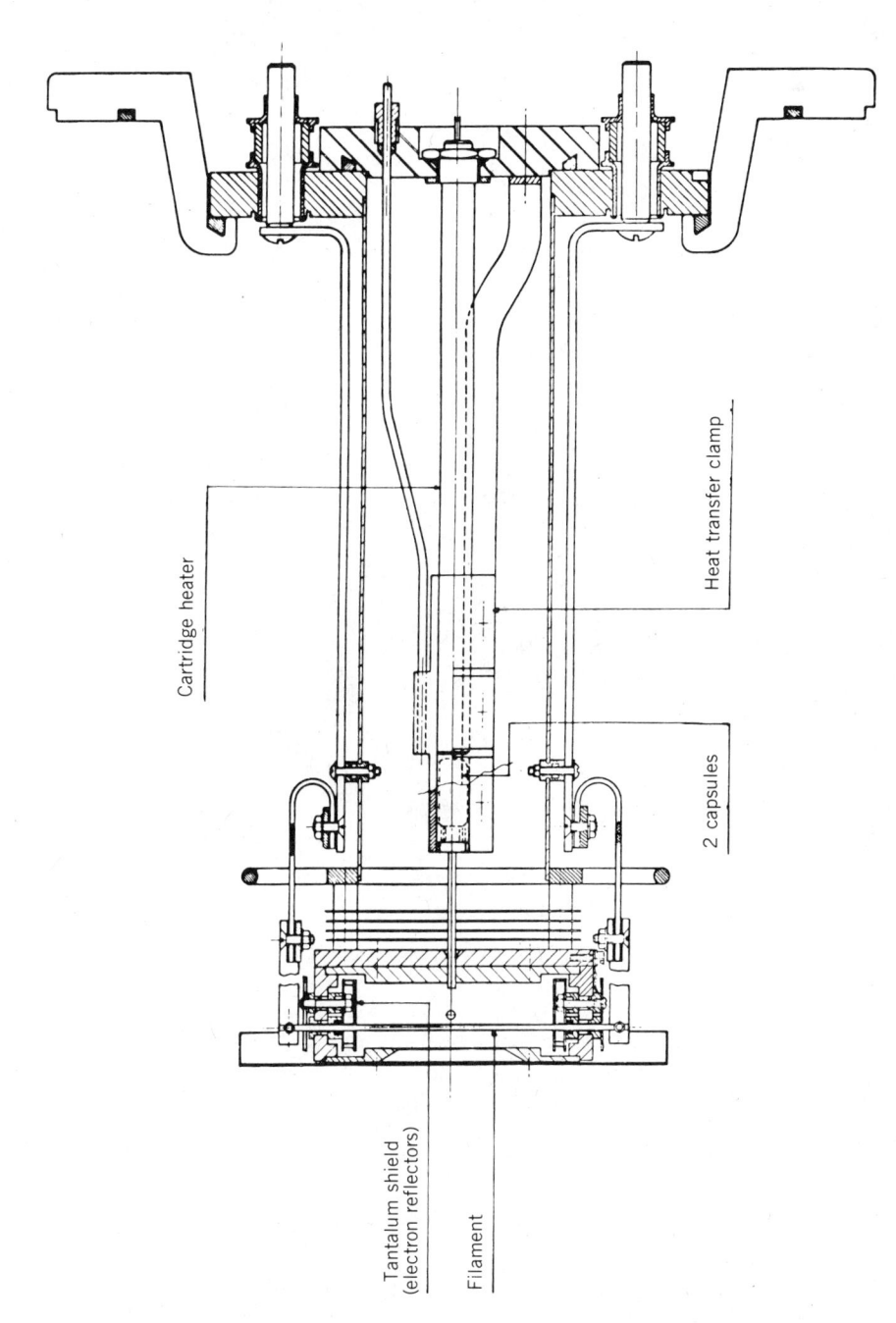

Figure 9.4 The Lintott Series III Freeman source.

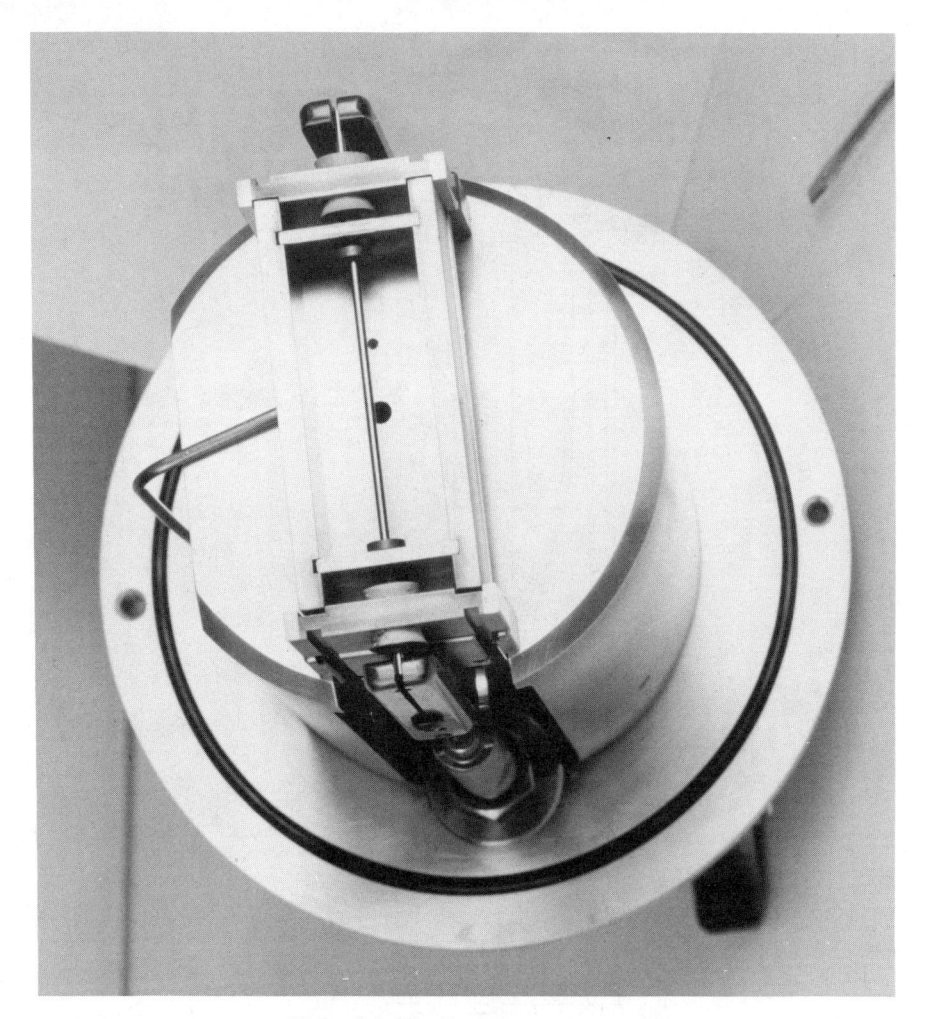

Figure 9.5 The Nova Freeman source.

which can lead to poor control, particularly at high arc currents (above 2 A). A successful control system must cope with the instabilities that result from the sluggish temperature response of such a large filament. An example of a technique for overcoming this problem will be described later.

3.2 Nonuniformity of the Arc Discharge along the Length of the Filament

The Freeman source is an intrinsically asymmetric source. The voltage is different at the two ends of the filament, the difference being the filament voltage. The electrical current along the filament produces a magnetic field that interacts with the externally applied field to produce a resultant helical field. Although this has

3 *Source Characteristics*

Figure 9.6 The PI 9000 Freeman ion source.

some desirable attributes from the point of view of ionization efficiency, it does cause the primary electrons leaving the filament surface to undergo some complex and asymmetric paths. Hinkel [10] has calculated the electron trajectories that occur under various operating conditions. Figure 9.7 shows the primary electron trajectories calculated by Hinkel for 60 V arc and 130 A filament current. These calculations are for a free electron and collisions are not taken into account. The results are sensitive to the assumptions made for the relation between the potential in the

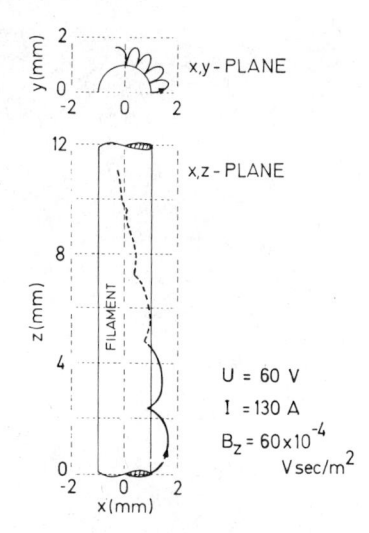

Figure 9.7 Electron trajectories. (Source: ref. 10.)

discharge as a function of the distance from the cathode, but the general conclusions are not affected.

The most significant conclusion from this work is that electrons emitted from the hot filament which do not collide with gas atoms do not reach the arc chamber wall over the central region of the arc chamber from which the beam is extracted. The electrons spiral in pseudo-cycloids around the filament and propagate toward the positively biased end. Therefore, electrons emitted from the negative end of the filament contribute to ionization at the positive end. Consequently, the discharge will tend to be more intense at this end and the erosion of the filament at this end will be particularly enhanced as a result. The only mitigating consideration is the arc voltage, which will be lower at the positive end and will, therefore, tend to reduce this effect. The direction of the external magnetic field has no effect on this argument, but reversing the polarity of the filament supply will change the end to which the elecrons migrate. The filament erosion would be more uniform if the arc voltage was increased or the external magnetic field reduced to a level that allowed electrons to escape from the filament. The former increases the erosion rate and therefore decreases filament life; the latter reduces the available beam current. These approaches are therefore of no value when maximum output is required. Another possibility is to reduce the diameter of the filament so that the necessary filament current for thermionic emission is reduced with a consequent reduction of the filament magnetic field. This does work; the performance of the source with a 1-mm-diameter filament is generally better than with the usual 2-mm-diameter filament but there is an adverse effect on filament life. The best solution is to periodically reverse the direction of the filament current to even out the erosion at the two ends of the filament. Some other solutions to this problem will be mentioned in the next section.

4 FREEMAN SOURCE VARIANTS

Since the Freeman source was first introduced there have been only a relatively small number of modifications made to the original design to optimize its use in modern commercial ion implanters.

It has been mentioned in the previous section that electrons find it difficult to escape radially from the filament but do not find any problem traveling along the filament to the positively biased end. This not only causes plasma nonuniformities but also limits the efficiency of plasma confinement by giving primary electrons an easy escape route. A number of techniques have been suggested for improving the situation.

4.1 Magnetic Bottle

Williams [11] has shown that the performance of the Freeman source, particularly for boron from boron trifluoride feed gas, can be significantly improved by incorporating iron poles at each end of the filament which create a magnetic bottle along the length of the filament. This prevents the easy escape of electrons at the ends of the arc chamber due to the converging and intensifying magnetic field and consequently increases the electron temperature in the plasma.

4.2 Electrostatic Electron Reflectors

In an early version of this source Freeman [1] had the facility to incorporate electron reflectors at each end of the emitting region of the filament. For practical reasons these were not used. Aitken [12] has shown that the incorporation of electron reflectors at each end of the filament, at filament potential, both improves the performance of the source and reduces the tendency to localized erosion at the positive end of the filament. The reflecting of primary electrons will return primary electrons to the plasma that would otherwise be lost. Ion bombardment of the reflectors will also create secondary electrons which, together with reflected primary electrons, can travel along the filament in either direction. The detailed situation is complex but the performance improvements are significant.

Figure 9.4 shows a source [12] with tantalum shields at the ends of the arc chamber which act as electron reflectors and also protect the filament insulators from metallization. A later version of this source had the filament insulators external to the arc chamber to further improve the insulator life.

4.3 AC Filament

Another variation is the use of an ac heating supply for the filament. This has been used by Tabata et al. [13] in machines designed for the surface modification of materials. This clearly is a technique that will remove the asymmetric output found with a dc filament on a time-averaged basis. This is suitable for the less stringent dose-uniformity requirements of these machines compared with implanters used for semiconductor doping. This technique is not widely used because the ac filament induces plasma oscillations and the source output is heavily modulated at mains frequency. This can lead to severe scan uniformity problems. It is nevertheless a simple way of making the beam more uniform for certain applications.

4.4 Source Magnet Control

In the Applied Materials Precision Implant 9000 [15] ion implanter the Freeman source is provided with an externally applied magnetic field that can be varied in both intensity and shape. The arrangement is shown in Figure 9.8. The two poles of the source magnet each have their own coil which is independently powered. If the current is greater in one coil than the other, there are two consequences:

1. The field strength will be biased so that one end of the Freeman source has a larger magnetic field strength than the other.
2. The magnetic bottle effect (mentioned in Section 4.1) will also be asymmetric.

The extra degree of freedom given by this technique allows some control over the distribution of plasma intensity along the length of the extraction slot. The optimum situation is achieved simply by maximizing the beam current through the ion implanter and this optimum is invariably asymmetric.

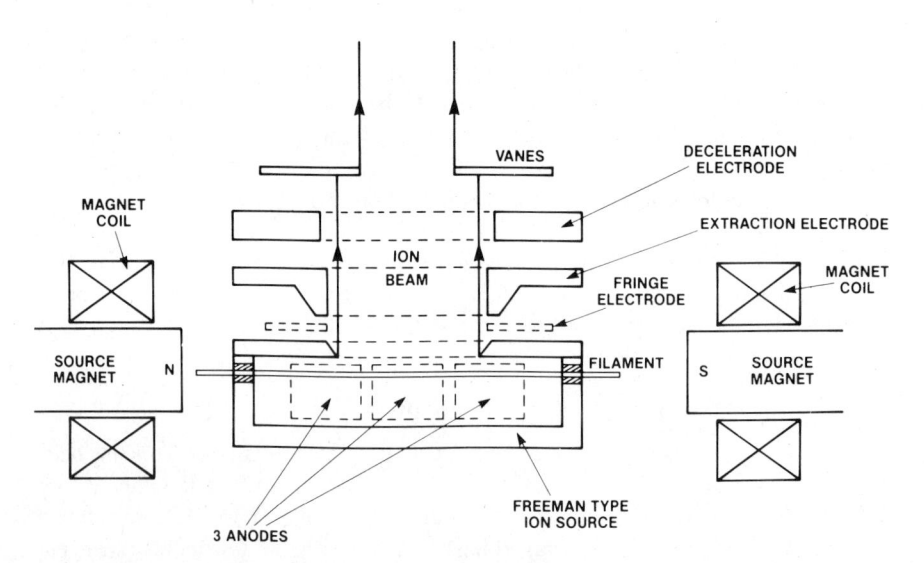

Figure 9.8 Schematic of the PI 9000 Freeman source.

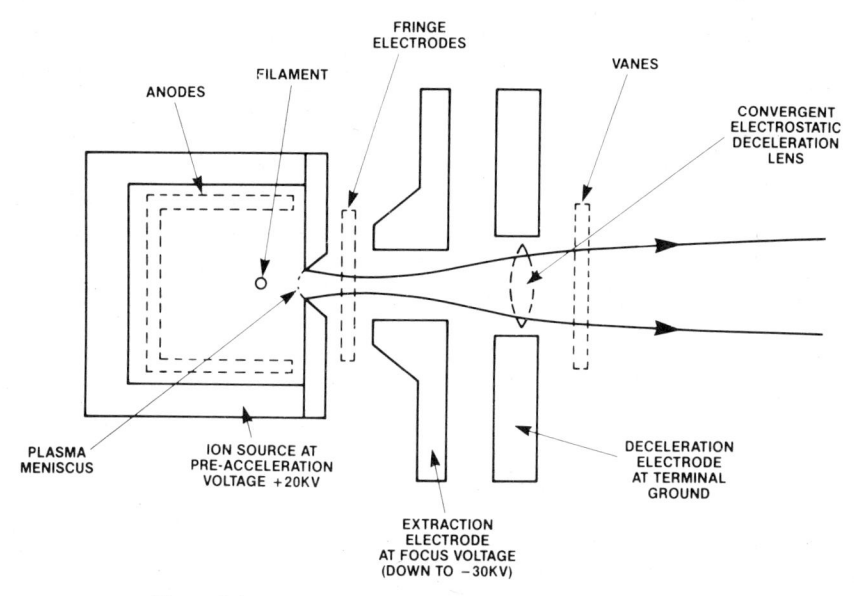

Figure 9.9 Ion extraction from the PI 9000 Freeman source.

4.5 Other Variants

Figures 9.8 and 9.9 show another technique that can be used to control the distribution of plasma intensity along the length of the source. This involves the use of independently powered anodes. The idea was found to be effective but was not incorporated in the PI 9000 machine because source magnet control was found to be sufficient [14].

5 ION BEAM EXTRACTION

The techniques mentioned in the previous section were aimed at either improving the efficiency of primary electron containment or influencing the uniformity of the plasma intensity along the length of the source. The main motive for the latter is to improve the efficiency of ion beam extraction. The extraction conditions can be optimized only for a particular beam perveance. If there is a current density variation along the length of the slot, then optimum extraction conditions cannot be achieved.

Another important beam extraction consideration is the difficulty experienced in extracting high beam currents (>20 mA). If a conventional extraction system is used, where virtually all the extraction voltage is supplied by the highly stabilized preanalysis acceleration supply (i.e., the potential applied to the ion source), then serious sparking problems are invariably encountered. If a spark occurs across the extraction gap, then the highly stabilized preanalysis acceleration supply "feeds" the discharge, attempting to maintain a constant voltage despite the rapidly increasing discharge current. This results in violent sparks that can damage local electronics and can limit the rate at which the extraction electrodes become conditioned to a relatively spark-free condition.

The solution to this problem adopted in the PI 9000 machine is to use a modest +20-kV preanalysis acceleration supply together with a generous down to −30-kV bias on the extraction electrode (see Figure 9.9). A negative bias on this electrode is, of course, generally applied to prevent electron backstreaming toward the ion source. In this case the negative bias, called the focus voltage, provides a significant fraction of the total extraction voltage (up to 50 kV) and is fed from a "soft" supply by using a series resistor between the power supply and the electrode. Any spark across the extraction gap is now quenched by the reduction in the focus voltage caused by the increased current but the energy of the beam entering the analyzing magnet remains constant at the preanalysis acceleration potential. There are many advantages to this type of extraction system which are described in detail elsewhere [14, 15].

6 ARC CONTROL

Stable control of the arc in a Freeman source is not a simple task. There is a range of arc currents, typically up to 2 A, where control is relatively easily achieved by the conventional control technique using the filament temperature as the primary control parameter. When maximum performance is required from the source, higher arc currents may be required and this can lead to a control regime where the thermal inertia of the filament can lead to stability problems.

The significant requirement is the need for a control technique where the response is not limited by the relatively sluggish relation between filament current and filament temperature. The technique used in the PI 9000 is to have three control loops:

1. A fast control achieved by increasing or decreasing the arc voltage to increase or decrease the arc current. It has been found that fast changes in arc impedance can be very effectively accomodated by this control technique.
2. In the longer term the above control will result in a drift in the arc voltage from the required setting (i.e., the arc voltage that gives the best yield of the required ion). It is therefore necessary to have a second slower control loop which sets the filament current to a value that gives the required arc voltage at the required arc current. In the PI 9000 this is software controlled.
3. A third control loop is a software loop that monitors the beam current on target and maintains it at the required value. This control is achieved in PI 9000 by either direct arc current control or by use of a variable width aperture.

With the above techniques an extremely stable, hash-free beam can be achieved with the degree of predictability necessary in a totally automated ion implanter.

7 FEED MATERIALS

Many factors determine the choice of feed materials. If high output is the major requirement, then the use of the element is the first choice, provided the chemical and/or vapor pressure properties are appropriate. Chemical compounds are the next alternative but the fraction of the required species in the output spectrum of the ion source is inevitably going to be lower. If a large number of species are required without removing the ion source, then gas feeds will have an advantage over solid feeds. Another factor that can influence the choice of feed material is the effect on the thermionic emission properties of the filament surface. These effects can be physical, such as activation or deactivation of the electron emission caused by condensation of material on the surface of the filament, or chemical, causing erosion or swelling. Figure 9.10 [16] shows an example of the filament deactivation and swelling problem found when using excessive boron trichloride as a feed material. Activation would appear to be advantageous, allowing a lower filament operating temperature, but in practice the balance between deposition of the activating species and removal due to evaporation and/or sputtering can lead to erratic behavior.

It is important when considering the appropriate feed material to make sure that the temperature of the arc chamber is sufficiently high to prevent condensation of the feed and subsequent flake formation, which can lead to erratic behavior. Sometimes this is not possible. A good example is the platinum group elements which have relatively low vapor pressures and in many cases lack convenient chemical compounds. The best solution is to use a sputter source (see Figure 9.3e). In this case deposition of unused feed material is inevitable.

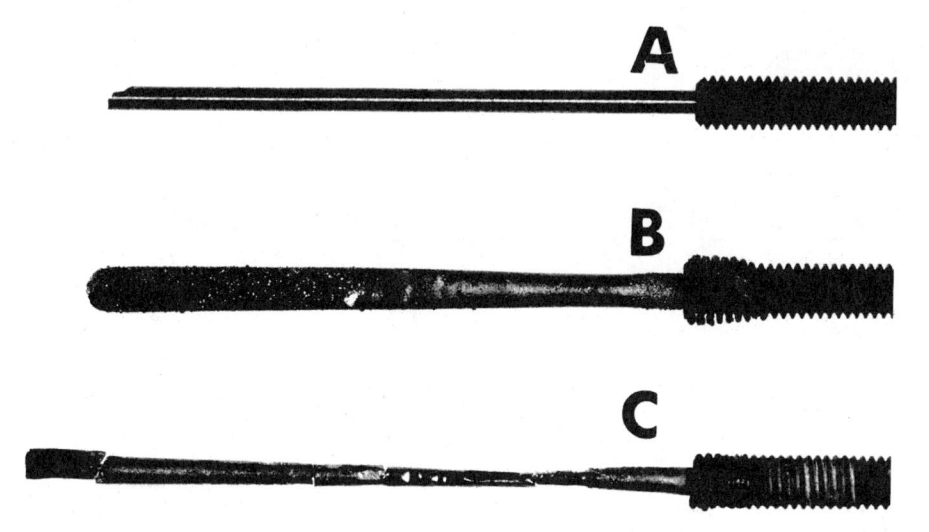

Figure 9.10 Freeman ion source filaments: (*a*) unused filament; (*b*) filament after brief operation with excess BCl_3 in discharge chamber; (*c*) filament after 35 h continuous operation (at 1 mA B^+) with correct vapor pressure of BCl_3. (Source: ref. 16.)

TABLE 9.1 Feed Materials for Semiconductor Ion Implantation

Species	Feed Material	°C for Vapor Pressure 10^{-2} Torr	Comments
Antimony	Element	533	Toxic
	Sb_2O_3	450	Toxic
	SbH_3	Gas	Very toxic
	SbF_3	Solid 760 Torr at 319 °C	Toxic
	SbF_5	Liquid, 760 Torr at 149 °C	Toxic
Argon	Element	Gas	—
Arsenic	Element	247	Toxic
	AsH_3	Gas	Very toxic
	AsF_3	Liquid, 100 Torr at 20 °C	Toxic
	AsF_5	Gas	Toxic
	$AsCl_3$	Liquid, 760 Torr at 130 °C	Toxic
Boron	Element	2027	Very high temperature required
	BF_3	Gas	Preferred feed
	BCl_3	Liquid 760 Torr at 13 °C	Deactivation problem
	B_2H_6	Gas	Spontaneously inflammable
Cadmium	Element	265	Convenient
Oxygen	Element	Gas	Severe filament oxidation
	CO_2	Gas	Preferred feed
Phosphorus	Element (red)	260	Convenient
	PH_3	Gas	Very toxic
	PF_3	Gas	
	PF_5	Gas	
	PCl_3	Liquid, 60 Torr at 10 °C	Deactivation problem
Selenium	Element	243	Convenient
	H_2Se	Gas	Very toxic
Silicon	Element	1632	Very high temperature required
	SiH_4	Gas	
	SiF_4	Gas	Preferred
	$SiCl_4$	Liquid, 100 Torr at 5 °C	Deactivation problem
Tellurium	Element	374	Convenient
	TeF_6	Gas	
Zinc	Element	344	Convenient

When a suitable feed compound exists, but its chemical properties or stability make it inconvenient to use, then a solution can be to synthesize the material in the ion source. A good example of this is the chlorination of heated oxides using carbon tetrachloride vapor [17]. The vaporizer becomes the chemical reactor and the reacting gas is passed over or through the heated solid. Precise temperature control is not usually necessary.

It is not the intent of this chapter to review the full range of feed materials. For detailed information the reader is referred to the excellent review by Freeman [18]. The details for a small number of species of particular interest for semiconductor ion implantation are shown in Table 9.1.

8 SUMMARY

The Freeman ion source has proved to be a reliable, predictable source for commercial ion implanters. Its practical advantages are its simplicity and the stable intense plasma fixed near the extraction slot by the filament.

REFERENCES

1. J. H. Freeman, *Nucl. Instrum. Methods* **22,** 306 (1963).
2. J. Koch, *Electromagnetic Isotope Separators and Applications of Electromagnetically Enriched Isotopes*, North-Holland, Amsterdam, 1958.
3. R. H. Dawton, *Electromagnetic Isotope Separators and Applications of Electromagnetically Enriched Isotopes*, North-Holland, Amsterdam, 1958.
4. I. Chavet and R. Bernas, *Nucl. Instrum. Methods* **51,** 77 (1967).
5. Lintott Engineering/Applied Materials, Foundry Lane, Horsham, West Sussex, England, RH13 5PY.
6. Extrion/Varian, Blackburn Industrial Park, Gloucester, MA 01930.
7. Nova/Eaton, 108 Cherry Hill Drive, Beverly, MA 01915.
8. J. H. Freeman, Atomic Energy Research Establishment Report, R6138, 1969.
9. J. H. Freeman, L. R. Caldecourt, K. C. W. Done, and R. J. Francis, Atomic Energy Research Establishment Report R6496, 1970.
10. H. Hinkel, *Nucl. Instrum. Methods,* **139,** 1 (1976).
11. N. Williams, Proceedings of the International Conference on Low Energy Ion Beams, September 5-8, 1977, University of Salford, UK, Institute of Physics, Bristol.
12. D. Aitken, *Radiation Effects* **44,** 159 (1979).
13. O. Tabata, S. Kimura, and M. Kunori, Proceedings of the International Ion Engineering Congress, ISIAT, 1983, Kyoto, 1983.
14. D. Aitken, *Vacuum* **36,** 953 (1986).
15. D. Aitken, *Nucl. Instrum. Methods, B* **21,** 274 (1987).
16. J. H. Freeman, G. A. Gard, and W. Temple, Atomic Energy Research Establishment Report R6758, 1971.
17. G. Sidenius and O. Skilbreid, in *Electromagnetis Separation of Radioactive Isotopes*, Higatsberger and Viehbock (Eds.), Springer-Verlag, Vienna, 1961, p. 243.
18. J. H. Freeman, *Ion Implantation*, North-Holland, Amsterdam, 1973.

10

Electron Cyclotron Resonance Ion Sources

Y. Jongen
Universite Catholique de Louvain
Louvain-la-Neuve
Belgium

C. M. Lyneis
Lawrence Berkeley Laboratory
University of California
Berkeley, California

ECR (Electron Cyclotron Resonance) ion sources are now in widespread use for the production of high-quality multiply charged ion beams for accelerators and atomic physics experiments, and industrial applications are being explored. In the last decade ECR sources have evolved from a single large, power-consuming, complex prototype (SUPERMAFIOS) [1] into a variety of compact, simple, reliable, efficient, high-performance sources of high charge state ions. Eight laboratories have one or more ECR sources in regular operation with cyclotrons. At least seven other ECR sources are under development or in the planning stage for use with cyclotrons. The CERN SPS accelerator has accelerated an oxygen beam, which was supplied by an ECR source, to 200 GeV/u [2]. A project is underway at Argonne to couple an ECR source to a superconducting heavy-ion linac [3]. Four ECR sources are in dedicated use for atomic physics and three of the ECR sources coupled to cyclotrons are also used for low-energy atomic physics research.

Several general characteristics of ECR sources explain their widespread acceptance. Most important is the ability to produce high charge state ion beams at useful intensities for nuclear physics experiments. Another characteristic of ECR sources is that the discharge is produced without cathodes, so that only the source material injected into an ECR source is consumed. As a result, ECR sources can be operated continuously for periods of weeks without interruption. Techniques have been developed in the last few years that allow these sources to produce beams from solid materials. The beam emittance from ECR sources is in the range of 50 to 200π mm-mrad at 10 kV, which matches the acceptance of cyclotrons and RFQs well. Maintenance required on ECR sources is also minimal, consisting mainly of occasional repair of vacuum equipment and electrical support equipment.

1 A BRIEF HISTORY OF THE DEVELOPMENT OF ECR SOURCES

The field of ECR ion sources has its roots in the plasma fusion developments in the late 1960s. The use of Electron Cyclotron Resonance Heating (ECRH) in plasma devices to produce high charge state ions was suggested in 1969 [4]. The first sources actually using the ECRH to produce multiply charged ions were reported in 1972 in France by Geller et al. [5] and in Germany by Wiesemann et al. [6]. Geller's MAFIOS source is illustrated in Figure 10.1. Although these devices, which used solenoid magnetic mirror configurations were capable of producing plasma densities of the order of 1×10^{12} cm^{-3} and keV electrons, typical operating pressures were 10^{-4} to 10^{-5} Torr, and the ion lifetimes were 10^{-4} s or less. This resulted in low charge state distributions (CSD) which, for example, peaked at N^{2+} for nitrogen and Ar^{2+} for argon. Similar sources were also developed later in Japan [7] and the Soviet Union [8].

A major step forward was made in 1974 when Geller transformed a large mirror device used for plasma research (CIRCE, 1973) [9] into an extremely successful ion source, SUPERMAFIOS [10], which is shown in Figure 10.2. Unlike the earlier ion sources using ECRH, the magnetic field of CIRCE, renamed SUPERMAFIOS, used a hexapolar field. This produced a minimum-B magnetic field

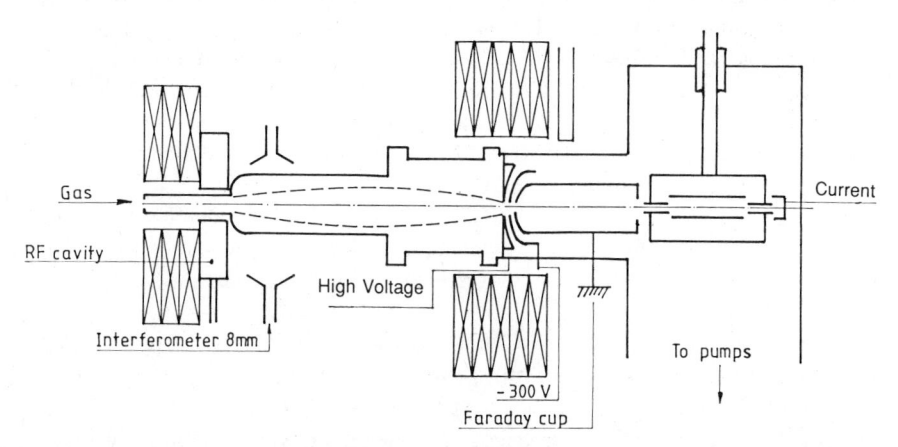

Figure 10.1 The MAFIOS source.

1 A Brief History of the Development of ECR Sources

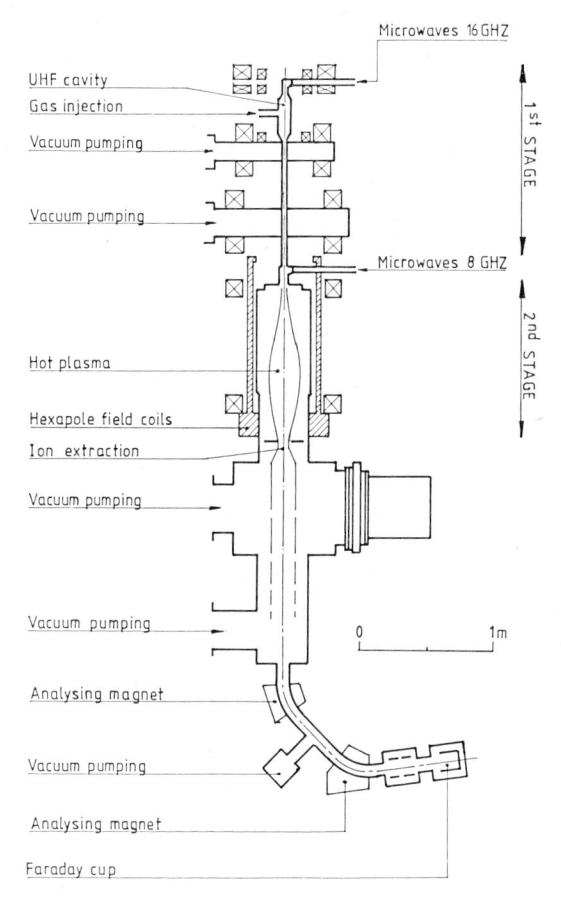

Figure 10.2 The SUPERMAFIOS source.

configuration which stabilized the plasma against MHD instability. In addition, CIRCE was a two-stage device. A cold plasma was generated in a first stage, operating at a higher pressure of about 10^{-3} Torr. This cold plasma flowed along the magnetic field lines, feeding the main confinement stage, which operated at much lower pressure, about 10^{-6} Torr. These two features, plasma stabilization in a minimum-B field configuration and main stage operating at low neutral pressure, produced an enormous improvement in the lifetime of the ions in the source ($\sim 10^{-2}$ s) and thereby also in the CSD. Several variations of this ECR source were tested between 1974 and 1977: SUPERMAFIOS-A, TRIPLEMAFIOS, and SUPERMAFIOS-B. However, the basic features of the design remained unchanged and are still the basis for present source design.

The INTEREM device at Oak Ridge was another plasma fusion device converted into an ion source. It used a combination of solenoid and quadrupole coils to provide a minimum-B geometry. Although this device succeeded in producing a nitrogen CSD peaked on N^{3+} with a small amount of N^{6+}, the extraction geometry was very inefficient and the resulting currents were too small for practical use [11].

The main drawback of SUPERMAFIOS was the 3-MW power consumed by the main stage. Different solutions were considered to reduce this power. The use of samarium cobalt (SmCo) permanent magnets was considered impractical because it was assumed that ECR sources should be as large as SUPERMAFIOS (diameter 30 cm, length 100 cm). Therefore, superconductivity was initially seen as the best way to solve this problem and large superconducting ECR sources were started in Louvain-la-Neuve [12], Karlsruhe [13], and Jülich [14].

In 1979 Geller transformed a reduced-scale permanent magnet model hexapole built in Louvain-la-Neuve into a successful source of much smaller size (diameter 7 cm, length 30 cm) called MICROMAFIOS [15]. The basic design of MICROMAFIOS is shown in Figure 10.3. After development and modification it was renamed MINIMAFIOS [16]. MINIMAFIOS sources have since been built by Geller for the following laboratories: KVI–Groningen (1982), SARA–Grenoble (1983), GANIL–Caen (1984), and GSI/CERN–Geneva (1985).

During the same period, small test sources using SmCo permanent magnet hexapoles were also built: PICOHISKA [17] in Karlsruhe and PRE-ISIS in Jülich. A small fully superconducting source, ECREVETTE [18], was built in Louvain-la-Neuve.

In 1983, ECR source development began in the United States at Berkeley [19] and Oak Ridge [20], followed by Michigan State University [21] and Argonne [3].

During 1984 and 1985, Geller demonstrated a new source, MINIMAFIOS-16

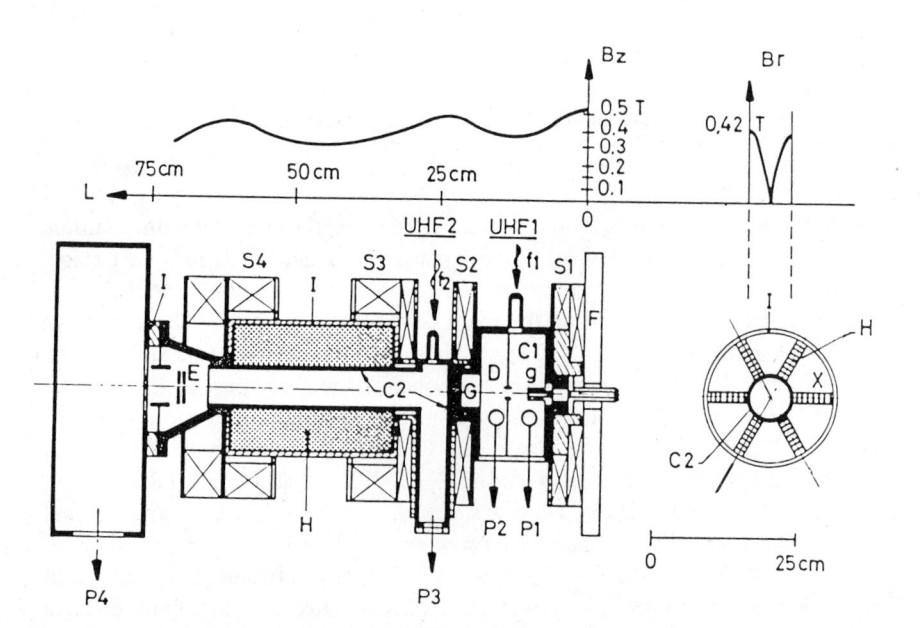

Figure 10.3 The MICROMAFIOS source.

GHz, which, using higher microwave frequencies, can improve both the total extracted current and the change state distribution of ECR sources [22].

2 BASIC PRINCIPLES OF ECR SOURCES

In recent years the design of ECR sources for multicharged ions has reached a certain maturity, and how to make an ECR source is well understood. In contrast, the detailed physical phenomena underlying the source operation are poorly understood. This is due in part to the difficulty of making unambiguous measurements of the plasma parameters in ECR sources. Therefore, the "basic principles" presented in this section are really a set of hypotheses generally accepted by the ECR source community rather than experimental facts.

2.1 Ionization by Electron Impact

It is generally accepted that in ECR sources multicharged ions are created mainly by a step by step ionization, caused by the impact of energetic electrons. Therefore, the electron impact ionization cross section is a significant parameter. A considerable amount of experimental data exists already in this field, and the use of ECR sources by atomic physics groups will extend these measurements to higher charge state ions. Semiempirical formulae, fitted to existing experimental results, have been proposed by Müller and Salzborn [23] and Lotz [24]. Figure 10.4 shows

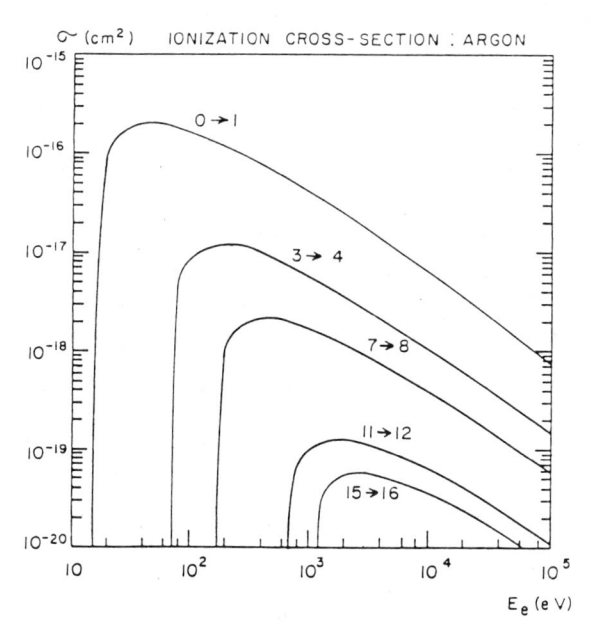

Figure 10.4 Computer calculation of some argon electron impact ionization cross sections using the Müller-Salzborn formula.

electron impact ionization cross sections calculated using the Müller-Salzborn formula. An excellent compilation was made in 1981 by Crandall [25].

The rate at which ions of charge state i are produced by electron impact ionization is given by

$$R_{\text{prod},i} = n_e \langle \sigma_{i-1,i} v_e \rangle n_{i-1} \tag{1}$$

where n_e is the electron density; $\langle \sigma_{i-1,i} v_e \rangle$, the rate coefficient, is the product of the electron impact ionization cross section from charge $i-1$ to i and the electron velocity averaged over the electron energy distribution; and n_{i-1} is the density of ions of charge state $i-1$. The rate coefficient calculated for argon with the Müller-Salzborn formula are illustrated in Figure 10.5.

2.2 Magnetic Confinement

Considering the small value of electron impact ionization cross sections, a large flux of high-energy electrons is necessary to produce a significant density of highly stripped ions. If the electrons travel only once through the system, as in an EBIS (Electron Beam Ion Source), the power flux needed is enormous. For instance, 1 to 10 MW/cm^2 would be required to produce 100 μA of Ar^{8+}. To avoid the

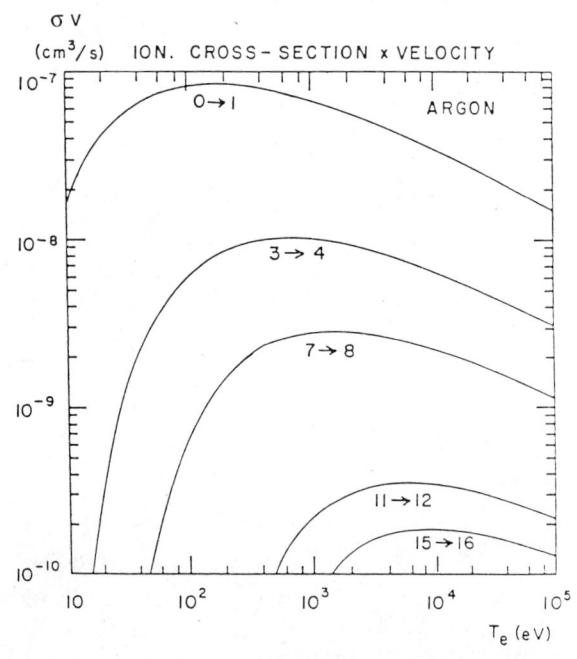

Figure 10.5 Computer calculation of some argon electron impact ionization rate coefficients using the Müller-Salzborn formula.

problems of using such large powers, it is possible to confine the electrons and ions, forming a neutral plasma. In this case, each electron makes many passes through the system, and the dissipated power is essentially limited to the energy carried away by the flux of electrons escaping from the confinement region.

To confine the plasma, a "minimum-B" magnetic field configuration can be used. Such a configuration, where the magnetic field is a minimum at the center of the device and increases in every direction from the center, stabilizes the plasma against MHD instabilities. The ratio of the maximum field strength at the magnetic mirrors to the field strength at the center of the device is called the "mirror ratio." It is an important parameter of the confinement geometry. Existing ECR sources use mirror ratios between 1.3 and 2.

In all of the successful ECR sources built so far, this minimum-B configuration is produced by combining an axial field, from solenoid coils, with a multipole field, either hexapole or octupole. These multipole fields can be produced by coils or permanent magnets.

2.3 ECR Heating

To optimize the rate of ionization by electron impact, electron temperatures between 1 and 10 keV are typically needed. On the other hand, an ion temperature as low as possible is desirable because the ion temperature is one source of the emittance and energy spread of the extracted beam. Therefore, a method to heat selectively the electrons in the plasma is desirable. The use of ECRH meets this condition. If we introduce into a plasma an electromagnetic wave whose frequency is equal to the cyclotron frequency of the electrons in the magnetic field, an extremely efficient energy transfer occurs between the wave and the electron population. The exact nature of this energy transfer is complex and has been discussed in numerous papers. A comprehensive discussion has been made by O. Eldridge [26]. In a minimum-B configuration, the magnetic field is not uniform but increases from the center to the outside. Therefore, the ECR condition is normally met only on a closed surface around the center, called the *ECR surface*. It has been found experimentally desirable to have the ECR surface closed inside the plasma chamber.

The propagation of electromagnetic waves in a plasma is quite complex. Using some oversimplification, one can describe the plasma as a high-pass filter. That is, waves with frequencies higher than a critical frequency, called the plasma frequency, can propagate while waves with frequencies lower than the plasma frequency are reflected. The plasma frequency ω_p is a function of the plasma density,

$$\omega_p = \sqrt{\frac{n_e e^2}{\epsilon_0 m}} \qquad (2)$$

where n_e is the electron density, e is the electron charge, m is the electron mass, and ϵ_0 is the vacuum dielectric constant. In practical units,

$$f_p = 8.91 \times 10^3 \sqrt{n_e} \tag{3}$$

where n_e is in electrons/cm^3 and f_p is in Hz. For a given microwave frequency ω, the critical density n_{ec} is defined by the density for which ω_p equals ω.

The production of overdense plasma ($n_e > n_{ec}$) has been demonstrated in some plasma physics experiments, but so far, attempts to use overdense plasma in actual ECR sources have been unsuccessful. Therefore, the use of higher frequencies seems the only practical way to reach higher plasma densities in ECR sources. In Table 10.1 the frequencies and resulting critical densities are given for some of the ECR sources presently in operation.

2.4 Losses by Charge Exchange

The loss of high charge state heavy ions from the plasma is dominated by two processes. These are charge exchange with neutral atoms in the plasma and loss of confinement. The cross sections for charge exchange between highly charged ions and neutrals are extremely large. An empirical formula, fitting some existing experimental data, has been proposed by Müller and Salzborn [27]. The cross section for charge exchange from initial charge state i to final charge state $i - 1$ is given by

$$\sigma_{i,i-1} = 1.43 \times 10^{-12} i^{1.17} V_{0,1}^{-2.76} \text{ cm}^2 \tag{4}$$

where $V_{0,1}$ is the first ionization potential (eV) of the neutral atom. Typical charge exchange cross sections are three to four orders of magnitude larger than corresponding electron impact ionization cross sections. Fortunately, the reaction rates are proportional to the projectile velocities, and neutral atoms are much slower than hot electrons. Still, to keep the rate of production by electron impact equal

TABLE 10.1 The Frequencies and Resulting Critical Densities for Some of the ECR Sources Now in Operation

Frequency (GHz)	Critical Density (10^{11}cm^{-3})	B_{reson} (kG)	Used at	Ref.
6.45	5.16	2.27	Berkeley, 2d stage	31
8.50	9.25	3.04	Louvain-la-Neuve, 2d stage	36
9.20	10.8	3.29	Berkeley, injector	31
10.00	12.8	3.5	MINIMAFIOS	37
14.30	26.2	5.12	Louvain-la-Neuve, injector	36
16.00	32.8	5.73	MINIMAFIOS 16 GHz	2

to the rate of loss by charge exchange for high charge state ions, it is necessary for the neutral atom density in the plasma to be two orders of magnitude lower than the electron density.

2.5 Ion Confinement

The second loss mechanism, the loss of confinement, although numerically less important, is of crucial importance for beam extraction. In ECR sources, unlike in other sources like EBIS, it is not possible to arbitrarily interrupt the ion confinement to extract the confined ions. All extracted ions have undergone a loss of confinement. That is, ions are available for extraction only to the extent that they are lost axially from the confinement region and stream toward the ion extractor. Consequently, the ion confinement time is a very critical parameter in ECR sources. If this confinement time is too short, ions do not have time to reach high charge states and if the confinement time is too long, the high charge state ions decay by charge exchange instead of being extracted. Unfortunately, the mechanisms of ion confinement are not clearly understood. Plasma potentials caused by the ambipolar diffusion are also present, but they are probably small.

Recently, interesting experiments have been carried out in Berkeley by comparing, in the same plasma, the charge-state distribution of two different isotopes of oxygen [28], but the results are still not conclusive. Therefore, ion confinement "tuning" in ECR source is essentially a blind procedure, carried out on a purely empirical basis. The use of gas mixing, as done presently in most ECR sources, is one such empirical optimization procedure [29].

2.6 The Plasma Injector Stage

The well-known Paschen law on avalanche discharge in gases [30] shows that it is not possible to break down a neutral gas to form a plasma if the gas pressure is too low. For this reason most ECR sources have been built as two-stage devices. The first stage is a cold plasma generator, operating at higher pressure. The plasma produced in the first stage flows along the magnetic field lines into the second stage. This flow is essentially governed by the density gradient between the first and second stages; experience has shown that the magnetic field gradient has little influence on the flow of this cold plasma. This can be explained by the fact that the first-stage plasma is highly collisional.

In principle, any method that generates a dense plasma can be used in the first stage, but all sources built up now use ECRH. To maximize the cold plasma flux into the second stage, a high-density gradient is necessary. Therefore, first stages generally operate close to the critical density defined by the heating frequency. For the same reason, the microwave frequency for the injector stage is frequently chosen to be higher than the frequency of the main stage. Finally, to increase the density gradient, the injector stage should be located as close as possible to the main stage.

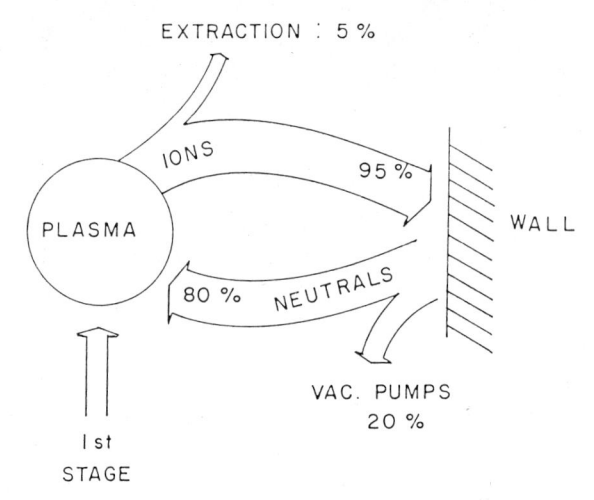

Figure 10.6 Plasma-wall gas recirculation.

2.7 Gas Recirculation in ECR Sources

Due to imperfect confinement of the plasma, an ion flux escapes continuously from the plasma. Only a small fraction of this flux (5 to 10%) escapes along the source axis, reaches the extraction hole, and is accelerated as beam. Most of the ion flux hits the walls, where the ions are either neutralized or scattered elastically back into the plasma. The walls appear, therefore, as an intense source of neutral gas. The neutrals generated at the walls are either reionized by the plasma or evacuated by the system vacuum pumps. Figure 10.6 shows the basic processes involved in plasma–wall recirculation. In ECR sources, the plasma pumping speed is much larger than the pumping speed of the external pumps. This means that each ion undergoes several plasma–wall cycles before escaping the system as an accelerated beam or as neutral gas evacuated by the vacuum pumps. One function of the first stage therefore, is to compensate for those losses by supplying singly charged ions instead of neutral gas.

3 WORLD ECR SOURCES

Since SUPERMAFIOS, more than 30 ECR sources have been built and used in different countries. Table 10.2 lists some of those sources with some key parameters.

Essentially all these sources have been successful, sharing common desirable features:

- ECR sources do not have consumable components. For this reason they are extremely reliable and stable, allowing unattended operation for days or weeks.

TABLE 10.2 ECR Ion Sources

Country Laboratory Source	Status	L (cm)	D (cm)	F (GHz)	Comments	Application
FRANCE						
Grenoble-CEN						
SUPERMAFIOS	Dism.	100	35	16, 8	3 MW used	Test
MICROMAFIOS	Dism.	30	7	10	Compact, copper, coils, SmCo	Test
MINIMAFIOS	Operat.	30	7	10	Now at KVI, SARA, GANIL and CERN-SPS	Test, A.P., and Cyc. Synchrotron
MINIMAFIOS-16 GHz	Test	30	7	16.6	Pulsed source	Test
CAPRICE	Test	13.5	8	10	Iron yoke, very compact	Test
MINIMAFIOS-18 GHz	Test	30	7	18	Highest frequency ECR	Test, Synchrotron
GERMANY						
KFZ-Karlsruhe						
p-HISKA	Dism.	28	7	14.5, 7.5	Compact, SmCo Sextupole	Test and Cyc
HISKA	Standby	70	10	14, 7.5	s.c. Solenoids, SmCo Sextupole	Cyclotron
LISKA	Operat.	25	7	7.5	Lithium ECR	Cyclotron
High-Efficiency ECR	Operat.	20	5	6.4	Isotope separation	Test
KFA-Jülich						
Pre-ISIS I	Dism.	25	5	2.5	Small, low-frequency, single-stage	Test
Pre-ISIS II	Operat.	25	5	5	Two-stage	Test and Cyc.
ISIS	Operat.	70	20	14.3, 14.3	Large s.c., high-frequency	Cyc
Giessen						
Pre-ISIS II	Operat.	25	5	5	Two-stage	Atomic physics

TABLE 10.2 *(Continued)*

Country Laboratory Source	Status	L (cm)	D (cm)	F (GHz)	Comments	Application
BELGIUM						
Louvain-la-Neuve						
ECREVETTE	Dism.	40	12	14.7, 8.5	1st superconducting ECR	Test and Cyc.
ECREVIS	Dism.	120	32	14.7, 8.5	Large superconducting ECR	Cyc. and A.P.
OCTOPUS	Operat.	70	18	14.3, 8.5	Iron yoke, open octupole	Cyc. and A.P.
UNITED STATES						
Oak Ridge National Lab						
ONRL-ECR	Operat.	40	8.5	10.6	Compact, SmCo	Atomic physics
Lawrence Berkeley Lab						
LBL-ECR	Operat.	33	9.	9.2, 6.4	Open SmCo sextupole	Cyc. and A.P.
NSCL-MSU						
RT-ECR	Operat.	50	14	6.4, 6.4	Iron yoke, 1st-stage, sextupole	Cyclotron
CP-ECR	Operat.	24	11	6.4	Single-stage, high-temp oven	Cyclotron
SC-ECR	Constr.	50	14	≤35	Superconducting magnets	Test and Cyc.
Argonne National Lab						
PIIECR	Constr.	53	10	10	On high-voltage platform	Linac
Texas A&M University						
ECR	Design			14, 6.4		Cyclotron
JAPAN						
RIKEN						
ECR1	Operat.	100	8	9, 2.5	Overdense experiments	Test
ECR2	Operat.	30	8	2.5, 2.5	Whistler mode heating	Test
ECR	Design	52	10	10		Cyclotron
SWEDEN/FINLAND						
The Svedberg Lab. (Uppsala)						
RT-ECR	Constr.	50	14	6.4, 6.4	Operation scheduled mid 88	Cyclotron

*a*Dism., Dismantled; A.P., atomic physics; Cyc., Cyclotron; s.c., superconducting.

- ECR sources have high electron temperatures (1 to 10 keV), and relatively low neutral pressure in the plasma (5×10^{-6} to 5×10^{-7} Torr). For these reasons, relatively high intensities of high charge state ions are obtained.
- Due to the selective heating of the electrons by the ECRH, the ions remain extremely cold in an ECR plasma, probably less than 1 eV. Consequently the energy spread of the beam from an ECR source is small and defined essentially by aberrations in the extraction process. With typical extraction holes, the energy spread is of the order of 5 eV $\times$ Q where Q is the charge state. With reduced diameter holes, energy spreads as low as 1 eV $\times$ Q have been observed.
- The emittance of an ECR source is dominated by the size of the extraction hole and conservation of the magnetic moment of the extracted ions. At 10 kV extraction voltage, with a 10-mm-diameter extraction hole, an emittance of 50 to 200π mm mrad is typical for existing ECR sources
- All the recent ECR sources, operating at frequencies between 6.4 and 10 GHz in the main stage, show surprisingly similar intensities and charge state distributions. This indicates that plasma densities and ion lifetimes are quite similar in all these sources. Although no precise measurement of plasma density has been carried out, indirect evidence suggests plasma densities of 1 to 3×10^{11} cm^{-3}.
- Past experience has shown that the ECR source technology is remarkably uncritical and forgiving. Even some poorly engineered ECR sources have worked well.

4 ILLUSTRATION OF A TYPICAL ECR SOURCE: THE LBL ECR

4.1 Source Design

The LBL ECR source [31] is a compact source, similar in size to MINIMAFIOS [13], using room temperature solenoid coils and a Sm-Co hexapole structure. The design is illustrated in Figure 10.7, and the main parameters are summarized in Table 10.3. The solenoid field is produced by 11 edge-cooled, tape-wound, copper coils, each individually powered by a 250-A SCR-regulated power supply. The tape-wound coils have a high packing fraction (75% including the edge cooling) and therefore allow the amount of copper to be high and the power requirements to be modest. Since not all coils are needed to produce the required solenoid field strength, the field configuration can be modified by adjusting the currents in the individual coils. In typical operation only 30 kW of magnet power are required. The sextupole field is produced by Sm-Co bars supported in an open copper fixture which allows for radial pumping between the bars. Microwave power is fed into the main stage by a circular waveguide, coming axially through the first-stage pumping chamber. This circular waveguide has an axial hole, to allow cold plasma to flow in from the injector stage. To contain the microwave power in the region of the plasma, a copper screen cylinder was formed around the sextupole. Before

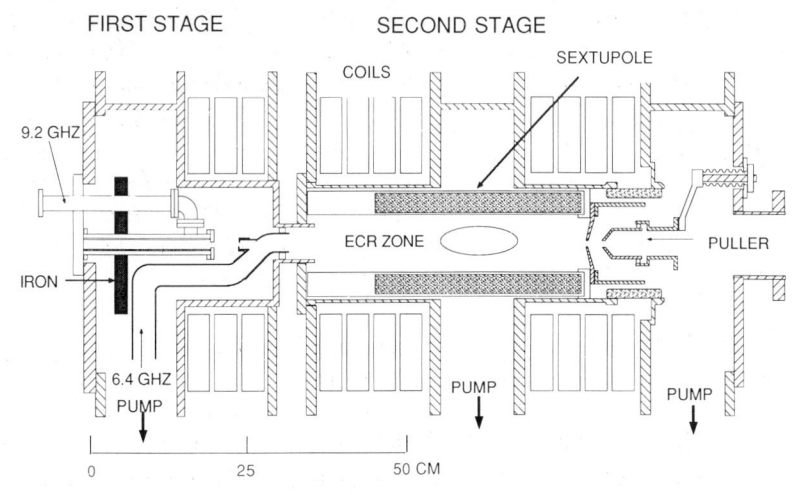

Figure 10.7 The LBL ECR source.

TABLE 10.3 LBL ECR Source Parameters

Parameter	Maximum	Typical
Magnetic field		
On axis	4.2 kG	3.5 kG
Mirror ratio	1.3–2.0	1.6
Sextupole at wall	3.1 kG	2.7 kG
Magnet power	110 kW	30 kW
Microwave Power		
Injector at 9.2 GHz	1.0 kW	0.150 kW
Main Stage 6.4 GHz	3.0 kW	0.300 kW
Vacuum	Base	Typical
Injector (ECR zone)	1×10^{-7} Torr	3×10^{-4} Torr
Injector vacuum chamber	1×10^{-7} Torr	8×10^{-6} Torr
Main stage	$<1 \times 10^{-7}$ Torr	6×10^{-7} Torr
Extraction	$<1 \times 10^{-7}$ Torr	$<1 \times 10^{-7}$ Torr
Extraction Geometry		
Plasma electrode hole	8 mm diam	8 mm diam
Puller hole	10 mm diam	10 mm diam
Gap	10–35 mm	30 mm

the screen was added, measurements on the source indicated that the microwave power was only weakly coupled to the plasma, apparently because the microwaves were flowing out between the bars of the sextupole. The cold plasma injector stage is illustrated in Figure 10.8. The location of the ECR zone in the first stage can be moved by adjusting the currents in the first three solenoid coils.

The basic design features of the extraction system and the Glaser lens [32] follow those used in ECREVIS at Louvain-la-Neuve. The details of the extractor geometry are given in Table 10.3. The puller is negatively biased to operate the extractor in an accel–decel mode. Additionally, the axial position of the extractor can be easily varied without stopping source operation. The beam analysis system, which is shown in Figure 10.9, consists of a Glaser lens, two sets of manually adjustable slits, and a double-focusing 90° analyzing magnet with a bending radius of 40 cm. The Glaser lens refocuses the diverging beam from the source onto the first set of slits. The 90° magnet coupled with the second set of slits allows a single charge state to be selected. The horizontal aperture of the slits must be closed down to 4 mm to attain a mass resolution of 1%. The extraction system and beam analysis system are quite efficient. The sum of the currents for all charge states collected at the Faraday cup just behind the second set of slits is typically at least 40% of the total source current, indicating excellent efficiency in the beam

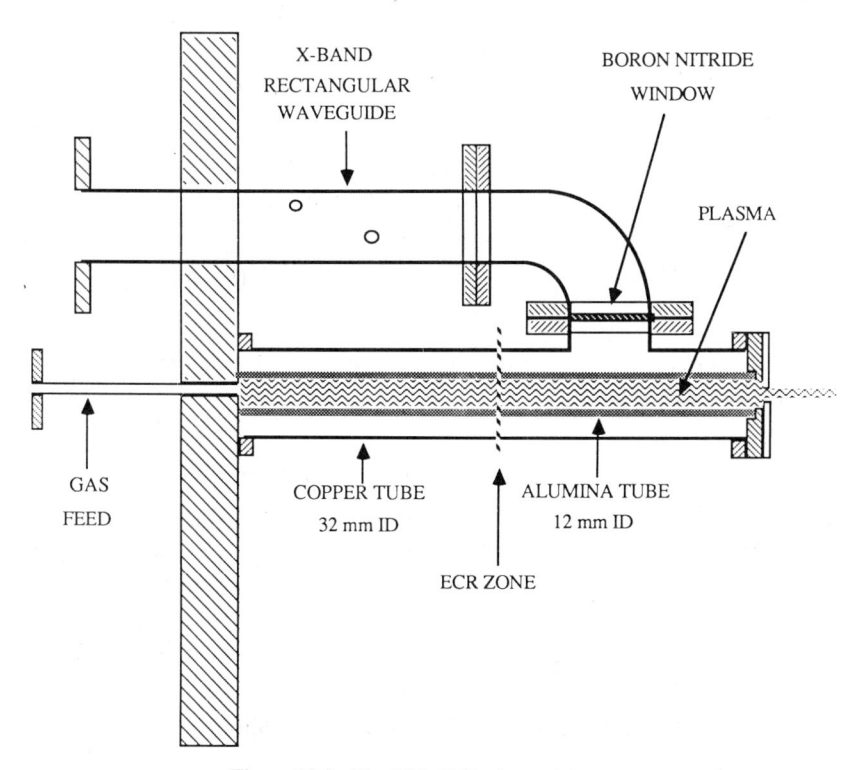

Figure 10.8 The LBL ECR plasma injector.

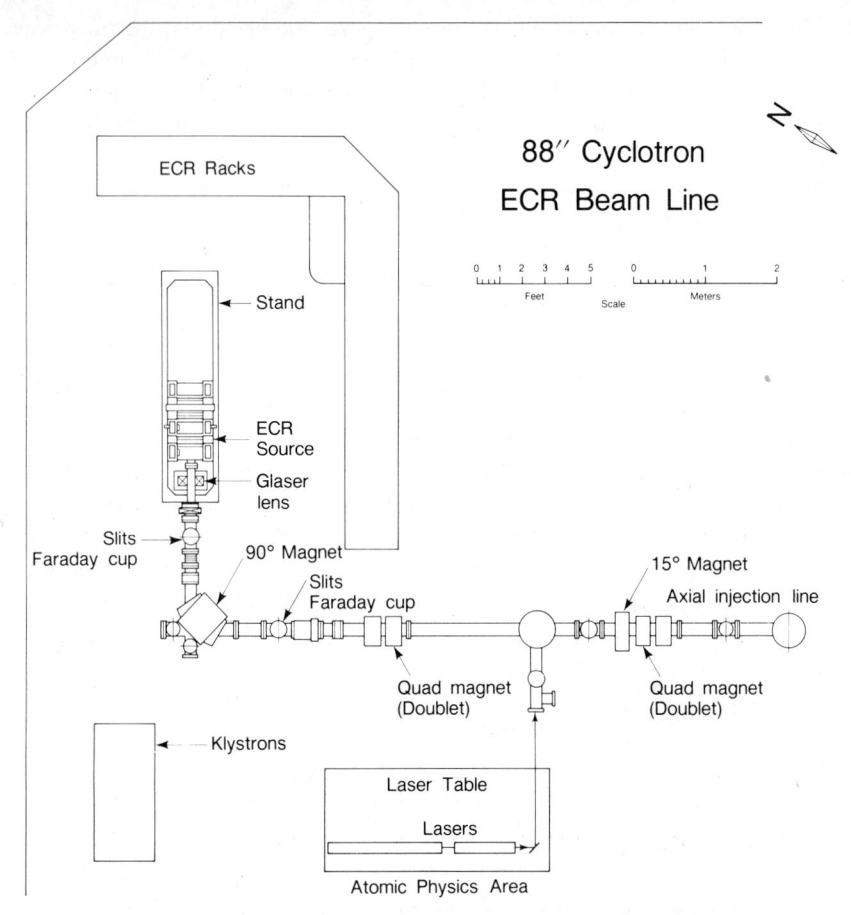

Figure 10.9 Plan view of the LBL ECR source and charge state analysis system.

analysis system. Figure 10.10 illustrates a typical oxygen charge state spectrum measured with the system.

4.2 LBL ECR Performance

The performance of the LBL ECR for a wide range of ions is summarized in Tables 10.4 and 10.5. The currents listed for gases are representative of the performance levels of a number of other ECR sources now in use with accelerators or for atomic physics, such as OCTOPUS, RT-ECR, MINIMAFIOS-10GHz, the ORNL-ECR, and ISIS. There is a much wider spread in the results for beams from solids.

4.3 ECR Source Beams from Solids

The development of techniques that allow the production of high charge state ions from solid materials has also been of great importance. Since a very high percentage

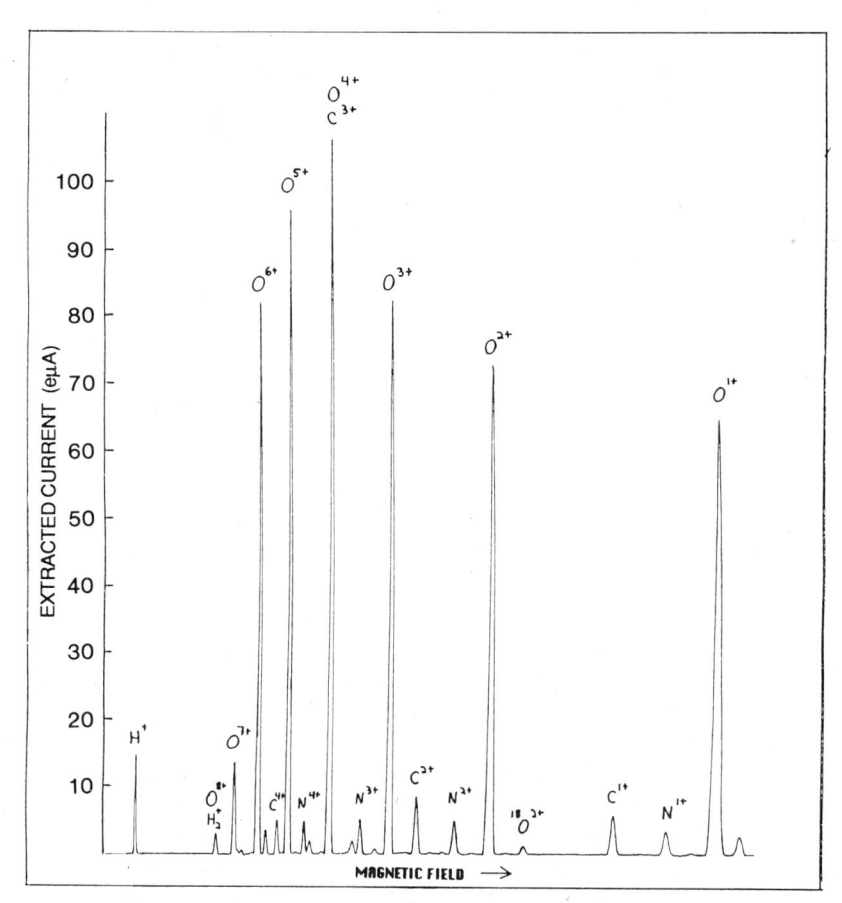

4 Illustration of a Typical ECR Source: The LBL ECR

Figure 10.10 Charge state distribution for oxygen measured on the LBL ECR. The plot was produced on an xy recorder by slowly sweeping the analyzing magnet.

of the elements more massive than argon are solids at room temperature, the ability to use solid materials as source feeds is vital for ion sources used with heavy-ion accelerators. The two main methods are direct insertion of solids into the plasma and use of ovens to vaporize solids. In some respects these two methods are complimentary. Direct insertion works well with high-temperature materials such as iron, nickel, niobium, tantalum, and tungsten. An oven works well with low-temperature materials such as lithium, magnesium, calcium, and bismuth.

Direction insertion has been studied in detail at Grenoble with the CAPRICE source for a wide variety of materials ranging from aluminum to gold [33]. In CAPRICE, a solid rod is positioned close to the ECR surface where it is vaporized by the plasma. The plasma is maintained by adding a support gas such as oxygen or nitrogen. To maintain a stable plasma the rod's position is automatically controlled with a feedback loop. In a one-week run using tantalum in CAPRICE, the average consumption was approximately 1 mg/h. Direct insertion has been

TABLE 10.4 Currents for the LBL ECR: Hydrogen through Silicon[a]

Q	^{1}H	^{3}He	^{12}C	^{14}N	^{16}O	^{19}F	^{20}Ne	^{24}Mg	^{28}Si
1+	300	300	27	82	118				
2+		200	37	117	143	43	51	32	20
3+			*	106	152	55	63	34	33
4+			31	110	*	53	78	28	69
5+			6.5	93	96	37	58	44	72
6+				19	82	17	45	34	47
7+					14	11	21	18	30
8+					0.95	1	11	8	17
9+						0.05	1.1	6.3	7
10+							0.04	2.2	2.7
11+								0.1	0.5
12+									0.2

[a]All currents in eμA measured at 10 kV extraction voltage. Natural isotopic abundance source feeds were used except for ^{3}He and ^{22}Ne^{10+}.
*Not measured because a mixture of two ions with identical charge to mass ratios were present.

used in the ORNL-ECR to produce iron, nickel, and chromium beams and in the LBL ECR to produce niobium beams for atomic physics. Even without feedback, stable beams can be produced for periods of several hours [34].

Several different metallic ion beams have been produced from the LBL ECR using a simple resistively heated oven, which is shown in Figure 10.11. Because the oven temperature is controlled externally it makes stable operation and efficient usage possible. The oven is inserted radially into the second stage so that vaporized metal atoms stream into the ECR plasma and are ionized by electron impact. With oven operation, the plasma is maintained by running a support gas such as oxygen or nitrogen in the first stage. This is similar to the use of a mixing gas when operating the source with gases heavier than oxygen. The amount of metal in the plasma is adjusted by varying the oven temperature. A proportional temperature controller is used to keep the oven temperature constant. The beam stability with the oven is quite remarkable. A number of cyclotron runs lasting several days have used metal ion beams from the ECR source, during some of which no adjustment of the ECR source or oven was required. When using rare isotopes such as ^{48}Ca, the usage can be minimized by lowering the operating temperature so that only the required intensity is produced. In this way a 68-h cyclotron run with ^{48}Ca consumed only 0.15 mg/h of an enriched sample of calcium (25% ^{48}Ca). ^{209}Bi was chosen to explore the performance of the LBL ECR source for very heavy elements because it is monoisotopic and its vapor pressure temperature characteristics are appropriate for the oven. In Figure 10.12 the currents produced using the oven with the LBL ECR for bismuth and iodine are shown. As listed in Table 10.5 the source produced 0.56 eμA of Bi^{31+} and 0.05 eμA of Bi^{34+}. Other materials, such as low-temperature chemical compounds, can also be used in the oven [35].

The successful coupling of ECR sources to cyclotrons, a synchrotron, and soon

4 Illustration of a Typical ECR Source: The LBL ECR

TABLE 10.5 Currents for the LBL ECR: Sulfur through Bismuth[a]

Q	^{32}S	^{39}K	^{40}Ar	^{40}Ca	^{48}Ti	^{84}Kr	^{127}I	^{129}Xe	^{209}Bi
3$^+$	10	4	38	23					
4$^+$	*	4.5	82	24					
5$^+$	20	5	*	*					
6$^+$	*	8.5	60	37		9			
7$^+$	63	11	66	38	2.4	12			
8$^+$	*	18	106	36	*	22			
9$^+$	36	37	72	31	12	25		4.1	
10$^+$	*	22	*	*	10	22	4.2	4.7	
11$^+$	5	12	18	22	8	19	4.9	5.1	
12$^+$	*	2.4	13	11	*	*	5.7	5.2	
13$^+$	0.4		5	3.2	1	21	7.5	5.2	
14$^+$	*		1.4	1.1		*	8.5	5	
15$^+$	0.001		*	*		16	11	4.3	
16$^+$			0.03	0.03		8	*	4.6	
17$^+$						7	12	4.3	
18$^+$						*	15	4.4	
19$^+$						2	15	4.8	
20$^+$						0.9	14	4.8	
21$^+$						*	*	4	2.2
22$^+$						0.1	11	3	2.6
23$^+$							10	3	3.1
24$^+$							8.3	2	3.7
25$^+$							5.6	2	3.6
26$^+$							2.1	1	*
27$^+$							0.83	0.3	3
28$^+$							0.2		2.5
29$^+$							0.05		1.6
30$^+$							0.009		*
31$^+$									0.56
32$^+$									0.26
33$^+$									0.1
34$^+$									0.05

[a]All currents in eμA measured at 10 kV extraction voltage. Natural isotopic abundance source feeds were used.
*Not measured because a mixture of two ions with identical charge to mass ratios were present.

to a heavy-ion linac makes it clear that continued development of ECR source technology is essential. This is a relatively young technology. ECR source performance has steadily progressed and further improvements are to be expected. These improvements may come in an incremental way as a result of refinements in extraction geometry, first-stage performance, source vacuum, or other areas. They could come in a more dramatic fashion, if higher frequency rf sources such as gyrotrons can be successfully used to drive ECR sources. Since the power density should

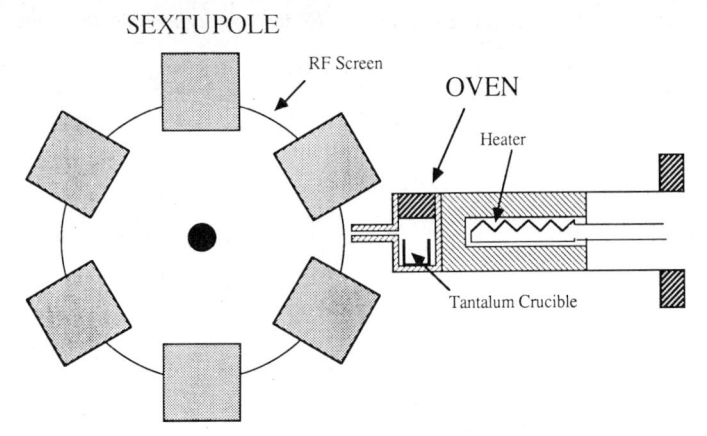

Figure 10.11 A cross-section view showing the radial position of the oven with respect to the sextupole structure. The source material is loaded into the tantalum crucible, which inhibits liquid film flow. The oven temperature is monitored and controlled using a type K thermocouple and a commerical proportional temperature controller.

increase with the plasma density, the size should decrease as the source frequency increases. Comparison of the performances of small and large sources give no clear indication that $n_e\tau_i$ scales with source size, although rf wavelength sets a minimum plasma chamber size. This is consistent with the trend to higher frequencies and smaller sources.

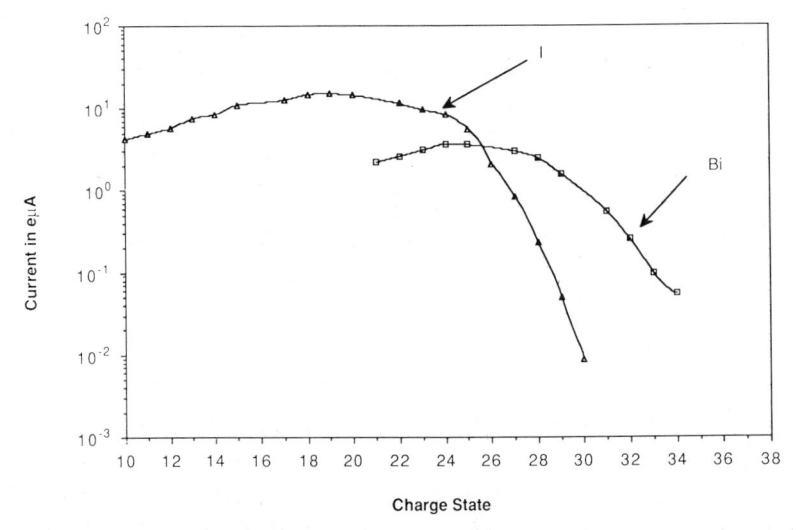

Figure 10.12 Performance of the LBL-ECR with ^{127}I and ^{209}Bi. The best iodine beams were produced from iodine that had been absorbed on the walls of the plasma chamber. The bismuth beams were produced by running the oven at 526°C.

REFERENCES

1. R. Geller, *IEEE Trans. Nucl. Sci.* **NS-23,** 904 (1976).
2. R. Geller, *Proceedings of the 11th International Conference on Cyclotrons and Their Applications*, Tokyo, October 1986, Ionics Publishing Co., p. 699.
3. R. Pardo, E. Minehara, F. Lynch, P. Billquist, W. Evans, B. E. Clift, and M. Waterson, Contributed Papers of the 7th International Workshop on ECR Ion Sources, Jülich, May 1986, KFA-Jülich, p. 223.
4. H. Postma, *Phys. Lett.* **31A,** 196 (1970).
5. S. Bliman, R. Geller, W. Hess, and B. Jacquot, *IEEE Trans. Nucl. Sci.* **NS-19,** 200 (1972).
6. K. Bernhardi and K. Wiesemann, *Plasma Phys.* **14,** 1073 (1972).
7. N. Sakudo, K. Tokiguchi, H. Koike, and I. Kanamata, *Rev. Sci. Instrum.* **49,** 940 (1978).
8. V. Dugar-Zhabov, K. Golovanevski, and S. Safonov, *Nucl. Instrum. Methods* **219,** 263 (1984).
9. R. Bardet, P. Briand, L. Dupas, C. Gormezano, and G. Melin, Proceedings of the European Conference on Controlled Fusion, Moscow, 1973, p. 247.
10. P. Briand, R. Geller, B. Jacquot, and C. Jacquot, *Nucl. Instrum. Methods* **131,** 407 (1975).
11. H. Tamagawa, L. Alexeff, C. M. Jones, and P. D. Miller, *IEEE Trans. Nucl. Sci.* **NS-23,** 994 (1976).
12. Y. Jongen, C. Pirart, G. Ryckewaert, and J. Steyaert, *IEEE Trans. Nucl. Sci.* **NS-26,** 3677 (1979).
13. V. Bechtold, H. P. Ehret, L. Friedrich, J. Möllenbeck, and H. Schweickert, *IEEE Trans. Nucl. Sci.* **NS-26,** 3680 (1979).
14. H. Beuscher, H.-G. Mathews, C. Mayer-Böricke, and J. Reich, Proceedings of the 9th International Conference on Cyclotrons and their Applications, Caen, September 1981, les Éditions de Physique, p. 285.
15. R. Geller, Proceedings of the 8th International Conference on Cyclotrons and Their Applications, Bloomington, Indiana, September 1978, IEEE, p. 2120.
16. F. Bourd, R. Geller, B. Jacquot, and M. Pontonnier, Proceedings of the 4th International Workshop on ECR Ion Sources and Related Topics, Grenoble, January 1982, Centre dÉtudes Nucléaires–Grenoble Press, p. 5.1.
17. Bechtold, L. Friedrich, and H. Schweickert, Proceedings of the 9th International Conference on Cyclotrons and Their Applications, Caen, September 1981, les Éditions de Physique, p. 249.
18. Y. Jongen, C. Pirart, and G. Ryckewaert, Proceedings of the 4th International Workshop on ECR Ion Sources and Related Topics, Grenoble, January 1982, Centre dÉtudes Nucléaires–Grenoble Press, p. 3.1.
19. D. J. Clark, J. G. Kalnins, and C. M. Lyneis, *IEEE Trans. Nucl. Sci.* **NS-30,** 2719 (1983).
20. F. W. Meyer, Proceedings of the 6th International Workshop on ECR Sources, Berkeley, January 1985, Lawrence Berkeley Laboratory, p. 37.
21. T. A. Antaya, H. G. Blosser, L. H. Harwood, and F. Marti, Proceedings of the 6th International Workshop on ECR Sources, Berkeley, January 1985, Lawrence Berkeley Laboratory, p. 107.
22. F. Bourg, P. Briand, J. Debernardi, R. Geller, B. Jacquot, P. Ludwig, M. Pontonnier, and P. Sortais, Contributed Papers of the 7th International Workshop on ECR Ion Sources, Jülich, May 1986, KFA-Jülich, p. 187.
23. A. Muller, E. Salzborn, R. Frodl, R. Becker, H. Klein, and H. Winter, *J. Phys. B* **13,** 1877 (1980).
24. W. Lotz, *Z. Phys.* **216,** 241 (1968).
25. D. H. Crandall, *Phys. Scr.* **23,** 153 (1981).
26. O. Eldridge, *Phys. Fluids* **15,** 676 (1972).
27. A. Müller and E. Salzborn, *Phys. Lett.* **62A,** 391 (1977).

28. C. M. Lyneis, Lecture Notes of the 1986 RCNP Kikuchi Summer School on Accelerator Technology, Osaka, October 1986, RCNP, p. 125; also LBL-22450.
29. Y. Jongen, Proceedings of the 10th International Conference on Cyclotrons and Their Applications, East Lansing, Michigan, April 1984, IEEE Catalog No. 84 CH 1996-3, p. 322.
30. See, e.g., J. D. Cobine, *Gaseous Conductors*, Dover, New York, 1941.
31. C. Lyneis, *Nucl. Instrum. Methods Phys. Res. B* **10/11,** 775 (1985).
32. W. Glaser, *Z. Phys.* **117,** 285 (1941).
33. F. Bourg, R. Geller, and B. Jacquot, *Nucl. Instrum. Methods A* **254,** 13 (1987).
34. F. W. Meyer and J. Hale, Proceedings of the 1987 Particle Accelerator Conference, Washington DC, March 1987, IEEE Catalog No. 87 CH 2387-9, p. 319.
35. C. M. Lyneis, Proceedings of the 11th International Conference on Cyclotrons and Their Applications, Tokyo, October 1986, Ionics Publishing Co. p. 707; also LBL-21860.
36. Y. Jongen and A. Chevalier, Contributed Papers of the 7th International Workshop on ECR Ion Sources, Jülich, May 1986, KFA-Jülich, p. 124.
37. V. Bechtold, N. Chan-tung, S. Dousson, R. Geller, B. Jacquot, and Y. Jongen, *Nucl. Instrum. Methods* **178,** 305 (1980).

11

Microwave Ion Sources

N. Sakudo
Hitachi, Ltd.
Tokyo, Japan

Ion sources using plasmas generated by a microwave discharge in a magnetic field have many advantageous features. Since they operate even with reactive source materials, they can provide long-life stable ion beams for a variety of ion species. This makes it possible to avoid frequent replacement of ion source parts, which interrupts machine operation.

Microwave ion sources can be classified into two types, according to operational conditions or purpose of application. One type is operated at the electron cyclotron resonance (ECR) to obtain multiply charged ions as described in a previous chapter, and the other uses off-resonance microwave plasma to obtain high-current singly charged ions, as required in industrial applications such as ion implantation. In general, the extractable ion current density is proportional to the product of electron density and the square root of electron temperature. Therefore, to achieve higher current ion beams, either or both of these parameters must be raised by increasing the absorbed microwave power.

1 PLASMA GENERATION BY MICROWAVE DISCHARGE IN A MAGNETIC FIELD

In an ECR ion source the magnetic field intensity is adjusted so that there is an electron cyclotron resonance zone within the plasma. The ion source is operated at relatively low pressure (10^{-5}–10^{-4} Torr). As a result, the collision frequency of electrons with neutrals and ions is very low compared to the microwave frequency. This allows continuous acceleration of electrons by the microwave electric field, resulting in high electron temperatures. Neutrals and ions are raised to higher charge states through multiple ionization by single impact of energetic electrons and through step-by-step ionization by multiple impacts of moderate-energy electrons.

However, for the purpose of obtaining a higher current of singly charged ions, higher electron density of the plasma is needed. Musil [1–3], Kopeckey [4], and Nanobashvili [5] have reported that higher electron density can be obtained by microwave discharges at higher magnetic fields and at somewhat higher pressures. In this case, electron cyclotron resonance is not a predominant condition for plasma generation at all.

According to the plasma wave theory of Stix [6], only the right-hand circularly polarized (RHCP) wave can propagate in a high-density plasma where (1) the electron plasma frequency is well above the wave frequency, $\omega_{pe} \gg \omega_{\mu w}$ and (2) the magnetic field strength is above the electron cyclotron resonance value and below the ion cyclotron resonance field, $B_{ce} < B < B_{ci}$ (or $\omega_{ce} > \omega_{\mu w} > \omega_{ci}$). When the RHCP wave propagates obliquely to the magnetic field in an inhomogeneous plasma, the energy in the microwaves is effectively absorbed as shown by Musil [2]. This means that the *critical density*, for which the electron plasma frequency equals the microwave frequency, 7.46×10^{10} cm^{-3} for 2.45 GHz, is not the upper limit of electron density realized by off-resonance microwave plasma at all, as it is for the ECR plasmas of Okamoto [7] and Geller [8], [9].

Microwave ion sources are distinguished from so-called rf ion sources, as shown by Cook [10] and Szuszczewicz [11], by the fact that their frequencies (one to several tens of GHz) are higher than those of rf ion sources (0.001–0.1 GHz), and also by the fact that the wavelengths are comparable to or smaller than the dimensions of the microwave ion sources. Brown [12] reported that ion temperatures in a cold microwave-produced plasma are 0.2 to 2 eV and decrease with increasing pressure. Microwave ion sources (Sakudo [13]) provide ion beams of higher current and smaller energy dispersion than rf ion sources, as shown by Tokiguchi [14], since ions in plasma are not accelerated by a microwave electric field as they may be by the rf electric field. The small dispersion of ion energy permits high mass resolution when combining a microwave ion source with an electromagnetic mass separator.

2 SOME PRACTICAL CONSIDERATIONS

One of the major problems in a practical ion source is how to couple the microwave power through the plasma boundary into the plasma, even though microwaves that have already entered the plasma are absorbed as described above. From a microwave circuitry point of view, the ion source containing plasma is considered a circuit element. However, the impedance of this element depends on the plasma characteristics, which themselves depend on the absorbed microwave power and the superimposed magnetic field. Therefore, if the incident microwave power and the magnetic field are fixed, the plasma parameters and microwave absorption that result from the degree of microwave impedance matching are also fixed by a negative feedback mechanism. Otherwise the plasma could not be maintained.

2 Some Practical Considerations

In the discharge chamber, the absorbed microwave power heats the electrons, in turn resulting in excitation and ionization of particles. The absorbed energy is then expended in heating the discharge chamber wall by charged particle bombardment and photon emission from excited particles.

One method of realizing a microwave absorption element is to let the resistivity vary gradually along the length of the element. This is accomplished in a practical ion source by using a magnetic mirror field. The field also helps to protect the ceramic used for vacuum sealing from plasma bombardment heating.

An off-resonance microwave ion source described by Sakudo [13] is represented schematically in Figure 11.1. The discharge chamber basically consists of a coaxial waveguide. Microwaves (2.45 GHz) are introduced into the discharge chamber via a coaxial line transition from the rectangular waveguide. The antenna is connected to the inner conductor of the coaxial transmission line, which uses a ceramic window for the vacuum feedthrough. A magnetic field with a mirror ratio of about 2 is superimposed on the entire discharge chamber, using three magnetic coils. The field intensity is higher than that needed for ECR (875 G) throughout the entire discharge region. Due to the magnetic field gradient, plasma bombardment of the ceramic is minimized.

Since the antenna is always in contact with the plasma, it is heated by charged particle bombardment. To avoid destruction of the ceramic, a water-cooled structure is employed. Usually, the coaxial waveguide is connected at right angles at a point $\lambda_g/4$ distant from the end of the rectangular waveguide, where λ_g is the wavelength in the rectangular waveguide. To cool the antenna, the inner conductor is made of a double tube with one end penetrating the end plate of the rectangular

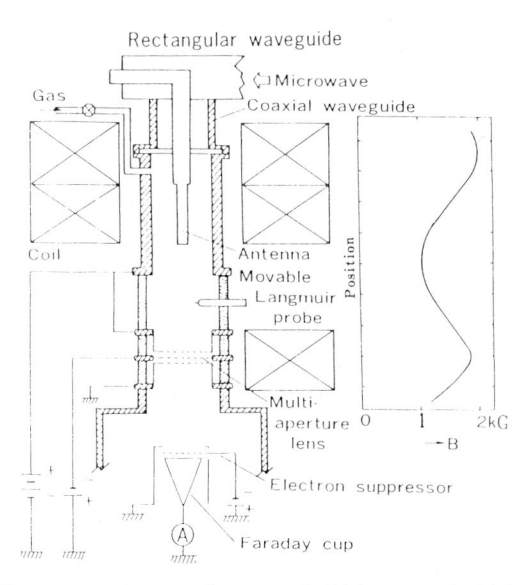

Figure 11.1 An off-resonance microwave ion source for high-current singly charged ions. The inner diameter of the discharge chamber is 60 mm. From Sakudo [13].

waveguide. The characteristic impedance of the coaxial waveguide is about 50 Ω. Therefore, the impedance of the plasma-filled discharge chamber should approximate this value for accurate impedance matching. Since the plasma is considered to be a resistive dielectric, the ratio of antenna diameter to chamber diameter should be smaller than that of the coaxial waveguide. The practical dimensions were determined empirically.

The impedance of a plasma-filled ion source depends on the electron density and electron temperature, as well as on the magnetic field intensity and its distribution. In addition, electron density and electron temperature change in a highly complicated manner, depending on the absorbed microwave power, gas pressure, and magnetic field. Therefore, it is practically impossible to examine the relation between any two of these parameters simply by holding the others constant.

In practice, when the geometric structure of the ion source, gas species, and magnetic field are determined, the plasma parameters are fixed depending on the absorbed microwave power. This in turn fixes the extent of impedance matching between the microwave circuit and ion source, thus determining the absorbed microwave power.

The relation between microwave absorption and magnetic field intensity is shown in Figure 11.2. In this case, the magnetic field intensity is varied while the field shape is maintained. Argon gas pressure is 8×10^{-3} Torr and incident microwave power is 1 kW. Magnetic field intensity is normalized to that shown in the inset in Figure 11.1, which was determined empirically so as to obtain ion beams with high currents. Microwave absorption is expressed as a percentage of incident microwave power. This absorption increases gradually with magnetic field intensity; its value exceeds 90% when the magnetic field intensity is above 95%. Below the 85% magnetic field, there should be one or two points that satisfy the ECR condition. However, no resonant property can be seen in Figure 11.2. This means that, in this case, the microwave power is mainly absorbed by off-resonance processes. This also suggests that even in a so-called ECR ion source, microwave absorption may be partially due to off-resonance effects, especially when the source is operated at a relatively high gas pressure.

When microwave absorption is plotted as a function of argon gas pressure, the change in absorption is gradual over the region from 5×10^{-4} to 8×10^{-3} Torr. Over the entire pressure region the absorption is above 75%. However, by employing an EH tuner between the ion source and the directional coupler, more than 90% of the microwave power can be absorbed.

Plasma parameters are measured with a movable Langmuir probe as shown in Figure 11.1. The magnetic field intensity at the position where the probe is placed is about 1300 G. Since the probe axis is perpendicular to the magnetic field, the effect of the magnetic field on probe measurement is negligible at the expected plasma parameters ($n_e > 10^{10}$ cm^{-3}, $T_e > 1$ eV) as shown by Dote [15]. The electron temperature of argon plasma and hydrogen plasma gradually increases from 3.7 to 6.8 eV and from 4.5 to 7.3 eV, respectively, as the absorbed microwave power varies from 200 to 1000 W. The electron densities are shown in Figure

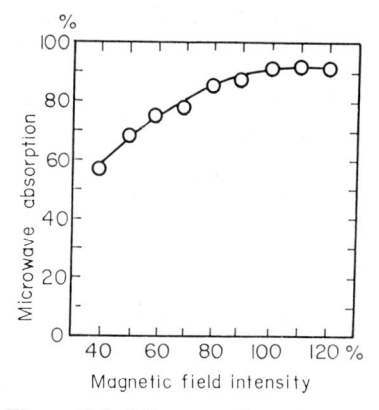

Figure 11.2 Microwave absorption versus magnetic field intensity. Argon gas pressure is 8×10^{-3} Torr and incident microwave power is 1 kW. From Sakudo [13].

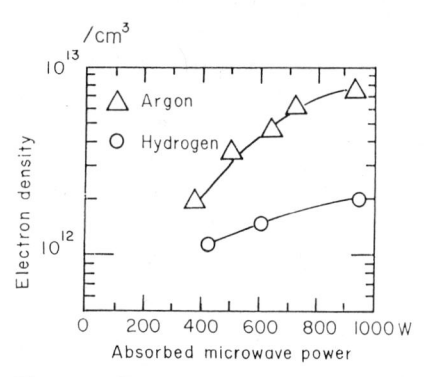

Figure 11.3 Electron densities versus absorbed microwave power. Argon pressure is 3×10^{-3} Torr and hydrogen pressure is 1×10^{-3} Torr. From Sakudo [13].

11.3. These are 10–100 times higher than the critical density, which is often the uppermost value obtained by ECR.

Any kind of waveguide can be used in place of the coaxial waveguide in the structure of the discharge chamber. A round waveguide, as in the microwave plasma etching apparatus described by Suzuki [16], is suitable for a large-volume discharge chamber.

3 VERSATILITY OF BEAM EXTRACTION

Microwave plasmas can be generated over a wide range of volume and in various forms. This permits a variety of beam extraction geometries.

3.1 Large Cross-Sectional Beam Extracted through a Multiaperture Lens

The current density extracted from the plasma J_i depends upon the plasma parameters as follows:

$$J_i = en_i(kT_e/M)^{1/2} \exp\left(-\tfrac{1}{2}\right) \qquad (1)$$

where k is Boltzmann's constant, e is electron charge, n_i is ion density, T_e is electron temperature, and M is ion mass. A high degree of microwave energy absorption must be realized since high electron density and/or high electron temperature are necessary for a high-current ion source. A small portion of the generated ions diffuses toward the extraction slit and is extracted as a beam. The power dissipated as heat in the discharge chamber walls is considerably greater than that used for ionization of the extracted ions.

A three-element multiaperture lens of 50 mm diameter is placed at the bottom of the discharge chamber as shown in Figure 11.1. The potentials of the upper element and discharge chamber are at the same level as the ion acceleration voltage. The middle element is supplied with negative voltage to prevent low-energy electrons in the beam from entering the discharge chamber. Argon and hydrogen ion currents are measured with a cone-shaped Faraday cup having an effective area of 40 mm diameter and equipped with a secondary electron suppressor.

The extracted argon ion currents are shown in Figure 11.4. Current increases with ion acceleration voltage. Maximum argon ion current is 200 mA at 5 kV, and that of hydrogen is around 400 mA. The critical voltage at which it is possible to extract argon ions as a beam is around 2.0 kV, as shown in the figure, and that for hydrogen ions around 800 V. Ion sheath widths have been obtained from measurements made by the probe. In argon and hydrogen plasmas, these widths are 0.38 and 0.62 mm, respectively, when the absorbed microwave power is 950 W and the probe potential is -70 V. These values also suggest a difference between the critical voltages for argon and hydrogen plasmas.

Oxygen ions, which are rather difficult to obtain with conventional hot-filament-type ion sources, can be stably produced with this ion source, as shown in Figure 11.5 by Tokiguchi [17]. At an ion extraction voltage of 5 kV, the oxygen ion current is 110 mA with a microwave power of 600 W. If the aperture radius of the lens and the electrode distance between the upper and middle elements are increased, ion extraction at the same current level can be optimized at a higher voltage, that is, several tens of kilovolts, as shown by Shimada [18].

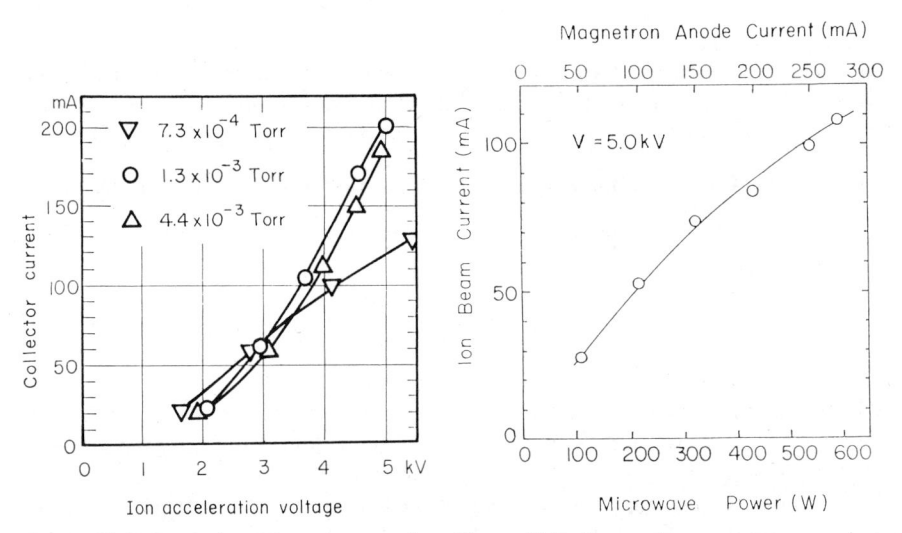

Figure 11.4 Argon ion currents versus ion acceleration voltage. From Sakudo [13].

Figure 11.5 Oxygen ion current versus microwave power. From Tokiguchi [17].

3.2 Slit-shaped Ion Beam for Ion Implanters

Sakudo [19] constructed a high-current ion source using an off-resonance microwave plasma, which provides a slit-shaped ion beam suitable for the electromagnetic mass separator of an ion implanter. This kind of separator requires a narrow ion beam to obtain high mass resolution. This is because the mass resolution Re in a symmetric-sector mass separator is (Jayaram [20])

$$Re = r/2S \tag{2}$$

where S is the beam width at the ion exit, r is the beam trajectory radius in the magnetic field, and ion energy spread is neglected. Therefore, ion beams for an electromagnetic mass separator should be extracted through a narrow slit.

The details of a microwave ion source for use as an implanter are shown in Figure 11.6 [21]. Microwaves are introduced into the discharge chamber through a vacuum-sealing dielectric plate placed between the rectangular waveguide and the tapered, ridged waveguide. Two iron blocks are placed just on the outer periphery of the latter waveguide. The cross section of the discharge region has a rectangular shape similar to the exit slit to effectively utilize the generated ions. The remaining space for the discharge region is filled with dielectric (boron nitride). Since the electric field between the ridges, that is, the discharge electrodes, is fairly uniform, this structure helps to realize a uniform plasma.

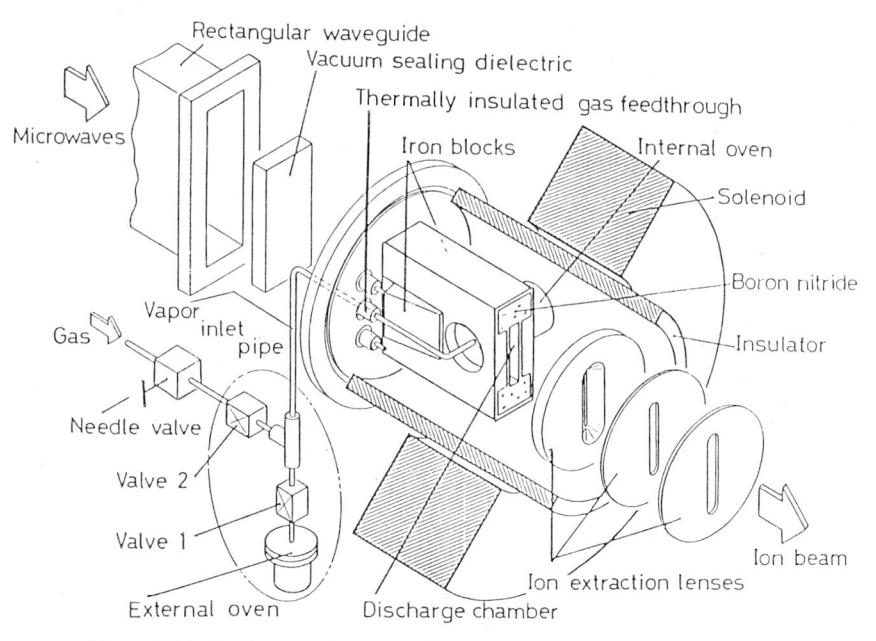

Figure 11.6 A microwave ion source for an ion implanter. From Sakudo [21].

An ion implanter equipped with this microwave ion source was constructed. The microwave ion source with its slit-shaped beam is combined with a 40-cm, 90° magnetic mass separator and a rotating disk target chamber, as shown in Figure 11.7 [22]. The ion energy can be varied from 20 to 80 keV. A high-current ion beam of 40 mA is extracted through the 2 × 40-mm^2 slit. Since the mass separator design is based on double-focusing optics with oblique entry and exit of ion beams, a high transmission level of 60% is attained. This permits 10 mA P$^+$ implantation (maximum 15 mA), and 4 mA B$^+$ implantation (maximum 6 mA). The implant current is continuously measured with a Faraday cup, which is composed of a rotating disk, an electron shield, and an electron suppressor, as shown in the inset in Figure 11.7.

The dependence of the mass peak ratio on the microwave power with PH$_3$ gas introduced into the ion source is shown in Figure 11.8. The P$^+$, H$^+$, and P^{2+} intensities increase, and those of H$_2^+$, P$_2^+$, and PH$^+$ decrease as the microwave power is increased. In the plasma, each PH$_3$ molecule is bombarded many times by electrons. The larger the number of collisions, the more the molecule decomposes. As microwave power increases, both the electron density and the electron temperature of the plasma increase. Usually, the plasma density of a microwave ion source is high, and so also is the percentage of atomic ions.

The lifetime of this ion source is 1–2 ampere-hours (implant current × duration) for 10 mA implantation of P$^+$ and As$^+$. This is several times longer than that of most conventional hot cathode type ion sources. The PH$_3$ gas consumption for these high-current operating conditions is 0.1–0.5 atm · cm^3/min, which is several times lower than that of conventional hot cathode-type ion sources. This long lifetime and low gas consumption rate leads to reduced maintenance frequency for

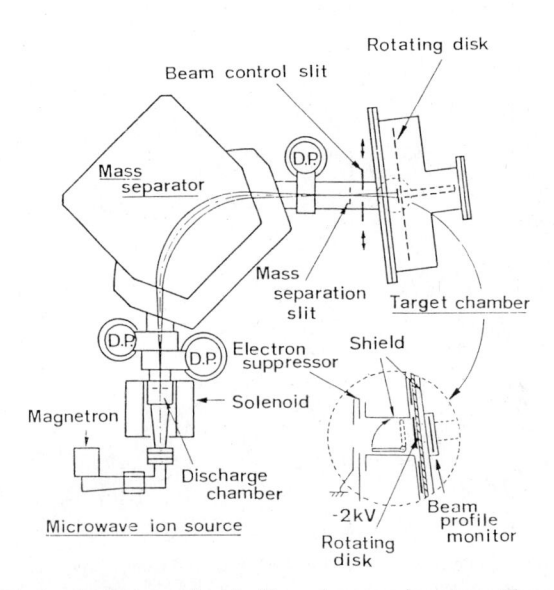

Figure 11.7 An ion implanter equipped with a microwave ion source. From Sakudo [22].

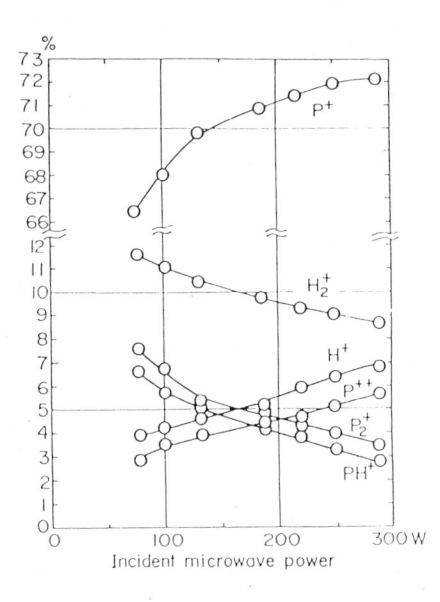

Figure 11.8 Dependence of mass peak heights for PH$_3$ on incident microwave power.

the ion source and vacuum system, and to an increased operating rate for the implanter.

3.3 Further Improvements in Slit-shaped Ion Beams

3.3.1 Increasing Ion Extraction Voltages

When a source is operated at a higher voltage simply by scaling up the standard microwave ion source with a slit-shaped beam (e.g., the source shown in Figure 11.6), a number of problems occur:

1. It is difficult to achieve sufficient voltage isolation between the high-potential discharge chamber and the magnetic field coil at ground potential, because their relative distance cannot be altered while the magnetic field intensity and distribution are maintained throughout the discharge region.
2. Sparks between electrodes in the vacuum are liable to occur in the ion acceleration space, since a magnetic field is superimposed on the accelerating electric field.
3. It is difficult to use a multistage extraction lens because of the magnetic coil at ground potential.

To avoid these problems, the ion source has been further modified [23]. The magnetic coil is positioned surrounding the discharge chamber at the high-voltage terminal, where it is also surrounded by a thick wall and an acceleration electrode, which are made of iron or high-permeability metal as shown in Figure 11.9. Thus,

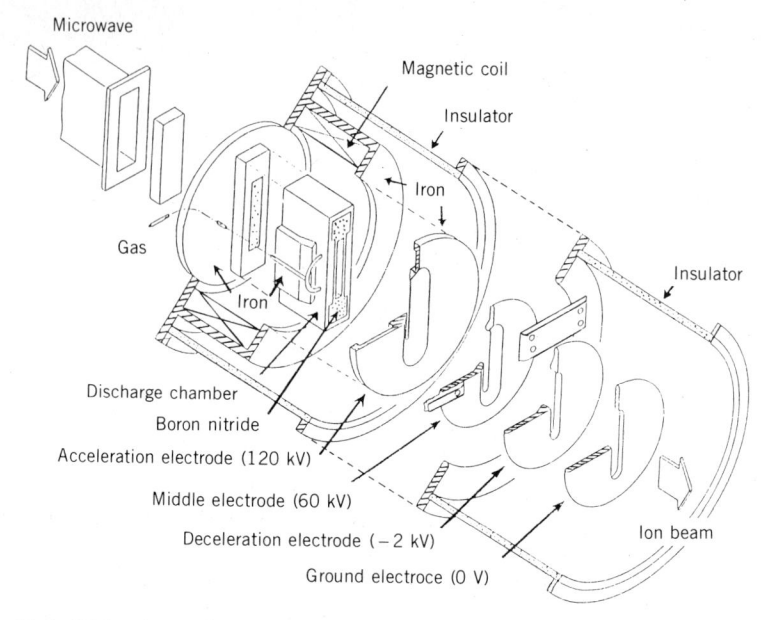

Figure 11.9 High-voltage microwave ion source with a closed magnetic field and a double-stage extraction lens. From Sakudo [23].

a magnetic circuit is formed to focus the magnetic field in the discharge chamber. Consequently, no magnetic field exists in the ion accelerating space in the new ion source. A double-stage extraction lens is also adopted. This ion source provides stable, mass-separated ion beams at a current of up to 15 mA and with extraction voltage of from 20 to 120 kV. A further increase in ion acceleration voltage can be realized by increasing the number of lens stages.

3.3.2 Increasing Extracted Currents

In general, to increase the current from an ion source in which the plasma density has reached a maximum, the ion-emitting area must be enlarged. It is very easy to lengthen the ion exit slit in the microwave ion source, although this may be somewhat difficult in a conventional (Freeman-type) ion source because the filament must also be lengthened. Extracted currents from a microwave ion source can easily be doubled by lengthening the slit without changing the ion optics in the narrower direction of the slit. If the extraction lens is curved along the longer direction to converge the ion beam into a magnet–pole gap of a conventional mass separator, mass-separated currents are also doubled without degrading the mass resolution, as shown in Figure 11.10. The ion exit slit is 80 mm high, although the effective height of the magnet–pole gap is 45 mm. A mass-separated P^+ ion of 30 mA has been obtained at the collector [23].

3 Versatility of Beam Extraction 239

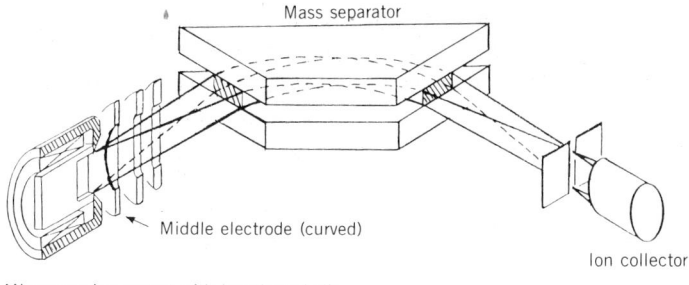

Middle electrode (curved)

Ion collector

Microwave ion source with lengthened slit

Figure 11.10 Mass-separated currents are doubled without deteriorating mass resolution, by combining a microwave ion source of lengthened slit with a conventional mass separator. From Sakudo [23].

3.4 Compact Microwave Ion Sources

Ishikawa [24] constructed a compact microwave ion source that extracts an ion beam of several milliamperes through a hole 2 mm in diameter as shown in Figure 11.11. To make the source small, a permanent magnet was used to replace the solenoids. The source size is 50 mm in diameter and 65 mm in height. The microwave frequency is 2.45 GHz and the power consumption is only 7 to 30 W. The ion beam has low emittance of 10^{-8} m · rad and high brightness of 10^{11} A · m^{-2} · rad^{-2}.

A microwave cavity plasma disk ion source was developed by Asmussen [25]. It uses a cylindrical microwave cavity operating in a hybrid mode associated with

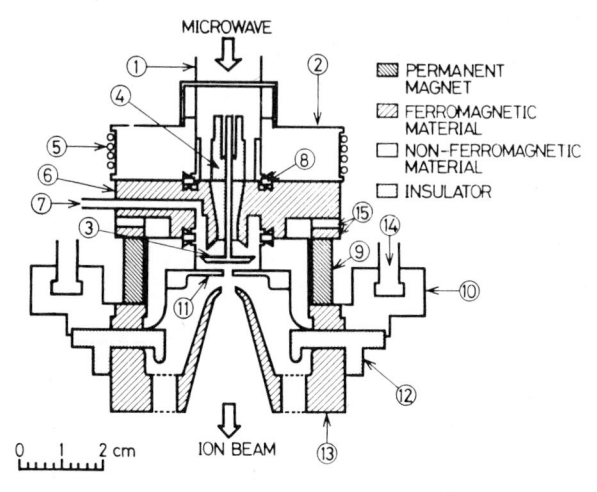

Figure 11.11 A compact microwave ion source using a permanent magnet: 1, sealed coaxial connector; 2, top flange; 3, antenna; 4, boron nitride; 5, sheath heater; 6, middle flange; 7, gas inlet; 8, copper gaskets; 9, permanent magnet; 10, base flange; 11, plate with an extraction aperture; 12, insulator; 13, extraction electrode; 14, coolant; and 15, spacers. From Ishikawa [24].

the TE$_{211}$ empty cavity mode. Without the aid of a static magnetic field, microwave fields produce a plasma in a disk-shaped quartz tube placed in the cavity. Ion beams are extracted with a multiaperture lens adjacent to the plasma. Since discharge losses per extracted ion are roughly proportional to the total surface area divided by the extraction area, the dish shape, where the height is much less than the radius, is an efficient discharge configuration. The performance of the ion source was improved by redesigning to surround the discharge zone with many closely spaced rare earth magnets producing a multicusp static magnetic field (Dahimene [26]).

Leung [27] constructed a small microwave ion source from a quartz tube with one end enclosed by a two-grid extraction lens. The source is also enclosed by a microwave cavity. An ion beam is extracted through a 0.8-mm-diameter extraction aperture. The current density exceeds 200 mA · cm^{-2} when the microwave power is 400 W.

4 DIVERSITY OF AVAILABLE ION SPECIES

General ion species for implantation into silicon semiconductors have traditionally been P$^+$, As$^+$, and B$^+$ ions. However, new silicon devices using silicon-on-insulator (SOI) technology require oxygen ion implantation at very high doses. Microwave ion sources can stably use even reactive materials as the source materials. Some of these sources, with slit-shaped beams, can provide a mass-separated O$^+$ ion beam of more than 10 mA for such implantation. In this case, the ratio of O$^+$ ions to the total ions in the mass spectrum is about 0.85 as shown in Figure 11.12. This high atomic fraction is due to the high electron density and/or the high electron temperature of the microwave plasma. A research group with

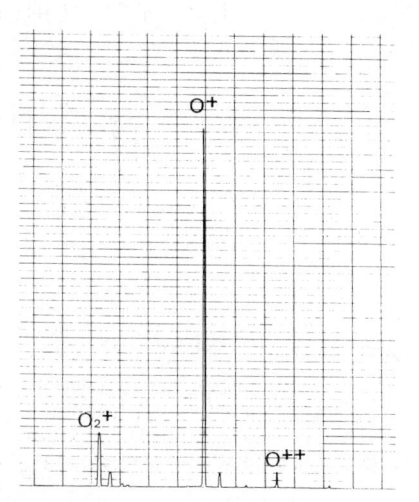

Figure 11.12 Mass spectrum for oxygen ions generated by microwave discharge in a magnetic field.

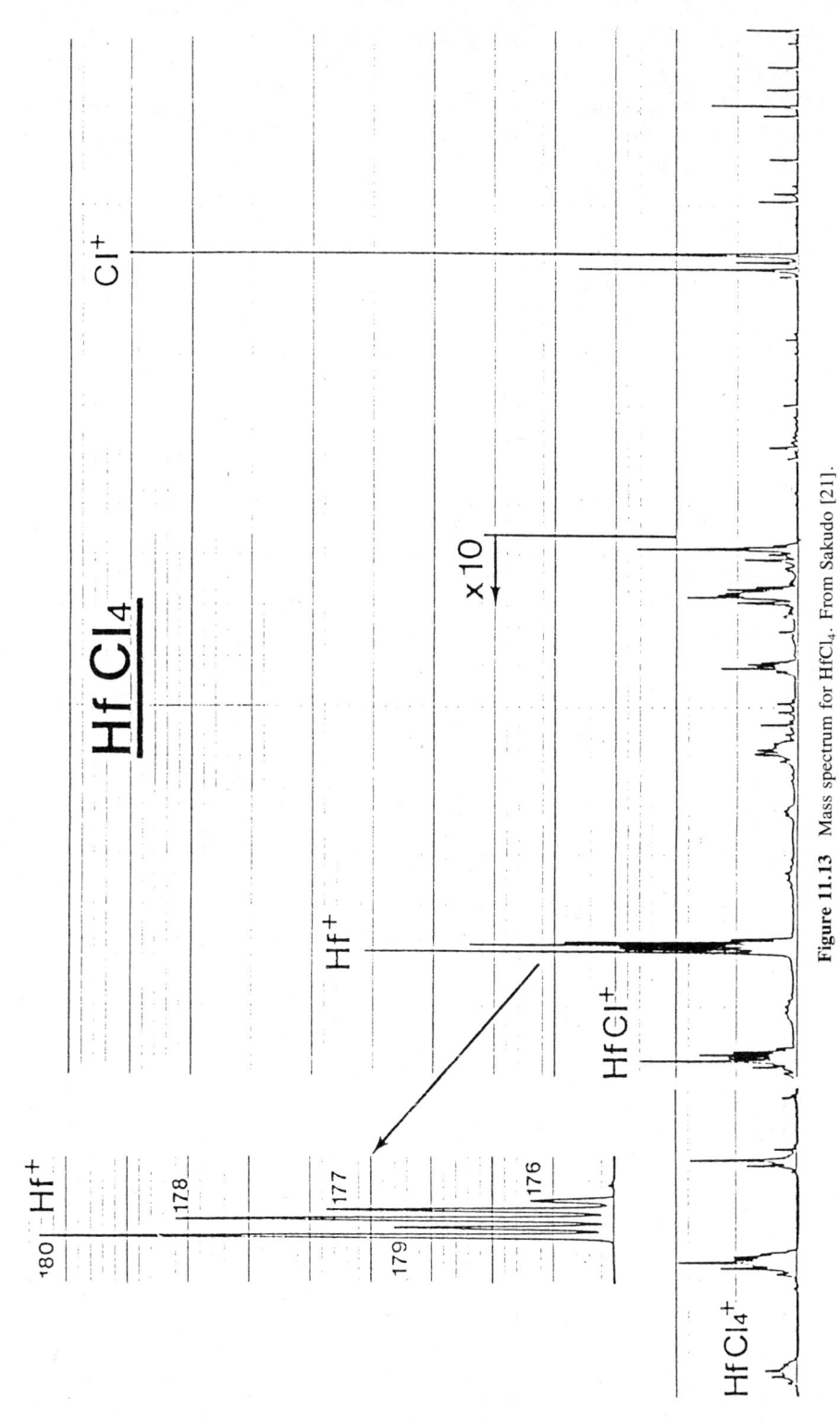

Figure 11.13 Mass spectrum for HfCl$_4$. From Sakudo [21].

NTT Japan has fabricated a 4-kb CMOS/SIMOX SRAM by O^+ ion implantation with a commercial implanter (Hitachi IP-815) as reported by Omura [28].

For implantation into compound semiconductors and for materials modification in metals and insulators, various other kinds of metal ions are required. Source materials for most metal ions are readily obtained in the form of halides, the evaporation temperatures of which are usually lower than those of the original metals. Halides sometimes cause problems in traditional hot-filament-type ion sources due to rapid erosion of the filament by the highly reactive halide plasma. However, the microwave ion source does not have any quickly consumed part. Furthermore, by bombarding and chemically cleaning the walls with halogen ions and radicals in the plasma, it is possible to prevent metal from being deposited on the inner walls of the discharge chamber; metallic deposition on the walls might prevent microwaves from entering.

To introduce liquid or solid materials into the chamber, the microwave ion source must have two types of ovens, an external one and an internal one, as shown in Figure 11.6, in addition to the usual gas inlet system. Low-vapor-pressure materials can be introduced with either oven depending on the evaporation temperature.

The mass spectrum for $HfCl_4$ extracted at 30 kV is shown in Figure 11.13. Hf^+ isotope ions with a 2-mA peak are obtained. The source feed material, $HfCl_4$, is put into the external oven and heated to 170 °C. Using this ion source, it is possible to obtain many other metal beams, such as Al^+, Ga^+, Ti^+, and Sc^+, at currents of several milliamperes.

An ion implanter designed by Iwaki [29] for materials modification in metals and insulators is shown in Figure 11.14. Ions are extracted from a microwave ion

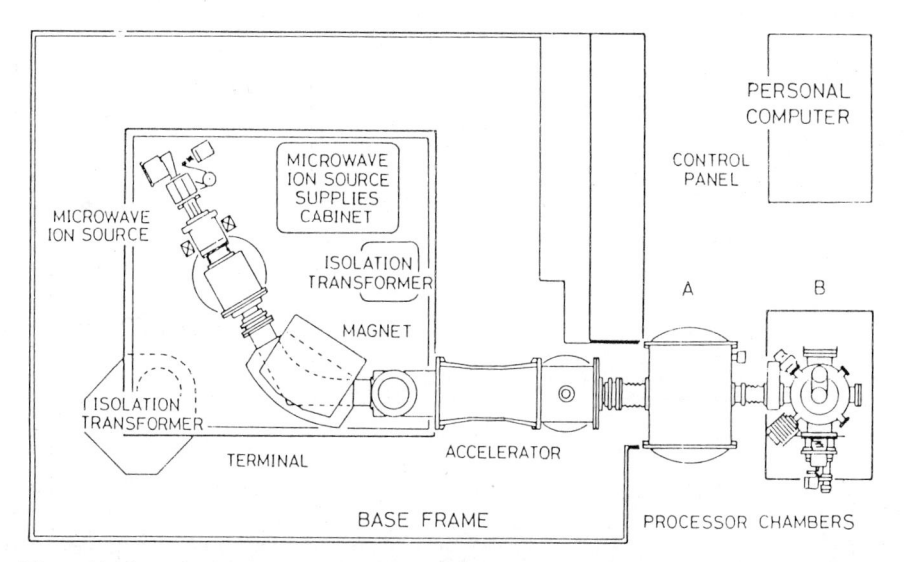

Figure 11.14 An ion implanter for materials modification in metals and insulators. From Iwaki [26].

source at 30 kV and then postaccelerated. The maximum acceleration voltage is 200 kV. To observe the modified surface composition, the target chamber has a secondary mass analyzer, an ion source for sample cleaning and for SIMS, and a current integrator. The in situ observation of surface composition is performed during ion implantation. Most metal ions are obtained with metal halides fed into the source.

5 SUMMARY

Microwave ion sources can provide long-life stable ion beams for a variety of ion species by using a wide range of source materials including reactive chemical compounds. In practice, an ion source for large cross-sectional beams of several hundred milliamperes has been developed, whose ion current can be further increased by increasing the discharge chamber volume and the area of the multi-aperture extractor. Another ion source, producing slit-shaped beams, has been constructed for application in ion implantation. An implanter equipped with this ion source can provide implant currents of 1–30 mA for most ion species including oxygen and metal ions. To increase ion acceleration voltages, this ion source can be easily combined with multistage extraction lenses. To increase the ion current, the ion exit slit can be easily lengthened. With versatility of beam extraction, many other kinds of modification can be made to suit particular industrial applications.

REFERENCES

1. J. Musil and F. Zacek, *Czeck. J. Phys.* B **22**, 133 (1972).
2. J. Musil and F. Zacek, *Czeck. J. Phys.* B **23**, 736 (1973).
3. J. Musil, F. Zacek, and P. Schmiedbergen, *Plasma Phys.* **16**, 971 (1974).
4. V. Kopecky, J. Musil, and F. Zacek, *Plasma Phys.* **17**, 1147 (1975).
5. S. Nanobashvili, G. Rostomashvili, and N. Tsintsadze, *Sov. Phys. Tech. Phys.* **20**, 280 (1975).
6. T. H. Stix, *The Theory of Plasma Waves*, McGraw-Hill, New York, 1962, p. 32.
7. Y. Okamoto and H. Tamagawa, *Rev. Sci. Instrum.* **43**, 1193 (1972).
8. R. Geller, *IEEE Trans. Nucl. Sci.* **NS-23**, 904 (1976).
9. R. Geller, B. Jacquot, and C. Jacquot, Proceedings, Int. Ion Engineering Congress, Kyoto, 1983, p. 187.
10. C. Cook, O. Heinz, D. Lorenz, and J. Peterson, *Rev. Sci. Instrum.* **33**, 649 (1962).
11. E. Szuszczewicz, *Phys. Fluids* **15**, 2240 (1972).
12. I. G. Brown, *Plasma Phys.* **18**, 205 (1976).
13. N. Sakudo, K. Tokiguchi, H. Koike, and I. Kanomata, *Rev. Sci. Instrum.* **48**, 762 (1977).
14. K. Tokiguchi, N. Sakudo, and H. Koike, *J. Vac. Sci. Technol.* A **2**, 29 (1984).
15. T. Dote, H. Amemiya, and T. Ichimiya, *Jpn. J. Appl. Phys.* **3**, 789 (1964).
16. K. Suzuki, S. Okudaira, N. Sakudo, and I. Kanomata, *Jpn. J. Appl. Phys.* **16**, 1979 (1977).
17. T. Tokiguchi, H. Itoh, N. Sakudo, H. Koike, and T. Saitoh, *Vacuum* **36**, 11 (1986); presented at 5th Int. Conf. on Ion and Plasma Assisted Techniques, Munich, May 1985.

18. M. Shimada, I. Watanabe, and Y. Torii, Proceedings, 10th Symposium on Ion Source and Ion-Assisted Technology, Tokyo, 1986, p. 131.
19. N. Sakudo, K. Tokiguchi, H. Koike, and I. Kanomata, *Rev. Sci. Instrum.* **49,** 940 (1978).
20. R. Jayaram, *Mass Spectrometry*, Plenum, New York, 1966, p. 25.
21. N. Sakudo, K. Tokiguchi, and H. Koike, *Vacuum* **34,** 245 (1984).
22. N. Sakudo, K. Tokiguchi, and H. Koike, *Rev. Sci. Instrum.* **54,** 681 (1983).
23. N. Sakudo, *Nucl. Instrum. Methods Phys. Res. B* **21,** 168 (1987).
24. J. Ishikawa, Y. Takeiri, and T. Takagi, *Rev. Sci. Instrum.* **55,** 449 (1984).
25. J. Asmussen and J. Root, *Appl. Phys. Lett.* **44,** 396 (1984).
26. M. Dahimene and J. Asmussen, *J. Vac. Sci. Technol. B* **4,** 126 (1986).
27. K. N. Leung, S. Walther, and H. W. Owren, *IEEE Trans. Nucl. Sci.* **NS-32** (5), 1803 (1984).
28. Y. Omura, S. Nakashima, and K. Izumi, Proceedings, Symposium on VLSI Technol., Kobe, 1985, p. 24.
29. M. Iwaki, K. Yoshida, N. Sakudo, and S. Satou, *Nucl. Instrum. Methods Phys. Res. B* **6,** 51 (1985).

12

Electron Beam Ion Sources

E. D. Donets
Joint Institute for Nuclear Research
Dubna, USSR

The Electron Beam Ion Source (EBIS) is a relatively new type of ion source. It was first proposed in 1967 [1] and from the beginning of its development the main goal was the creation of conditions for the production of extremely highly charged ions, including bare nuclei of heavy elements. Although a working EBIS was demonstrated in 1968 [2], the first really effective sources appeared in the mid 1970s. The last decade has been a period of rapid growth of activity in EBIS development and the ion charge states produced. In fact, there are more than 20 EBIS projects, completed or under construction, around the world now, and the highest ion charge states produced by means of the EBIS KRION-2 at the Joint Institute for Nuclear Research are Xe^{53+} and Xe^{54+}.

Nevertheless, the field of EBIS investigation and development is far from closed. Many problems in science and technology remain to be solved for the application of the EBIS to a large number of fundamental and applied areas of research in the physics of highly charged ions.

1 PHYSICAL BASIS AND MODE OF OPERATION OF EBIS [3]

The Electron Beam Ion Source is a device for the production of multiply charged ions [1] that involves (1) producing an extended electron beam of given energy and density, (2) creating an electrostatic ion trap along the beam, (3) injecting a definite number of low charge state ions of the working substance into the trap during a definite period of time, (4) containing the ions in the electron beam for a period of time sufficient for the ions to reach the desired charge state, and (5) extracting these highly charged ions from the trap along the electron beam and preparation for the next cycle.

In this section the above operations are considered without reference to specific techniques. Thus, a model of the electron beam method of multiple ionization of atoms is presented that can be useful before beginning an experimental EBIS project.

1.1 Physical Basis

The main physical process used in the EBIS to produce highly charged ions is ionization by electron impact. Since the probability of multiple ionization in one collision is small, sequential ionization is the main process leading to high ion charge states. This means that multiple electron–ion collisions are required, with removal of only one electron from an electron shell of the ion each time. In this case, the ionization factor $j\tau_i$ (the product of the electron flux density in the beam j and the time of bombardment of the ion by electrons τ_i) is the main quantitative characteristic of the process of production of highly charged ions if the energy of the beam electrons is high enough.

The probability of transition of the ion from charge state q to charge state $q + 1$ is

$$P_{q \to q+1} = \sigma_{q \to q+1} j\tau_i \tag{1}$$

where $\sigma_{q \to q+1}$ is the effective cross section for ionization of the ion with charge state q by electron impact. Thus, on the average, all ions of charge state q transform to ions of charge state $q + 1$ when $j\tau_i = 1/\sigma_{q \to q+1}$. This means that to obtain ions of mean charge k from singly charged ions, the ionization factor

$$j\tau_i = \sum_{q=1}^{k-1} \sigma_{q \to q+1}^{-1} \tag{2}$$

is required.

The effective cross sections $\sigma_{q \to q+1}$ can be estimated using Lotz's empirical formula [4], and the $j\tau_i$ required to obtain any ion charge state thus estimated.

The calculated values of $j\tau_i$ needed to obtain Ne, Ar, Kr, Xe, and U ions of all charges are given in Figure 12.1 [5], where it has been assumed that ionization occurs at electron energy $E_e = 2I_{q-1}$, where I_{q-1} is the binding energy of the last electron in the shell of the ion of charge state $q - 1$. One can see that to produce bare nuclei, about 90% of the required $j\tau_i$ factor is spent in removing the K-shell electrons.

The production of U^{92+} nuclei is generally regarded as the most difficult ionization problem. In this case, an electron beam energy E of about 200 keV is needed, and since the cross section for ionization of U^{91+} by electron impact is about 1.1×10^{-24} cm^2, a $j\tau_i$ factor of over 10^{24} cm^{-2} is required. This means that if the beam density is about 10^4 A/cm^2, a containment time in the beam of more than

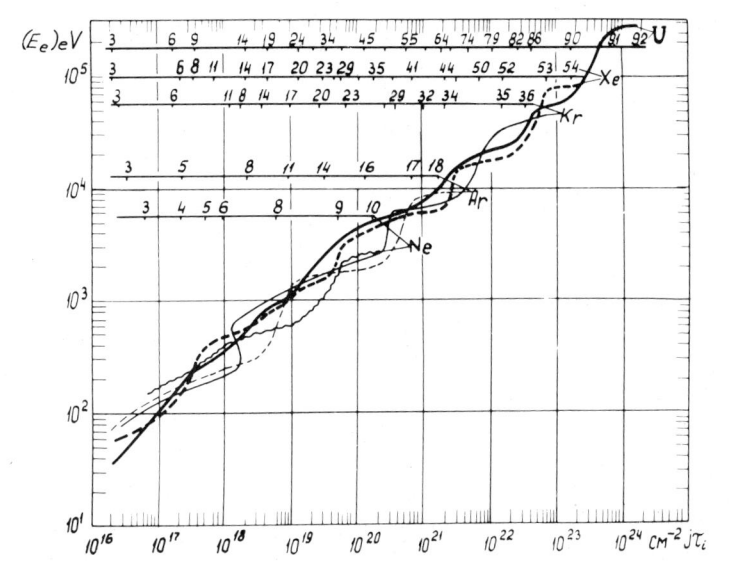

1 Physical Basis and Mode of Operation of EBIS 247

Figure 12.1 Calculated values of $j\tau_i$ needed to produce Ne, Ar, Kr, Xe, and U ions of the indicated charge states, for the corresponding energies of the bombarding electrons.

30 s is required. This example shows that if an electron beam of the necessary energy and density is produced, the task of ionization is solved by sufficiently long containment of ions in this beam. Of course, recombination processes must be suppressed during the containment time.

Let us consider the electric potential of the space charge of the cylindrical electron beam passing inside a drift tube with a definite potential. It is known that

$$U(r, \phi) = U_{dt} + s^- \ln \frac{R^2 - 2ar\cos\phi + a^2r^2/R^2}{r^2 - 2ar\cos\phi + a^2} \qquad (r > r_0)$$

and

$$U(r, \phi) = U_{dt} + s^- \left\{ \ln(R^2 - 2ar\cos\phi + a^2r^2/R^2) - [(r^2 - 2ar\cos\phi + a^2)/r_0^2 + \ln r_0^2 - 1] \right\} \qquad (r < r_0) \quad (3)$$

where U_{dt} is the potential of the drift tube,
$\quad\quad s^-$ is the linear charge density of the electrons of the beam,
$\quad\quad R$ is the internal radius of the drift tube,
$\quad\quad a$ is the distance between axes of the beam and the drift tube,
$\quad\quad r_0$ is the radius of the beam,
r and ϕ are cylindrical coordinates.

In the case of axial symmetry, then

$$U(r) = U_{dt} + 2s^- \ln \frac{R}{r} \qquad (r > r_0)$$

and

$$U(r) = U_{dt} + s^- \left[2 \ln \frac{R}{r_0} + \left(1 - \frac{r^2}{r_0^2}\right) \right] \qquad (r < r_0) \qquad (4)$$

The potential minimum coincides with the axis of the system and the potential well is axisymmetric in the latter case (4). But in the more general case (3), the space-charge potential is not axisymmetric, the potential minimum and the axis of the drift tube are displaced, and the magnitude of the potential minimum depends on a. This is shown in Figure 12.2.

A simple estimate shows that in case (4) the depression in the potential U produced by the space charge of an electron beam of 10 keV energy and 1 A current, is equal to about 150 V. Such a depression can be used for radial containment of ions in the electron beam. To vary the potential along the axis of the system, there are several possibilities, the simplest of which is to make the drift tube in sections and apply different potentials to different sections. There are other possibilities also; for example, variation of R if r_0 is constant, or the variation of r_0 if R is constant, and the most difficult way is the variation of s^- along the axis.

If in an ion-free electron beam system, a low-energy ion of charge state q^+ is created at a point $r < R$, it cannot reach the wall of the tube without gaining energy; the potential depression of the beam space charge provides radial trapping of the ion. To close the electrostatic trap in the axial direction, it is sufficient to

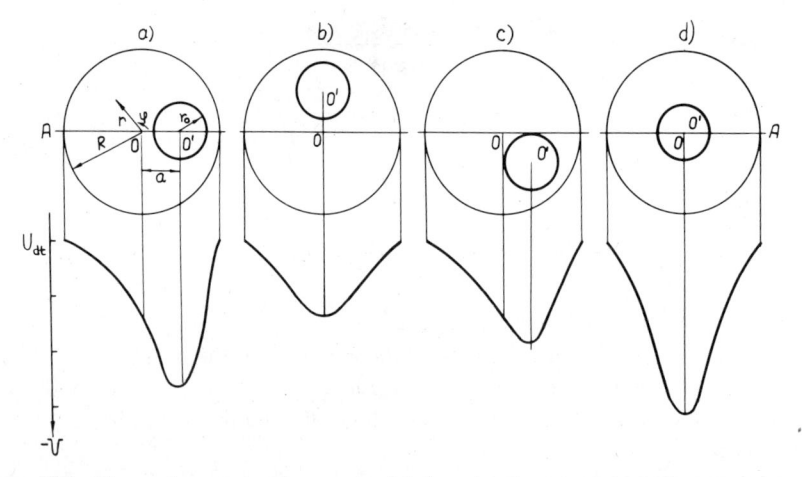

Figure 12.2 Electron beam space charge potential along the diameter of the drift tube cross section, for various positions of the beam inside the tube.

1 Physical Basis and Mode of Operation of EBIS

apply a potential $U_t \geq |s^-(2 \ln R/r_0 + 1)|$ at the end sections of the system. Note that because of the acceleration of the beam in the end sections by the applied U_t, s^- is decreased slightly there, and if the electron beam moves off axis with respect to the drift tube with a variable parameter a, then the level of the bottom of the trap follows a, and the bottom cannot be flat in the axial direction.

The radial containment forces decrease as $1 - \eta = 1 - |s^+/s^-|$, where s^+ is the linear charge density of the contained ions and η is the level of compensation of the electron beam space charge by the ion space charge. The radial confinement completely disappears if $\eta = 1$. Then

$$C^+ = 3.36 \times 10^{11} I_e L E_e \tag{5}$$

where C^+ is the ion capacity of the trap (the maximum number of elementary positive charge in the trap),
I_e is the beam current in amperes,
L is the length of the trap in meters, and
E_e is the electron energy in keV.

A closed electrostatic trap configuration cannot be maintained indefinitely because the electron beam is always formed in a vessel with some residual gas pressure. Gas is ionized by electron impact, and ions can stay in the trap, compensating the space charge of the beam. After a period of time τ_c, the positive space charge of the residual gas ions completely compensates the electron beam space charge.

Let us assume that all the ions that arise from the residual gas stay in the trap, remaining singly charged. Under this assumption,

$$\tau_c = (\sigma_{0 \to 1} n v)^{-1} \tag{6}$$

and using Lotz's formula for $\sigma_{0 \to 1}$,

$$\tau_c = 7.5 \times 10^4 \frac{E_e^{1/2} I_e}{n \ln (E_e/I_e)} \tag{7}$$

where $\sigma_{0 \to 1}$ is the cross section for ionization of residual gas atoms,
τ_c is in seconds,
n is the density of the residual gas (atoms per cm^3),
v is the velocity of the electrons in the beam (cm/s),
E_e is the electron energy (eV), and
I_e is the ionization potential of the residual gas atoms (eV).

Equation (7) provides only an approximate estimate of τ_c, because it does not take into account the possibility of ions leaving the trap, nor their charge state evolution during τ_c. But these processes tend to compensate each other, and one can use Eq. (7) for practical estimation of τ_c. For example, if the residual gas is nitrogen, $E = 10$ keV, $n = 3.6 \times 10^6$ cm^{-3} (10^{-10} Torr), then τ_c is approximately equal to 4.5 s. Naturally, in this consideration τ_c does not depend on j. Thus, if the electrostatic trap is used to contain ions of the working substance for multiple ionization, the necessary value of the factor $j\tau_i$ can be attained for $\tau_i \ll \tau_c$. It should be noted that the ion residence within the electron beam leads to an increase in charge state only when recombination with the residual gas is suppressed. Unfortunately, there is insufficient information on effective charge exchange cross sections of low-energy, highly charged ions on various atoms. Nevertheless, one can take $\sigma_{q \rightarrow q-1} \simeq 10^{-13}$ cm^2 for ions of type U^{90+} as an approximation. This means that if $\tau_i = 30$ s, the residual gas pressure must be less than 10^{-12} Torr in experiments to obtain bare uranium nuclei.

1.2 Basic Operation of the Electron Beam Method of Producing Highly Charged Ions

Methods for producing dense extended electron beams represent a highly developed and complicated field of science and technology, and we will not describe them in detail. We merely point out that to obtain a good quality electron beam it is necessary to synthesize anew in each case the total system of beam formation and focusing. To ensure that it has a constant shape and cross section over the drift length, a longitudinal magnetic field is usually applied. In this case, there are three main modes of beam formation, providing essentially different values of beam density for a given magnetic field induction (B) in the drift space:

1. *Magnetic ducting:* j is constant and is limited by the emissivity of the cathode of the gun.
2. *Magnetic compression:* j is proportional to B.
3. *Brillouin focusing:* j is proportional to B^2.

The shape of the electric potential trap along the beam can be created by application of appropriate voltages to various parts of the drift tube structure.

A description of other basic operations—injection of ions into the trap, ion containment during ionization, and axial extraction—can now be made in the framework of several different models:

1. The simplest one is the single-ion model. The criterion for validity of this model is $s^+ \ll |s^-|$, or rather, $s^+ \leq 10^{-2} |s^-|$, which means that the ion space charge does not influence their motion in the field of the electron space charge.
2. When the ion space charge is comparable to the electron beam space charge, that is, when $s^+ \simeq |s^-|$, one can use the self-consistent field model, in

which the ion space-charge field has a strong influence on ion motion but does not influence the motion of electrons.
3. If an interaction between the electron and ion fields appears, then a collective model must be used.

1.2.1 Injection of Ions Into the Trap

The most straightforward method of filling the trap with ions is to generate them directly in the trap from the atoms of a cloud of the working substance by electron impact in the electron beam. In this method of injection, ions are born with a very small kinetic energy compared to $-q_i s^-$, which is the maximum potential energy of a newly born beam ion of charge state q_i. We can then consider ion motion within the trap in the single-ion model. When the electron beam passes through the cloud of the working substance, mostly singly charged ions are produced with equal probability at any point r_i. These ions then execute oscillatory motion about the point of minimum potential. In the axisymmetric case and without magnetic field, the ion radial motion can be described by the expression

$$r = r_i \cos \left(\frac{q_i |\rho|}{2 M \epsilon_0} \right)^{1/2} t \qquad (8)$$

where ρ is the density of the negative charge of the electron beam,
q_i is the charge state of the ion,
M is the mass of the ion, and
ϵ_0 is the dielectric constant.

If a magnetic field is applied, the equation of radial motion is more complicated. Figure 12.3 shows part of the trajectory of an O^+ ion for $B = 2$ T, $\rho = 3.2 \times$

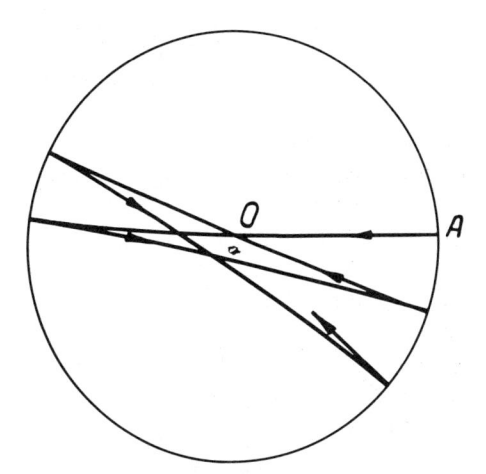

Figure 12.3 Part of the trajectory of O^+ ion.

10^{-8} C/cm^3 (2×10^{11} electronic charges/cm^3), $r_i = 0.15$ cm. The ion moves about the point $r = 0$ with frequency about 1.05×10^8 rad/s. This example corresponds to the real case of a beam of current 0.13 A, current density 200 A/cm^2, and 10 keV electron energy.

Thus, in the single-ion model, there is a range of ion energies that lie between 0 and $-q_i s^-$ with a constant number of ions per unit energy range. In their motion they do not leave the beam; this is very important, for example, in the consideration of ionization. But in this case, the number of ions in the beam is relatively small ($s^+ \lesssim 10^{-2} |s^-|$).

For a description of injection of more ions, one can use the self-consistent field model. Since the real injection time is always appreciably greater than the ion oscillation period, the characteristics of the ion motion slowly change with the slow variation of the self-consistent field. Exact solution of the ion motion is complicated, and we can indicate only the main qualitative features of the ion behavior during injection. Since the oscillation frequency ω_i is determined by the rigidity of the system, and growth of the ion space charge can be expressed as a decrease in rigidity, ω_i decreases during the injection process, and the energy of oscillation also decreases. The amplitude of the oscillation increases, leading ultimately to a loss of the most energetic ions at the wall of the drift tube. It is interesting to consider the motion of ions produced from atoms at the point $r_i = r_0$. Calculations show that if $R/r = 2$, during the increase of s^+ due to ion injection, r increases, and when $s^+/|s^-| = 0.3$, the most energetic ions strike the wall of the tube.

Thus, in the injection process, new ions are produced under conditions of decreasing self-consistent field, and when $s^+ = |s^-|$ equilibrium is established between the current of ions striking the wall and the ion charge generated in the beam per unit time. Then the mean ion energy becomes very small, near thermal. If the injection is now stopped, then during subsequent ionization of the injected ions the growth in q_i will be accompanied by a growth of s^+, which results in ions moving out of the beam. This can introduce a significant uncertainty into the analysis of the ionization process.

There are two possible ways to produce ions within the axial region of the electron beam following injection, while avoiding their subsequent loss during the ionization process.

One can inject ions into a beam of relatively low density, s_{in}^-, until $s^+ = |s_{in}^-|$ is satisfied, and then increase s^- to a final value equal to about $s_{in}^- q_{if}/q_{ii}$, where q_{if} is the expected final ion charge state and q_{ii} is the mean ion charge state before the increase of s^-. If the time to increase s^- is much greater than $1/\omega_i$, then the ion motion is opposite to the motion considered earlier: ω_i and E_e increase while r_i decreases, leading to a concentration of ions in the axial region of the beam. Another possibility is to inject the ions in the presence of axial potential barriers $U_t < |s^-(2 \ln R/r_0 + 1)|$. In this, the preferred case, the more or less energetic ions can leave the trap in the axial direction, and the column of ions in the axial region of the beam is formed of ions of thermal energy.

1 Physical Basis and Mode of Operation of EBIS

To provide pulsed injection of ions into the electron beam, one can use a pulse-generated cloud of atoms of the working gas. But almost always this is a very difficult operation. Alternatively, the working gas enters the injection region continuously and the electron beam intercepts the gas cloud in this region, with an appropriate axial potential distribution to ensure that the actual position of the interception is within the electrostatic trap for only a controlled interval of time.

The arrangement of the drift tubes with the gas puff in region *A* and the corresponding potential distributions are shown in Figure 12.4. The distribution in Figure 12.4a precedes the beginning of ion injection into the trap B. The directions of motion of the positive ions of the working gas and of the ions of the background gas in region B generated by the electron beam are shown by the arrows.

When injection begins, the distribution of potentials (see Figure 12.4b) combines regions A and B into a single trap, so that the ions generated in region *A* are freely distributed throughout the region A + B.

Once the necessary number of ions has been introduced into the electron beam, the potential distribution shown in Figure 12.4c is established. In it, ions that then are generated in region A move away in the direction shown by the arrows, while the ions intended for subsequent ionization are contained in the trap in region B.

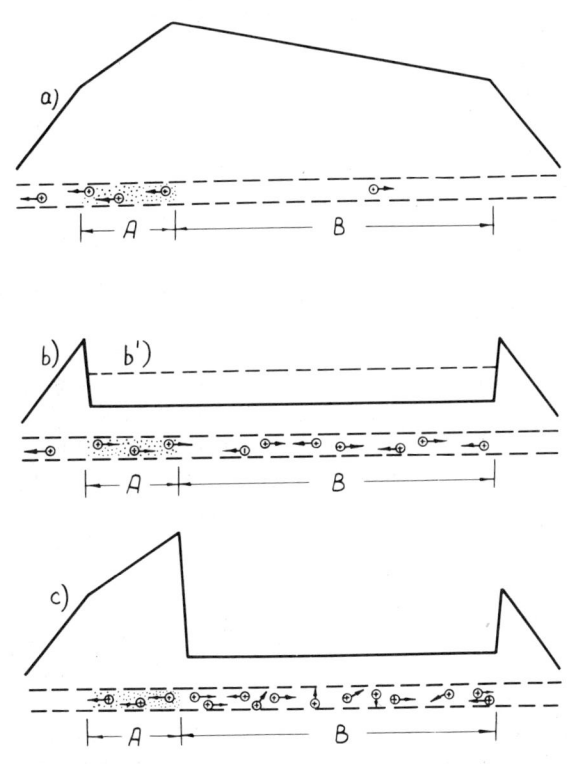

Figure 12.4 Injection scheme, for moving ions from region A into the trap B.

In this method, there are four different parameters that can be used to regulate the number of ions introduced into trap B: the electron current during the injection time, the injection time τ during which region A and B are united, the level of the axial trapping barrier during the injection time, and the density of atoms of the working gas in region A.

A necessary condition for performing pulsed ion injection is, of course, elimination of gas passing directly from region A to region B; this is not always easily satisfied.

Another way is the pulsed injection of a low charge state ion beam of the working substance, produced with an additional ion source, into the EBIS electron beam [6]. Such a beam can be introduced into the region of the electron beam along its axis, reflected at one side from an applied potential barrier, and contained in the trap by application of a potential barrier at the other side. Apparently this mode of injection has numerous significant advantages. There are no problems with neutral gas flowing into the trap region; there is no continuous generation of slow electrons in the region of interception of the beam and the gas cloud; there is the possibility of electromagnetic mass analysis of the ion beam before injection, and others. But, of course, external injection requires significantly better technology.

1.2.2 Containment of the Ions: Ionization

Ions with initial charge state q_{in} injected into the trap must be contained in the electron beam for the ionization time

$$\tau_i = \frac{1}{j} \sum_{q_{in}}^{q_f - 1} \sigma_{q \to q+1}^{-1} \tag{9}$$

to reach the mean final charge state q_f.

Let us consider the motion of ions during ionization. In the single-ion model, ions moving along the trajectory as shown in Figure 12.3 can be ionized at every point, but the greatest probability of ionization is near the equilibrium positions, where the potential energy is maximum and the ion velocity is approximately zero. If q increases at the point of production of the ion, that is, at r_{imax}, then in the absence of a magnetic field the trajectory does not change but the energy E_i of the ion increases in proportion to q, $\omega_i \sim q$, and r_{imax} is conserved, since the rigidity changes nonadiabatically. Similar changes take place in the presence of a magnetic field. But if q increases at $r_i < r_{imax}$, the ion cannot subsequently reach $r_i = r_{imax}$, although E_i and ω_i increase.

Thus, by the end of the ionization process, the energy spread of the ensemble of ions increases ($E_i - q_i s^- \geq 0$) with a smaller increase in the most probable energy, ω_i increases with q with a smaller increase in the most probable frequency, and the most probable value of r_i decreases, that is, there is a more or less significant concentration of ions in the direction of the beam axis.

An exact treatment, including a radial distribution of the ensemble of ions during their ionization, can be made by the method described in [7].

1 Physical Basis and Mode of Operation of EBIS

Now we consider three cases in the self-consistent field model.

1. Following injection, the electron beam is filled with slow ions, so that $s^+ \simeq |s^-|$. Due to multiple ionization, s^+ becomes greater than $|s^-|$, and forces of radial loss appear. Ions move out to the drift tube wall, with a delay because of their motion across the magnetic field if applied. In this case, there are some disadvantages. For example, since the more highly charged ions have a greater velocity they leave the beam more rapidly and after recombination on the wall of the drift tube some of them can return into the beam. Also, in the trap there is now an unconserved number of ions, which makes a description of the ionization process much more complicated.

2. A core of ions with near-thermal energies is formed as a result of injection in the axial region of the beam so that in this region the charge of the beam is fully compensated, while the main part of the beam is free of ions. The region with ions will be overcompensated by ion charge during the ionization process and, if at the beginning of the ionization process the containment voltage is increased (see the transition from Figure 12.4b to Figure 12.4c), then the ions cannot leave the trap in the axial direction, and their charge compensates the region of the beam that is free of ions. In this case, it is possible to obtain the maximum number of ions of the final charge state, since at the end $s^+ = |s^-|$.

3. The simplest case—ions moving during their containment time in the field of the electrons or of ion–electron space charge—was considered above. But there are various sources of heating that can change the picture of ion motion, leading to their "evaporation" from the beam. In particular, heating of ions by Coulomb collisions with beam electrons has been considered by Becker [8]. Since the collision frequency is proportional to j and the number of collisions during the containment time is proportional to $j\tau_i$, some limitation in obtaining super high charge states can appear because the ions can "evaporate" from the trap before they reach the desired charge state. There are other ways ions can be heated also.

In the previous consideration, in each ionization event secondary electrons appear that can influence the self-consistent field. Thus, the treatment will be correct for the case of rapid removal of these electrons from the trap. How can this be done? The energy spread of the secondary electrons is very narrow. After creation at a point with potential U_0, the secondary electrons move along a complicated trajectory to the region with potential $U > U_0$. But loss from the trap in the radial direction is small because of the strong magnetic field, and electrons move to the section of the drift tube where the barrier voltage is applied. If the secondary electrons are not caught here, their motion is oscillatory within the trap region. To catch the electrons inside the barrier section of the tube, one can apply there an electric field perpendicular to the magnetic field. Then the drift of the secondary electrons to the catch electrodes can be very fast. If the special catch cell is not installed, a fast drift can appear because of possible asymmetric motion of the electron beam within the drift tube.

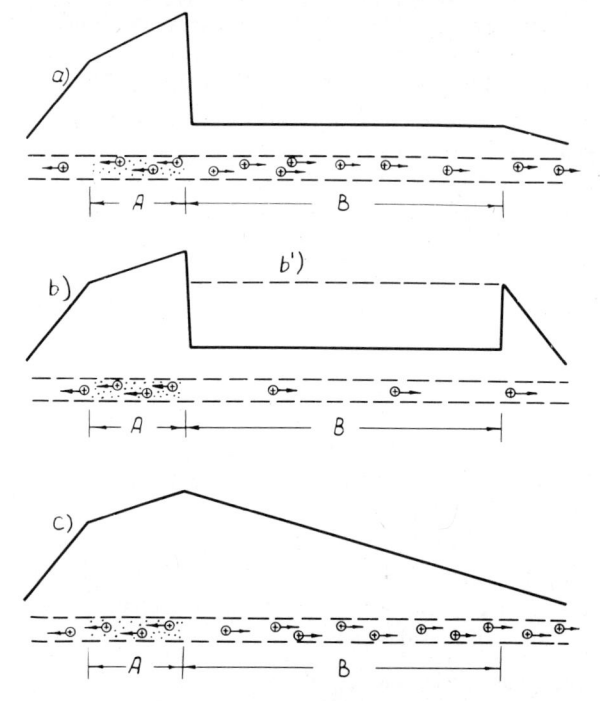

Figure 12.5 Extraction schemes, for moving ions out of the trap B.

1.2.3 Ion Extraction

Ions are extracted from the trap B in the axial direction by the creation of an appropriate distribution of potentials along the beam (Figure 12.5). Three modes of ion extraction can be used.

1. Ion extraction can be passive, when the right-hand barrier is removed (see Figure 12.5a), and ions leave the trap by virtue of their kinetic energy (in the single-ion model). In the self-consistent field model, the axial gradient of the field also appears in the direction of extraction. When passive extraction is used, the energy spread of the ions outside the trap is the same as inside the trap, that is, $0 < E_i < |s^-|(2 \ln R/r_0 + 1)$. The extraction time, τ_{ex}, is different for different cases, but one can take the transit time for ions with charge state q and mass M through the length of the trap L in the electric field of gradient

$$\frac{\partial U}{\partial z} = |s^-|\left(2 \ln \frac{R}{r_0} + 1\right)/L \qquad (10)$$

as being typical for τ_{ex}. In this case

1 Physical Basis and Mode of Operation of EBIS

$$\tau_{ex} = \left[\frac{2M}{q} \frac{L^2}{|s^-|\left(2\ln\frac{R}{r_0} + 1\right)} \right]^{1/2} \tag{11}$$

For example, if $q/M = 0.5$, $I_e = 1$ A, $R/r = 10$, $E_e = 10$ keV, $L = 1$ m, then $\tau_{ex} = 10$ μs.

2. The energy spread of ions outside the trap can be made very small by a slow change in the potential distribution b' (see Figure 12.5), when the right-hand barrier is constant and the level of the bottom of the trap is increased slowly.

3. Fast active extraction occurs when the distribution b is transformed into the distribution c by the rapid application of an external electric field $\partial U/\partial z$ along the system. Then,

$$\tau_{ex} = \left(\frac{2M}{q} L \Big/ \frac{\partial U}{\partial z} \right)^{1/2} \tag{12}$$

if the electric field is applied instantaneously.

1.2.4 Some Remarks About the Collective Model

In some cases an instability can occur in the ion–electron beam system of EBIS. In particular, the growth and stabilization of the Budker-Buneman [9] instability has been considered by Bonch-Osmolovskiy [10]. This instability can be stabilized by the same longitudinal magnetic field that is also used for the electron beam focusing. It was shown that the minimum value of the magnetic field necessary for the stabilization of resonant harmonics is given by

$$H = fq_i \frac{j_e \lambda}{c\beta^2} \tag{13}$$

where f is the ratio of the number of ions to the number of fast electrons in the trap, j_e is the electron current density in the beam, $\beta = v/c$ (v is the electron velocity), and λ is the wavelength of the perturbation.

For example, in the case of 10% compensation of the electron beam space charge by the ion space charge and a magnetic field of about 5 T, it is possible to use an electron beam of current density about 10^5 A/cm². That is, the ionization faction $j\tau_i = 10^{24}$ cm^{-2} needed to obtain uranium nuclei U^{92+} can be achieved in 1.5 s.

An ideal system has been assumed [10], with ions of the same q and monoenergetic fast electrons only. A real system has many differences from these assumptions, and other modes of instability can occur. This means that the collective model must include a consideration of all possible collective ion–electron motions.

In constructing the above model of the electron beam method of ionization, we have used only a simple electrostatic consideration and analyzed an idealized system of slow ions and a beam of fast electrons. The actual experimental situation is unavoidably more complicated with respect to the number of species of participating particles and fields as well as with respect to the processes that can take place. Thus, the effectiveness of the method and the limits of its applicability can only be established experimentally.

2 TYPES OF EBIS CONSTRUCTION

EBIS is an electron beam device consisting mainly of the following components and systems: vacuum chamber, electron gun, drift tube section, electron collector, ion extractor, focusing solenoid, working gas injector or external ion source with injection system, electric power supplies and electronic control systems, and diagnostics and vacuum pumping systems.

All EBIS devices can be divided into two classes, according to the solenoid construction:

1. "Warm" EBIS, where a solenoid of normal conductivity is used.
2. Cryogenic EBIS, with a superconducting solenoid.

In general, a "warm" EBIS is like a developmental test bench for a powerful electron beam rf facility, except that the device for the excitation of oscillation is replaced by a device for their suppression. This means that for the construction of a "warm" EBIS, one can use a lot of the scientific and technical construction techniques used in such electron beam facilities, especially the diagnostics. But "warm" EBIS devices have several disadvantages, a most important one of which is the relatively low focusing magnetic field strength (up to 1 T) and the relatively large power needed to provide it.

Cryogenic EBIS devices have the following advantages:

1. Use of a superconducting focusing solenoid provides a much stronger magnetic field (up to 10 T), and if the solenoid is wound with a fine superconducting wire, it is simpler to achieve the necessary magnetic field homogeneity.
2. The technique for providing the superconductivity is readily coupled to techniques for obtaining super high vacuum.
3. Cooling the drift tube to 4.2 K or lower allows the pulsed injection of working gas into the electron beam.

Perhaps the main disadvantage of the cryogenic EBIS is the difficulty in application of diagnostics.

All electron beam ion sources can be divided into two types with respect to the formation and focusing of the electron beam:

1. EBIS devices with the electron gun immersed in the magnetic field of the solenoid (IG EBIS).
2. EBIS devices with an external electron gun that is fully shielded from the magnetic field (EG EBIS).

The IG EBIS has a relatively simple construction and a fairly rigid electron beam, because the electron beam is strongly magnetized in this case. The main disadvantage of the IG EBIS is the relatively low electron beam current density that can be produced. In fact, the use of the highest emissivity cathode material, for example, LaB_6 [11], and of a magnetic field in the drift region of about 5–6 T, can provide a beam density of no more than a few hundred A/cm^2. Also, in this case the electron beam diameter is always oscillatory (spatially modulated), which affects the dynamics of the confined ions.

The main advantage of the EG EBIS is the possibility of producing and using for ionization a very high current density electron beam, up to about 10^4 A/cm^2; the beam obtained in such a device can have a steady or only weakly modulated diameter [12]. On the other hand, producing a beam of the necessary quality is significantly more complicated technologically.

To illustrate the construction of an EBIS, let us consider the structure of the cryogenic IG EBIS KRION-2 at the Joint Institute for Nuclear Research (Dubna, USSR) [13], and the cryogenic EG EBIS CRYEBIS at the Institute of Nuclear Physics (Orsay, France) [14].

2.1 Construction of KRION-2

KRION-2 was constructed in 1974 for the production and investigation of the physics of highly charged ions, after obtaining record charge states, for that time, on the first cryogenic EBIS KRION [15, 16]. Since then KRION-2 has been upgraded several times, and now it is a working device for the production and investigation of ions up to Kr^{35+}, Kr^{36+} [17], Xe^{52+} [18], Xe^{53+}, and Xe^{54+}.

A general view of the cryomagnet system of KRION-2, which is very similar to the KRION system, is shown in Figure 12.6. The main parameters of the system are:

Solenoid winding length—1.2 m
Internal winding diameter of the solenoid—0.05 m
Maximum magnetic induction—2.25 T
Heat flow to the liquid helium—0.125 w

The KRION-2 vacuum system consists of a mechanical pump and an oil diffusion pump, which are used only for a preliminary pumpdown; cryogenic pumping by internal surfaces at temperatures of 78 and 4.2 K produce a residual gas pressure of less than 10^{-12} Torr.

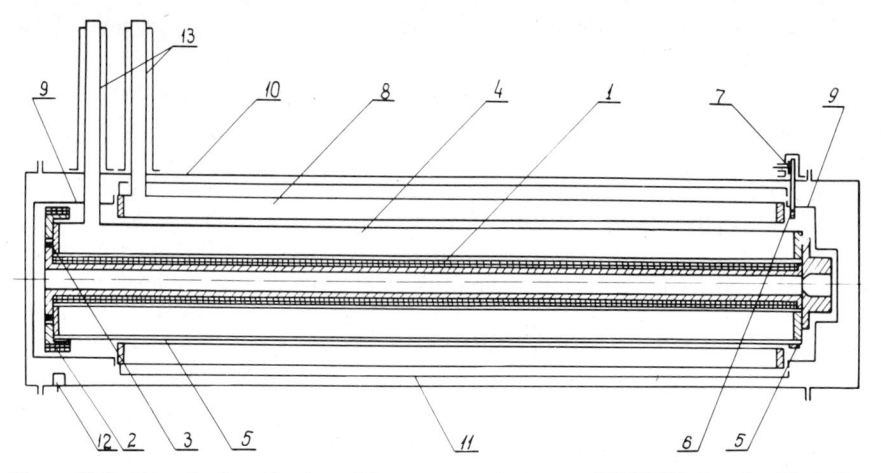

Figure 12.6 General schematic view of the cryomagnetic system of KRION-2: 1, solenoid winding: 2, superconducting "key"; 3, junctions of superconducting wires; 4, cryostat for liquid helium; 5, 6, current-carrying bars; 7, bar contact device; 8, cryostat for liquid nitrogen; 9, 11, radiation screens; 10, vacuum jacket; 12, vacuum gauge; 13, tubes for filling the cryostats with cryoliquids.

The electron optical and ion optical systems of KRION-2 are shown in Figure 12.7 [13]. Here the shape of the magnetic field induction B along the axis of the source and the distribution of the electric potential U for ion motion control are also shown.

There are three temperature terminals inside the ion source: 300, 78, and 4.2 K. The electron gun (EG), electron collector (EC), and electron extractor (EE) are connected to the room temperature terminal; the first five sections of the drift tube and the last (the 25th one) are connected to the liquid nitrogen temperature terminal; the sections of the drift tube from number 6 through number 24 are connected to the liquid helium temperature terminal. The working gas flows into the third section of the tube through a channel at 78 K. Cryosorbtion of working gas on the walls of drift tube sections 6 and 7 provides a gas pressure in the region of the trap (sections 8–22) of less than 10^{-12} Torr, which is sufficiently low for obtaining extremely highly charged ions.

The main diagnostic device used on KRION-2 is the time-of-flight mass spectrometer. There are also X-ray spectrometers for hard X rays.

2.2 Construction of CRYEBIS [14]

The CRYEBIS electron beam ion source was built after KRION and KRION-2 were commissioned; it, thus, is a member of the new series of cryogenic EBISes. A schematic outline of CRYEBIS is shown in Figure 12.8. The installation includes a source of neutral atoms of the working substance that can be injected as a beam into the electron beam through a hole in the cathode of the electron gun.

The electron gun is fully shielded from the magnetic field. The electron beam is injected into the magnetic field of a short normal conductivity solenoid, Solin,

2 Types of EBIS Construction

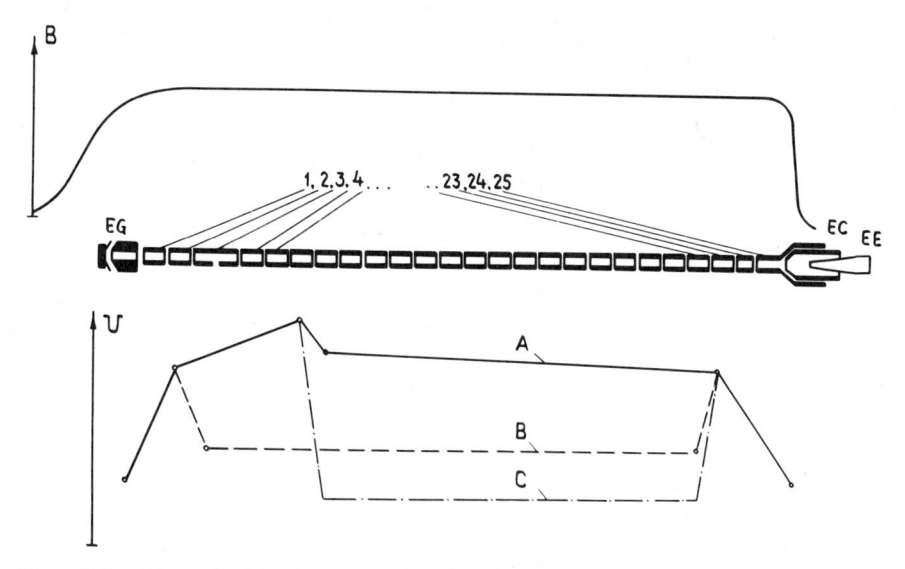

Figure 12.7 Schematic of the electron- and ion-optical systems of KRION-2: EG is the electron gun, 1–25 are the drift tube sections, EC is the electron collector, and EE is the extracting electrode. The distribution of the magnetic induction B along the source axis is shown as well as the distribution of the electric potentials U along the drift tube structure.

and then in Brillouin flow into the magnetic field of the main superconducting solenoid, Supersolo, where the electron beam is further magnetically compressed. Another short normal conductivity solenoid, Solex, is used at the extraction end, to adjust the electron beam blowup position.

The drift tube structure is located in the vacuum chamber where cryopanels are installed, and is isolated from the vacuum of the cryomagnetic system, unlike KRION-2 where there is a common vacuum chamber.

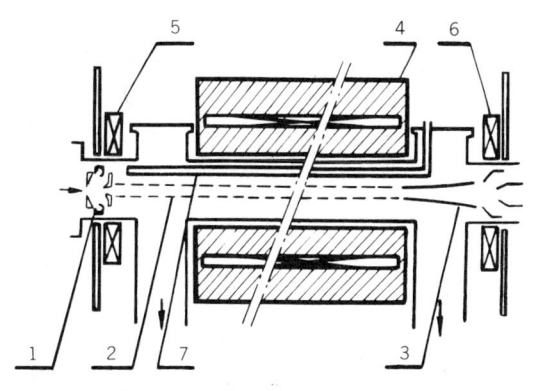

Figure 12.8 Schematic of the CRYEBIS source: 1, the electron gun; 2, the drift tube structure; 3, output electron and ion optics; 4, Supersolo; 5, Solin; 6, Solex; 7, the cryopanel.

The main parameters of the CRYEBIS system are:

Solenoid winding length—1.5 m
Inner solenoid diameter—0.2 m
Maximum magnetic field induction—3 T
Electron beam current—up to 3 A
Electron beam density—more than 10^3 A/cm^2
Electron beam energy—more than 10 keV

The time-of-flight mass spectrometer for A/q analysis is also used as the main diagnostic device.

3 EXPERIMENTAL DEVELOPMENT OF EBIS

Although there are at present more than ten electron beam ion sources in operation, there is only one more or less systematic description of the experimental investigation of the ion injection process and the containment and ionization of the ions to high charge states [3]. These studies were done on KRION-2.

3.1 Capacity of the Ion Trap

The capacity C^+ of the electrostatic ion trap is the limiting value of the ion charge Q^+ accumulated in the electron beam over the length of L (sections 2–22), with variation of any of three variables: q_f, the gas flux in the region of the third section; τ_{in}, the injection time; and U_t, the height of the barrier at section 23. Figure 12.9 shows the family of curves $Q^+ = f(U_t)$ for different values of I_e at $E_e = 8$ keV. All the curves are similar: There is an approximately linear increase and a transition to saturation at $U_t = U_s$ in accordance with estimates based on the self-consistent field model (the broken line for $R/r_0 = 10$, $I_e = 140$ mA). The dependence of U_s on I_e, which is a straight line because the natural potential depression on the beam axis is proportional to the current for unchanged electron beam energy, is shown in Figure 12.10.

If a charge Q^+ less than C^+ is introduced into the electron beam (for example by decreasing q_f), a saturation occurs in the function $Q^+ = f(U_t)$ at a U_s that is smaller than for $Q^+ = C^+$. The dependence of U_s on Q^+ for $I = 150$ mA and $E_e = 8$ keV is shown in Figure 12.11. An anomaly occurs at $Q^+ \simeq 3.0 \times 10^{-2} C^+$, when U_s ceases to decrease with further decrease in Q^+. The value of U_s at this point is equal to the potential difference between the boundary of the beam and its axis. This picture corresponds to the single-ion model and indicates that the ions do not pass outside the electron beam.

In the self-consistent field model, C^+ is equal in magnitude to the number of fast electrons within the trap, and so is proportional to I_e for E_e = constant. The family of curves $C^+ = f(I_e)$ for several values of E_e is shown in Figure 12.12. It

3 Experimental Development of EBIS

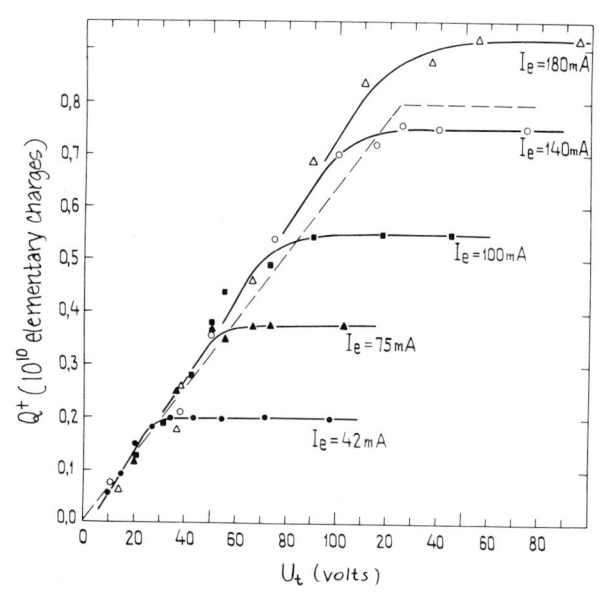

Figure 12.9 Dependence of the total ion charge Q^+ in the trap on the trapping potential U_t.

can be seen that linear growth of C^+ holds only up to a definite value of I_e, at which a deviation from linear growth of C^+ begins, then followed by a decrease in C^+ with increasing I_e. This is called the critical value of beam current, I_{ecr}. It has been found that for a range of Q^+, I_{ecr} does not depend on the number of ions in the trap. In a study of the dependence of I_{ecr} on the various source parameters, it was established that I_{ecr} depends only on E_e, increasing linearly with beam energy (Figure 12.13). This, in particular, indicates that the process leading to ejection

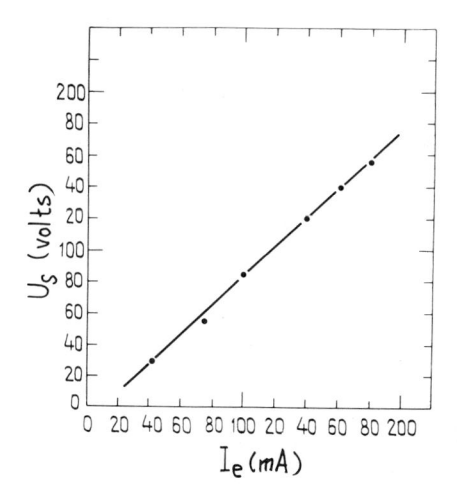

Figure 12.10 Dependence of the trapping potential at saturation on I_e.

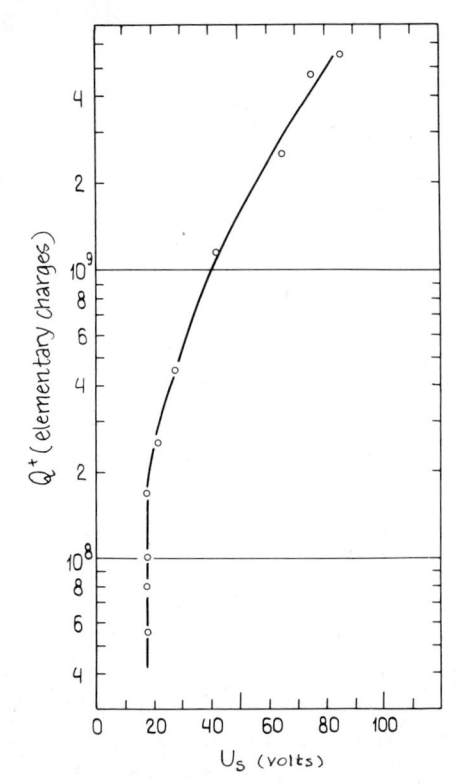

Figure 12.11 Values of the trapping potential at saturation for various values of ion charge in the trap.

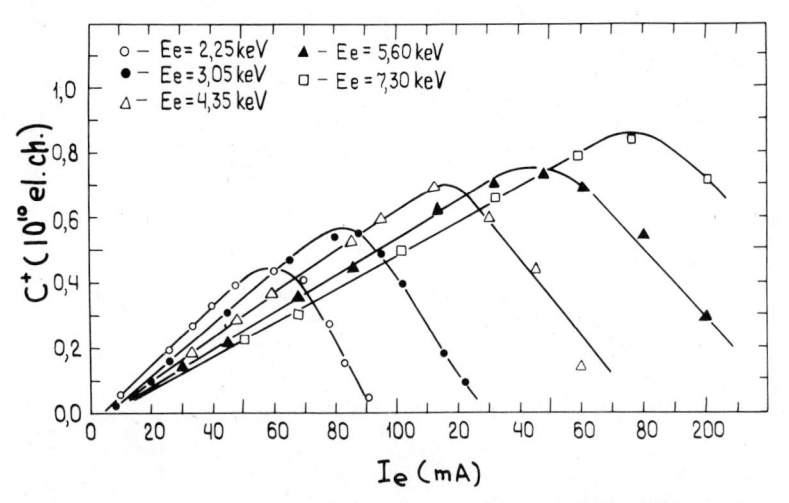

Figure 12.12 Dependence of the trap capacity C^+ on the beam current for different values of the electron energy.

3 Experimental Development of EBIS

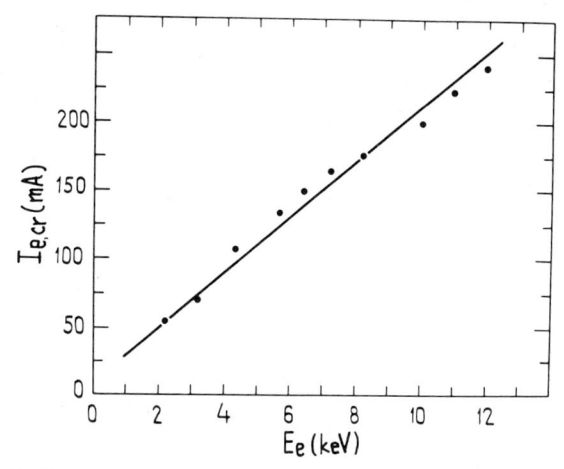

Figure 12.13 Dependence of the critical beam current on the electron energy.

of ions from the trap begins at a definite value of s^-/v (where v is the electron velocity), this value being constant for all energies and currents of the electron beam. It follows from the family of curves in Figure 12.12 that, in accordance with the self-consistent field model, C^+ is proportional to $E_e^{-1/2}$ for I_e = constant and for $I_e < I_{ecr}$.

A characteristic feature of the experimental dependence $Q^+ = f(\tau_{in})$ for $U_t > U_s$—this dependence is rather difficult to describe theoretically—is the transition to saturation, when $Q^+ = C^+$, corresponding to the self-consistent field model. Study of the dependence of C^+ on the length of the trap over the range 0.15 to 0.95 m revealed a linear growth.

3.2 Containment of Ions in the Beam

Containment within the electron beam of the ions introduced into it during the complete time τ_i is the most important condition for success of the electron beam method of ionization. The containment efficiency has been studied as follows: Initially, n_0 nitrogen ions were injected into the beam, this number being measured by $Q^+/\bar{q}_i$, where Q^+ is the ion charge measured directly at the end of injection and $\bar{q}_i$ is the mean ion charge measured by means of a time-of-flight spectrometer. The value of n was measured as a function of the containment time for different values of E_e and I_e and for $U_e \geq U_s$. The dependences $Q^+ = f(\tau_i)$ and $n = f(\tau_i)$ for three values of n_0 for the same I_e and E_e are given in Figure 12.14. One can see that for small n_0, all the ions are contained in the trap until the end of the ionization cycle (τ_i = 100 ms). But once a certain value n_0 is reached, ions begin to be lost after a certain time τ_i, although Q^+ still continues to grow because of the increase in $\bar{q}_i$. If the compensation level is already near unity when $\tau_i = 0$, the loss of ions begins when $\tau_i = 0$, and $\partial Q^+/\partial \tau_i$ is maximum at $\tau_i = 0$.

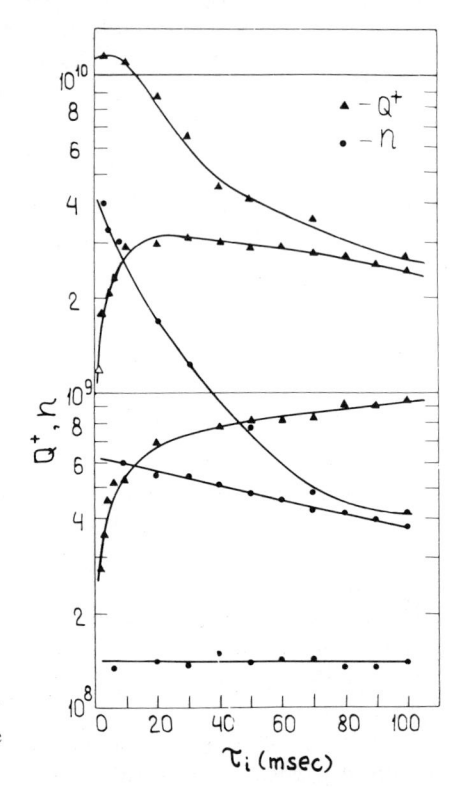

Figure 12.14 Dependence of the ion charge Q^+ and the number of ions n in the trap on τ_i for three different pairs of initial values of Q^+ and n.

The dependence $Q^+ = f(\tau_i)$ for full initial compensation was investigated for different values of I_e and E_e. These data, transformed into the dependence $\eta = f(\tau_i)$, where η is the level of compensation, are shown in Figure 12.15. It is interesting that for all $I_e < I_{ecr}$, the curves almost coincide. A difference appears when $I_e = I_{ecr}$, and the larger I_e, the steeper is the drop in the curve $\eta = f(\tau_i)$. The loss rate approaches 0 at $\eta \simeq 0.15$, which means that the ionization process can be extended over a prolonged period without loss of ions from the trap.

Thus, the investigations showed that there are definite parameter ranges of the ion–electron system within which the processes of injection and containment of the ions can be satisfactorily described by the single-ion model and the self-consistent field model. But if $I_e > I_{ecr}$, processes appear in the ion–electron system that lead to a fast loss of ions from the electron beam and failure to obtain highly charged ions.

3.3 Production of Highly Charged Ions

Experiments to obtain multiply charged ions and bare nuclei in KRION-2 were made for $I_e < I_{ecr}$. In Figure 12.16 such experiments are illustrated by means of the evolution of the charge state spectra of neon ions. This figure demonstrates

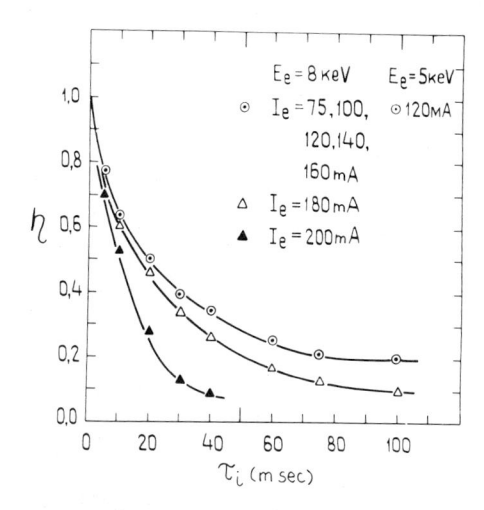

Figure 12.15 Dependence of the compensation level on τ_i.

with great clarity one of the important differences between EBIS and other sources of multiply charged ions: During the ionization process, ions of low charge states are completely depleted, transforming into ions of higher charge state. As a result, there is a more or less narrow charge state spectrum at the EBIS exit, and in the limit there are only nuclei completely stripped of their electron shells. The main features of all evolutions are as follows:

1. The increase in the ion charge state continues throughout the entire containment time (up to 40 s for the case of xenon ionization, for example) if the beam energy is high enough and the nuclear state is not reached.
2. The spectra evolve more slowly for large η than for small η, indicating radial loss of ions from the beam during containment.
3. The spectra differ somewhat depending on the time in the ion pulse at which the spectrum is sampled.

Thus, it has been experimentally demonstrated by use of KRION-2 that it is possible to obtain ions of any charge state in the EBIS.

3.4 Other Experimental Results

Among the results obtained on other EBIS devices, most interesting was the observation of the production of highly charged ions, with anomalously short containment in the electron beam, in the EG EBIS CRYEBIS [19, 20]. The measured value of the ion-free electron beam current density was about 10^3 A/cm^2 in these experiments, but using a containment time of only about 5 ms, bare nuclei of nitrogen, neon, and argon, and Kr^{34+} and Xe^{44+} were obtained. A consideration of the evolution of krypton ions, for example, showed that sequential ionization takes place, but the effective electron beam current density needed to explain such

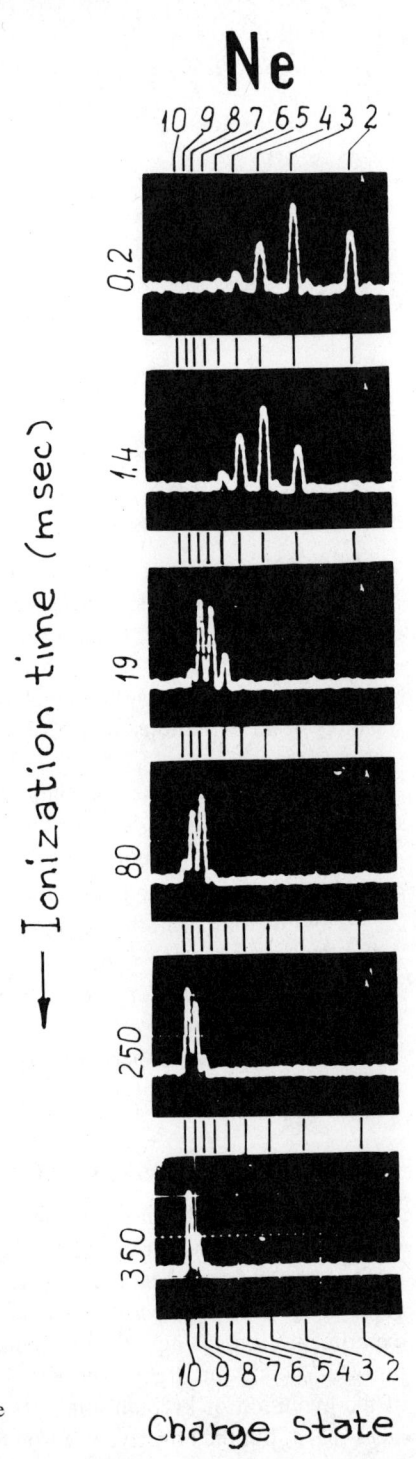

Figure 12.16 Dependence of the charge state spectrum of Ne ions on the ionization time.

a short ionization time is over 10^5 A/cm^2. It was also shown that the effective beam density increases by a factor of more than 100 when the beam is filled by ions, and decreases if the ion compensation of the beam decreases.

This feature of the Brillouin electron beam, called ion supercompression, was not known before the experiments of Arianer [19, 20]. The investigation and use of this phenomenon is very important for the development of the electron beam method of ionization, especially for increase in the efficiency of EBIS. Unfortunately, the conditions in CRYEBIS under which the phenomenon of ion supercompression was observed have been lost and not yet recovered.

The behavior of the ion–electron beam system was studied on the warm EG EBIS at the Lawrence Berkeley Laboratory by the detection of X-ray emission arising in the region of the trap [21, 22]. As a rule, some increase in the diameter of the beam was observed when the electron beam was filled with ions. That is, a phenomenon contrary to ion supercompression was reported.

In some experiments, an overcompensation of the electron beam space charge by the ion space charge has been observed [23, 24], indicating an accumulation of secondary electrons in the trap region. It is then possible for the EBIS to develop longitudinal oscillation of the primary electrons [25].

In the Saturne National Laboratory (France), a system for external injection of ions of the working substance into the electron beam has been developed [6]. A duoplasmatron was used as the external ion source. The system is very reliable and provides good reproducibility for highly charged ion production in the cryogenic EBISes of this laboratory.

4 SHORT REVIEW OF PROJECTS AND DEVICES

More than 20 EBIS projects, either completed or under construction, have been described up to the present time. It is not possible to consider them all in detail here, but we will nevertheless give a short description of most of the projects, with their main results, ideas, problems, and with references. For each project, the name and type of EBIS, the institution and year when the project was proposed or the installation first operated, is also given.

1. IEL [2]—"warm" IG EBIS, JINR (Dubna, USSR), 1968. This was the first test device on which the EBIS principles were experimentally demonstrated.
2. IEL-2 [26]—"warm" IG EBIS, JINR (Dubna, USSR), 1970. It was shown that the EBIS efficiency increases in proportion to the length of the trap. The necessity for super high vacuum in the electron beam region turned out to be a difficult problem.
3. SILFEC [23]—"warm" IG EBIS, IPN (Orsay, France), 1971. Ne^{9+}, Ar^{14+} and Xe^{21+} ions were obtained, which was excellent for a "warm" EBIS. Pulsed injection of neutrals into the electron beam through a hole in the cathode of the electron gun was proposed.

4. KRION [15, 16]—cryogenic IG EBIS, JINR (Dubna, USSR), 1972. This was the first cryogenic EBIS with which the full sequence of operations of the electron beam method of ionization was realized. Bare nuclei of C, N, O, and Ne as well as Ar^{16+} and Xe^{28+} ions were first obtained. This was the first EBIS applied to a synchrotron [27]. The first studies of positive ion ionization by electron impact inside the EBIS electron beam were done. Overcompensation of the EBIS electron beam was observed [24].

5. KRION-2 [13, 28]—cryogenic IG EBIS, JINR (Dubna, USSR), 1974. Magnetic compression of the electron beam was used. Super high vacuum in the trap region, better than 10^{-12} Torr, was obtained. Bare nuclei of Ar, Kr, and Xe were first produced. Studies of the basic processes of the electron beam method of ionization and systematic investigations of positive ion ionization by electron impact, including cross-section measurements, were carried out. Processes leading to a limitation of EBIS efficiency were observed.

6. CRYEBIS [19, 23, 29]—cryogenic EG EBIS, IPN (Orsay, France), 1975. This was the first EG EBIS with Brillouin flow electron beam ($j_e \simeq 10^3$ A/cm^2). The very important phenomenon of ion supercompression of the electron beam was discovered. The first attempt to use EBIS as an ionizer-accumulator for polarized particles was undertaken. This was the second EBIS to be applied to a synchrotron.

7. Texas EBIS [30, 31]—"warm" EG EBIS, Cyclotron Institute (College Station, Texas, USA), 1975. This was the first "warm" EG EBIS. The first attempt to build an EBIS for cyclotron use was undertaken.

8. TOFEBIS [32]—"warm" EG EBIS, Institut fur Angewandte Physik (Frankfurt/Main, FRG), 1975. Dc mode of operation of EBIS was proposed. The ions reach moderately high charge states during a single transit time in the electron beam from the electron collector, where they are created from the working gas cloud, to the cathode of the electron gun, where multiply charged ions are extracted through a hole in the cathode. An electric field can be applied along the axis of the electron beam to lead the ions from the collector to the cathode.

9. Frankfurt EBIS [25]—cryogenic IG EBIS, Institut fur Angewandte Physik (Frankfurt/Main, FRG), 1977. This was the first attempt to create an EBIS with longitudinal oscillation and accumulation of the primary electrons so as to increase the efficiency of use of the electron beam power. Studies and possible suppression of instabilities connected with oscillation and accumulation of the electrons were foreseen.

10. LBL EBIS [21, 22, 33]—"warm" EG EBIS, LBL (Berkeley, California, USA), 1979. This was a "test-stand" for investigation of EBIS processes. Record axial symmetry of the magnetic field was achieved by use of a special *magnetic homogenizer*. X rays produced in electron–ion interactions first applied for EBIS diagnostics. An increase in the interaction region when the beam is filled by ions was found.

11. Cornell EBIS [34, 35]—"warm" EG EBIS, Cornell University (Ithaca, New York, USA), 1981. This EBIS is for atomic physics studies. The axial trap voltage is microprocessor controlled.

12. CRYEBIS-2 [36, 37]—cryogenic EG EBIS, IPN (Orsay, France), and RIP (Stockholm, Sweden), 1981. Two identical ion sources for atomic physics were constructed. The electron beam energy and other parameters are sufficient to produce U^{90+} ions. The electron beam energy can be increased in the future.

13. NICE EBIS [38, 39]—cryogenic IG EBIS, IPP (Nagoya, Japan), 1979. EBIS for atomic physics experiments with bare nuclei of relatively light elements; operates in dc mode. Very low energy spread (about 0.8q eV) is obtained, allowing the study of atomic energy levels by translation energy spectroscopy.

14. Novosibirsk EBIS [40]—"warm" EG EBIS, INP (Novosibirsk, USSR), 1980. This is a bare-nucleus source for the synchrotron, cycling with a 50-Hz repetition rate. Fast extraction of ions, about 1 μs, was proposed.

15. KRION-3 [28]—cryogenic EG EBIS, JINR (Dubna, USSR), 1983. This is the Synchrophasotron ion source project. The energy of the electron beam is high enough for the production of krypton nuclei. Anomalously short ionization time for the production of highly charged ions was observed.

16. DIONE [29]—cryogenic EG EBIS, NLS (Saclay, France), 1983. This is the ion source for the synchrotron "Saturne." External injection of low charge state ions into the electron beam is used. Decrease in internal temperature to lower than 4.2 K, to prevent desorption of hydrogen, is foreseen. Superb technology is used in the source. Bare nuclei of Ne and Ar are obtained. The electron beam energy is sufficient for production of Kr nuclei.

There are three more recent EBIS projects also. These are the Kansas EBIS (Kansas State University), the SNLL EBIS (Sandia National Laboratories, Livermore), and the LLNL EBIS (Lawrence Livermore National Laboratory). Information about these projects was presented at the Third Workshop on EBIS (Cornell University, 20–23 May, 1985, Ithaca, New York, USA).

5 SOME SPECIAL APPLICATIONS

We consider here some applications of EBIS that are closely related to its mode of operation and construction, namely investigation of positive ion ionization by electron impact, ion-at-surface spectroscopy, and the experimental study of the nuclear properties of highly charged ions.

5.1 Ionization of Positive Ions by Electron Impact [3, 41]

Experiments in this field with the EBIS KRION-2 have resulted in the creation of a new way to measure cross sections for extremely highly charged ions. Let us consider this technique and some results.

The ionization cross sections of ions were measured as follows [41]. Ions of the investigated element with relatively low charge state values ($1 < q < 4$) were introduced into an electron beam. Immediately after injection, the charge state spectrum of this sample of ions was measured. Since the ion bunch at the output of the source usually had a half-width of about 50 μs, and a narrow (about 100 ns) pulse is separated for the time-of-flight (TOF) analysis, five or six cycles were needed to obtain reliable information, so different time sections of the ion bunch were analyzed each time. The analyzer was operated in the current mode, so that the amplitude of a spectral line was proportional to the total electric charge of ions of given charge state in the TOF pulse. The charge state spectrum was then transformed into the distribution, normalized to unity, of the number of ions with given charge states (the charge state distribution) as follows:

$$n_q = \frac{1}{q} \sum_{k=1}^{a} A_q^{(k)} \bigg/ \sum_{q_{\min}}^{q_{\max}} \left(\frac{1}{q} \sum_{k=1}^{a} A_q^{(k)} \right), \quad \sum_{q_{\min}}^{q_{\max}} n_q = 1 \qquad (14)$$

where n is the number of ions of charge state q, normalized to unity, $A_q^{(k)}$ is the amplitude of the line of the TOF pulse of ions of charge state q in test k, a is the number of tests, and $q_{\min}$ and $q_{\max}$ are the minimum and maximum charge states of the ions in the bunch.

At various times, measured from the end of injection, the new charge state distribution was measured in other cycles. The form of this distribution is determined by the value of τ_i and the effective ionization cross sections, which are not known but which can be extracted from the results of the measurements. The kinetic equation for the number n_q in all cases except for ionization of metastables is

$$\frac{dn_q}{d(j\tau_i)} = - \sum_{f=1}^{f_{\max}} n_q \sigma_{q \to q+f} + \sum_{r=1}^{r_{\max}} n_{q-r} \sigma_{q-r \to q} \qquad (15)$$

where f and $f_{\max}$ are the current number and maximum possible number of electrons that can be simultaneously stripped from an ion of charge state q, $\sigma_{q \to q+f}$ is the cross section for this process, r and $r_{\max}$ are the current number and the maximum possible number of electrons that can be simultaneously stripped from an ion of charge state $q - r$, and $\sigma_{q-r \to q}$ is the cross section for this process.

In the special case of successive ionization,

$$\frac{dn_q}{d(j\tau_i)} = -n_q \sigma_{q \to q+1} + n_{q-1} \sigma_{q-1 \to q} \qquad (16)$$

5 Some Special Applications

To find all the unknown values of σ's, it is convenient to measure the dependencies $n_q = f(j\tau_i)$ for all q in Eq. (15), that is, to obtain the evolution of the charge state distribution of the ions.

The experimental evolution of the charge state distribution of nitrogen ions is given in Figure 12.17 [41].

From these data one can make as many vertical sections as necessary to obtain all coefficients of n_q and $dn_q/d(j\tau_i)$ for the system of x equations, if x unknown σ's are to be found. In this case, the system is canonical and it has a unique solution, but it is difficult to find the probable deviations $\Delta\sigma$. As a rule, however, the information content of the evolution is very rich, and one can make more vertical sections and so construct a multitude M of systems of x equations for x unknown σ's. The solution for each system gives a corrsponding set of values for the σ's. If all the systems have equal weight, the result is a set of values of the σ's averaged over the complete experiment (for all $j\tau_i$) and the probable deviations $\Delta\sigma$ as well.

There is another way of extracting solutions from the experiment. Using regularized iterative processes of the Gauss-Newton type, Bochev et al. [42, 43] developed a program for solving the inverse ionization problem. In it, one finds a set of σ's that when substituted into the direct problem give the smallest difference between the calculated and experimental charge state distribution evolutions. The difficulty here is in determining whether the solution is unique.

As an illustration, Figure 12.18 shows the histograms of σ values obtained from the experimental charge state distribution evolution for nitrogen ions by using the first of the two methods described above. Above the top of each histogram, the values of σ (in 10^{-20} cm^2) obtained by the first method (a) and by the second method (b) are given. One can see that the two methods give values of σ that agree to within the errors.

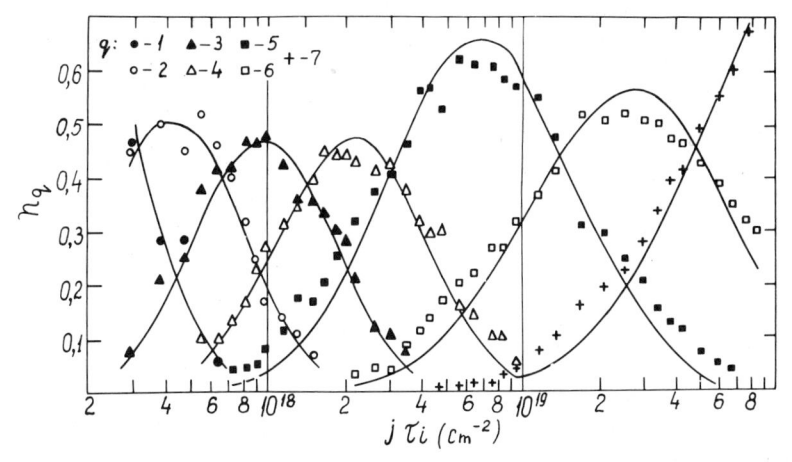

Figure 12.17 Evolution of the charge distribution for nitrogen ions for $E = 5.45$ keV.

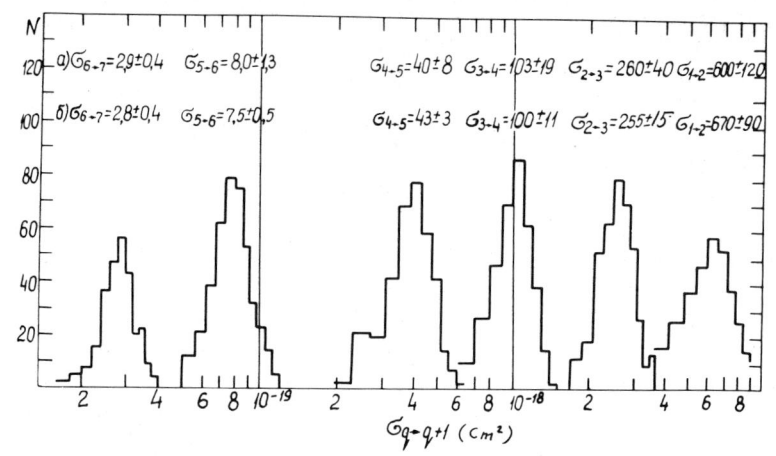

Figure 12.18 Histograms of $\sigma_{q \to q+1}$ for nitrogen ions.

A model for the ionization process can be chosen in each case, but the best way is by solving the system of equations, Eq. (15), with all energetically possible values of f_{max} and r_{max}.

Naturally, to find the right values of the cross sections, the evolution should be obtained when $I_e < I_{ecr}$ and the ion compensation of the electron beam is very small (about 10^{-2}), that is, under conditions when there is no loss of ions from the trap and all the ions stay in the beam for the full containment time.

An advantage of the above method for obtaining cross sections is the possibility of measuring extremely small values (to 10^{-24} cm^2) for very highly charged ions (up to U^{91+}).

Cross sections for many ions have been measured in this way [18, 41]. For those cases for which there are results available from other methods, there is good agreement. As an illustration of this method of cross-section measurement using the EBIS, the energy dependencies of the cross sections for ionization of the hydrogen-like ions of C, N, O, Ne, and Ar are given in Figure 12.19 [18].

5.2 Dependence of Nuclear Properties on the Ion Charge State

The investigation of the influence of the electron shell structure on nuclear properties is one of the most interesting applications of the EBIS. The problem was considered long ago, but there have been no experimental results up to the present time. One can now try to begin such experiments, by the use of a good EBIS.

Let us consider the possible experimental study of the dependencies of α-decay energy (E_α) and α-decay half-life ($T_{1/2}$) on the ion charge state, for example. A suitable nucleus for beginning the experiment is ^{220}Rn, with $T_{1/2} = 51$ s and $E_\alpha = 6.28$ MeV. As a continuous source of ^{220}Rn atoms, it is convenient to use a container with a fine powder of the oxide of natural thorium, and as α-particle detectors to use surface-barrier detectors of cylindrical shape with the sensitive

5 Some Special Applications

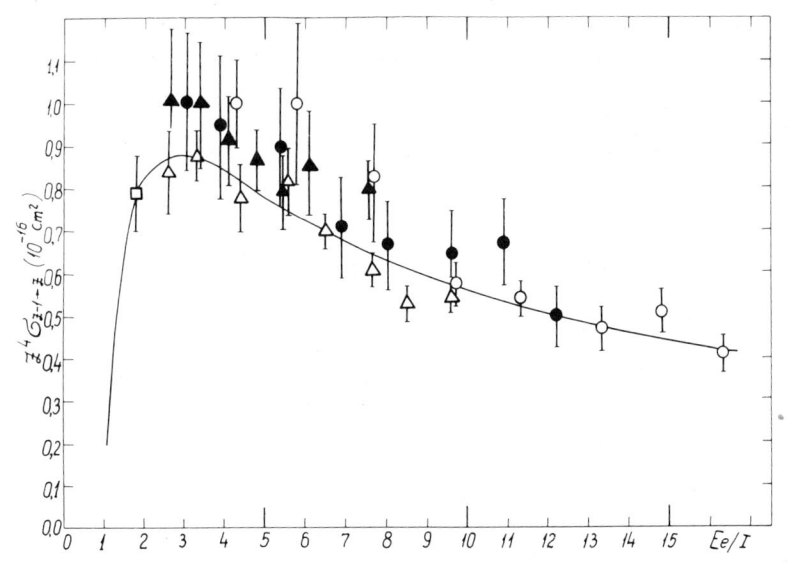

Figure 12.19 Dependence of the reduced cross sections for ionization of hydrogen-like ions on the electron energy: ○, C^{5+}; ●, N^{6+}; △, O^{7+}; ▲, Ne^{9+}; □, Ar^{17+}. The continuous curve is calculation.

layer on the inside facing the electron beam of the EBIS. In considering the method to be used in the measurement, it must be borne in the mind that during the ionization process, ions may be lost from the electron beam in the axial and radial directions with subsequent sorption (in the latter case) on the walls of the drift tube or detector; this results in background α-radiation when both E_α and $T_{1/2}$ are being determined. For the determination of E_α, the operation of subtracting the background is very simple, but the measurements associated with the determination of $\Delta T_{1/2}$ appear at present to present much greater problems. The following procedure for such a measurement has been proposed [18]. At a certain time τ, which is measured from the time of introduction of low charge state Rn ions into the electron beam, the counting rate n_τ of the α-particles is measured for all radon ions and atoms in the electron beam and on the wall of the drift tube. The electron beam is then switched off directly after the measurement, but the axial containment of the ions in the trap is maintained. The disappearance of the space charge of the electrons has the result that the Rn ions move to the walls of the drift tube, where they are neutralized. Directly after the beam has been switched off, a new measurement is made of the α-particle counting rate n'_τ for all atoms on the wall of the drift tube. The difference $n_\tau - n'_\tau = \Delta n_\tau$ arises because of the difference $\Delta \lambda = \lambda_0 - \lambda_q$, where λ_0 is the decay probability for nuclei of the ^{220}Rn atoms, and λ_q is the decay probability for nuclei of the ^{220}Rn ions. To determine the number of ions associated with the observed Δn_τ, the Rn ions in the beam at the time τ in the following cycle of measurements are extracted from the trap in the axial direction directly after the measurement of n_τ. The α-particle counting rate for the Rn atoms on the walls of the drift tube, n''_τ, is measured directly after extraction. The difference $\delta n_\tau = n'_\tau - n''_\tau$ gives the number of ions of charge state q that must be associated with the observed change Δn_τ in the counting rate. The expected shape of the dependence of the α-particle counting rate on τ is plotted in Figure 12.20. Note that in this method of measurement the main observed effect is the jump Δn_τ, which must be measured with the greatest possible accuracy in one cycle. The number of atoms corresponding to the effect is measured in two neighboring cycles. The accuracy is then lower, but the requirements are appreciably lower. Making measurements for different values of τ, it is possible to find the dependence $\Delta \lambda = f(q)$.

Calculations show that at the present time, such a method could yield an experimental determination of a change of order $\Delta \lambda / \lambda_0 \gtrsim 10^{-3}$.

Similar experiments to study K capture and $\beta^\pm$ decay can be considered, especially since the predicted effects in these cases are much larger.

5.3 Spectroscopy of Slow Highly Charged Ions on Solid Surfaces

Investigations in this field of atomic physics have commenced recently with the observation of hard X rays from the recombination of slow Ar^{17+} ions produced in KRION-2 on the surface of a solid target [28, 44]. After this experiment, it soon became clear that a method for the generation and observation of the decay of previously unseen superexcited states of atoms had been found. In fact, present thinking about the recombination process is that slow highly charged ions are neutralized on the surface by the population of very high electronic levels, with principal quantum number equal to about the q of the ion, if it is slow enough.

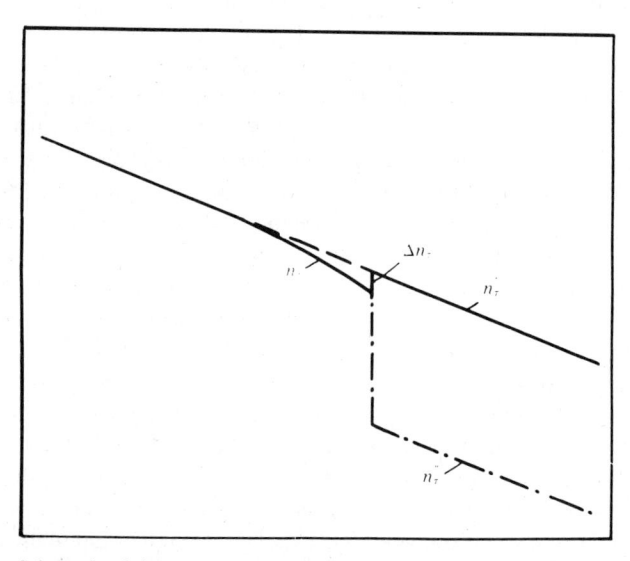

Figure 12.20 Schematic of the expected α-particle counting rates as a function of the time that the Rn ions stay in the electron beam.

5 **Some Special Applications**

This means that a neutral atom with a large excitation energy is obtained at the first step of the process. The decay of superexcited states occurs via the production and decay of various atomic and ionic levels having, as a rule, many inner shell vacancies. In radiative transitions into these vacancies, hard X rays are emitted, and by means of X-ray spectroscopy very interesting information can be obtained about the energies of the various levels of the superexcited atoms, the transition probabilities, and the charge states of the recombining ions as well.

The spectrum of K-series X rays, emitted when slow Kr^{35+} ions with a weak Kr^{36+} nuclei contamination are incident upon an aluminum target, and obtained by a semiconductor X-ray spectrometer, is shown in Figure 12.21 [17]. One can see that the components of this spectrum are very different from the components of the spectrum of krypton with only one K-shell vacancy. This allows one to use ion-at-surface spectroscopy for EBIS diagnostics as well. In fact, for example, if highly charged ions leave the electron beam in the radial direction and then are incident upon the surface of the drift tube, it is possible to detect this process by ion-at-surface X-ray spectroscopy.

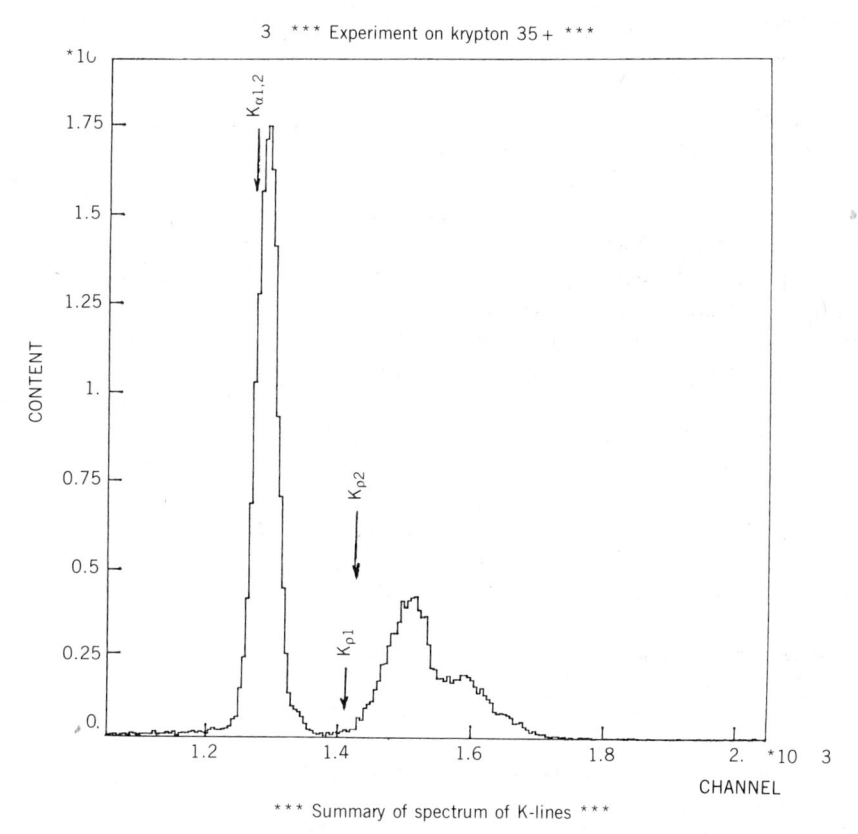

Figure 12.21 Spectrum of K-series X rays for Kr^{35+} ion recombination on the surface of a solid target. The values of energy of the diagram lines for krypton are given.

6 PROBLEMS AND PROSPECTS OF EBIS DEVELOPMENT

Up to now, electron beam ion sources have provided the highest charge state ions of all types of ion sources, and doubtless in the near future, by using EBIS, the most difficult problem of ionization—the production of slow uranium nuclei—will be solved.

A significant advantage of the EBIS is the very small emittance of the ion beam produced. The normalized ion beam emittance of this type of source is about 10^{-7} m · rad, and it can be smaller. This small emittance is due to the fact that during extraction ions acquire a considerable longitudinal velocity in a region with radial focusing by the electron beam space charge.

Undoubtedly, the possibility of investigating, within the EBIS itself, various fundamental phenomena that were previously inaccessible, is another advantage of the EBIS.

The relatively severe technical complications of EBIS, in its present state of development, are a disadvantage, of course. In fact, only the simultaneous use in the same device of superconductivity, super high vacuum, super dense electron beam, super low temperature, and some other supers, can produce super high charge states in EBIS. This makes construction of a modern EBIS a rather complicated and expensive activity.

A relatively low intensity of highly charged ions is the main disadvantage of the EBIS at present. The value of 10^{11} elementary positive charges for the quantity of ions per pulse can be taken as what is achievable for nuclear beams of light and moderate elements, and for heavy elements the value is smaller by approximately an order of a magnitude.

The prospects for development and broadening of EBIS applications will depend mainly on progress in producing progressively more dense electron beams. In the first step toward solving this problem, by using classical methods of electron beam formation, one can expect an increase in beam density up to 10^4 A/cm^2. But future progress will be connected with the investigation and application of the discovered phenomenon of ion supercompression of the electron beam [23].

REFERENCES

1. E. D. Donets, USSR Inventor's Certificate No. 248860, 16 March 1967, Byull. OIPOTZ No. 23, 1969, p. 65.
2. E. D. Donets, V. I. Ilushchenko, and V. A. Alpert, Proc. Premiere Conf. sur les Sources d'Ions, INSTM, Saclay, France, 1969, p. 625.
3. E. D. Donets, *Phys. Elementary Particles At. Nucleus* **13,** 941 (1982).
4. W. Lotz, *Z. Phys.* **216,** 341 (1968).
5. E. D. Donets, *Proceedings, Fifth All-Union Conference on Charged-Particle Accelerators*, Vol. 1, Nauka, Moscow, 1977, p. 346.
6. J. Faure and B. Feinberg, *Nucl. Instrum. Methods* **219,** 449 (1984).
7. G. S. Janes et al., AVCO-Everett Research Laboratory, Res. Report 235, 1965.

8. R. Becker, Proceedings, II EBIS Workshop, Saclay-Orsay, 1981, J. Arianer and M. Olivier (Eds.), p. 185.
9. G. I. Budker, *Atomic Energy* **5,** 9 (1956).
10. A. G. Bonch-Osmolovskiy, Preprint JINR P9-8379, Dubna, 1974.
11. H. Ahmed and A. N. Broers, *J. Appl. Phys.* **43,** 2186 (1972).
12. K. Amboss, Proceedings, II EBIS Workshop, Saclay-Orsay, 1981, Arianer and M. Olivier (Eds.), p. 59.
13. E. D. Donets and V. P. Ovsyannikov, Preprint JINR, P7-9799, Dubna, 1976.
14. J. Arianer et al., GSI-P-3-77, Darmstadt, 1977, p. 65.
15. E. D. Donets, *IEEE Trans. Nucl. Sci.* **NS-23,** 897 (1976).
16. E. D. Donets and A. I. Pikin, *JETP* **45,** 2373 (1975).
17. E. D. Donets, S. V. Kartashov, and V. P. Ovsyannikov, JINR Rapid Comm. No. 20-86, Dubna, 1986, p. 27.
18. E. D. Donets, *Phys. Scr.* **T3,** 11 (1983).
19. J. Arianer et al. *IEEE Trans. Nucl. Sci.* **NS-26,** 3713 (1979).
20. J. Arianer and C. Goldstein, Rep. IPNO NO. 79-02, Orsay, 1979.
21. I. G. Brown and B. Feinberg, *Nucl. Instrum. Methods* **220,** 251 (1984).
22. M. A. Levine, R. E. Marrs, and R. W. Schmieder, *Nucl. Instrum. Methods A* **237,** 429 (1985).
23. J. Arianer and C. Goldstein, *IEEE Trans. Nucl. Sci.* **NS-23,** 979 (1976).
24. V. P. Vadeev, Preprint JINR, P9-81-660, Dubna, 1981.
25. R. Becker, H. Klein, and M. Kleinod, GSI-P-3-77, Darmstadt, 1977, p. 35.
26. V. A. Alpert et al., Preprint JINR D7-5769, Dubna, 1971.
27. V. P. Vadeev et al., Preprint JINR, P7-10823, Dubna, 1977.
28. E. D. Donets, *Nucl. Instrum. Methods B* **9,** 522 (1985).
29. B. Gastineau, J. Faure, and A. Cortois, *Nucl. Instrum. Methods B* **9,** 538 (2985).
30. R. W. Hamm, L. M. Choate, and R. A. Kenefick, *IEEE Trans. Nucl. Sci.* **NS-23,** 1013 (1976).
31. R. A. Kenefick and R. W. Hamm, GSI-P-3-77, Darmstadt, 1977, p. 21.
32. R. Becker and H. Klein, *IEEE Trans. Nucl. Sci.* **NS-23,** 1017 (1976).
33. R. W. Schmieder, EBIS TN-017, SNLL, 1986.
34. V. O. Kostroun et al., Proceedings II EBIS Workshop, Saclay-Orsay, 1981, J. Arianer and M. Olivier (Eds.), p. 30.
35. V. O. Kostroun et al., *Phys. Scr.* **T3,** 47 (1983).
36. J. Arianer et al., Proceedings II EBIS Workshop, Saclay-Orsay, 1981, J. Arianer and M. Olivier (Eds.), p. 240.
37. J. Arianer et al., *Phys. Scr.* **T3,** 36 (1983).
38. T. Twai et al., *Phys. Rev. A* **26,** 105 (1982).
39. S. Ohtani, *Phys. Scr.* **T3,** 110 (1983).
40. W. G. Abdulmanov et al., Proceedings, Tenth Int. Conf. on HEPA, Vol. 1, Serpukhov, 1977, p. 345.
41. E. D. Donets and V. P. Ovsyannikov, *JETP* **80,** 916 (1981).
42. B. Bochev, T. Kutsarova, and V. P. Ovsyannikov, Preprint JINR, P5-11566, Dubna, 1978.
43. B. Bochev, T. Kutsarova, and V. P. Ovsyannikov, Preprint JINR P9-11567, Dubna, 1978.
44. E. D. Donets et al., Preprint JINR, P7-83-627, Dubna, 1983.

13

Beam–Plasma Ion Sources

J. Ishikawa
Kyoto University
Kyoto, Japan

Ion sources in which a controlled electron beam with an energy of several keV is used for plasma production can be classified into two groups. The first group is the EBIS (electron beam ion source) for multiply charged ion production [1–3]. This ion source is operated at a very low gas pressure. The electron beam produces ions only by collisional ionization, and does not actively interact with a generated plasma. In the second group, the main ionization mechanism is a beam–plasma discharge that Ishikawa and Takagi have proposed as an effective plasma production method for ion source [4–12]. Experiments on beam–plasma discharges were also investigated by Demirkhanov et al. [13, 14], but they did not extract ions as a beam.

In the beam–plasma ion source [4–12] an electron beam is effectively utilized for the dual purpose of high-density plasma production, as a result of the beam–plasma discharge, and high-current ion beam extraction with ion space-charge compensation.

The plasma of a high-current ion source should have both a high plasma density and a high electron temperature, because the ion saturation current is proportional to the product of the plasma density and the square root of the electron temperature. At the same time, the ion temperature should be low, so as to obtain an extracted ion beam with a low emittance. The combination of microwave discharge and electron cyclotron resonance is suitable for the production of an ion source plasma, since the plasma electrons are selectively heated. When both the generation of the microwaves and ionization by heated plasma electrons take place in the same region, an effective plasma production is realized. By using beam–plasma interactions, microwave power suited for the ion source plasma can be generated within the plasma, when the electron beam conditions and the interaction dimensions are appropriate. This kind of plasma is called a *beam–plasma discharge* [12].

TABLE 13.1 Main Parameters for Beam–Plasma Ion Sources C to F

Parameter[a]	Source				
	C	D	D'	E	F
V_{ex}	8–10	8–10	5–8	1–5	1–8
B_{max}	0.6	4	4	4	3
b	0.55	0.9	0.9	0.9	2.0
L_i	30	18	18	27	63
V_{cd}	0.2	0.2	0.2	1–5	1–8

[a] V_{ex} (kV), ion extraction voltage (=primary electron beam energy); B_{max} (kG), maximum magnetic field in the drift space; b (cm), inner radius of the drift tube; L_i (cm), length of the drift tube (=length for the beam–plasma interactions); V_{cd} (kV), secondary-electron acceleration voltage.

When ions are extracted from a high-density plasma, space-charge effects between the ions are an obstacle to high-current extraction. Therefore, normally an ion current of hundreds of miliamperes cannot be extracted from a single aperture at an extraction voltage of several kilovolts. One method for solving this problem is ion space-charge compensation in the extraction region by externally imposed electron space charge [8]. For this purpose, a primary electron beam is injected into the plasma from the ion extraction electrode, and a thermalized secondary electron beam from the plasma region is introduced into the ion extraction region. In this way an ion current of more than several tens times the normal space-charge-limited current can be extracted from a single aperture under the ion space-charge compensation condition.

1 ION SOURCE CHARACTERISTICS

Beam–plasma ion sources (versions A to F) have been developed by Ishikawa and Takagi [4–12]. In Figures 13.1 and 13.2, the beam–plasma ion sources E and F are shown, respectively. The main parameters for ion sources C to F are described in Table 13.1. The extracted ion current as a function of the primary electron beam current for these ion sources is shown in Figure 13.3. A continuous hydrogen ion current of 0.6 A was extracted from the ion source F. The extraction voltage was about 8 kV; thus, ion extraction with space-charge compensation factor (ratio of ion current with space-charge compensation to the normal space-charge-limited ion current) of about 20 was realized. In this ion source a nitrogen ion current of the same order was extracted, the compensation factor then was about 100.

This kind of ion source can be operated at a low gas pressure, in the range 10^{-4} to 10^{-3} Torr. The gas efficiency (fraction of gas atoms that are extracted as ions) is high, 40 to 90% for the ion sources E and F. The energy spread for the ion source F was 20 to 40 eV at the hydrogen ion extraction current of 0.05 to 0.5 A.

2 Source Construction

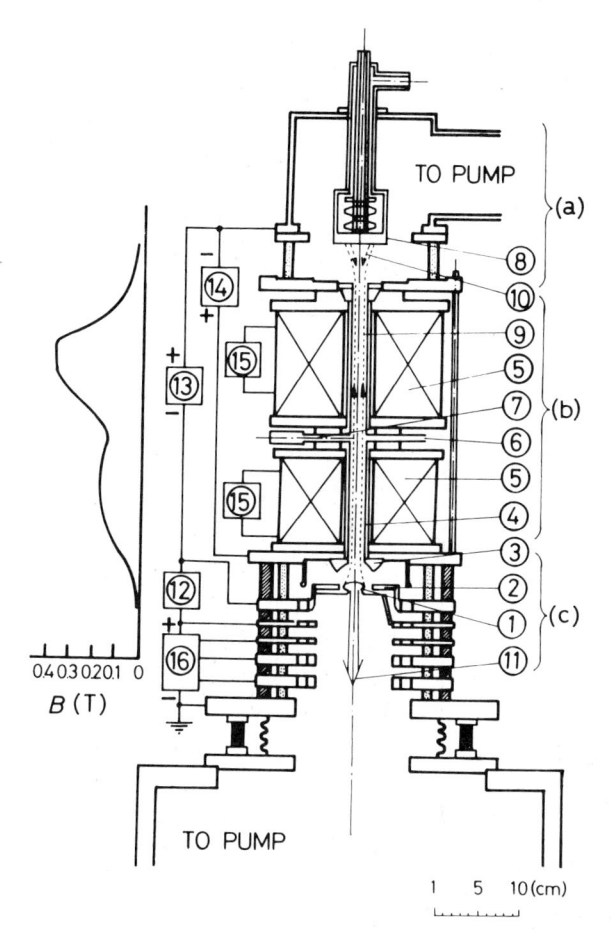

Figure 13.1 Schematic diagram of the beam–plasma ion source, version E. (*a*) Electron beam collector region, (*b*) drift space, (*c*) ion extraction and electron gun region. 1, Filament cathode; 2, ion extraction electrode (Wehnelt); 3, plasma expansion cup (anode); 4, drift tube; 5, solenoid; 6, gas inlet; 7, probe; 8, collector; 9, primary electron beam; 10, secondary electron beam; 11, ion beam; 12, filament power supply; 13, primary electron beam acceleration power supply; 14, secondary electron beam acceleration power supply; 15, solenoid power supply; 16, ion acceleration power supply.

The normalized emittances for the ion sources A and B, C and D, and E and F were about $4 \times 10^{-8} \pi$ m · rad for about 1 mA, $2-3 \times 10^{-7} \pi$ m · rad for 2–10 mA, and $1-2 \times 10^{-6} \pi$ m · rad for 0.1 to 0.6 A, respectively. The normalized brightness for these ion sources was 10^{10} to 2×10^{11} A rad^{-2} m^{-2}.

2 SOURCE CONSTRUCTION

Figure 13.4 shows the operating principle of the ion source. The ion source consists of an ion extraction region with an electron gun, a drift space, and an electron

284   *Beam–Plasma Ion Sources*

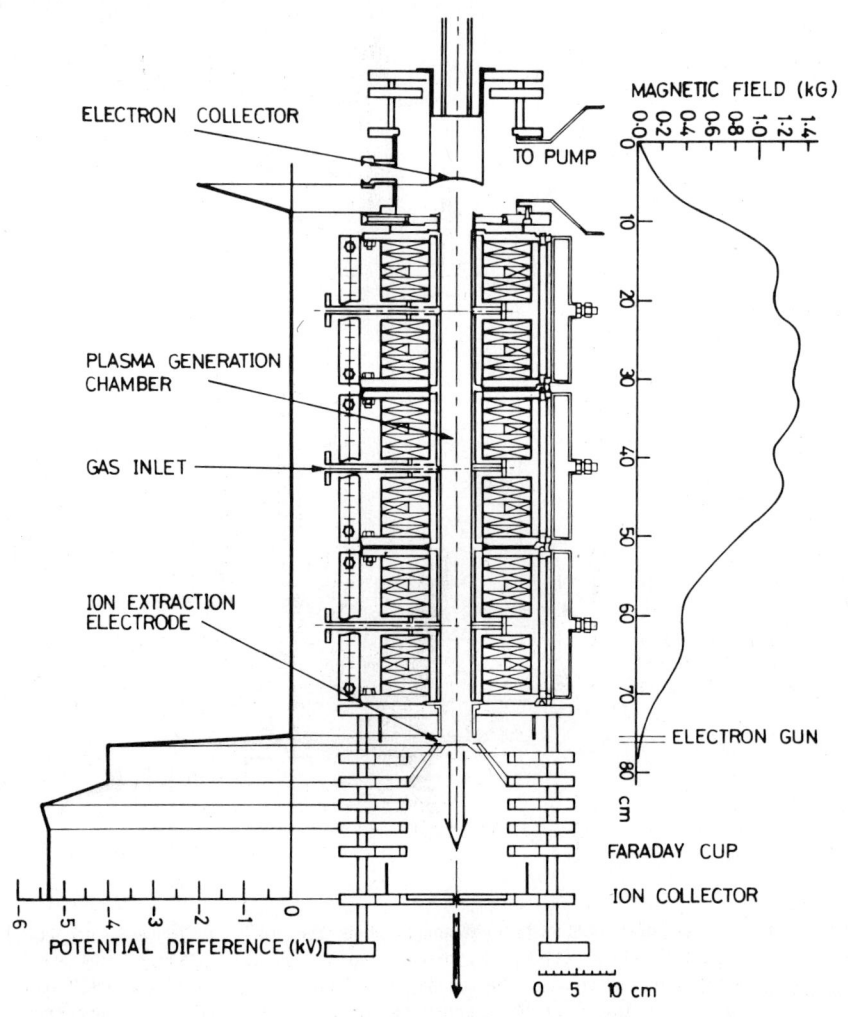

Figure 13.2 Schematic diagram of the beam–plasma ion source, version F. The axial electrostatic potential distribution is shown.

beam collector region. The electron beam formed by the electron gun is injected into the drift tube and passes through the drift tube to the collector. The electron beam interacts with a thin background plasma in the drift space, and a high-density plasma (10^{11}–10^{13} cm^{-3}) is produced by the beam–plasma discharge. The ions are extracted by the same electric field that accelerates the electron beam. The space charge of the ion beam is compensated for by the negative space charge of the primary electron beam and the backward secondary electron beam.

2 Source Construction

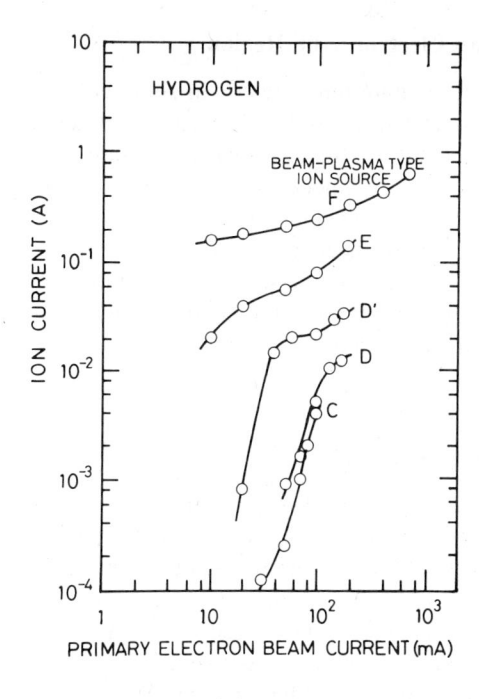

Figure 13.3 Typical ion extraction current as a function of the primary electron beam current for the beam–plasma ion sources C, D, D′, E, and F.

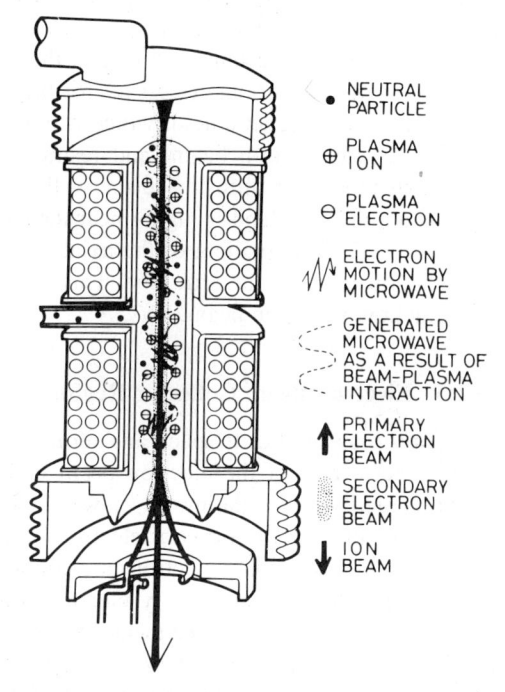

Figure 13.4 Illustration of the operational principle of the beam–plasma ion source.

2.1 Ion Extraction and Electron Gun Region

A directly heated filament cathode made of tungsten wire, or in some cases a cylindrical cathode made of tantalum or tungsten, is used. Electrons are extracted from the cathode by the focusing electric field of a high-perveance structure with an aspect ratio of about 1.0, which is formed by anode, Wehnelt, and cathode electrodes. The electron beam thus extracted is injected into the drift space to form a thin beam due to the focusing magnetic field of the solenoid (the beam diameter is typically about 1 cm). A high positive voltage, with respect to the cathode, is applied to the anode, drift tube, and electron beam collector electrodes, whereby the ions produced in the drift tube are extracted from the plasma that has diffused into the plasma expansion cup.

2.2 Drift Space

The plasma production chamber is a long, thin tube (drift tube) made of electrically conductive materials such as stainless steel or copper. The gas flow conductance is low enough to restrain the neutral particles from flowing out to the high-vacuum region and to maintain the proper gas pressure for ionization in the drift space. The gaseous material to be ionized is usually introduced from a gas inlet located in the middle of the drift tube. When metal vapor is ionized, the whole drift tube should be heated to a high temperature [10]. Measured gas flow rate for hydrogen was 1.15×10^{-2} Torr L/s at a mean gas pressure in the drift tube of 1×10^{-3} Torr, for the ion source E. The inner radius and the length of the drift tube should be determined by the conditions necessary for the occurrence of the beam–plasma discharge. The axial magnetic field, which is produced by water- or oil-cooled solenoids, focuses the electron beam, and restrains the high-density plasma produced in the drift tube from diffusing to the wall. The distribution of the axial magnetic field is not uniform, and is higher near the collector region, as shown in Figures 13.1 and 13.2. The gradient of the axial magnetic field causes plasma ions to move to the lower end of the drift tube. This magnetic field configuration enhances the extraction of the ion beam.

2.3 Electron Beam Collector Region

The electron beam passes through the drift space and terminates at the electron beam collector electrode. As a part of the electron beam energy is lost at the collector, it needs to be cooled with insulating oil or water. When the electron beam from the electron gun, that is, the primary electron beam, bombards the surface of the collector electrode, secondary electrons are emitted. The secondary electron emission factor δ depends on the primary electron beam energy at the collector and the collector surface. Typically, the measured δ was 0.9 for a copper collector at a primary electron beam energy of 5 keV. A negative voltage is applied to the collector with respect to the drift tube, controllable from zero up to the primary electron-beam acceleration voltage. Secondary electrons emitted from the collector are accelerated back and form a secondary electron beam within the drift tube. This secondary electron beam can produce a relatively high-density background plasma for initiation of the main beam–plasma discharge. When the high-energy secondary electron beam is effectively used, the behavior of both the beam–plasma discharge and also the ion beam extraction is greatly improved. The collector region should be kept at high vacuum ($P < 10^{-4}$ Torr) to allow high-voltage hold-off between collector and drift tube.

3 BEAM–PLASMA DISCHARGE [12]

The beam–plasma interaction phenomenon is essentially nonlinear. However, most of the results derived from a linear theory can be useful in the case where the oscillation amplitude is small. As the oscillation power is increased, by increasing the electron beam current, the energy spectrum of the electron beam tends to broaden as shown in Figure 13.5; in this case the beam–plasma system should be analyzed by a quasilinear theory.

When the energy distribution of the electron beam is a δ-function, hydrodynamic instability occurs due to the active coupling between a space-charge wave on the electron beam and a high-frequency plasma wave. In conjunction with the occurrence of the instability, a high-frequency oscillation with comparatively small amplitude is also generated. When the beam and plasma parameters satisfy the condition for strong growth of the oscillation amplitude, the velocity distribution

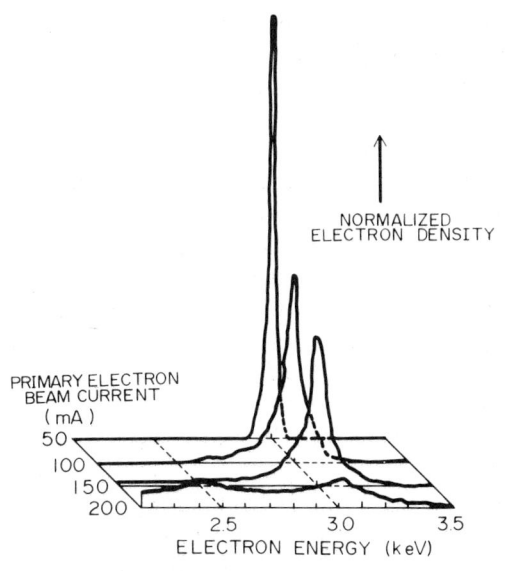

Figure 13.5 Energy distribution in the vicinity of the electron beam acceleration voltage 3 kV with variation in the electron beam current in beam–plasma system.

of the electron beam becomes wider. Then, the hydrodynamic instability weakens, and a kinetic instability due to the shape of the electron velocity distribution grows. This instability shows strong growth when $v_{th}/v_b > (n_b/n_i)^{1/3}$, where v_{th} and v_b are the thermal and directed velocities of the electron beam, respectively, and n_b and n_i are the electron beam density and the plasma ion density, respectively [15]. The thermal spread of the electron beam increases, and when the electron velocity distribution function eventually becomes flat, the velocity distribution cannot change any more. This state is called quasilinear relaxation of the beam–plasma system [16, 17].

3.1 Beam–Plasma Discharge Due to Hydrodynamic Instability

The waves that exist in the beam–plasma system are space-charge waves on the electron beam, slow and fast cyclotron waves, and plasma waves [18]. By comparing the values of the imaginary parts of the propagation constant and the angular frequency, it can be shown that the interaction between the space-charge wave on the electron beam and the high-frequency plasma wave (i.e., an absolute instability) takes place most easily. The interaction frequency is extremely close to the upper hybrid frequency, $f = (f_{pe}^2 + f_{ce}^2)^{1/2}$, where f_{ce} is the electron cyclotron frequency and f_{pe} is the plasma frequency. The imaginary part of the frequency changes drastically at the point where the electron cyclotron frequency is nearly equal to the plasma frequency.

Figure 13.6a shows the variation of the imaginary part of the frequency ω_{si} as a function of the plasma density n_i. The ω_{si} curve calculated without collisions has a maximum value at a plasma density of n_{max}. Meanwhile, the collision frequency ν, corresponding to the collision effects by neutral gas and Coulomb force which restrain the beam–plasma system from oscillation, increases with plasma density. In general, the collision term reduces the value of ω_{si} in proportion to the collision frequency [19], and therefore the curve due to collisions can be shown on the same figure. The difference between those two curves, indicated by the hatching, shows the virtual value of ω_{si}, corresponding to growth of the oscillation amplitude with time. By the above considerations, the two cross points, n_{min} and n_{i0}, at which the ω_{si} curve without collisions and the curve with collisions intersect, mean the following. When the plasma density of the beam–plasma system is less than n_{min}, oscillations cannot start. When the plasma density is greater than n_{min}, oscillations will start, and the plasma density begins increasing. Moreover, the amplitude of the oscillation is increased due to the positive feedback, and the plasma density ultimately obtained becomes n_{i0}. Therefore, to raise the plasma density in the drift tube, the external conditions should be changed to increase the value of n_{i0}, which is strongly dependent on electron beam current, voltage, and magnetic field. Figure 13.6b shows the plasma density measured as a function of the voltage between collector and drift tube in the beam–plasma ion source C. In the voltage region greater than 30 V, secondary electrons have a sufficient ionization cross section to produce a threshold plasma density in the drift tube. Then the plasma density increases suddenly because of the occurrence of the beam–plasma discharge. The

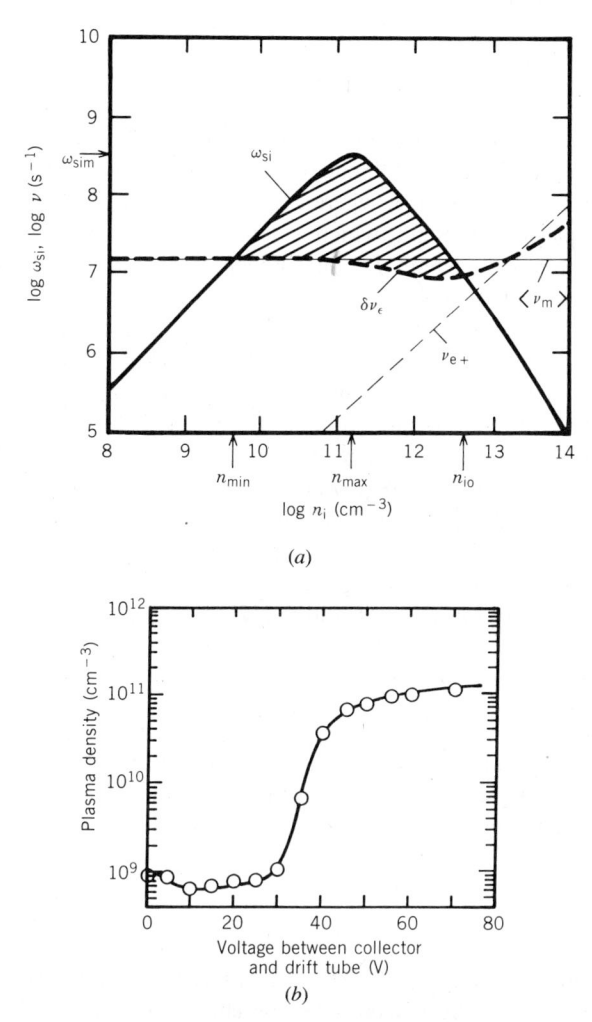

Figure 13.6 (a) Variation in the imaginary part of angular frequency ω_{si} as a function of the plasma density n_i. Collision terms ($\delta\nu_e$, which reduces ω_{si} due to collision frequency; ν_{e+}, collision frequency between plasma electrons and plasma ions; and $\langle \nu_m \rangle$, effective collision frequency between plasma electrons and neutral particles) are also indicated. The curves for ω_{si} and $\delta\nu_e$ have two intersections of n_{min} and n_{i0}. The ω_{si} curve has a maximum value (ω_{sim}) at n_{max}. Conditions of the calculation: electron beam energy $V_p = 5$ kV, current $I_p = 100$ mA, magnetic field $B = 0.1$ T, hydrogen gas pressure $P = 1 \times 10^{-3}$ Torr, inner radius of the drift tube $b = 1$ cm, and plasma electron temperature $T_e = 5$ eV. (b) Plasma density measured as a function of the voltage between collector and drift tube in the beam-plasma ion source C. The conditions of experiment: $V_p = 10$ kV, $I_p = 90$ mA, $B = 0.1$ T, $b = 0.55$ cm, $P = 5 \times 10^{-3}$ Torr (hydrogen).

3.2 Quasilinear Relaxation

Using a simplified one-dimensional model where the oscillation frequency is close to the plasma frequency, the growth rate of the wave γ_i due to the kinetic instability can be given as $\gamma_i = \omega_{pe}(n_b/n_i)$, where ω_{pe} is the angular plasma frequency [16, 17]. When the quasilinear length, which is defined as v_b/γ_i, is nearly equal to the beam–plasma interaction length L_i, quasilinear relaxation of the beam–plasma system takes place completely [16]. In this state the oscillation amplitude of the nonlinear beam–plasma interaction is strongest. Then, a plasma suitable for an ion source is produced because of the effective heating of the plasma electrons. A radial boundary condition for the beam–plasma interaction should be considered because the beam–plasma interaction takes place in a drift tube with a wall at a finite radial distance. An empirical relation $v_b = \omega_{pe}/p$ between the radial wave number p, the angular plasma frequency ω_{pe}, and the electron beam velocity v_b should be satisfied so that the energy of the electron beam can be most efficiently converted into the axial electromagnetic energy of high-frequency wave [20].

From the optimum condition for quasilinear relaxation one obtains the electron beam current necessary for its optimization [12],

$$I_b = BL_i^{-1}V^{3/2} \qquad (1)$$

where V is the electron beam voltage and B is a proportionality coefficient. On the other hand, in the case of finite length of the drift tube one obtains the condition for the start of hydrodynamic instability [18],

$$I_b = AL_i^{-3}V^2 \qquad (2)$$

where A is a proportionality coefficient. A measurement of the microwave power in the drift tube measured by a coaxial probe is shown in Figure 13.7. Figure 13.7a shows the relation between the primary electron beam current and the detected microwave power when the primary electron beam voltage was varied and the secondary electron beam was set to zero. Both the beginning of the oscillation due to hydrodynamic instability and the strong oscillation due to quasilinear relaxation were observed. The dependence of the observed currents corresponding to oscillations on the electron beam voltage is illustrated in Figure 13.7b. The electron beam current for strong oscillation is proportional to the three-halves power of the voltage, as predicted by Eq. (1).

The plasma density n_i and the electron temperature T_e in the case of quasilinear relaxation are given by

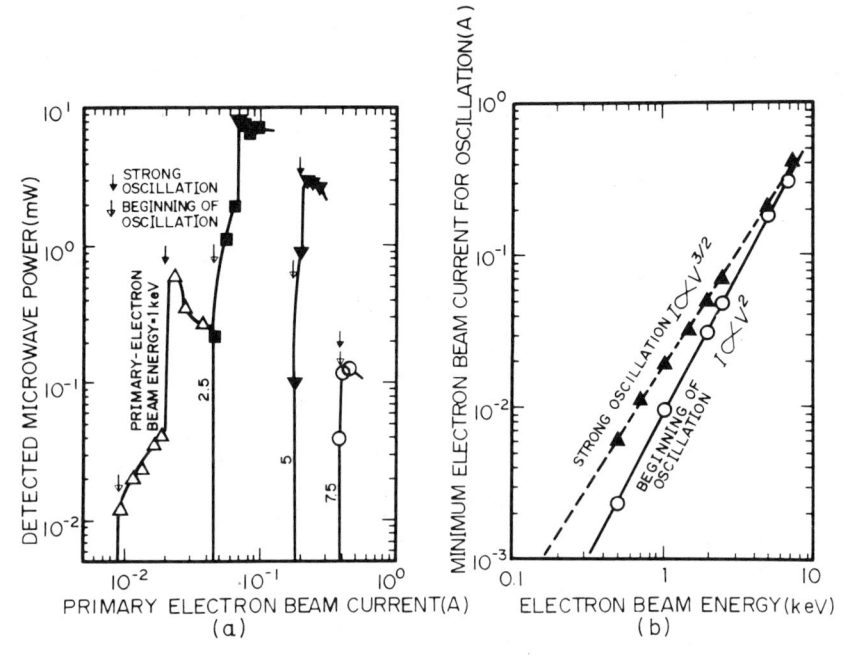

Figure 13.7 (a) Measurement of the microwave power in the drift tube detected by a coaxial probe as a function of the primary electron beam current. (b) The relation between electron beam current and voltage for strong oscillations and the beginning of oscillation.

$$n_i = \left(\frac{\epsilon_0 m_e \, \epsilon_{nm}^4}{\pi^2 \, e^4}\right)^{1/3} \left(\frac{L_i}{b}\right)^{2/3} \frac{1}{b^2} I_b^{2/3} \quad (3)$$

$$\frac{kT_e}{e} = \frac{2}{3} \frac{n_b}{n_i} \frac{m_e v_b^2}{e} \propto I_b^{2/3} \quad (4)$$

where b is the inner radius of the drift tube and ϵ_{nm} the mth zero of Bessel function $J_n(x)$ [12, 17].

4 ION EXTRACTION WITH ION SPACE-CHARGE COMPENSATION [8]

The influence of ion space charge is disadvantageous to both the quality and quantity of the extracted ion beam. In beam–plasma ion sources, the ion beam is extracted with the ion space charge neutralized by electrons to ameliorate the undesirable influence of the ion space charge. The electrons used for ion space-charge compensation can be divided into three groups: the primary electron beam, the secondary electron beam, and the plasma electrons.

292   *Beam–Plasma Ion Sources*

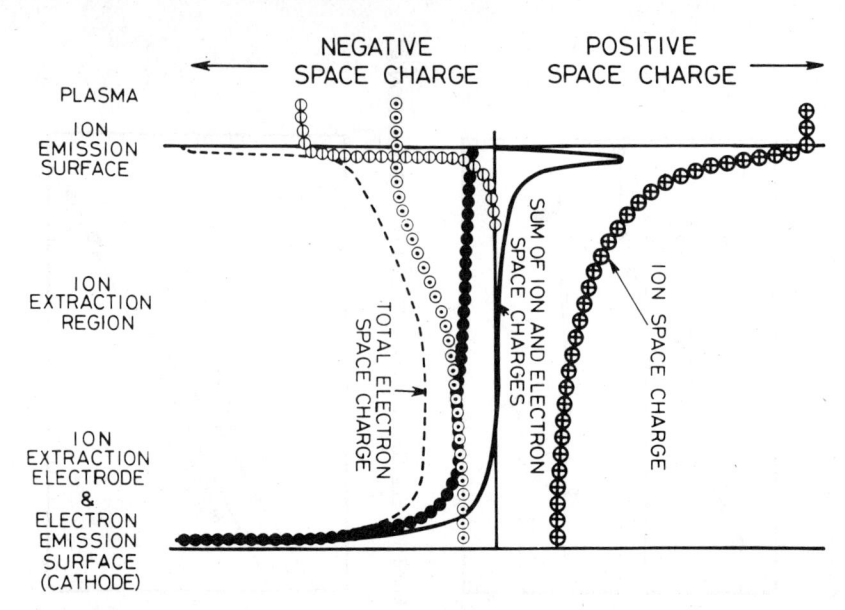

Figure 13.8 Illustration of ion space-charge compensation by the various energetic electrons distributed in the ion extraction region of the beam–plasma ion source. ⊕, Ion space charge; •, space charge of primary electron beam; ⊖, space charge of low-energy plasma electrons; ⊙, space charge of high-energy plasma electrons and secondary electron beam.

Figure 13.8 illustrates ion space-charge compensation by the various energetic electrons that are distributed in the ion extraction region of this ion source. The ion space-charge density near the ion emission surface is high, because the ion velocity is very low, and it decreases gradually with proximity to the ion extraction electrode. The low-energy electrons neutralize much of the ion space charge near the ion emission surface. The secondary electron beam and the high-energy plasma electrons neutralize the ion space charge within the middle region. The primary electron beam neutralizes the ion space charge near the ion extraction electrode. Thus, these groups of electrons in all can effectively neutralize the entire ion space charge of the extraction region.

4.1   Relaxation of the Space-Charge Limitation of Extractable Ion Current

For simplicity, we assume that the electrons in the ion extraction region consist of the following three groups: (1) the primary electron beam extracted from the cathode electrode; (2) the group of high-temperature plasma electrons consisting of the secondary electron beam together with the high-energy plasma electrons; and (3) the normal low-temperature plasma electrons with Maxwellian velocity distribution. Then, Poisson's equation in the ion extraction region can be written

4 Ion Extraction with Ion Space-Charge Compensation

as [8, 13, 14]

$$\frac{d^2U}{dz^2} = \frac{1}{\epsilon_0}\left[J_i\left(\frac{m_i}{2e(U+U_0)}\right)^{1/2}\right.$$
$$- J_e\left(\frac{m_e}{2e(V-U)}\right)^{1/2} - en_{e1}\exp\left(-\frac{eU}{kT_{e1}}\right)$$
$$\left. - en_{e2}\exp\left(-\frac{eU}{kT_{e2}}\right)\right] \quad (5)$$

where m_i and m_e are the ion and the electron masses, J_i and J_e are the ion and the electron current densities, U_0 is the initial plasma ion energy, and n_{e1} and n_{e2} are the plasma electron densities with low (T_{e1}) and high (T_{e2}) electron temperatures. To facilitate understanding, the potential U is represented in a reverse sign. To obtain the extractable current, Eq. (5) should be solved using proper boundary conditions. The extractable ion current density in the presence of the space-charge-limited primary electron beam is given by [8]

$$J_i = J_{si}\gamma \quad (6)$$

where J_{si} is the normal space-charge-limited ion current,

$$\gamma = \left\{\frac{3}{4}\int_{y_0}^{1}\frac{dy}{\left[(y+y_0)^{1/2} - y_0^{1/2} + \frac{(1-y)^{1/2} - 1}{y_0^{1/2}} + \frac{y_0+1}{y_0^{3/2}}C(y)\right]^{1/2}}\right\}^2 \quad (7)$$

and

$$C(y) = \frac{\frac{\alpha_1}{2}(1-\lambda)(e^{-y/\alpha_1} - 1) + \frac{\alpha_2}{2}\lambda(e^{-y/\alpha_2} - 1) + 1 - (1-y)^{1/2}}{\frac{2}{\alpha_1}(1-\lambda) + \frac{2}{\alpha_2}\lambda + 1} \quad (8)$$

where

$$y = \frac{U}{V} \quad y_0 = \frac{U_0}{V} \quad \alpha_j = \frac{kT_{ej}}{eV} \quad \lambda = \frac{n_{e2}}{n_{e1} + n_{e2}} \quad (9)$$

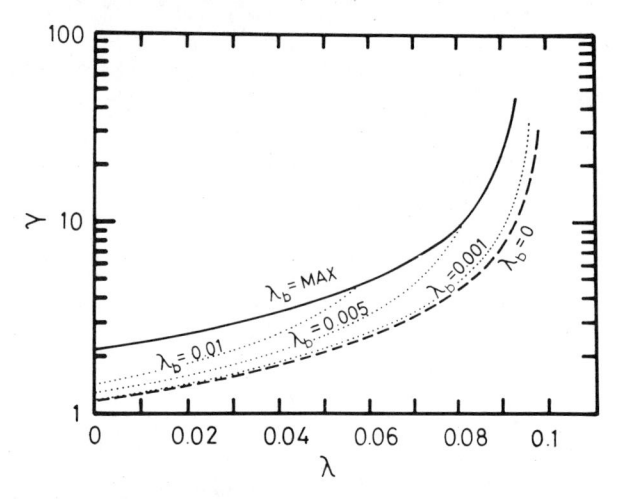

Figure 13.9 Relation between the space-charge compensation factor γ and the ratio of high-temperature plasma electron density to total plasma electron density λ. λ_b is the ratio of primary electron beam density to total plasma electron density. The conditions in the calculation are an ion extraction voltage of 5 kV and plasma electron temperatures of 2.5 keV (high) and 5 eV (low).

The space-charge compensation factor γ, indicated in Eq. (7), means the degree of the relaxation of ion space-charge limitation due to the space-charge compensation by electrons in the ion extraction region. Figure 13.9 shows the dependence of the space-charge compensation factor γ on the ratio λ of high-temperature plasma electron density to total plasma electron density. In the figure, the solid curve labeled λ_b = max shows the case where the primary electron beam is injected at its space-charge limitation, the thick broken curve labeled λ_b = 0 shows the case without the primary electron beam [13, 14], and the thin broken curves show the case where the primary electron beam is injected in the temperature limitation. As the high-temperature plasma electron density is increased, the space-charge compensation factor grows quickly, reaching a few tens near λ = 0.1. When the primary electron beam is space-charge limited, the compensation factor is about twice as large as that without the primary electron beam [21, 22]. Thus, with a primary space-charge-limited electron beam and adequate high-temperature plasma electron density, such as provided by microwaves and the thermalized secondary electron beam, a high-current ion beam can be obtained from a single-aperture extraction system because of the drastic relaxation of the ion space-charge limitation.

4.2 Extracted Ion Current

Figure 13.10 shows the extracted ion current and classification of the discharge region as functions of gas pressure and secondary electron beam voltage. In the beam–plasma discharge regime (Region I in the figure), the extracted ion current

4 Ion Extraction with Ion Space-Charge Compensation   295

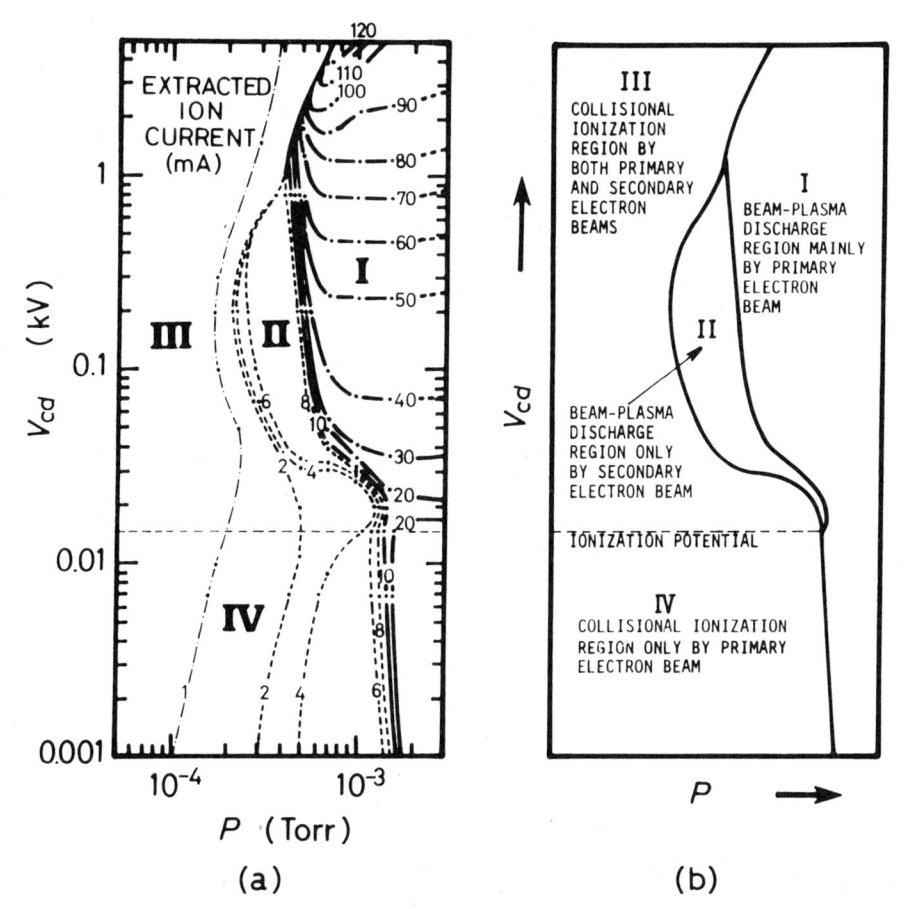

Figure 13.10 (*a*) Extracted hydrogen ion current, and (*b*) classification of the discharge region, as a function of the gas pressure in the drift tube P, and the secondary electron beam voltage V_{cd}, under the condition of a primary electron beam voltage of 5 kV and a current of 300 mA.

increases with increasing secondary electron beam energy. The normal perveance of the ion extractor in this ion source was 6×10^{-8} AV$^{-3/2}$ for H$_2^+$, so that the space-charge-limited ion current without ion space-charge compensation was calculated as 21.2 mA for an extraction voltage of 5 kV. The extracted ion current at the low secondary electron beam voltage in Region I was about 20 mA. It was found that most of the ion space charge was not compensated by electrons. In the case of the high-energy secondary electron beam, the thermalized secondary electron beam and the high-energy plasma electrons had neutralized the ion space charge of the extractor effectively, the space-charge limitation was relaxed, and ion optics were improved. When the energy of the secondary electron beam was near that of the primary electron beam, a current several times larger than that which was produced without the ion space-charge compensation, was extracted.

5 SUMMARY

A high-density plasma of 10^{11}–10^{13} cm^{-3} is produced by the beam–plasma discharge. The electron beam current necessary for the beam–plasma discharge increases with electron beam voltage, but decreases with beam–plasma interaction length. For the case of low electron beam current (< 100 mA) and low secondary electron beam voltage (< 200 V), a stable plasma with a density of around 10^{11} cm^{-3} is produced by a hydrodynamic instability. When the electron beam current is increased (100–1000 mA) along with the secondary electron beam voltage (around the primary electron beam voltage), a strong oscillation is generated due to quasilinear relaxation, and a high-density plasma (10^{12}–10^{13} cm^{-3}) is produced. Then, both the primary and secondary electron beams are thermalized into plasma energy. The thermalized secondary electron beam is effectively distributed in the ion extraction region, and neutralizes the space charge of the extracted ion beam. In this way the space-charge limitation of extractable ion current is greatly relaxed (space-charge compensation factor: 5–100), and a high-current ion beam (20–600 mA for hydrogen or nitrogen) can be extracted at relatively low extraction voltage (5–8 kV) from a single aperture.

This ion source is suitable for low-voltage, high-current ion beam application fields.

REFERENCES

1. E. D. Donets, V. I. Ilyuschenko, and V. A. Alpert, Proceedings of the International Conference on Ion Sources, Saclay, France, 1969, p. 635.
2. E. D. Donets, *IEEE Trans. Nucl. Sci.* **NS-23**(2), 897 (1976).
3. J. Arianer and C. Goldstein, *IEEE Trans. Nucl. Sci.* **NS-23**(2), 979(1976); see also Chapter 12 on Electron Beam Ion Sources, by Donets.
4. T. Takagi, I. Yamada, and J. Ishikawa, 11th Symposium on Electron, Ion and Laser Beam Technology, Boulder, CO, 1971, p. 579.
5. T. Takagi, I. Yamada, J. Ishikawa, and H. Iwao, Proceedings of the 2d International Conference on Ion Sources, Vienna 1972, p. 367.
6. T. Takagi, I. Yamada, J. Ishikawa, F. Sano, and N. Kusano, Proceedings of the 2d Symposium on Ion Sources and Formation of Ion Beams, Berkeley, CA, 1974, p. III-2-1.
7. J. Ishikawa, F. Sano, H. Tsuji, and T. Takagi, Institute of Physics Conference Series, No. 38, 1978, p. 84.
8. J. Ishikawa, F. Sano, and T. Takagi, *J. Appl. Phys.* **53**, 6018 (1982).
9. J. Ishikawa, A. Motamed Ektessabi, and T. Takagi, *Jpn. J. Appl. Phys.* **22**, 309 (1983).
10. J. Ishikawa and T. Takagi, *Jpn. J. Appl. Phys.* **22**, 534 (1983).
11. J. Ishikawa, A. Motamed Ektessabi, and T. Takagi, *Nucl. Instrum. Methods* **207**, 487 (1983).
12. J. Ishikawa and T. Takagi, *J. Appl. Phys.* **54**, 2911 (1983).
13. R. A. Demirkhanov, Yu. V. Kursanov, and L. P. Skripal', *Sov. Phys. Tech. Phys.* **15**(7), 1047 (1971).
14. R. A. Demirkhanov, Yu. V. Kursanov, and L. P. Skripal', *Sov. Phys. Tech. Phys.* **18**(5), 684 (1973).

15. A. B. Mikhailovskii, *Theory of Plasma Instabilities*, Vol. 1, Consultants Bureau, New York, 1974.
16. M. Seidl and P. Sunka, *Nucl. Fusion* **7,** 237 (1967).
17. V. D. Shapiro, *Sov. Phys. JETP* **17,** 416 (1963).
18. R. J. Briggs, *Electron-Stream Interactions with Plasmas*, MIT, Cambridge, MA, 1964.
19. S. A. Self, *J. Appl. Phys.* **40,** 5217, 5232 (1969).
20. J. R. Conrad, *J. Appl. Phys.* **47,** 4859 (1976).
21. I. Langmuir and J. M. Mott-Smith, *Gen. Electr. Rev.* **27,** 449, 538, 616. 762, 810 (1924).
22. H. M. Mott-Smith and I. Langmuir, *Phys. Rev.* **28,** 727 (1926).

14

Laser Ion Sources

R. H. Hughes and R. J. Anderson
University of Arkansas
Fayetteville, Arkansas

Relatively low-cost, high-intensity lasers are now available to most laboratories. Irradiation of solid targets by these lasers can produce plasmas from which ions can be extracted and focused into a useful beam. This simple ion source should have a basic appeal because of its simplicity of operation, its production of copious quantities of highly charged states in a wide variety of elemental species, and its relatively small dimensions. However, it is still considered a novel source and its use is not yet widespread. It is hoped that the following discussion will familiarize the reader with the techniques necessary to produce and extract from dense laser-generated plasmas and to demonstrate the capability of this unique ion source.

1 BACKGROUND

Laser ion source is a generic term. For example, it may be used to describe the application of lasers to a collection of atoms to selectively ionize a wanted species by photoionization through resonance excitation (resonance ionization spectroscopy [1], RIS). The resulting ions can then be extracted as a beam. In fact, this technique is actively being pursued to produce isobarically and possibly isotopically pure ion beams that can be injected into accelerators, storage rings, or ion traps [2]. However, the most common usage of the term probably describes the generation and extraction of ions produced by the interaction of a high-power laser with solids.

A high-power laser beam focused upon the surface of a solid target at a power density of greater than 10^8 W/cm^2 will penetrate the surface where the electron density is low until the electron plasma frequency matches the laser frequency.

Since the light cannot penetrate past this cutoff point where rapid electron heating takes place, intense ionization results and a dense hot plasma is produced. The target material is explosively ablated as an extremely dense plasma plume along the direction of the greatest hydrodynamical pressure gradient, usually perpendicular to the surface. The rapidly expanding plasma blowoff may be further heated by the laser pulse when the electronic density drops sufficiently to allow further laser light penetration. Although electron recombination occurs during the initial phase of the plasma expansion, a freeze in the charge state distribution occurs later in time as the plasma density decreases in the advancing plume [3]. The character of the plasma is a strong function of the laser power density incident upon the target [4]. For example, the charge state multiplicity and the ion plasma flow velocity scale with the laser power density on the target, and the plasma directivity also increases to a point where the highest charge states are found only in plasma filaments projected along the normal to the target surface [4, 5]. For power densities less than 10^{12} W/cm^2, thermalized plasmas are produced with the highest charge states possessing the highest initial kinetic energies. For power densities in the 10^{10}–10^{11} W/cm^2 range, depending on the laser, charge states requiring 500 eV ionization energy can be formed in the plasma. Thus, Fe^{16+} and fully stripped carbon (C^{6+}) are observed in such plasmas. Figure 14.1 shows an iron spectrum obtained by Phaneuf [6] using a 10-J, 80-ns, pulsed CO$_2$ laser to irradiate a solid iron target at an estimated power density of 3×10^{10} W/cm^2. Ion kinetic energies in the keV range are typical for the higher charge states [5].

For laser power densities greater than 5×10^{12} W/cm^2 the electron temperatures are sufficiently large to produce charge separation within the plasma. The hot electrons precede the ion component in the plasma plume and an accelerating potential for the ions is created [7, 8]. This effect has been observed for laser power

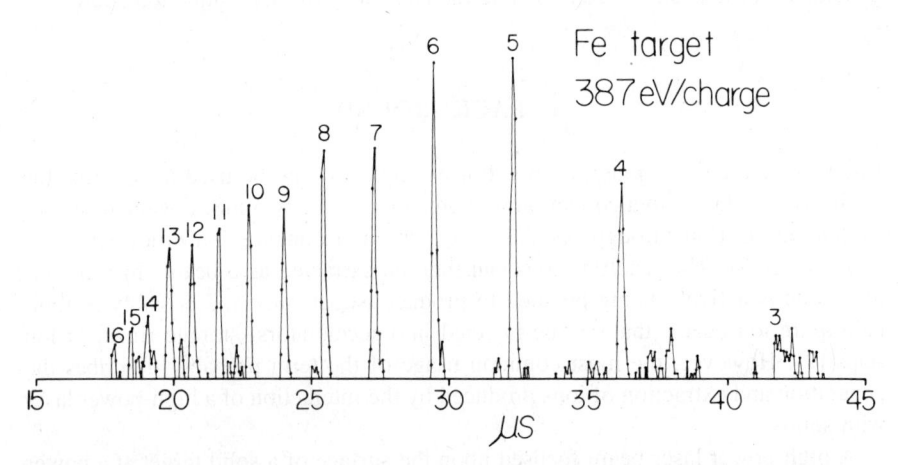

Figure 14.1 Charge state spectrum produced by electrostatic analysis of a CO$_2$ laser-produced iron plasma. The power density is about 3×10^{10} W/cm^2. The figure is taken from Phaneuf [6], © 1981 IEEE.

densities of 10^{15} W/cm^2 with plasmas formed containing ions of very high charge state (Au^{38+}) and MeV kinetic energies [9, 10]. The introduction of relativistic self-focusing effects at sufficiently high laser power densities [11, 12] has resulted in the observation of ions with energies as high as 100 MeV [13].

The potential use of laser generated plasmas as a source of highly charged heavy ions was recognized early. The literature in the late 1960s and in the 1970s is rich in the analysis of laser plasmas. A bibliographic review of the effects of high-power laser interaction with solids published by De Michelis in 1970 [14] contains 212 references pertaining to plasma production by high-power lasers. Most of this work was driven by the interest in developing intense ion sources for accelerators and by the laser fusion program. Tonon's review [15] of the subject of laser plasmas as a potential ion source for accelerators, published in 1972, contained some 70 references describing a variety of techniques and experiments. Work continues today on the development of intense laser ion sources for ion inertial fusion [16], for cyclotrons [17] and small accelerators [18], and for atomic physics experiments [19, 20]. In addition, laser produced plasmas are being combined with pulsed power technology to produce intense ion beams [21].

2 LASER PLASMAS USED AS AN ION SOURCE

Many of the properties of the laser produced plasma plumes are potentially useful for an ion source. Among these are [23] (1) a copious supply of ions per pulse; (2) high states of ionization; (3) short plasma generation times useful for time-of-flight measurements; (4) directional plasma plumes which can be oriented along the extraction axis (to provide low emittance); (5) versatility in producing a variety of nuclear species since *any* solid material is a potential plasma source; (6) simplicity in design and construction since, in principle, only the solid target need be at accelerator terminal potential because it is optically connected to the laser at ground potential; (7) the possibility of extraction of ions directly from the plasma plume without the application of an additional extractor potential because of its directed expansion velocity, thus making the laser ion source unique in producing slow multicharged ions; and (8) the absence of a carrier gas since the source operates most efficiently in a high vacuum. Continuing improvements in laser technology including increased repetition rates and power output make this technique a relatively inexpensive method of producing multicharged ions, an important consideration for limited-budget research.

The choice of a laser, however, may pose a problem for some investigators. The development of laser ion sources has centered around the use of Nd:YAG and CO$_2$ lasers operating at 1.06 and 10.6 μm, respectively, with the physics of plasma heating [15] favoring the longer wavelength of the CO$_2$ laser. This wavelength advantage is primarily caused by the fact that a lower electronic plasma density is required for optical resonance absorption of the longer CO$_2$ laser radiation; that is, 10^{19} el/cm^3 *versus* 10^{21} el/cm^3 for 10.6 μm *versus* 1.06 μm, respectively. However, unless a special laser is to be developed one must compare the characteristics of commercial off-the-shelf Nd:YAG and CO_2 lasers in the near one-joule/pulse range before making a selection. To produce sufficiently high power densities, both types of lasers must be Q-switched. Typical pulse widths for active Q-switched Nd:YAG systems are 10 to 15 ns at repetition rates in the range of 10 to 30 pps, while commercial CO_2 lasers generally employ unstable optical resonators to produce pulse widths of near 100 ns at comparable repetition rates. Use of this data in estimating the anticipated power output, however, may be misleading since the output energy per pulse may be quoted by the manufacturer over the fluorescence lifetime rather than that under the Q-switched pulse. This affects CO_2 lasers to a greater degree since they tend to have a longer fluorescence lifetime, depending on the gas mixture. In addition, the beam divergence for multi-mode lasers is considerably better for the Nd:YAG laser (0.6 vs. 2 mR) so that its focal spot is smaller than that of the CO_2 laser and, therefore, is capable of producing a larger power density. Thus, specified in terms of energy per pulse the capability for efficient plasma heating by the Nd:YAG system can become comparable to that of the CO_2 laser in the case of off-the-shelf lasers.

The small crater diameter produced by the Nd:YAG laser results in a small hole being drilled into the target that follows the direction of the laser. In the current Arkansas Laser Operated Ion Source (LOIS) described in Ref. 20, the crater resulting from a single laser shot from an actively Q-switched Nd:YAG laser of 800 mJ per pulse is about 100 μm in diameter using a 12-cm focal length lens. Since the laser direction is generally off the extraction axis, plasma production directed along the axis degrades rapidly with the number of laser shots; that is, the plasma blowoff direction now follows the laser beam direction. This effect necessitates repositioning the laser target surface to avoid deep hole drilling in repetitive operation. To circumvent this problem the present LOIS target is a (aluminum) cylinder 7.6 cm in diameter and 2.5 cm in length which can be externally both rotated and translated offering 180 cm^2 of target surface. It can be rotated many rounds before a noticeable drop occurs in the higher stages of ionization. Subsequent translation of the target allows at least a million laser shots to be taken before installation of a new target is required.

Since an average power of only approximately 5 W is dissipated in the ion source there is no need to cool the target. Gas generation, however, does occur in the laser source. For example, LOIS is presently operated in a vacuum system without differential pumping and with a base pressure below 10^{-8} Torr. On a single shot basis, a vacuum gauge cannot detect an appreciable change in the background pressure; however, at a laser repetition rate of 6 Hz the 280-L/s pumping system cannot keep up with the gas generation and the pressure rises above 5×10^{-8} Torr. It is clear that ultrahigh vacuum techniques such as an extensive high-temperature bakeout of the system and/or differential pumping of the target area would be desirable if high repetition rates and ultrahigh vacuums are desired. With low repetition rates or less stringent vacuum requirements, however, it is not at all necessary. In LOIS much of the gas generation is attributed to irradiation of the chamber walls by XUV plasma light causing photodissociation of wall adsorbants. Photoelectrons having energies of 100 eV, produced by XUV photons reflecting their way around the curved surfaces of a 180° electrostatic ion energy analyzer, have been observed in the apparatus. (X-ray emission by laser generated plasmas is well known [24]).

3 PERFORMANCE OF LASER ION SOURCES

3.1 The Dubna Laser Sources

An active developmental program has been carried out for some time [20] at the Laboratory for Nuclear Reactions at the Joint Institute for Nuclear Research, Dubna (USSR). (Reference 25 describes the injection and the acceleration of laser-produced carbon ions by the Synchrophasotron there.) Development of laser multicharged ion sources for cyclotrons has been a high-priority item at the Dubna laboratories. In the case of a cyclotron the ion source plasma is created in a magnetic field with the field direction perpendicular to the ion extraction direction. Figure 14.2 shows one version of the Dubna ion source [17] that used a 4-J output CO_2 laser with a pulse width of 100 ns to generate the plasma [26]. Laser light was directed between the pole faces of a magnet used in an ion source test station and reflected into the target chamber. The laser was focused at 45° to the target

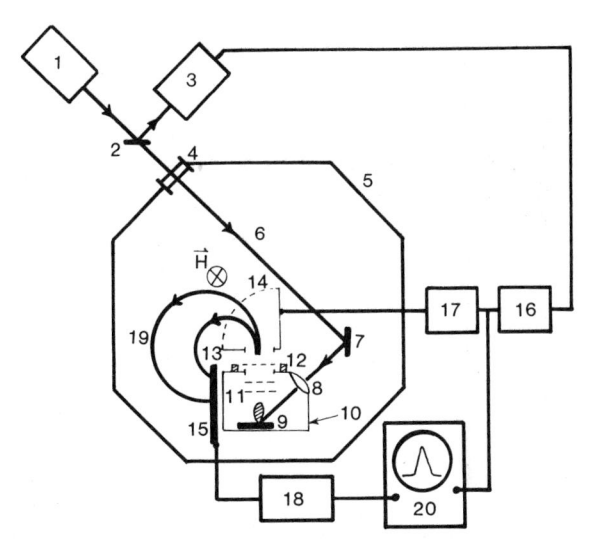

Figure 14.2 Schematic of one version of a laser ion source designed for cyclotrons at the Dubna ion source test facility. The figure is taken from Anan'in et al. [17], © 1983 APS. Here a CO_2 laser beam is reflected by mirror and focused by lens onto a target. The plasma is formed in a magnetic field H (see text). 1, Laser; 2, beam splitter; 3, photoresistor; 4, window in vacuum chamber; 5, chamber housing the ion sources; 6, laser beam; 7, mirror; 8, lens; 9, target; 10, box housing the ion laser source; 11, attenuating grids; 12, insulated electron stripping grid; 13, high-voltage electrode; 14, shielding grid; 15, ion collector; 16, electronic time delay; 17, high-voltage pulse generator; 18, amplifier; 19, ion orbit; 20, dual beam oscilloscope.

normal with power densities estimated to be from 1 to 4 × 10^9 W/cm^2. The resulting plasma plume expanded across a 1- to 2-kG field to a 5 × 10-mm^2 gridded ion emission slit located 6 cm away. A delayed 100-ns-wide, 5-kV extraction pulse was then applied across the extraction gap. The extracted ions were directed into a nearly 360° orbit where they were collected. For a given elemental species and magnetic field strength the time-of-flight (TOF) of the individual charge packets in orbit was inversely proportional to the charge and allowed the charge states to be determined. Figure 14.3 shows the number of each ion species to reach the detector per ion pulse. For the light atoms bare Li^{3+}, Be^{4+}, and C^{6+} were produced. Using heavier atom targets [26], maximum charge states of Si^{8+}, V^{11+}, Zr^{13+}, In^{11+}, Ta^{9+}, and Bi^{8+} were detected.

The work has also made use of a 2-J, 1-Hz CO_2 laser with a 200-ns-wide Q-switched pulse which contained about 30% of the total output energy. The plasma generated by this laser has been projected at an angle of 45° with respect

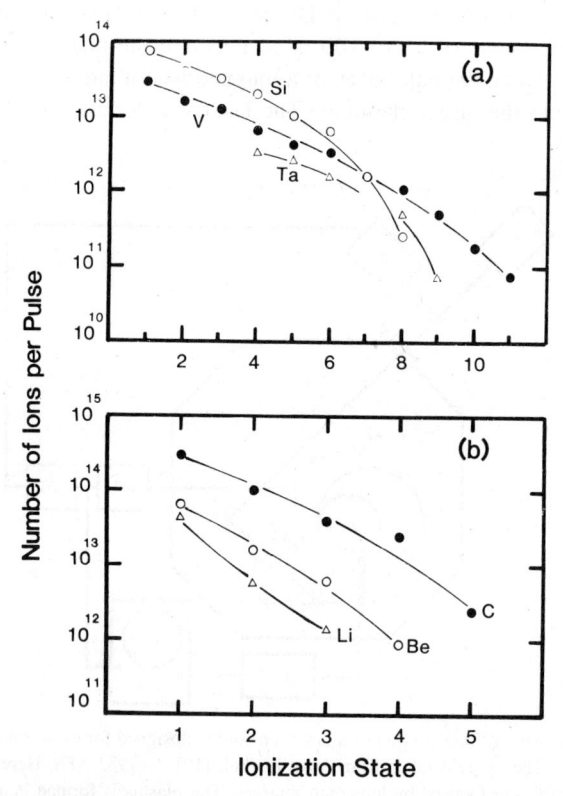

Figure 14.3 Number of ions per laser pulse extracted from the CO_2 laser ion source shown in Figure 14.2. Figure 14.3a is taken from O. B. Anan'in et al. [27], © 1982 APS. Figure 14.3b is taken from O. B. Anan'in et al. [17], © 1983 APS. Power densities are a very few times 10^9 W/cm^2 and magnetic fields are a very few tenths of teslas.

3 Performance of Laser Ion Sources

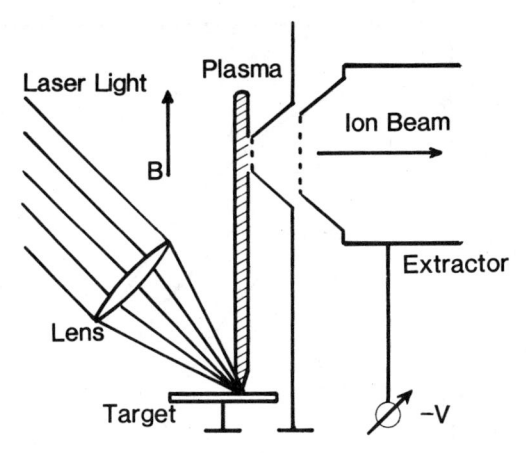

Figure 14.4 A Dubna laser source with plasma plume projected along the magnetic field direction (see text). The figure is taken from Bykovskii et al. [28].

to the magnetic field orientation [28]. In yet another version [28, 29] the plasma plume is projected parallel with the field as shown in Figure 14.4. The diamagnetic plasma is compressed by the magnetic field into a filament that streams across the anode which has an emission slit area of 3.5×22.5 mm^2. The slit width is larger than the diameter of the compressed plasma plume. A Nd:YAG laser producing 40-mJ pulses at a rate 25 Hz is used in this investigation. The laser pulse width is about 12 ns and produces target craters having diameters near 300 μm. The maximum power density is estimated to be about 5×10^9 W/cm^2 per laser pulse. Figure 14.5 shows extraction results when an aluminum laser target is used. The ordinate is expressed as particles per second considering the laser operating at 25 Hz. Maximum charge states extracted from the laser plasmas produced by irradiation at maximum power densities are Li^{3+}, B^{3+}, C^{4+}, Al^{6+}, Si^{5+}, Ti^{5+}, Mo^{4+}, and Bi^{4+}. It is suggested that from 100 to 1000 laser shots can be used on a static target before significant degradation of the ion output occurs.

Some comments may be worthwhile here. An ion source internal to a cyclotron has its plasma created in a magnetic field which is oriented perpendicular to the electric field lines present at the extractor. Thus the extractor gap becomes a magnetically insulated diode [30] by its very nature. This fact makes possible ion extraction from high-density plasmas, such as those generated in the configuration shown in Figure 14.2. Without the presence of the magnetic field, secondary electrons produced by ion impact on the cathode surfaces would cross the extractor gap initiating a vacuum arc. The presence of grids on the extractor apertures is also a required feature when dealing with high-density plasmas. The grid spacing must be smaller than the plasma Debye shielding length [31] to allow penetration of the plasma by the extraction field to affect the ions. If shielding is allowed to occur, then electrons can breach the extractor gap causing a short circuit.

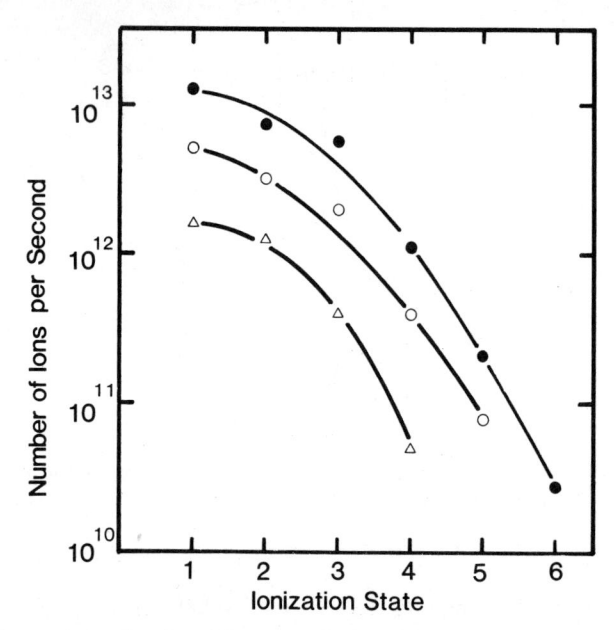

Figure 14.5 Ion output as a function of charge state from the ion source configuration shown in Figure 14.4. The ordinate is the ion output per second based on the use of a 25-Hz Nd:YAG laser focused onto an aluminum target. Power densities of △, 1.2×10^8 W/cm^2; ○, 1.1×10^9 W/cm^2; ●, 4.4×10^9 W/cm^2. The plasma plume (see Figure 14.4) travels a distance of 6.6 cm before flowing over the anode aperture. The magnetic field is 0.4 T.

3.2 The Arkansas Experiments

The original laser system used at the University of Arkansas was a commercial Nd:YAG system consisting of an oscillator and one stage of amplification. It produced 15-ns-wide, 300-mJ pulses at a nominal 50-Hz rate. This laser system has been upgraded in energy to about 800 mJ with the addition of a second amplifier stage. The source is currently operated at 6 Hz, computer control limited. It would probably be laser limited now to about 20 Hz because of the additional amplifier.

The early experiments, designed to test the capability of a laser generated ion source, made use of the 300-mJ version of the laser system. Figure 14.6 shows the first experimental arrangement [18] using a 20-MW Nd:YAG laser focused at a power density near 10^{11} W/cm^2 on target. (*Note:* The cover glass plate, C, shown in Figure 14.6 is essential to repetitive operation of the source. It keeps ablation products from the laser optics. The system is self-cleaning. The cover glass becomes opaque except where the laser passes.) At a target-to-extractor distance of 8 cm, a peak current of over 200 mA could be extracted from the 6.4-mm-diameter anode aperture using 100-mesh/1-mil screens, a 6.4-mm-diameter cathode aperture, an extractor gap of 6.4 mm, and 15-kV extraction potential ($J > 0.6$ A/cm^2). The resulting extracted ion current exceeded space-

3 Performance of Laser Ion Sources

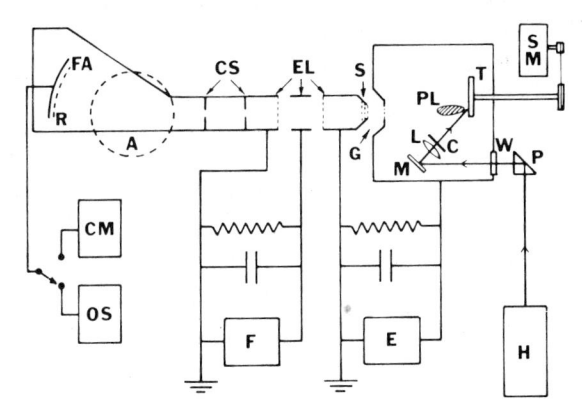

Figure 14.6 Schematic diagram of experimental apparatus taken from Ref. 23, © 1980 APS. H, 20-MW Nd:YAG pulsed laser; P, right-angle prism; W, vacuum window; M, totally reflecting mirror; L, 4-cm focal length lens; C, cover plate to protect lens surface; T, rotatable laser target; PL, laser plasma plume; SM, stepping motor with belt drive to rotate target; G, ion extraction gap and screens; S, space-charge neutralizing screens; EL, gridded Einzel lens assembly; CS, collimating slits; A, analyzing magnet; R, repeller grid; FA, Faraday cup array; CM, coulomb meter; OS, dual channel oscilloscope; E, high-voltage extractor power supply; F, high-voltage focus power supply.

charge-limited expectations. When the target-to-extractor distance was subsequently increased to 30 cm, the peak current dropped to 44 mA. Space-charge effects were still severe. Further investigation of the extracted beam showed that the faster, higher charge state ions suffer the most severe loss in beam blowup near the extractor gap. Two more screens were added in back of the anode grid to reduce the beam blowup problem. With the screens in place, a 15-mA aluminum ion beam could be focused with a gridded strong focusing Einzel lens through a 1-cm^2 aperture with an efficiency of 70%. Friichtenicht et al. [32], using a 1.5-J ruby laser with a 80-ns Q-switched temporal width, extracted an aluminum ion current of 6 A over an anode aperture area of 44 cm^2. Their extraction gap was also 6.4 mm and their extraction potential was 10 kV.

At these moderate laser power densities, the extracted ions show their thermal origin. Figure 14.7, from Ref. 23, shows a computer fit of a Maxwellian distri-

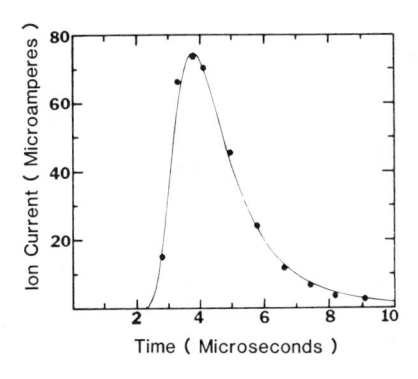

Figure 14.7 Computer fit of a C^{3+} ion signal produced by a near-10^{11}-W/cm^2 Nd:YAG laser burst on a graphite target. The figure is taken from Ref. 23, © 1980 APS. The ion signal is represented by the closed circles while the line represents a computer fit to a single Maxwell-Boltzmann velocity distribution. The ion temperature is 237 eV and the flow velocity is 5.2×10^4 m/s.

bution to a C^{3+} ion component selected by the magnetic analyzer shown in Figure 14.6. The curve fit demonstrates thermal equilibrium of the ion component. An interesting result is that the ion temperatures measured for different charge states produced by the same laser shot were not the same and the charge states did not necessarily scale with the temperature, but correlated better with the flow velocity. (Up to four stages of ionization could be monitored simultaneously.) The Maxwellian fit was excellent for all targets studied [33] which included C, Al, Cu, Pb, Ta, and Au, although it was found that often the velocity distribution for 1^+ and 2^+ ions had to be fit with two Maxwellian distributions, which suggested the presence of recombination in the plasma plume. Higher stages of ionization never exhibited such double Maxwellian distributions. It was also observed that the actual number of ions produced for a given charge state roughly scaled inversely with the sum of the ionization potentials necessary for stepwise ionization. Representative numbers of ions estimated to be actually extracted from the source aperture after a plasma flight of 30 cm from the target to extractor for the first five ionization stages (in 10^9 ions) were, respectively, C: 45, 28, 10, 2, 1; Al: 48, 32, 25, 9, 1; Cu: 18, 11, 4, 1, 0.3; and Pb: 315, 81, 7, 2, 1. These numbers are of the order of 10^3 smaller than those obtained in Ref. 17 after extracting in a transverse magnetic field from a plasma with a flight distance of only 6 cm to the extracting gap.

To improve the performance of the source and its duty cycle, the ion pulse width was increased by increasing the plasma flight distance to the extractor from an aluminum target to 60 cm, with the plasma passing through a confining axial magnetic field [34]. Extensive use of screens was made to reduce the beam blowup problem. Analysis showed that at least 40 μC of charge or about 2×10^{14} ions per laser pulse were captured in the magnetic field and transported to extractor gap plane. (Because of the attenuation screens, only a charge of about 20 μC was measured in the extractor plane.) The ion number of 2×10^{14} is probably representative of the total useful ion output per laser pulse.

To extract ions by applying a 9-kV potential across the 12.7-mm extractor gap (12.7-mm-diameter gridded apertures were used in both the anode and cathode), a massive soft-iron magnetic field terminator was used to avoid secondary electrons produced by ion impact on the cathode from funneling back to the anode and causing breakdown of the extractor gap. The extracted ions were subsequently magnetically analyzed. Figure 14.8 shows the time-of-flight analysis of the magnetically collected ions. The times shown in the figure essentially represent the ion flight times to travel 70 cm to the extractor gap. The data show the extremes in energy available in the ion source. Multiply charged states are produced with energies ranging from tens of eV to low keV. The singly charged ions, meanwhile, are produced with energies from a few eV to a few hundred eV. Figure 14.8 also demonstrates that the ion output from this thermal source can produce ion beams with large energy spreads that scale with the charge state, ranging from a few hundred eV for the low charge states to a few keV for the higher charge states.

Our attempts and attempts of others [35–37] to confine a laser produced plasma in a magnetic field were not particularly successful. A confined plasma would

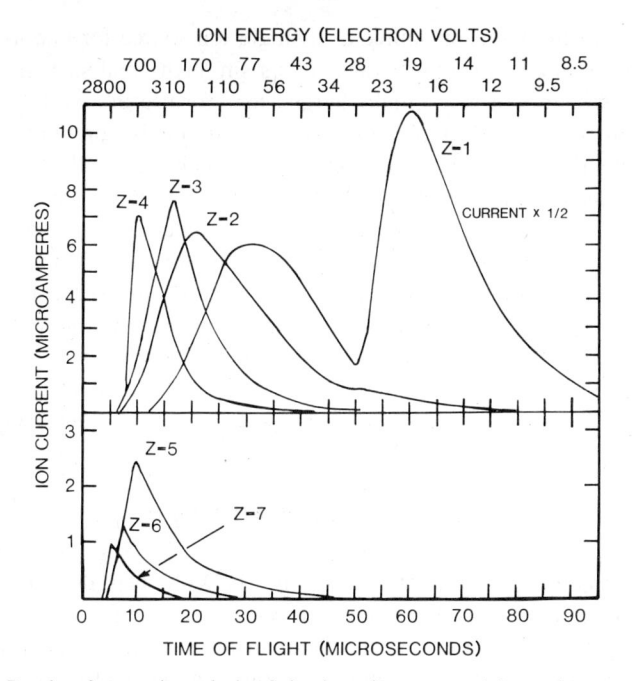

Figure 14.8 Results of magnetic analysis of aluminum ions extracted from a laser generated plasma plume transported 70 cm through an axial magnetic field to the extractor gap. The plasma is generated by a Nd:YAG laser beam focused at near 10^{11} W/cm^2 and incident on a planar target at 45°. The figure is taken from Ref. 34, © 1982 APS.

result in an increased ion pulse width and higher duty cycle and would open the possibility of further plasma heating; however, at present only singly charged ions from a laser plasma have been successfully injected into an electrostatic trap for storage [38]. A simple method of extending the duty cycle has been suggested by Gray et al. [34]: Simply drift the plasma, confined by an axial magnetic field, through a large distance. They also suggested that properly phased pulsing of the extractor potential could also be used to give monoenergic ions.

4 FUTURE OF LASER ION SOURCES

The laser ion source has a sufficient number of unique properties to ensure its survival in a field where numerous alternatives exist: (1) it is essentially a thermal source that can supply its own extraction mechanism; hence, it is unrivaled as a source of *slow*, multicharged ions; (2) it stands alone among current ion sources in simplicity of design and operation; (3) it can easily provide a wide range of ion species—simultaneously if necessary by using laser beam splitting and multiple targets; and (4) its cost, size, construction, and maintenance are minimal when compared with other high charge state sources such as the electron cyclotron resonance (ECR) source and the electron beam ion source (EBIS).

Future use of the laser ion source as a single-ion source for internal mounting in a cyclotron seems promising because of its simplicity and size. In addition, it is an excellent source of slow, multicharged ions for use in a variety of time-of-flight experiments. For example, the Arkansas group is continuing its use by extracting directly from the laser plasma *without* acceleration. The first four stages of ionization have been extracted at 100 eV/charge from an aluminum plasma. Experiments are being done with 40-eV/charge aluminum ions originally extracted as 280-eV/charge ions. It also seems to have potential use in ion implantation [39]. Effective ion plating and implantation over a large area seems possible. Here its lower duty cycle may not be an important factor since a total of 10^{14} ions or more are available in each laser pulse. Although electrostatic focusing can provide high-density ion implantation and the source can provide simultaneous ion implantation of different nuclear species, this technique has not been extensively explored.

ACKNOWLEDGMENTS

The authors greatly appreciate the help of B. H. Wolf, Gesellschaft fur Schwerionenforschung mbH (GSI), Darmstadt, Germany, in sending us his collection of material on laser ion sources and the German translations of the Russian 1986 preprints from Dubna. They also appreciate the material sent by R. A. Phaneuf, Oak Ridge National Laboratory, which included Figure 14.1. The Arkansas work was supported by the National Science Foundation, the Nuclear Science Division, and is currently supported by National Science Foundation Grant DMR-8516109 (jointly funded by the Atomic, Molecular, and Plasma Program and the Solid State Physics Program).

REFERENCES

1. G. S. Hurst, M. G. Payne, S. D. Kramer, and J. P. Young, *Rev. Mod. Phys.* **51,** 767 (1979).
2. H.-Jurgen Kluge, F. Ames, W. Ruster, and K. Wallmeroth, invited talk given at the Accelerated Radioactive Beams Workshop, Vancouver Island, Canada, 4–7 Sept. 1985, CERN-EP/85-162, Oct. 4, 1985.
3. F. E. Irons and N. J. Peacock, *J. Phys. B* **7,** 2084 (1974).
4. Yu. A. Bykovskii, N. N. Degtyarenko, V. F. Elesin, Yu. P. Kozyrev, and S. M. Sil'nov, *Sov. Phys. JETP (USA)* **33,** 706 (1971).
5. G. A. Doschek, U. Feldman, P. G. Burkchalter, T. Finn, and W. A. Feibelman, *J. Phys. B* **10,** L745 (1977).
6. R. A. Phaneuf, *IEEE Trans. Nucl. Sci.* **NS-28,** 1182 (1981).
7. A. W. Ehler, *J. Appl. Phys.* **46,** 2464 (1975).
8. T. S. Pearlman and G. H. Dahlbacka, *Appl. Phys. Lett.* **31,** 414 (1977).
9. B. Luther-Davis and J. L. Hughes, *Opt. Commun.* **18,** 351 (1976).
10. M. Siegrist, B. Luther-Davis, and J. L. Hughes, *Opt. Commun.* **18,** 605 (1976).
11. H. Hora, E. L. Kane, and J. L. Hughes, *J. Appl. Phys.* **49,** 923 (1978).

References

12. H. Hora, D. A. Jones, E. L. Kane, and B. Luther-Davis, *Laser Acceleration of Particles*, AIP Conf. Proceed. No. 91, P. J. Channell (Ed.), American Institute of Physics, New York, 1982, p. 112.
13. A. M. Sessler, *Laser Acceleration of Particles*, 1982 AIP Conf. Proceed. No. 91, P. J. Channell (Ed.), American Institute of Physics, New York, 1982, p. 10.
14. C. De Michelis, *IEEE J. Quant. Electronics* **QE-6,** 630 (1970).
15. G. F. Tonon, *IEEE Trans. Nucl. Sci.* **NS-19,** 172 (1972).
16. L. Z. Barabash, D. G. Koshkarev, Yu. I. Lapitskii, S. V. Latyshev, A. V. Shumshurov, Yu. A. Bykovskii, A. A. Golvbev, Yu. P. Kosyrev, K. I. Krechet, R. T. Haydarov, and B. Yu. Sharkov, *Laser and Particle Beams* **2,** 49 (1984).
17. O. B. Anan'in, Yu. A. Bykovskii, V. P. Gusev, Yu. P. Koznev, I. V. Kolesov, A. S. Pasyuk, and V. D. Peklenkov, *Sov. Phys. Tech. Phys.* **28,** 54 (1983).
18. G. Korschinek and J. Sellmair, *Rev. Sci. Instrum.* **57,** 745 (1986); G. Karschinek, *Bull. Am. Phys. Soc.* **31,** 1277 (1986).
19. R. A. Phaneuf, I. Alvarez, F. W. Meyer, and D. H. Crandall, *Phys. Rev. A* **26,** 1892 (1982).
20. R. H. Hughes, D. O. Pederson, and X. M. Ye, *Appl. Phys. Lett.* **47,** 1282 (1985).
21. T. Ohmori, M. Katsurai, and T. Sekiguchi, *Jpn. J. Appl. Phys.* **19,** L728 (1980).
22. T. Ohmori, M. Katsurai, and T. Sekiguchi, *Jpn. J. Appl. Phys.* **22,** 728 (1983).
23. R. H. Hughes, R. J. Anderson, C. K. Manka, M. R. Carruth, L. G. Gray, and J. P. Rosenfeld, *J. Appl. Phys.* **51,** 4088 (1980).
24. For example, H. C. Gerritsen, H. van Brug, F. Bijkerk, and M. J. van der Wiel, *J. Appl. Phys.* **59,** 2337 (1986).
25. O. B. Anan'in, A. M. Baldin, Yu. D. Beznogikh, Yu. A. Bykovskii, A. I. Guvorov, L. P. Zinov'ev, Yu. P. Kozyrev, L. G. Makarov, A. Monchinskii, I. K. Novikov, V. D. Peklenkov, A. M. Raspopin, and I. N. Semenyushkin, *Sov. J. Quantum Electron.* **7,** 873 (1977).
26. O. B. Anan'in, Yu. A. Bykovskii, V. P. Gusev, Yu. P. Kozyrev, A. S. Pasyuk, and I. V. Kolesov, *Proceedings of the Seventh All-Union Conference on Charged-Particle Acceleration* (in Russian), Vol. 1, Ob'edin. Inst. Yad. Issled., Dubna, 1981, p. 98.
27. O. B. Anan'in, Yu. A. Bykovskii, V. P. Gusev, Yu. P. Kozyrev, I. V. Kolesov, A. S. Pasyuk, and V. D. Peklenkov, *Sov. Phys. Tech. Phys.* **27,** 903 (1982).
28. Yu. A. Bykovskii, V. P. Gusev, Yu. P. Kozyrev, I. V. Kolesov, V. B. Kutner, A. S. Pasyuk, V. D. Peklenkov, S. G. Stetsenko, K. G. Suvorov, and D. A. Uzienko, Joint Institute for Nuclear Research, Report No. P9-86-2, Dubna, 1986 (in Russian).
29. Yu. A. Bykovskii, A. N. Oblizin, Yu. P. Kozyrev, I. V. Kolesov, V. B. Kutner, A. S. Pasyuk, V. B. Peklenkov, S. G. Stetsenko, K. G. Suvorov, and D. A. Uzienko, Joint Institute for Nuclear Research, Report No. P9-86-3, Dubna, 1986 (in Russian).
30. R. N. Sudan and R. V. Lovelace, *Phys. Rev. Lett.* **31,** 1174 (1973).
31. L. Spitzer, *Physics of Fully Ionized Gases*, 2nd ed., Interscience, New York, 1962, p. 22.
32. J. F. Friichtenicht, N. G. Utterback, and J. R. Valles, *Rev. Sci. Instrum.* **47,** 1489 (1976).
33. J. P. Rosenfeld, Masters Thesis, University of Arkansas, Fayetteville, 1980.
34. L. G. Gray, R. H. Hughes, and R. J. Anderson, *J. Appl. Phys.* **53,** 6628 (1982).
35. E. W. Sucov, J. L. Pack, A. V. Phelps, and A. G. Englehardt, *Phys. Fluids* **10,** 2035 (1967).
36. A. E. Haught, D. H. Polk, and W. J. Fader, *Phys. Fluids* **13,** 2842 (1970).
37. J. Bruneteau, E. Fabre, H. Laman, and P. Vasseur, *Phys. Fluids* **13,** 1796 (1970).
38. R. D. Knight, *Appl. Phys. Lett.* **38,** 221 (1981).
39. Yu. K. Al'ludov, T. A. Basova, Yu. A. Bykovskii, V. G. Degtyarev, Yu. N. Kolosov, I. D. Laptev, and V. N. Nevolin, *Sov. Phys. Tech. Phys.* **24,** 1077 (1979).

15

Liquid Metal Ion Sources

L. W. Swanson and A. E. Bell
Oregon Graduate Center
Beaverton, Oregon

The liquid metal ion source (LMIS) consists of a liquid metal coating on a needle substrate. Application of a potential of several kilovolts in a vacuum $<1 \times 10^{-7}$ Torr to a nearby extractor electrode distorts the liquid metal at the needle apex to form a conical shape. At the apex of the stabilized liquid cone the electric field is sufficiently high to produce ions at currents in the range of 1 to 100 μA.

The liquid metal ion source is unique among all ion sources because of its extraordinary high brightness which can exceed 1×10^6 A/cm^2 sr. It is the latter property, coupled with a reasonably small energy spread (typically <10 eV), that makes this ion source extremely attractive for use in fixed or scanning beam focusing columns where beam sizes <5000 Å with current densities of 1 to 10 A/cm^2 are desired. Such focused ion beam (FIB) systems are being used for maskless implantation as a step in IC fabrication [1, 2] and for a variety of micromilling applications, many of which are associated with submicron IC fabrication [3, 4]. Other submicron resolution, surface analytical applications such as scanning microscopy and secondary ion mass spectroscopy [5, 6, 7] have also been a part of LMIS FIB applications.

The technology of LMIS has now developed to the point that a number of pure metal and alloy sources have been developed with lifetimes in excess of several hundred hours. Table 15.1 lists a number of these sources along with their respective substrates. LMIS of pure elements with a reasonably long life are restricted to those elements that are conductors and that have a low vapor pressure (typically $<10^{-7}$ Torr) at the melting point. These criteria are obeyed admirably by Ga, which has a vapor pressure of less than 10^{-10} Torr at its melting point of 29.8 °C. LMIS source technology has been extended to include elements not meeting the above criteria by incorporating the element of interest in a suitable binary alloy such as Pd$_2$As which has been used to create a LMIS of As [20], an element that

TABLE 15.1 Various Commonly Used Single and Multicomponent Liquid Metal and Substrate Combinations

Liquid Metal	Substrate	Principal Species	Reference
Ga	W	Ga$^+$	[8, 9]
In	W	In$^+$	[10]
Bi	Ni/Cr	Bi$^+$	[10, 11]
Al	C	Al$^+$	[12, 13]
Sn	W	Sn^{2+}, Sn$^+$	[14]
Cs	W	Cs$^+$	[15]
Au	W	Au$^+$, Au^{2+}	[16]
Au$_{0.80}$Si$_{0.20}$	W	Au^{2+}, Au$^+$, Si^{2+}	[17]
Au$_{0.60}$Be$_{0.40}$	W	Au$^+$, Au^{2+}, Be^{2+}	[18]
Au$_{0.59}$Si$_{0.26}$Be$_{0.15}$	W	Au$^+$, Be^{2+}, Si^{2+}	[19]
Pd$_{0.66}$As$_{0.34}$	W	Pd$^+$, As^{2+}	[20]
Ni$_{0.45}$B$_{0.45}$Si$_{0.10}$	C	Ni^{2+}, B$^+$, Ni$^+$, Si^{2+}, Si$^+$	[21]
Pd$_{0.70}$As$_{0.16}$B$_{0.14}$	W	Pd$^+$, As^{2+}, B$^+$	[20]
Pd$_{0.64}$As$_{0.11}$B$_{0.09}$P$_{0.16}$	W	Pd$^+$, Pd^{2+}, P$^+$, As^{2+}, P^{2+}, B$^+$	[20]

could not be used in its pure state because of its very high vapor pressure at its melting point. Of course, the alloy LMIS produces ions of all the alloy species, some of which must be removed. This is usually accomplished by incorporating a Wien mass filter in the ion optical column used to deliver the focused ion beam to the target [22]. Alloy LMIS providing ions of As, B, Be, Si, and P have been developed and used in FIB systems to carry out maskless implantation of submicron features in semiconductor devices [18].

In this chapter we review the development and current understanding of the mechanism of LMIS operation. A brief discussion of the LMIS emission characteristics in relation to focused beam applications is also presented.

1 PRINCIPLE OF OPERATION

The operation of the LMIS consists of a delicate balance of electrostatic f_e and surface tension f_s forces on the liquid surface so as to form a stable cone-shaped structure. In terms of the surface tension γ, electric field strength F, and principal radii of curvature r_1 and r_2 the forces can be written as

$$f_s = \gamma/(1/r_1 + 1/r_2) \tag{1}$$

and

$$f_e = \tfrac{1}{2}\epsilon_0 F^2 \tag{2}$$

where ϵ_0 is the vacuum dielectric constant. In 1964 G. Taylor [23] showed that when subjected to an electric field a liquid surface tended toward a conical surface with a cone half angle of ~49°. He then determined mathematically that one shape of the liquid surface that satisfied the condition $f_s = f_e$ was an infinite cone with half angle $\alpha = 49.3°$ and a counterelectrode whose shape was given by

$$r = R_0[P_{0.5}(\cos\theta)]^{-2} \qquad (3)$$

where $P_{0.5}$ is a Legendre polynomial of fractional order 0.5, θ and r are polar coordinates measured from the cone apex, and R_0 is the axial separation between the cone apex and the counterelectrode. Taylor further derived the following relation for the voltage V_s necessary to stabilize the liquid cone:

$$V_s = 4.52 \times 10^5 (R_0\gamma)^{1/2} \quad (V) \qquad (4)$$

for R_0 in meters and γ in newtons/meter.

Mahoney et al. [24] and Swatik and Hendricks [25] in the late 1960s applied the ideas of Taylor to explain their experimental ion emission results for low-melting liquids flowing through small nozzles under the influence of a high electric field. Since that time the LIMS geometry has evolved as shown in Figure 15.1 to a wetted needle configuration. The above authors also proposed a field evaporation mechanism for singly charged ion formation and, using a simplified model for field evaporation, they calculated the field F at the cone apex to be

$$F = 0.069(H_0 + I - \phi)^2 \quad (V/\text{Å}) \qquad (5)$$

where H_0 is the heat of evaporation, I is the first ionization potential, and ϕ is the work function all in eV units. Using the Taylor theory, they were able to show that the radius at the cone apex was in the range of 2 to 10 Å.

In the intervening years the liquid in a LMIS was assumed to exist in the shape of a Taylor cone with some rounding off at the apex as seen in SEM pictures of

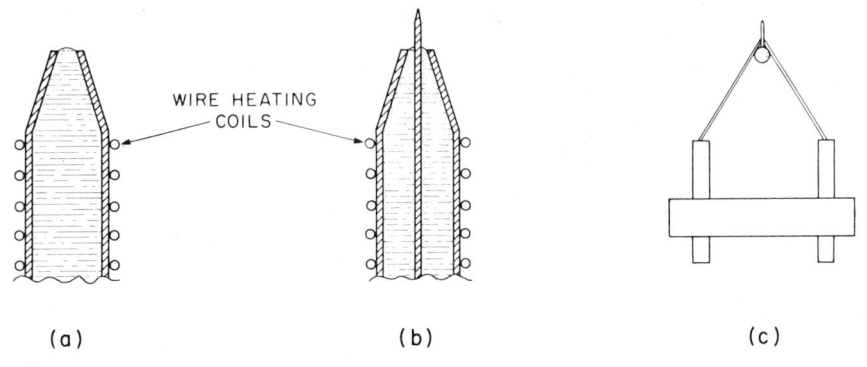

Figure 15.1 Various geometrical configurations of the LMIS: (*a*) nozzle type; (*b*) needle/nozzle type; (*c*) needle/filament type.

"frozen-in" cones, as shown in Figure 15.2 for Au. This led Gomer [26] to conclude that the effect of space charge is so significant that a surface field high enough to allow field evaporation could not be sustained at currents greater than 10 μA and that field ionization must be the major current formation mechanism. This was disputed by Prewett et al. [27] who favored a field evaporation mecha-

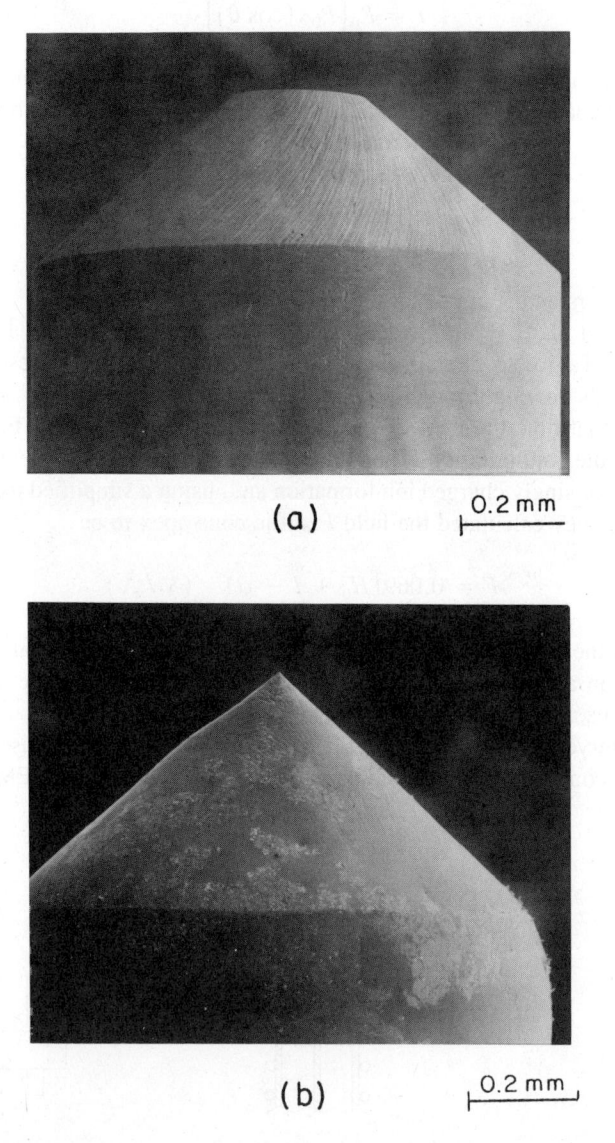

Figure 15.2 SEM photographs of a wetted needle type Au LMIS: (*a*) bare substrate; (*b*) substrate with frozen Au formed during operation. A larger and truncated needle substrate was used to emphasize visualization of liquid cone.

nism, but who did not produce a counter argument to Gomer's result that space-charge effects prevent field evaporation.

This disagreement can be resolved if the assumption of a rounded-off Taylor cone shape is discarded, as suggested by Kang and Swanson [28]. They adopted a model of a conical protrusion on the end of a Taylor cone shape and found that, for a protrusion of length 3000 Å and diameter 300 Å, currents in excess of 50 μA could be produced by field evaporation. Later, experimental evidence for a jetlike protrusion was presented by Benassayag [29].

The identification of the LMIS liquid geometry with the Taylor cone shape has generated considerable controversy with respect to the mathematical approach by Taylor [30, 31]; however, high-voltage TEM photographs of an operating LMIS presented by Benassayag et al. showing the presence of a liquid metal cone of approximately the correct shape for an operating LMIS supports the general conclusions of Taylor.

More recently, it has been realized that the static solution provided by Taylor may not be appropriate for the region near the apex of the liquid cone where mass flow velocities become very high. The latter becomes especially true at higher currents. A dynamic model including the effect of space charge was put forth by Kingham and Swanson [32] (KS) in an attempt to justify the existence of the jetlike protrusion which has been shown experimentally by TEM measurements to increase with current. The basis of their approach was to approximate the dynamic situation by inclusion of a Bernoulli term $\rho v^2/2$ (where v and ρ are the fluid velocity and density, respectively) in Taylor's balance equation for the static case:

$$\gamma(1/r_1 + 1/r_2) + \rho v^2/2 = \epsilon_0 F^2/2 \qquad (6)$$

where the approximations that have been made are discussed in Ref. 32. Ideally one would like to solve Eq. (6) to find a stable shape, which would be possible if F were known over the whole surface. Unfortunately, this is not the case, so instead a particular shape was assumed and the surface field F_s required to stabilize the shape was calculated from Eq. (6). Next the Poisson field F_p for the assumed shape was calculated numerically, including the effects of space charge. Good agreement between F_s and F_p over the liquid surface (except at the emitter apex where emission of ions occurs) was taken as an indication of a stable emitter shape. At the emitter apex the field was assumed to be the evaporation field given by Eq. (5). In this way the dynamic stability of the Taylor cone shape with a jetlike protrusion was justified. One interesting result was the predicted small values of the apex radius r_a, not much different from the early predictions of Mahoney et al. [24], and the approximate proportionality between the emission current I and r_a^2. This implies that the current increase with applied voltage is mainly through an increase in emitting area, and the current density and applied field remain approximately constant. On the basis of the predicted and experimentally observed values of r_a, the current density for a Ga LMIS is in excess of 1×10^8 A/cm^2 for all values of I.

2 SOURCE CHARACTERISTICS

The group IA and IIIA elements principally form singly charged monomer ions. Table 15.2 lists the principal species observed in the Ga and Al LMIS; also shown in Table 15.2 are the observed species in LMIS of the Group VA element Bi and several alloys. The large percentage of M^{2+} species observed in some LMIS has been explained quite adequately by the postionization model put forth by Kingham [34]. The formation of various dimer, trimer, etc., species as shown in Table 15.2 is less well understood.

From the energy distribution and energy deficits one can learn a great deal concerning the ion formation mechanism. For example, from the Figure 15.3 potential energy diagram depicting the field evaporation and post ionization model for the formation of M^+ and M^{2+} species, one can determine that the energy deficit ΔE_1 for a singly charged ion is given by

$$\Delta E_1 = H_0 + I_1 - \phi - Q(F) \qquad (7)$$

where $Q(F)$ is the activation energy for field evaporation. Close agreement between the predictions of Eq. (7) and experimental results has provided strong support for field evaporation as the primary ion formation mechanism [35].

TABLE 15.2 Relative Abundance of Various Species for the Indicated LMIS at 10 µA Current

LMIS	Primary Species in Beam	Reference
Al	Al^+(99.8), Al^{2+}(0.23), Al_2^+(0.011)	[33]
Ga	Ga^+(99.7), Ga_2^+(0.23), Ga_3^+(0.019)	[33]
Bi	Bi^+(60.4), Bi_3^+(14.5), Bi_2^+(12.6), Bi_3^{2+}(10.0), Bi_5^{2+}(1.6)	[33]
$Au_{0.60}Be_{0.40}$	Au^+(56.9), Au^{2+}(22.0), Be^{2+}(8.7), Au_2^+(5.0), Be^+(3.0), Au_5^{2+}(2.4), $AuBe^+$(1.9)	[33]
$Au_{0.69}Si_{0.31}$	Au^{2+}(63.6), Au^+(22.6), Si^{2+}(6.3), Au_2^+(3.1), $AuSi^+$(1.6), Si^+(1.4)	[33]
$Au_{0.59}Si_{0.26}Be_{0.15}$	Au^+(53), Si^{2+}(15), Si^+(8), Au^{2+}(6), Be^{2+}(6), Be^+(6), Au_2^+(4), Si_2^+(2)	[19]
$Pd_{0.66}As_{0.34}$	Pd^+(48.6), Pd^{2+}(19.6), $AsPd^{2+}$(7.9), As^{2+}(7.3), As^+(7.3), $AsPd^+$(2.5), As_2^+(2.3), $AsPd_2^{2+}$(1.6), As_2Pd^+(1.1)	[20]
$Pd_{0.70}As_{0.16}B_{0.14}$	Pd^+(46.0), Pd^{2+}(21.1), As^{2+}(7.3), As^+(6.1), $AsPd^{2+}$(4.9), PdB^+(3.4), B^+(2.8), $AsPd^+$(1.9), As_2^+(1.4), Pd_2B^+(1.3), $AsPd_2^+$(1.0)	[20]

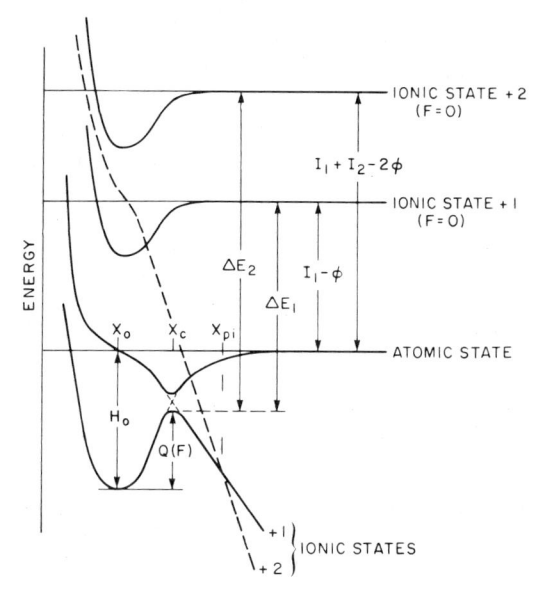

Figure 15.3 Potential energy diagram showing field evaporation ion formation of M^+ and subsequent postionization formation of M^{2+}. The energy deficits ΔE_1 and ΔE_2 for M^+ and M^{2+} are indicated.

Figure 15.4 shows a pictorial summary of the various mechanisms of ion formation including the possibility of ion formation from thermally evaporated atoms via gas phase field ionization. Evidence for this comes primarily from the total energy distributions (TED) of M^+ species, which occasionally show secondary low-energy peaks or long low-energy tails [36, 37]. The latter are indicative of ion formation farther out from the emitter surface. It has been shown for the jetlike protrusion at high currents that plausible mechanisms exist to raise the temperature sufficiently to cause nonnegligible thermal evaporation and subsequent ion formation mechanism via field ionization [38].

Further evidence of the existence of evaporated neutrals comes from the observation of optical emission from the space immediately in front of the emitter. Spectral analysis shows most of the optical emission to emanate from excited neutrals [39]. This has lead to the prediction of the following charge exchange reactions, not only to account for the excited neutrals, but also for the long tails frequently observed on both the low as well as the high-energy side of the TEDs of M^+ species:

$$M^+(\text{fast}) + M(\text{slow}) = M^+(\text{slow}) + M^*(\text{fast}) \qquad (8)$$

and

$$M^{2+}(\text{fast}) + M(\text{slow}) = M^+(\text{fast}) + M^+(\text{slow}) \qquad (9)$$

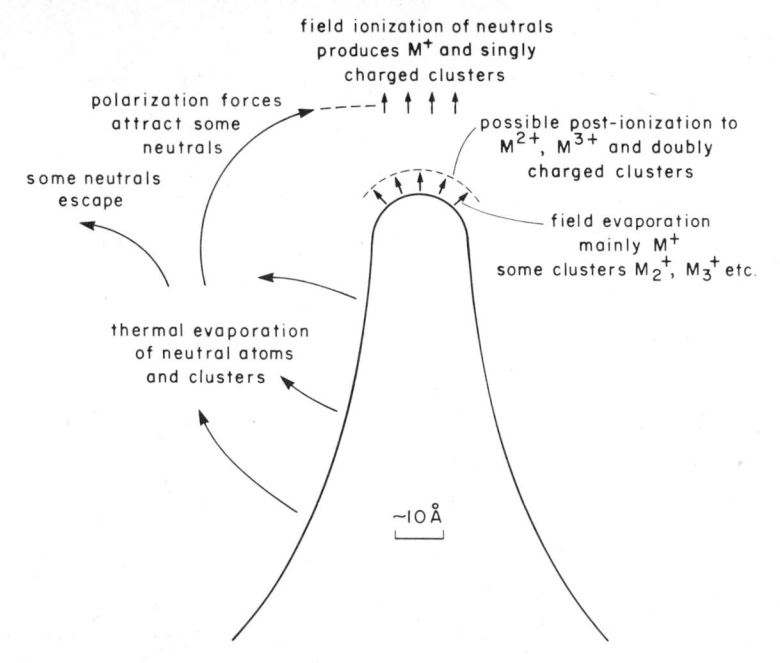

Figure 15.4 A schematic view of the proposed model of LMIS operation.

where M* represents an electronically excited state. Reaction (9) would lead to tails on the high-energy side of the TED when the M^{2+} species is dominant (see for example the Si^+ TED results in Ref. 35).

Accompanying the emission of monomeric ions are polymeric ions, the proportion and size of which increase with current. At sufficiently high currents a substantial fraction of the mass loss is due to charged microdroplets [40]. It has been shown that the droplet angular distribution is much narrower ($\sim 2°$ half angle) than the ion distribution. Rudenauer [41] found that In droplets at 25 μA emission current from a nozzle source were about 2500 Å in diameter. On the other hand, D'Cruz et al. [42] found that for a Au wetted needle type LMIS operating at 150 μA the average droplet size was 1 μm. Because the microdroplets are charged, they can be focused along with the monomeric charged particles [40, 43]. In this regard, Benassayag et al. [44] achieved a focused deposit of Au droplets of 2.5 μm diameter with volume flow rates of 1 to 1.5 $\mu m^3/s$.

2.1 Total Energy Distribution and Angular Intensity of the LMIS

The TED curves for the Ga^+ species of a Ga LMIS are shown in Figure 15.5 as a function of emission current. In addition to a dependence on total current we also note a strong dependence on particle mass as shown in Figure 15.6 where the

2 Source Characteristics

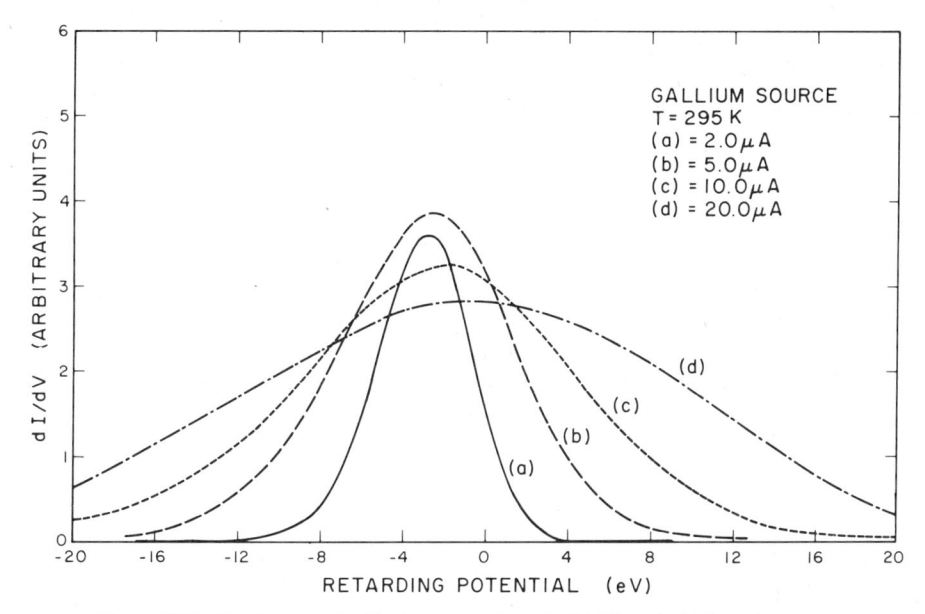

Figure 15.5 Total energy distribution curves for a Ga LMIS at the indicated currents.

full widths at half maximum (FWHM), ΔV, of the TEDs of several pure metal LMIS are given. Extrapolation of the Figure 15.6 results to $I = 0$ suggests a minimum $\Delta V_0 \sim 5$ eV independent of mass. Unfortunately, because of the threshold nature of the Taylor cone formation, it is difficult to achieve stable operation of the LMIS below ~ 1 μA. However, it has been shown that by the use of smaller needle radius the energy spread at threshold can be reduced to 2 to 3 eV [45]. In view of the accepted field evaporation mechanism of ion formation, it is expected that the zero current value of ΔV_0 would be ~ 2 eV if other energy broadening mechanisms are not involved.

The Figure 15.6 results can be fitted reasonably well to the following empirical equation:

$$\Delta V' = k m^{1/2} I^{0.6} \tag{10}$$

where $\Delta V' = (\Delta V^2 - 25)^{1/2}$, m is the ion mass, and k is a constant. The Eq. (10) relation is believed to arise in part from an increasing contribution of field ionization and charge exchange reactions along with random density fluctuations in the beam, which has been shown by Knauer [46] to cause an increase in ΔV through Coulomb interactions among the emitted particles.

The angular intensity I' (defined as the current per unit solid angle in the central portion of the beam) is generally observed to increase linearly with total current. In the case of the pure metal LMIS, where M^+ is the dominant species, the Figure 15.7 results show that the angular divergence of the beam increases not only with

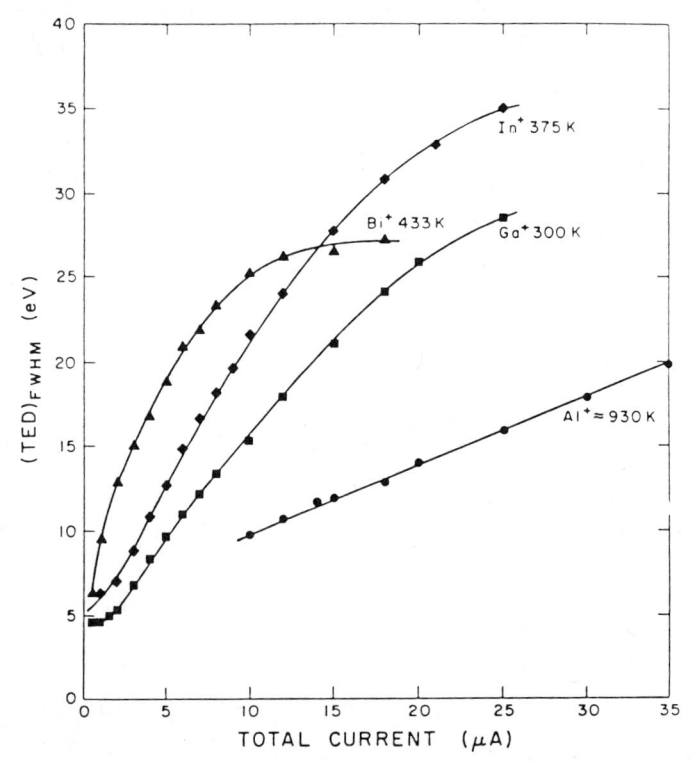

Figure 15.6 Full width at half maximum (FWHM) of the total energy distribution vs. total current for the indicated LMIS.

I but also with particle mass m. Detailed analysis shows that I' decreases as $m^{1/2}$ for $I = 10$ μA. Furthermore, it is easily shown that

$$I' = Jr_a^2/m_\alpha^2 \tag{11}$$

where m_α is the angular magnification of the source, r_a is the apex radius, and J is the current density. Since the jetlike protrusion model of KS [32] shows that J and r_a are nearly independent of m for $I < 10$ μA, then the mass dependence of I' can be due only to a variation in m_α with mass. Again the KS model supports this deduction by predicting that protrusion elongation significantly increases with mass at a constant current. This, in turn, will modify the trajectories such that m_α will increase with mass and thus account for the observation that

$$I' \propto I^n m^{-1/2} \tag{12}$$

The variation of the angular intensity distribution with I, shown in Figure 15.8 for an In LMIS, is typical for most pure metal LMIS. One of the unusual features of the distribution is the extreme uniformity and the sharp fall off at the beam edge.

2 Source Characteristics

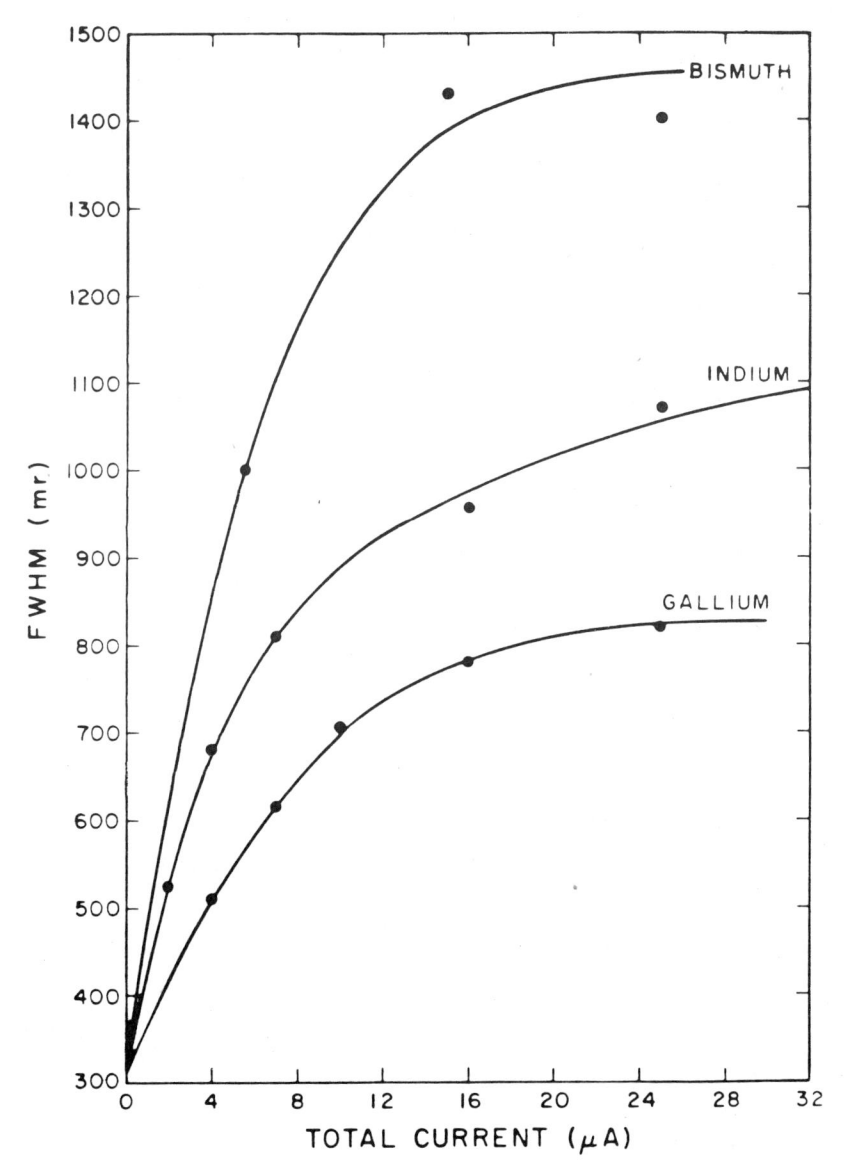

Figure 15.7 Full width at half maximum of the beam angular divergence (see Figure 15.8) vs. total current of the indicated LMIS.

For the wetted needle LMIS geometry shown in Figures 15.1b and c, it has been shown that substrate geometry parameters such as the needle shaft diameter D_s, cone angle α_c, and apex radius of curvature r_n can dramatically alter the value of I' for a specific I as well as the voltage V_t for current onset [47]. From experimental results and model calculations it has been shown that V_t increases with D_s, α_c, and r_n [47]. If the value of V_t to maintain a constant I increases due to variation

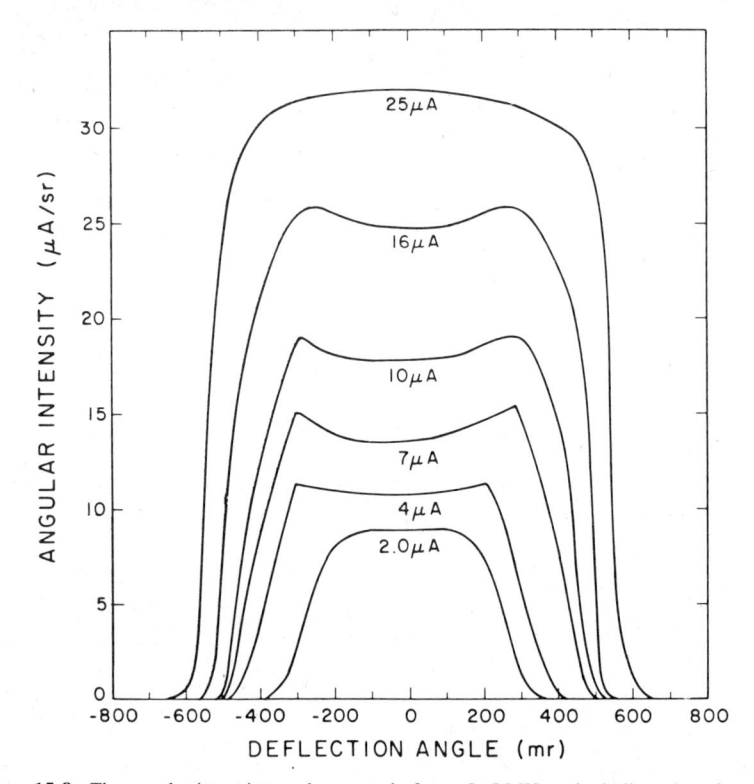

Figure 15.8 The angular intensity vs. beam angle for an In LMIS at the indicated total currents.

in one or more of the aforementioned geometric parameters, it can be shown that m_α decreases as $V_t^{-1/2}$. Thus, substituting for m_α in Eq. (11) the value of I' is seen to increase as V_t at constant I. Since ΔV is a function of I, and not I', one can expect to achieve higher values of I' for a specific ΔV simply by proper variation of the substrate geometrical parameters.

2.2 Virtual Source Size of the LMIS

As will be shown below, one of the important source parameters for fine focus applications is the virtual source size d_v. The value of d_v is obtained as shown in Figure 15.9 by extrapolating the trajectory tangents, at a distant plane from the surface, to a common crossover that usually occurs behind the physical emitter. The size of the circle formed at the crossover point is usually referred to as the virtual source and is the object for focusing applications. In other words, the minimum size of the focused beam cannot be less than Md_v, where M is the overall focusing column magnification. From the known atomic resolution of the field ion microscope one would conclude that d_v for a LMIS should be of the same order of magnitude.

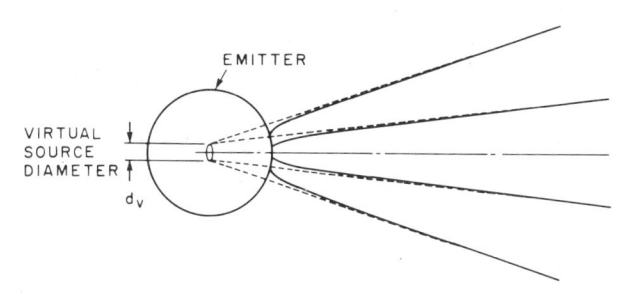

Figure 15.9 Beam trajectories and trajectory tangents. Extrapolated trajectory tangents define the indicated virtual source size d_v.

The measurement of d_v, although difficult, has been achieved both by experimental and theoretical approaches. By reducing the acceptance angle sufficiently to eliminate chromatic and spherical aberration contributions to the focused beam, some workers have been able to place an upper limit of ~ 500 Å on d_v [35, 48]. Using the latter approach Komuro [49] has experimentally determined d_v for a LMIS of Ga and In to be 400 to 500 and 670 Å, respectively. A Monte Carlo model developed by Ward et al. [50] showed that stochastic Coulomb interactions within the beam were the primary cause of the larger than expected value of d_v, which was in the range of 500 to 1000 Å.

Nevertheless, in spite of the larger than expected value of d_v, the source brightness B which determines the beam size and current density values in the focused beam, is quite large. The specific brightness of an LMIS can be given as

$$B = 4I'/\pi d_v^2 \tag{13}$$

If one assumes $d_v = 500$ Å and $I' = 20$ μA/sr, which is a typical experimental value obtained at 5 to 10 kV, a value of $B = 4 \times 10^6$ A/cm^2 sr is realized. In spite of the larger than expected d_v, B is much larger than any other known source of heavy metal ions.

3 ALLOY SOURCES

Considerable interest has been shown in the development of ion sources capable of producing a variety of technologically important ions such as B$^+$, As^{2+}, Be^{2+}, and Si^{2+}. In the case of highly volatile or excessively high melting point elements a suitable binary or ternary alloy and compatible substrate can often be found to form a long-lived alloy LMIS provided that a number of materials requirements can be met [51]. Besides the need for achieving a sufficiently low vapor pressure and melting point of the alloy, the element of interest must appear in the beam at a reasonably high level. As seen in Tables 15.2 and 15.3 the beam composition of A$^+$ in an AB alloy is different from the alloy composition because of the forma-

TABLE 15.3 A Comparison of the Beam and Alloy Compositions at I = 10 μA for Several Alloys

Alloy Composition	Beam Composition	Reference
$Au_{0.69}Si_{0.310}$	$Au_{0.90}Si_{0.10}$	[33]
$Au_{0.60}Be_{0.40}$	$Au_{0.88}Be_{0.12}$	[33]
$Au_{0.59}Si_{0.26}Be_{0.15}$	$Au_{0.55}Si_{0.16}Be_{0.28}$	[19]
$Pd_{0.66}As_{0.33}$	$Pd_{0.70}As_{0.30}$	[20]
$Pd_{0.70}As_{0.16}B_{0.14}$	$Pd_{0.73}As_{0.20}B_{0.07}$	[20]

tion of unwanted AB_n^{m+}, A^{m+}, and A_n^+ species and because of selectivity in the field evaporation process. The latter is often a serious problem because the alloy composition and hence melting point of the alloy varies with time. Thus, if a low melting eutectic composition is used, the melting point will increase and often lead to unacceptable alloy/substrate interaction.

It can be shown from a simplified model of field evaporation for a AB alloy, where A and B charged monomers are the principle beam species, that congruent ion formation (i.e., beam and alloy compositions are equal) will occur if [52]

$$H_A + I_A = H_B + I_B \qquad (14)$$

where H_A and H_B are the respective partial heats of vaporization of the alloy species. Equation (14) is obtained by equating the activation energies $Q_A(F) = Q_B(F)$ for field evaporation of the alloy species. Table 15.3, which compares the beam and alloy stoichiometries for several alloys, shows that congruent ion formation can be achieved for some of the alloy systems investigated. Even without congruent ion formation, alloy LMIS such as AuSi have been operated stably for several hundred hours at low currents, and the PdAsB alloy source, which supplies ions of both As and B, has operated stably for up to 150 h [20]. Ishitani et al. [21] have reported 250 h of continuous operation for the NiSiB alloy LMIS. Most important in alloy LMIS is the achievement of low volatility at the melting point, good wetting of the substrate without excessive attack by one of the alloy components, and a sufficient reservoir of the alloy for the desired source life.

4 APPLICATIONS OF LIQUID METAL ION SOURCES

That the LMIS is an ideal source for submicron focusing applications was recognized early in its development [8] and has been the driving force for research activity during the past decade. A review of such applications particularly as they relate to various aspects of IC fabrication has been given by Melngailis [53]. In contrast, one of the earliest and still active applications involves its use for electrostatic space propulsion [54]. Another early application that still receives current

4 Applications of Liquid Metal Ion Sources

interest is thin-film deposition using the LMIS in the high-current mode where ion emission is accompanied by copious, charged, microdroplet emission [55]. As mentioned in Section 2 the charged droplets can also be focused to beam sizes as small as 2.5 μm thereby allowing micron sized features to be patterned. Other applications include surface analytical applications such as scanning ion miroscopy and SIMS [5–7].

A number of focusing systems designed for both pure metal and alloy LMIS have been constructed. Typically, the primary consideration in focusing applications at a particular beam voltage is the minimum beam size, its current density, and working distance (i.e., distance from final lens or deflection plates to target). Because of the relatively large energy spread and angular intensity of the LMIS, the focused beam size d is limited by the chromatic aberration $d_c = \alpha C_c (\Delta V/V)$ (where α is the aperture half angle) and the virtual beam size d_v. It can be shown that the current density is related to these parameters and the column magnification M as follows [35]:

$$J = \left(M^{-2} - (d_v/d)^2\right)(4\pi V^2/C_c^2)(I'/\Delta V^2) \tag{15}$$

where V is the beam voltage, ΔV is the energy spread per unit charge, and C_c is the effective chromatic aberration coefficient of the column. The minimum achievable beam size is $d = M d_v$. Using a two-lens system at 40 kV, Levi-Setti et al. [48] have reported a beam size of 430 Å at a beam current density of 0.11 A/cm² and a working distance of 3 cm. With a single lens focusing column operating at 25 kV and a working distance of 75 mm a beam size of 2000 Å and a current density of 1.0 A/cm² has been obtained [35].

The last factor in Eq. (15), $I'/\Delta V^2$, contains only source parameters and can be referred to as the source figure of merit. Figure 15.10, which compares the figure of merit for various single component LMIS, clearly shows a strong mass and current dependence. One arrives at the somewhat paradoxical conclusion that to increase the focused beam current density one must reduce the source total current. The figure of merit for the various species of alloy LMIS also shows a similar dependence on m and I. That this is an expected result can be shown by combining Eqs. (10) and (12) with the definition of source figure of merit $I'/\Delta V^2$, which gives

$$I'/\Delta V^2 \propto m^{-3/2} I^{n-1.2} \tag{16}$$

Energy broadening of the beam places a limit not only on the size of the focused beam, but also on its current density distribution. It has been shown [56] that in case of chromatically limited focused beams, the shape of the energy distribution curve is manifested in the current density distribution of the focused beam. Thus, long tails in the energy distribution will show up as long tails on the current density distribution. The latter have been observed both for pure metal [57] and alloy sources [58]. It should be noted that when both M^+ and M^{2+} species exist in the

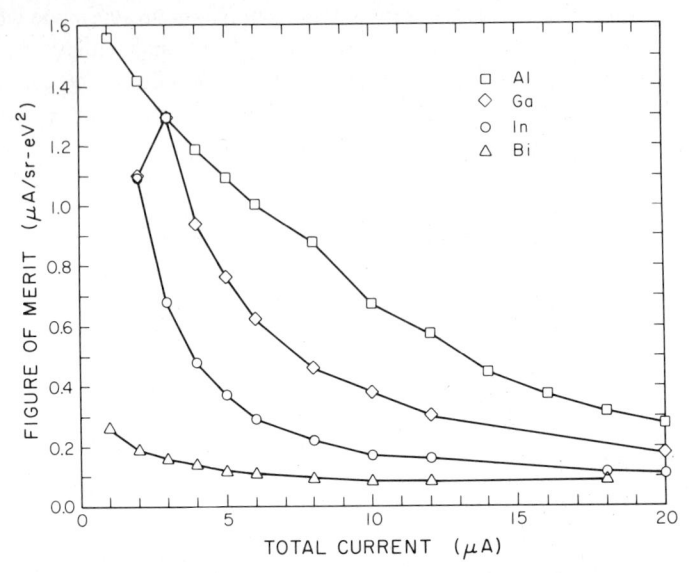

Figure 15.10 Source figure of merit vs. total current for the indicated LMIS are shown. Since I' for various source geometries varies linear with source voltage [Ref. 47], the I' values are normalized to a common source voltage of 12 kV.

beam that the M^{2+} species tend to show narrower energy distributions and therefore should yield a narrower current density distribution in the focused beam.

5 SUMMARY

The understanding of the mechanism of operation of the liquid metal ion source has advanced considerably during the past 15 years, although more work remains to be done. The formation mechanisms of large cluster ions and microdroplets is not well understood. Some LMIS such as group IVA and VA metals exhibit a large contribution of cluster ions, while others such as group IIIA metals give a relatively small contribution. A more comprehensive model of the role of hydrodynamic and space-charge forces in the operation of the field stabilized cone would be useful. The role of such parameters as work function, ionization potential, heat of vaporization, and composition on the relative ion formation of alloy components remains to be rigorously elucidated.

In spite of the lack of full understanding of all aspects of LMIS operation, a great deal of experimental studies of emission characteristics now exists. This has allowed rapid utilization of the LMIS in focused beam applications where extremely high-current densities and small beam sizes are required. Commercial focused beam systems and relevant LMIS designed for applications ranging from photo mask repair to maskless ion implantation are now available from JEOL, Ltd. and Seiko of Japan, Vacuum Generators of the United Kingdom, and FEI Co., Micro Beam Corp. and Micrion Corp. of the United States.

REFERENCES

1. R. L. Kubena, C. L. Anderson, R. L. Seliger, R. A. Jullens, and E. H. Stevens, *J. Vac. Sci. Technol.* **19,** 916 (1981).
2. V. Wang, J. W. Ward, and R. L. Seliger, *J. Vac. Sci. Technol.* **19,** 1158 (1981).
3. R. L. Seliger, R. L. Kubena, R. D. Olney, J. W. Ward, and V. Wang, *J. Vac. Sci. Technol.* **16,** 1610 (1979).
4. D. B. Rensch, R. L. Seliger, G. Csanky, R. D. Olney, and H. L. Stover, *J. Vac. Sci. Technol.* **16,** 1897 (1979).
5. A. R. Bayly, A. R. Waugh, and K. Anderson, *Nucl. Instrum. Methods* **218,** 375 (1983).
6. P. D. Prewett and D. K. Jefferies, *Inst. Phys. Conf.* **54,** 316 (1980).
7. R. Levi-Setti, Y. L. Young, and G. Crow, *J. Phys. Colloq. (France)* **45,** 197 (1984).
8. V. E. Krohn and G. R. Ringo, *Appl. Phys. Lett.* **27,** 479 (1975).
9. R. Clampitt and D. K. Jefferies, Inst. Phys. Conf. Ser. No. 38, (1978), p. 12.
10. L. W. Swanson, *Microcircuit Engineering 80,* R. P. Kramer (Ed.), Delft Univ. Press, Delft, 1981, p. 267.
11. L. W. Swanson, G. A. Schwind, A. E. Bell, and J. E. Brady, *J. Vacuum Sci. Tech.* **16,** 1864 (1979).
12. A. E. Bell, G. A. Schwind, and L. W. Swanson, *J. Appl. Phys.* **53,** 4602 (1982).
13. Y. Torii and H. Yamada, *Japan. J. Appl. Phys.* **22,** L444 (1983).
14. R. Clampitt and D. K. Jefferies, *Nucl. Instrum. Methods* **149,** 739 (1978).
15. C. Bartoli, H. Von Rohden, S. P. Thompson, and J. Blommers, *J. Phys. D* **17,** 2473 (1984).
16. P. Sudraud, C. Colliex, and J. van de Walle, *J. Phys. (France)* **40,** L207 (1979).
17. K. Gamo, T. Ukegawa, and S. Namba, *Jpn. J. Appl. Phys.* **19,** L379 (1980).
18. R. L. Kubena, C. L. Anderson, R. L. Seliger, R. A. Jullens, and E. H. Stevens, *J. Vac. Sci. Technol.* **19,** 916 (1981).
19. K. Gamo, T. Matsui, and S. Namba, *Jpn. J. Appl. Phys.* **22,** L692 (1983).
20. M. W. Utlaut, W. M. Clark, R. L. Seliger, A. E. Bell, L. W. Swanson, G. A. Schwind, and J. B. Jergenson, *J. Vac. Sci. Technol. B* **5,** 197 (1987).
21. T. Ishitani, K. Umemura, S. Hosoki, S. Takayama, and H. Tamura, *J. Vac. Sci. Technol. A* **2,** 1365 (1984).
22. V. Wang, J. W. Ward, and R. L. Seliger, *J. Vac. Sci. Technol.* **19,** 1158 (1981).
23. G. I. Taylor, *Proc. R. Soc. (London) Ser. A* **280,** 383 (1964).
24. J. F. Mahoney, A. T. Yahiku, H. L. Daley, R. D. Moore, and J. Perel, *J. Appl. Phys.* **40,** 5101 (1969).
25. D. S. Swatik and C. D. Hendricks, *A.I.A.A.J.* **6,** 1596 (1968).
26. R. Gomer, *Appl. Phys.* **19,** 365 (1979).
27. P. D. Prewett, G. L. R. Mair, and S. P. Thompson, *J. Phys. D.* **15,** 1339 (1982).
28. N. K. Kang and L. W. Swanson, *Appl. Phys. A* **30,** 95 (1983).
29. G. Benassayag, P. Sudraud, and B. Jouffrey, *Ultramicroscopy* **16** (1985).
30. N. Sujatha, P. H. Cutler, E. Kazes, J. P. Rogers, and N. M. Miskovsky, *Appl. Phys. A* **2,** 55 (1983).
31. D. R. Kingham and A. E. Bell, *J. Phys. Colloq. (France)* **45,** 139 (1984).
32. D. R. Kingham and L. W. Swanson, *Appl. Phys. A* **34,** 123 (1984).
33. L. W. Swanson and G. A. Schwind (unpublished).
34. D. R. Kingham, *Surf. Sci.* **116,** 273 (1982).
35. L. W. Swanson, *Nucl. Instrum. Methods* **218,** 347 (1983).

36. L. W. Swanson, G. A. Schwind, and A. E. Bell, *J. Appl. Phys.* **51,** 3453 (1980).
37. A. R. Waugh, *J. Phys. D* **13,** L203 (1980).
38. L. W. Swanson and D. R. Kingham, *Appl. Phys. A* **41,** 223 (1986).
39. T. Venkatesan, A. Wagner, and D. Barr, *Appl. Phys. Lett.* **38,** 943 (1981).
40. A. Wagner, T. Venkatesan, P. M. Petroff, and D. Barr, *J. Vac. Sci. Technol.* **19,** 1186 (1981).
41. F. G. Rudenauer, W. Steiger, R. Grotzschel, and F. Nahring, *Vacuum* **35,** 315 (1985).
42. C. d'Cruz and K. Pourrezaei, *J. Appl. Phys.* **58,** 2724 (1985).
43. C. Mahony and P. D. Prewett, *Vacuum* **34,** 301 (1984).
44. G. Benassayag, J. Orloff, and L. W. Swanson, *J. Phys. Colloq. (France)* **7,** 389 (1986).
45. H. P. Mayer, *Appl. Phys. Lett.* **47,** 1247 (1985).
46. W. Knauer, *Optik* **59,** 335 (1981).
47. A. E. Bell and L. W. Swanson, *Appl. Phys. A* **41,** 335 (1986).
48. R. Levi-Setti, Y. L. Wang, and G. Crow, *Appl. Surf. Sci.* **26,** 249 (1986).
49. M. Komuro, H. Hiroshim, H. Tanque, and T. Kanayama, *J. Vac. Sci. Technol. B* **1,** 985 (1983).
50. J. W. Ward, *J. Vac. Sci. Technol. B* **3,** 207 (1985).
51. M. J. Bozack, L. W. Swanson, and A. E. Bell, *J. Phys Colloq. (France)* **47,** 95 (1986).
52. M. J. Bozack, L. W. Swanson, and J. Orloff, *SEM IV*, 1139 (1985).
53. J. Melngailis, *J. Vac. Sci. Technol. B* **5,** 469 (1987).
54. C. Bartoli, H. von Rohden, S. P. Thompson, and J. Blommers, *J. Phys. D* **17,** 2473 (1984).
55. C. Mahony and P. D. Prewett, *Vacuum* **34,** 301 (1984).
56. D. W. Tuggle, L. W. Swanson, and M. A. Gesley, *J. Vac. Sci Technol. B* **4,** 131 (1986).
57. J. W. Ward, M. W. Utlaut, and R. S. Kubena, *J. Vac. Sci. Technol. B* **5,** 169 (1987).
58. Y. Wada, S. Shukuri, M. Tamura, H. Masada, and I. Ishitani, Proceedings, Electrochem. Soc. Meeting, Las Vegas, NV, October, 1986, p. 133.
58. Y. Wada, S. Shukuri, M. Tamura, H. Masada, and I. Ishitani, Proceedings, Electrochem. Soc. Meeting, Las Vegas, NV, October, 1986, p. 133.

16

The Metal Vapor Vacuum Arc Ion Source

I. G. Brown
Lawrence Berkeley Laboratory
University of California
Berkeley, California

While there has been steady progress in the development of ion sources for the production of beams of gaseous species, techniques for the generation of high-current beams of metal ions have been more limited. Metallic ion sources have for the most part employed vaporization of the solid material [1–4], or surface ionization [1, 5–7], or sputtering of metallic ions from a solid electrode by a gaseous "carrier" plasma [1, 8, 9]. Beams of up to several tens of milliamperes of metal ions have been obtained on a long-pulse or dc basis.

The MEVVA ion source is a new kind of source in which a *me*tal *v*apor *v*acuum *a*rc is used as the plasma medium from which the ions are extracted. With this source, pulsed beam currents of over 1 A of metallic ions have been produced from a wide range of solid electrode materials.

The study of metal vapor vacuum arc plasma discharges—also called vacuum arcs, or metal vapor arcs—had its origin in the high-power switching field, and research in this discipline has remained largely the domain of the high-power electrical engineering community. The metal vapor arc provides a means of switching high currents at high voltage in a vacuum environment—this is the vacuum switch, invented by R. A. Millikan and R. W. Sorensen. One of the earliest publications on this work is that of Sorensen and Mendenhall in 1926 [10]. A historical survey of work up to about 1960 has been given by Cobine [11]. More recently, a very complete review of the entire field of metal vapor vacuum arcs has been given by Lafferty [12].

A closely related kind of plasma discharge, the vacuum spark, is fairly well developed and has been used as a spectral source [13, 14] and ion source [15, 16] for some time. Characteristic of vacuum sparks is the production of very highly stripped species of the electrode material; for example, the heliumlike spectral lines of Ti^{20+}, Fe^{24+}, and Co^{27+} have been observed [14]. These sources, however, are inherently of submicrosecond duration.

The production of ions in the metal vapor vacuum arc plasma has been investigated by a number of authors for at least the last two decades [17-26]. One of the earliest attempts to incorporate the arc as the method of plasma production for use in an ion source is the work done as part of the Manhattan Project in World War II; however, this source suffered from several drawbacks and was not pursued [27]. Revutskii et al. [28], in 1968, investigated a cylindrically symmetric arc geometry; their configuration employed ion extraction through a hole in the cathode, and this work also seems to have not been pursued. More recently, the source has been developed by Brown and coworkers [29-32], Humphries and coworkers [33-37], and Adler and Picraux [38]. These recent versions have produced impressive results, and this is the kind of source described here.

1 THE METAL VAPOR VACUUM ARC PLASMA

The metal vapor vacuum arc is a plasma discharge between two metallic electrodes in vacuum. The ambient pressure should be sufficiently low that the gas does not affect the discharge phenomena; a pressure of about 10^{-4} Torr might be considered a rough upper limit, and the 10^{-6}-Torr range is usual.

The physics involved in the mechanism by which the arc is established in the first place is poorly understood. Certainly the concept of arc growth by an electron avalanche in the gaseous medium between the electrodes—the usual picture of gaseous breakdown—is not applicable. Here, conduction of the arc current is supported by metal plasma that is evolved from the solid electrode (cathode) material itself. Thus, the problem is taken back to the initial step of establishing plasma evolution from the cathode. Triggering of the vacuum arc has been accomplished by using a discharge to the cathode across an insulating surface [39-42], a laser-initiated plasma discharge [25], and by physically separating the two electrodes while current is being conducted between them [18-20, 24].

A basic characteristic of this kind of discharge is the formation of "cathode spots" [12, 17-26, 43-45]. These are minute regions of intense current concentration—micron-sized spots in which the current density can be over 10^6 A/cm^2—on the cathode surface, at which cathode material is vaporized and ionized. Individual spots have been observed to move around on the cathode surface, and the lifetime of a spot may be only microseconds; on the other hand, small surface irregularities like edges or protuberances will tend to anchor the spots to these areas. The plasma pressure near the solid surface is high, and a strong pressure gradient is established which causes the plasma generated within the spot to plume away from the surface in a manner similar to the behavior of the plasma generated by the interaction of an intense focussed laser beam with a solid surface. The current carried by a cathode spot is typically of the order of a few amperes,

Figure 16.1 Photograph of the dense metal plasma plume streaming through the 1-cm-diameter anode hole (left) and ducted along the magnetic field to the extractor (right). Uranium plasma in MEVVA I. Courtesy Lawrence Berkeley Laboratory, University of California.

depending on the metal, and if the arc is caused to conduct a higher total current, then more cathode spots are formed; thus, a typical metal vapor arc discharge of several hundred amperes arc current might involve the participation of several tens of cathode spots. The assemblage of cathode spots thus gives rise to a dense plasma of cathode material. This quasi-neutral plasma plumes away from the cathode, initially normal to the cathode and in the general direction of the anode, thereby allowing the arc current to flow and the arc to persist.

It is this expanding plasma plume that constitutes the medium from which the ion beam is to be extracted. The plasma consists entirely of material evolved from the cathode, since that is where the cathode spots are located. The plasma plume may be magnetically ducted, and the plasma size and density can to some extent be controlled. The plasma plume is quasi-neutral and flows from the cathode to the anode. The plasma streams through a hole in the anode toward a set of extractor grids, where, in the normal fashion, the ion component is extracted from the plasma and the beam generated. A photograph of the intense uranium plasma produced in the MEVVA I ion source is shown in Fig. 16.1. Here the plasma is flowing through the central hole in the anode on the left and streaming along the magnetic field lines toward the extractor on the right. Note that the plasma stream grows in diameter not because of space charge blowup (the plasma is charge-neutral), but because of the diverging magnetic field lines of the flux tube defined by the anode hole.

2 SOURCE DESIGN

Several versions of the MEVVA concept have been developed at the Lawrence Berkeley Laboratory (LBL) [46]. A schematic of the MEVVA IIb device is shown in Figure 16.2 and a partially disassembled source in Figure 16.3. A cylindrically symmetric configuration is used in this embodiment; this arc geometry has been studied by Gilmour and Lockwood [47], and the LBL design has drawn extensively upon this work. Plasma created at the cathode flows through a central hole in the anode, of diameter about 1 cm, and through a drift space of several centimeters to the extraction grids. Coolant (low-conductivity water, or Freon for higher voltage work) is carried to the cathode and anode regions for heat removal. The cathode is a simple cylindrical plug of the material of interest, and the trigger electrode surrounds the cathode tip, separated from it by a thin (about 1 mm thick) alumina insulator, as shown in Figure 16.4. Since only the front surface of the cathode is eroded by the vacuum arc, it is not essential that the entire mass of the cathode be fabricated from the element required; thus, the front surface of the cathode may be a thin wafer only several millimeters thick that has been secured to, for example, a stainless steel *cathode mounting slug*. Since the plasma is formed only from the cathode material, where the cathode spots are located, and there is virtually no contribution to the plasma from other components in the discharge region, the materials of which the trigger electrode, trigger-cathode insulator, and anode, for example, are fabricated are not critical; nonetheless, they should be materials that

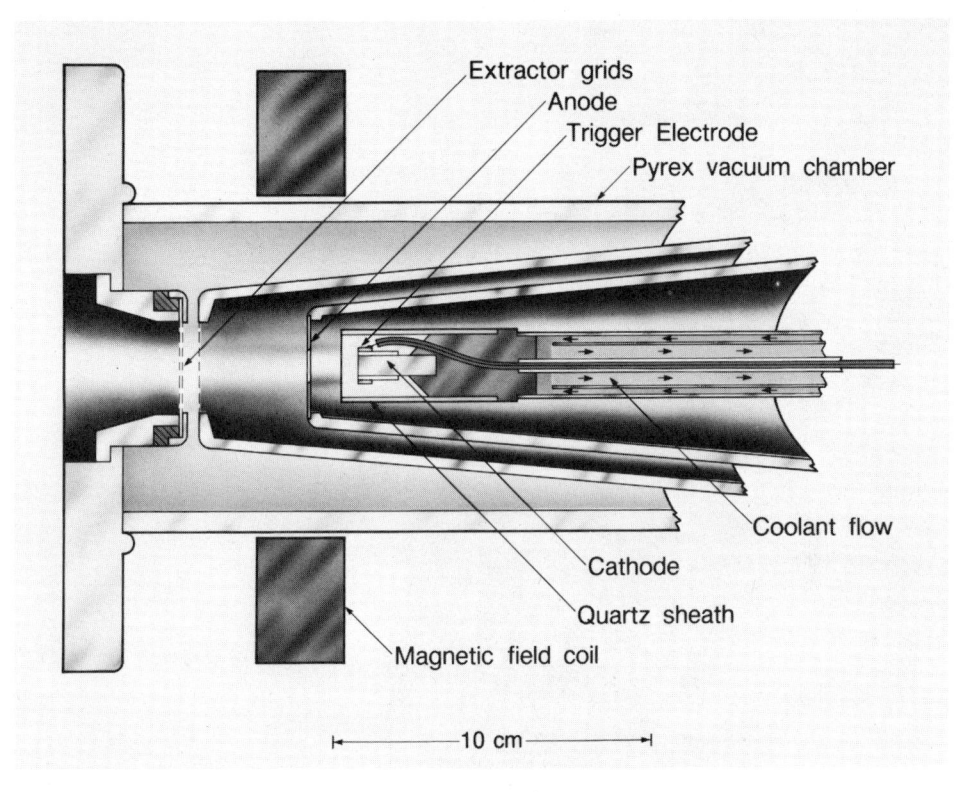

Figure 16.2 Outline of the MEVVA ion source. The embodiment shown is that of MEVVA IIb. Courtesy Lawrence Berkeley Laboratory, University of California.

are good for vacuum and plasma application. The precise geometry of the cathode-trigger assembly is the subject of present development. Cathodes of diameter between about 0.05 and 0.50 in. and of various detailed shapes have been tried. One version that works well uses a 0.250-in.-diameter cathode with the alumina insulator glued to it with an alumina-based cement such as Aremco [48] or Sauereisen [49]. The purpose of the cement is to ensure that the trigger spark occurs as a surface discharge rather than across a vacuum gap. The front surface of the cathode-insulator-trigger assembly can be cleaned and smoothed, to remove excess material after the glue has dried, using quite ordinary abrasive cloth.

The magnetic field coil that establishes the field in the arc region is of minimal design. The field itself is not necessary for the source operation, but serves to increase the efficiency (ratio of beam current to arc current) with which the source operates. The field can be varied up to several hundred gauss. The effect of this magnetic field on the transport of the metal plasma generated at the cathode to the extractor grids has been studied in detail [50]. The metal vapor arc is such a prolific generator of plasma, however, that the additional plasma that is in this way

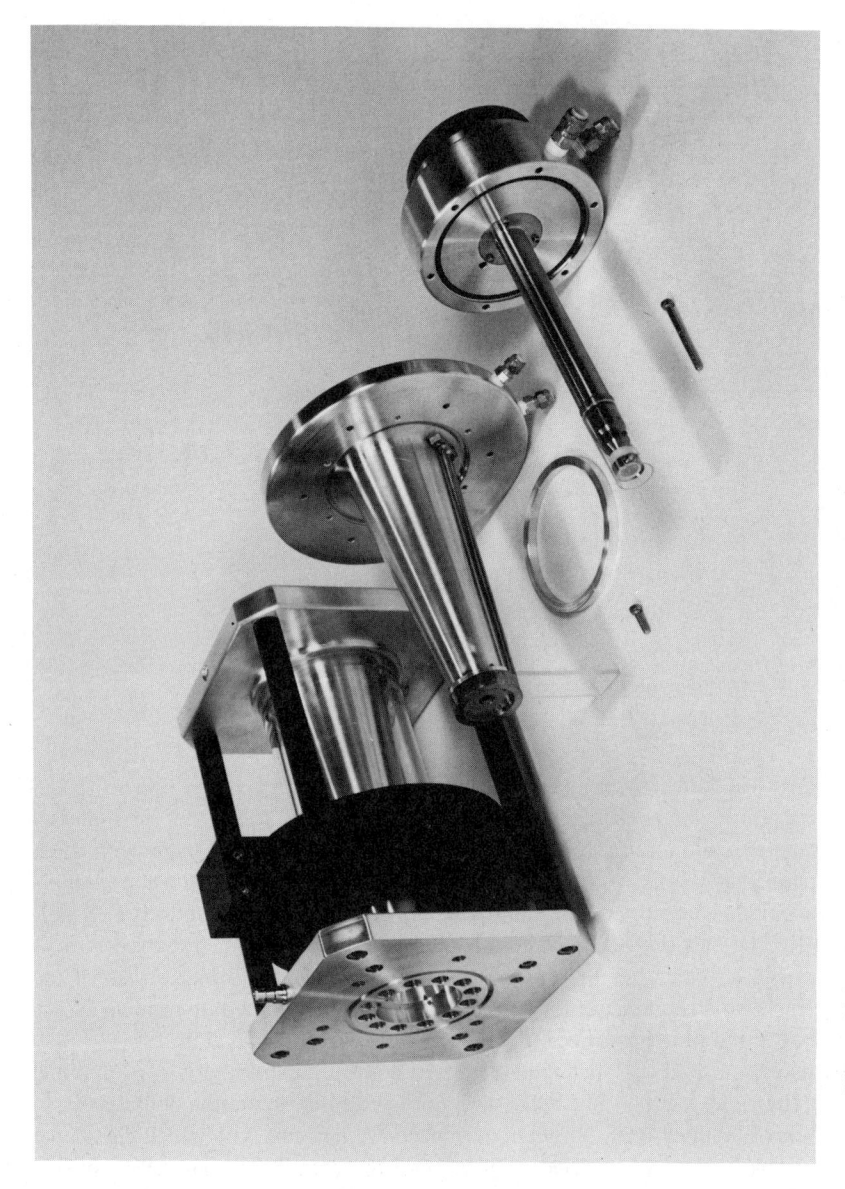

Figure 16.3 The partially disassembled MEVVA IIb source. Courtesy Lawrence Berkeley Laboratory, University of California.

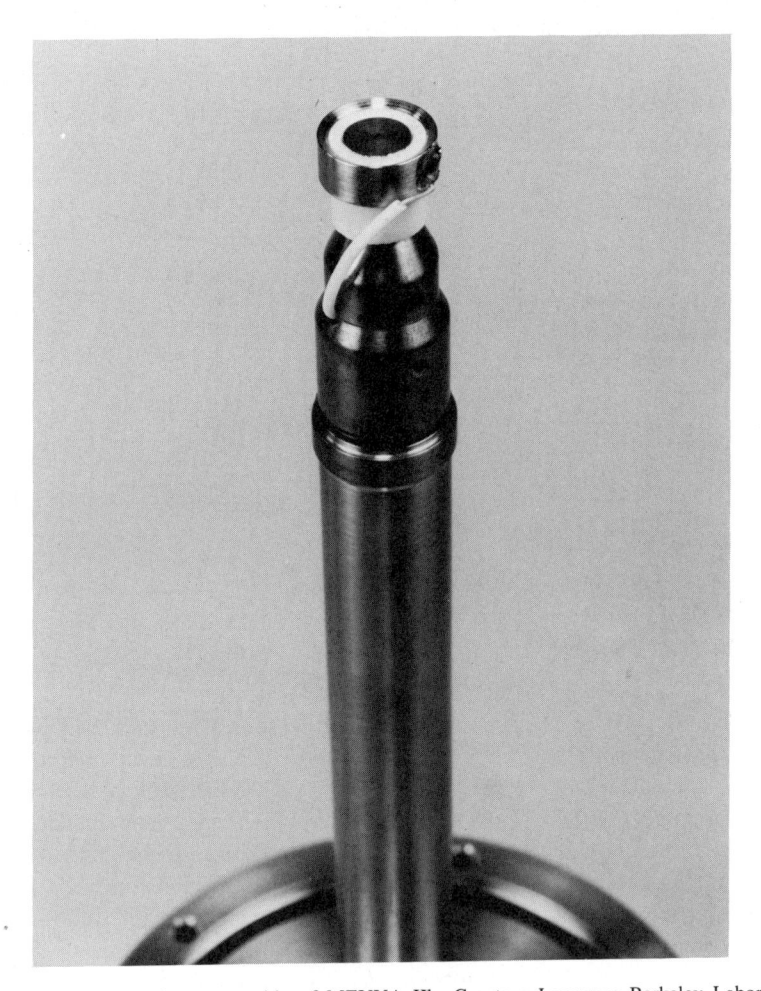

Figure 16.4 Cathode-trigger assembly of MEVVA IIb. Courtesy Lawrence Berkeley Laboratory, University of California.

presented to the extractor is rarely required, and in actual use, the magnetic field has not normally been used.

The extractor configuration is a multiaperture, accel–decel design. The three grids are fabricated from about 0.030-in.-thick stainless steel or aluminum and have anywhere from 7 to 200 individual extraction holes, each of diameter in the range 1 to 5 mm. Extractor design is a field in itself, and is covered elsewhere in this book. The maximum beam current that can be produced by the source is determined by the extractor, and is not limited by the available plasma. Thus, if it is important for a given application to produce the highest possible beam current, then the extractor design is critical and determines the limiting current.

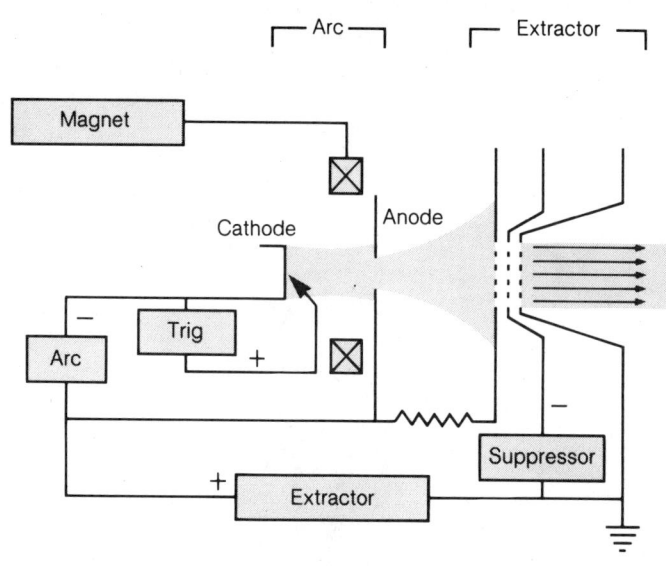

Figure 16.5 Schematic of the source electrical configuration. Courtesy Lawrence Berkeley Laboratory, University of California.

The electrical systems necessary to drive the source are indicated in Figure 16.5. The trigger pulse may be 5–10 kV in amplitude and of several microseconds duration, conveniently delivered across the trigger-cathode by a step-up transformer which also serves as high-voltage isolation. The minimum applied voltage for which triggering will occur depends largely on the design and condition of the cathode-trigger assembly, particularly on the degree to which the insulator presents a surface path rather than a vacuum gap. For a well-prepared cathode-trigger assembly, the minimum trigger voltage can be as low as a few hundred volts. The arc power supply can be a low-impedance LC pulse line if the required arc pulse length is not much more than a few milliseconds. Typical might be a line of length 250 μs, of impedance 1 Ω, that is charged to a voltage of several hundred volts. This is a very simple way of providing the 50–500 A necessary for the arc, and the problem of floating the supply up to extractor voltage is minimized also. If the pulse line charging voltage is too low, so that the pre-arc voltage across the anode–cathode gap is too low, then triggering can be difficult. This situation can be aided by using a line of higher impedance (higher open-circuit voltage with still not too high arc current) or by using a series resistor of 1 Ω or so in the arc circuit. The other components of the system are quite standard.

Further details of the source construction and operation can be found in references [29–32, 50]. This source is a new development and is still undergoing much evolution. Thus, the embodiments and performance described here are in this sense still preliminary.

3 EMBODIMENTS

Several embodiments of the source concept have been made and tested at LBL [46]. These different versions of the source are briefly described here.

3.1 MicroMEVVA

The MicroMEVVA source is a miniature embodiment of the MEVVA concept. One version of this source is less than "thumb size"—1.5 cm in diameter and 6 cm in overall length. Beam currents of up to 15 mA at 15 kV have been produced. The power dissipation capability of this source is minimal, since there is very poor heat removal. Nonetheless the source can operate at, for example, a pulse length of about 100 μs and a repetition rate of about 1 pps. This source is shown in Figure 16.6a, and the components from which it is assembled are shown in Figure 16.6b; clearly it is a simple source! It is fabricated from an array of metal and alumina tubes one within another. The beam composition is determined by the cathode material; to change the ion species the cathode cylinder, or at least its exposed tip, should be changed. This embodiment and some results obtained with it have been described in the literature [51]. A MicroMEVVA II version has been made also. In this second iteration of the microsource, the cathode-trigger design was changed to facilitate changing the cathode material and to improve the triggering reliability and cathode lifetime, and the voltage hold-off capabilities were improved. This source is shown in Figure 16.7a, and its components are shown in Figure 16.7b. The cathode is the central rod and is copper in the source shown. An additional electron between anode and ground has been added, improving the extraction optics. Miniature MEVVA embodiments are finding application as injectors of metal ions into EBIS (*E*lectron *B*eam *I*on *S*ource) devices, where it is important to inject the "ion feedstock" while still maintaining a very high-quality vacuum [52]. Some preliminary work has also been done using the MicroMEVVA as a tool for Fourier transform ion cyclotron resonance mass spectrometry [53].

3.2 MEVVA II

This is the version that has been used to describe the basic MEVVA design and performance, above, and is shown in Figures 16.2–16.4. The fully assembled MEVVA IIb device is shown in Figure 16.8. Note that the magnetic field coil is not essential, and the source performs well without it (see below for further discussion on the effect of the magnetic field). In the MEVVA IIa version the trigger electrode was a central thin rod or pin, about 1 mm diameter, sheathed in an alumina tube with the cathode surrounding it coaxially. The MEVVA IIb design reversed this configuration, with the cathode being the central conductor—a simple cylindrical rod—surrounded coaxially by the alumina insulator and then the trigger electrode. This second configuration seems to give better performance in a number of ways, including cathode lifetime and triggering reliability; it is also a simpler cathode geometry to manufacture.

340   *The Metal Vapor Vacuum Arc Ion Source*

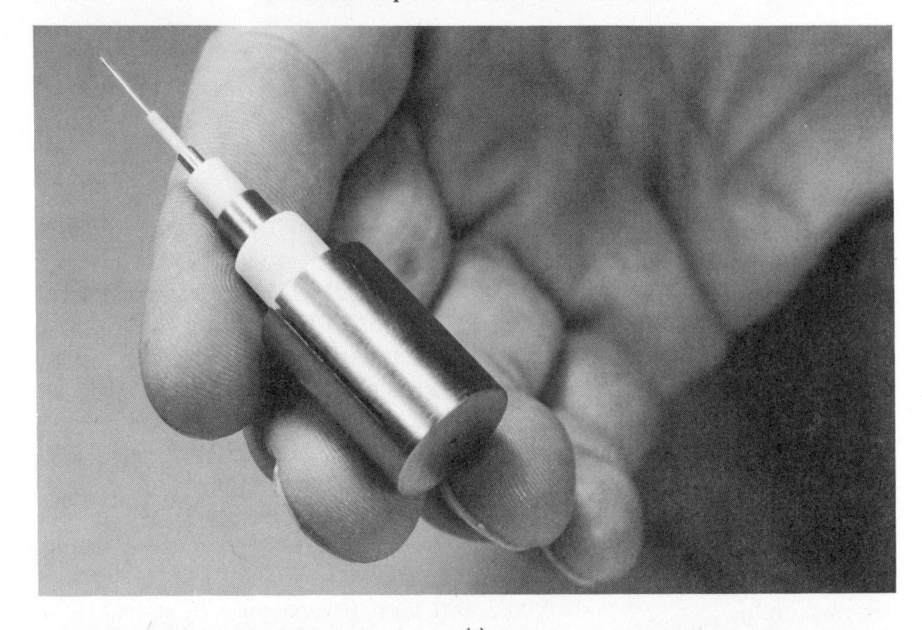

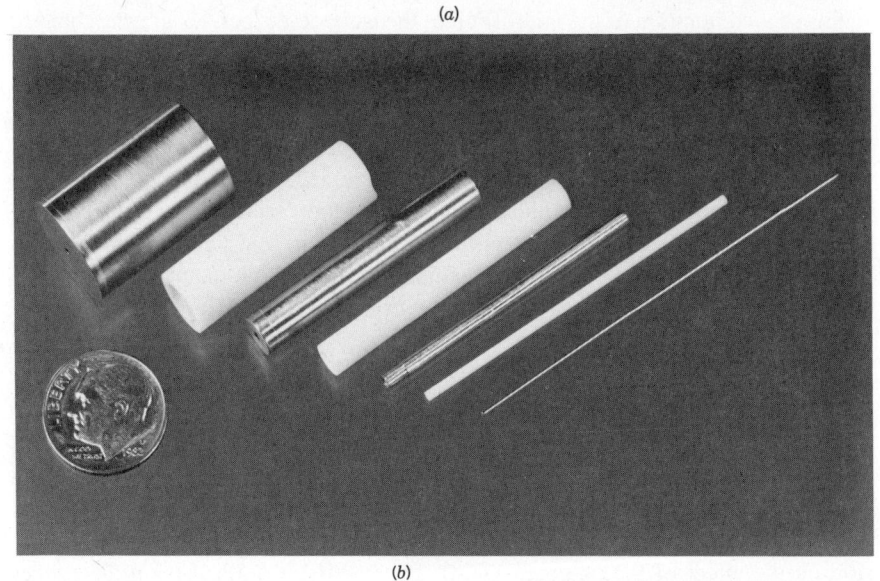

Figure 16.6 (*a*) MicroMEVVA I, assembled; (*b*) MicroMEVVA I, component parts. Courtesy Lawrence Berkeley Laboratory, University of California.

Figure 16.7 (*a*) MicroMEVVA II, assembled; (*b*) MicroMEVVA II, component parts. Courtesy Lawrence Berkeley Laboratory, University of California.

(a)

(b)

Figure 16.8 MEVVA IIb; overall length is 16 in. Courtesy Lawrence Berkeley Laboratory, University of California.

3.3 MEVVA III

This embodiment is about the size of a closed fist. Its arc and extractor components are about the same size as in the MEVVA II, and the source can deliver pulsed beams of hundreds of milliamperes. It resides entirely within the vacuum chamber, and because the cooling is quite minimal, so the average power dissipation capability is modest also; that is, the duty cycle is limited. This source is shown in Figure 16.9.

3.4 MEVVA IV

This version incorporates a number of novel features, the most significant of which is the multicathode design. A circular cathode assembly houses an array of 16 separate cathodes in a Gatling gun-like fashion. The cathode assembly can be rotated so as position any one of the 16 cathodes into the firing position. The rotation can be done while under high vacuum and in a time of a second or less. While the first rendition of MEVVA IV calls for manual operation of this feature, it is, of course, a simple extension to add a remote electromechanical positioning device. With the multiple cathode source, the lifetime of the source (operating time between scheduled maintenance) is increased by a factor of 16, or alternatively, different cathode materials can be installed and thus the beam species changed simply and swiftly.

The source operates well at an extraction voltage of 100 kV, and has been tested up to 120 kV. It is interesting to note that this high extraction voltage, in conjunction with the fact that in general the ion species produced are multiply ionized with mean charge state up to $\bar{Q} = 3$ (see below), means that the ion energy can be up to about 300 keV.

That beams can be produced from 16 different cathode materials in a single experimental run is quite an advantageous operational feature, and makes possible a range of experiments that would otherwise be difficult or impossible. Characteristics of a wide range of beam species can be intercompared while maintaining the same experimental conditions [54].

This source, partially disassembled, is shown in Figure 16.10, which also shows the multiple cathode assmebly.

4 SOURCE PERFORMANCE

4.1 Ion Species

The source has been operated using a wide range of cathode materials spanning the periodic table and including Li, C, Mg, Al, Si, Ti, Cr, Fe, Co, Ni, Cu, Zn, Y, Zr, Nb, Mo, Rh, Pd, Ag, In, Sn, La, Gd, Ho, Hf, Ta, W, Pt, Au, Pb, Th, U, LaB_6, CdSe, FeS, PbS, SiC, TiC, and WC. All of these materials produce intense beams—hundreds of milliamperes under typical operating conditions. Soft materials like Li, Sn, and Pb tend to have a shorter lifetime before triggering problems due

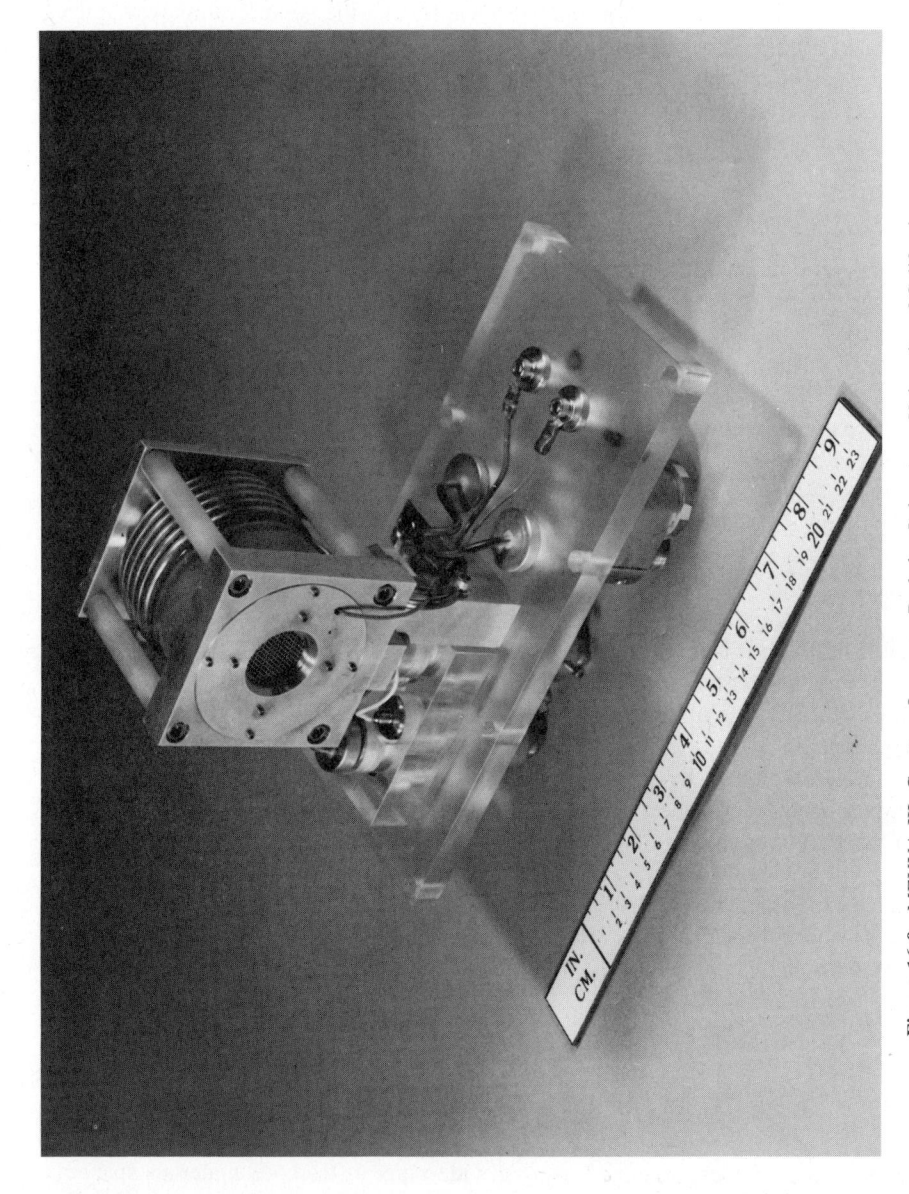

Figure 16.9 MEVVA III. Courtesy Lawrence Berkeley Laboratory, University of California.

Figure 16.10 The MEVVA IV source is a multicathode embodiment of the concept, in which any one of 16 separate cathodes can be switched into operation. Courtesy Lawrence Berkeley Laboratory, University of California.

to plating over of the cathode-trigger insulator necessitate maintenance, but this depends on how the source is run (e.g., the arc current and pulse length). Cathodes that are chemical compounds rather than pure elements produce beams containing ions of the constituent species, and it is interesting to note that beams containing nonmetallic elements, such as B and S, can be made by using conducting compound electrodes in which the nonmetal is a constituent.

4.2 Beam Current

The maximum ion beam current that has been measured in the embodiments of the source concept described above is approximately 1 A (electrical), and currents of several hundred milliamperes can be produced with only modest arc power, for example, 100 A arc current and 20 V arc voltage. Not surprisingly, the beam shape is roughly Gaussian at some distance from the source, and half the total beam current is contained within the Gaussian half-width, at which the beam divergence is typically about 3° and the emittance 0.05π cm mrad (normalized). When the extraction is optimally tuned up, the emittance can be less than 0.02π cm mrad (normalized) (at, roughly, the 80% beam intensity level). Operationally, the arc current is varied so as to maximize the beam current measured into the acceptance of a Faraday cup, which occurs when the plasma density is best matched to the extractor parameters [55–57]. Keller has developed a high-voltage extractor for the MEVVA source that has produced an extremely bright beam; the emittance-normalized brightness was $B_{\epsilon n} = 21$ A/$(\pi$ mm mrad$)^2$ [57, 58].

4.3 Extraction Voltage

The maximum voltage at which beam has been extracted in the present source embodiments is approximately 100 kV. This was achieved with the MEVVA IV source. Since the charge state distribution of the ions produced contains multiply ionized species with charge state up to about $Q = 5+$ for some metals and with mean charge state of typically $\overline{Q} = 2$ to 3, the mean energy of the beam ions produced can be 200–300 keV.

4.4 Beam Noise

Beam noise is a parameter of some importance to good transport and it is vitally important to some applications. For optimal beam extraction, the plasma density must be matched to the grid geometry, so that the plasma meniscus is correctly located; for the MEVVA, the plasma density is directly and simply controlled by the arc current. We have found that when the plasma density is properly matched to the extractor grids employed, not only is the beam then maximum in current (measured into the fixed divergence defined by the Faraday cup), but also the beam noise is minimum. This may be a feature common to other kinds of sources also. Figure 16.11 shows an oscillogram of the beam current and arc current for this optimized condition; here the beam noise is well less than 5% rms.

4.5 Duty Cycle

The upper limit to duty cycle at which the source can be run has been steadily increased. The limitations are set by the electronics, including the trigger generator, arc supply, and extractor supply, as well as heat removal from the arc region. Clearly a higher duty cycle can be expected when the arc current is lower. A high repetition rate triggering system, employing spark gaps instead of thyratron, has been developed which can operate at up to several hundred pulses per second, and the arc supply efficiency has been increased. With these changes, and using the better cooling capacity of the MEVVA IIb or MEVVA IV, the source was operated at 30 pps for periods of several hours. The pulse width was 250 μs, and so the

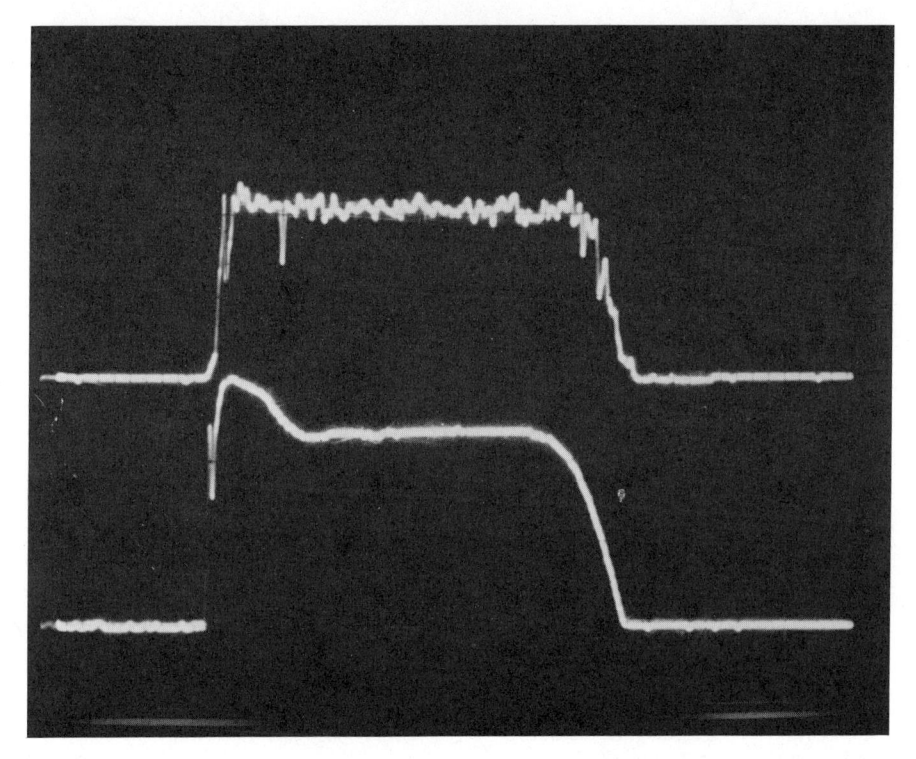

Figure 16.11 Oscillogram of the ion beam current when the plasma density is optimized to the extractor. Ion species: iron. Upper trace: beam current, 40 mA/cm. Lower trace: arc current, 50 A/cm. Sweep speed: 50 µs/cm. Courtesy Lawrence Berkeley Laboratory, University of California.

duty cycle was 0.75%. The peak beam current was moderate, in the 100-mA range. The limit to still higher rep rate and duty cycle was again set by the power supplies.

It is important to recognize that in the present embodiments of the source, the arc plasma is vastly underutilized. It has been shown by several workers that the total amount of ion current generated by a vacuum arc discharge is equal to approximately 10% of the current used to sustain the arc [19, 40, 59]. Measurements made on the MEVVA source configuration [50] have shown that about half of this ion current can be ducted to the extractor location for the case when the magnetic field is maximally effective; that is, for example, with an arc current of 200 A, about 20 A of metal ion current is generated at the cathode spots, and about 10 A of ion current is available within the plasma at the extractor. Figure 16.14 shows the results of measurements of the plasma ion current at the extractor as a function of arc current, for various magnetic field strengths. In Figure 16.13 some of these data are plotted as a function of magnetic field, and it can be seen that the magnetic field becomes maximally effective at about 100 G, at which field strength the *source efficiency factor*, I_{ion}/I_{arc}, is over 5%. Thus, there is 10–100 times as much plasma generated by the arc than is formed into beam by the extractor configuration

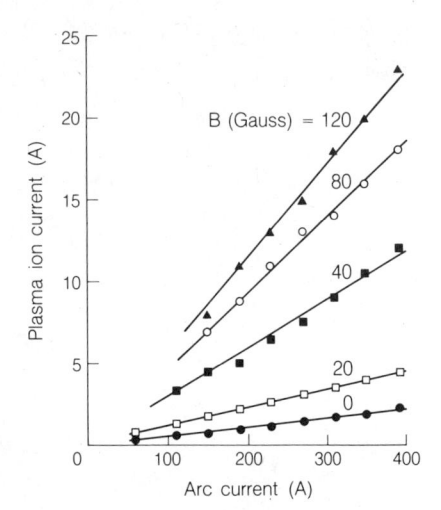

Figure 16.12 Plasma ion current measured at the extractor location, as a function of arc current for several different magnetic field strengths. Courtesy Lawrence Berkeley Laboratory, University of California.

employed to date, even for minimal arc current. Most of this plasma is simply dumped within the source prior to beam formation: The present source embodiments are inherently operating very inefficiently. An important future development is to utilize (i.e., extract from) a greater fraction of the plasma that is created, perhaps by greatly increasing the diameter of the extractor and appropriately ducting the plasma to this enlarged extractor. In this case, the peak beam current could be as high as tens of amperes, and the mean current in the range 10–100 mA or more.

4.6 Source Lifetime

The source lifetime is presently limited by cathode erosion. As an example, after several hundred thousand shots each of 250-μs duration and with a beam current of several hundred milliamperes, approximately 1 g of material is eroded away from the cathode. Triggering then becomes difficult and erratic. It may be that a different trigger-insulator-cathode design might improve the lifetime, as deter-

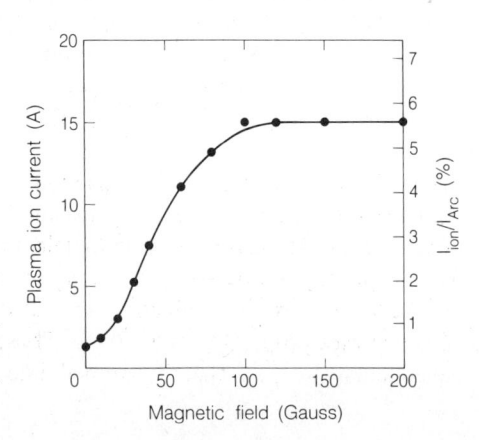

Figure 16.13 Plasma ion current and source efficiency factor, I_{ion}/I_{arc}, as a function of magnetic field strength. I_{arc} = 270 A. Courtesy Lawrence Berkeley Laboratory, University of California.

4 Source Performance

mined by this limitation, significantly. There is no significant deterioration in either the trigger or the trigger insulator even after as many as one million shots.

4.7 Charge State Distribution

Measurements of the charge state distribution of ions generated by the vacuum arc plasma discharge have been reported by several workers [17, 18, 20, 36, 60], and it is well recognized that the distribution in general contains a high fraction of multiply stripped species. The charge state spectrum of the ion beam produced by the MEVVA ion source has been measured for a wide range of cathode materials using a time-of-flight system [61], and these results have been reported in detail [54]. The measured spectra are shown in Figure 16.14. These data were all taken

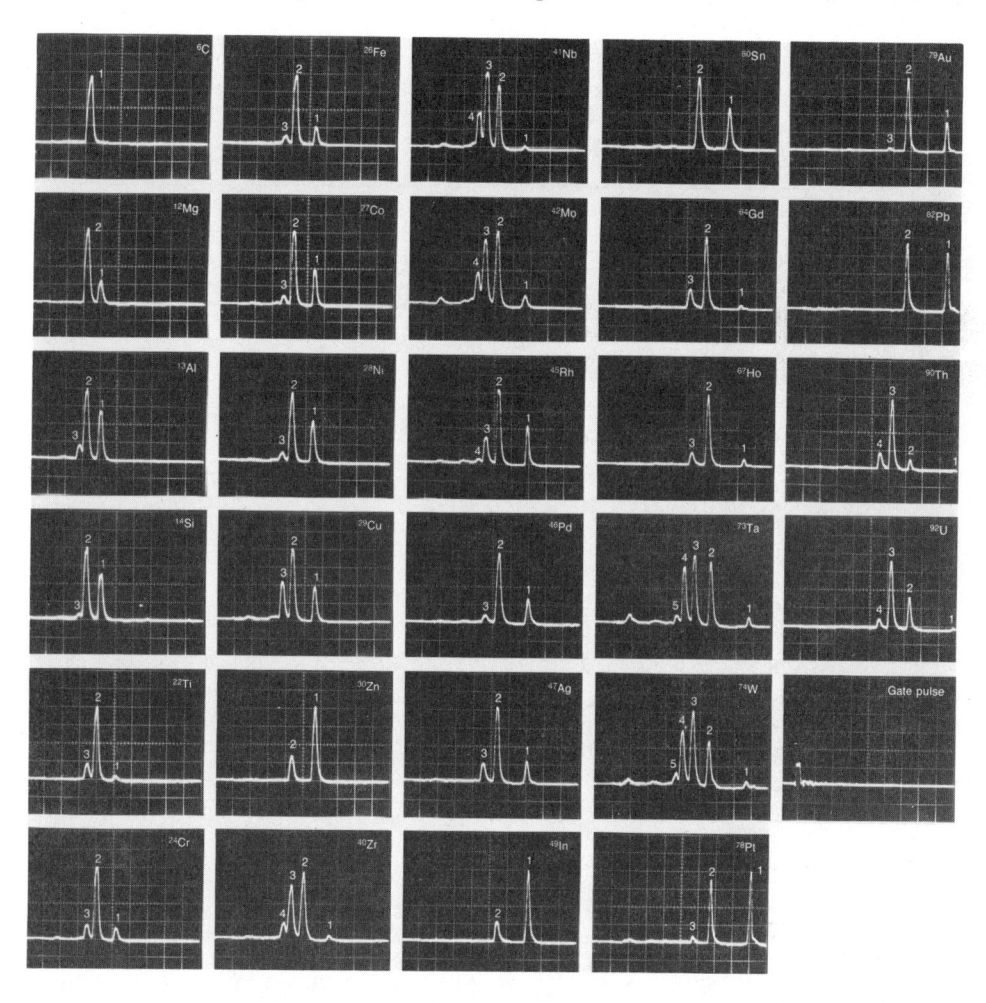

Figure 16.14 Measured charge state distributions obtained using the MEVVA IV ion source. Vertical scale: current collected by Faraday cup, gain approximately 100 μA/cm. Courtesy Lawrence Berkeley Laboratory, University of California.

at the same arc current, 200 A, and for a beam extraction voltage of 60 kV; the oscillogram sweep speed is 1 μs/cm. The spectra were obtained as ion current measured into a Faraday cup, and the amplitudes of the charge state peaks in the oscillograms are proportional to electrical current; the electrical current is greater than particle current by the charge state Q, $I_{elec} = QI_{part}$. To obtain spectral data that would be visually intercomparable, the oscilloscope gain was adjusted for each cathode material, and the vertical current scale in Figure 16.14 is not the same for different materials; nonetheless, the current scale is always within a factor of several of 100 μA/cm. The charge state distributions shown in Figure 16.14 are summarized numerically in Table 16.1. Here, $\overline{Q}$ is the mean charge state, $\overline{Q} = \Sigma fQ/\Sigma f$, where f is fraction of the total beam current in a given charge state (i.e., the charge state fraction); $\overline{Q}_p$ and $\overline{Q}_e$ are the mean charge states expressed in terms of particle

TABLE 16.1 Charge State Fractions and Mean Charge State, Expressed in Terms of Electrical Currents and Particle Currents

Element	\multicolumn{6}{c}{Electrical Current Fractions (%)}	\multicolumn{6}{c}{Particle Current Fractions (%)}
	$Q=1$	2	3	4	5	$\overline{Q}_e$	$Q=1$	2	3	4	5	$\overline{Q}_p$
C	100					1	100					1
Mg	23	77				1.77	37	63				1.63
Al	38	52	10			1.72	56	39	5			1.48
Si	38	58	4			1.66	56	42	2			1.46
Ti	3	80	17			2.14	6	82	12			2.05
Cr	14	73	13			1.99	25	67	8			1.82
Fe	18	74	8			1.90	31	64	5			1.73
Co	30	62	8			1.78	47	49	4			1.57
Ni	35	58	7			1.72	53	44	3			1.51
Cu	26	49	25			1.99	44	42	14			1.70
Zn	76	24				1.24	86	14				1.14
Zr	4	47	38	11		2.56	9	55	30	6		2.33
Nb	2	36	43	19		2.79	5	46	37	12		2.56
Mo	6	40	36	18		2.66	14	47	28	11		2.35
Rh	28	52	18	2		1.94	46	43	10	1		1.65
Pd	24	69	7			1.83	39	57	4			1.64
Ag	18	66	16			1.98	32	59	9			1.77
In	79	21				1.21	88	12				1.12
Sn	36	64				1.64	53	47				1.47
Gd	3	78	19			2.16	6	81	13			2.07
Ho	8	79	13			2.05	15	76	9			1.93
Ta	5	30	33	28	4	2.96	13	39	28	18	2	2.58
W	3	25	39	27	6	3.08	8	34	36	19	3	2.74
Pt	52	44	4			1.52	69	29	2			1.33
Au	28	69	3			1.75	44	54	2			1.58
Pb	47	53				1.53	64	36				1.36
Th	1	10	72	17		3.05	3	15	70	12		2.92
U	1	29	62	8		2.77	3	38	54	5		2.62

currents and electrical currents, respectively. For the case of cathodes that are compounds rather than elements, the spectra produced are mixtures of the spectra from the individual constituents. The beam can then contain ions of nonconducting species also; for example, a PbS cathode produces a beam containing S^+ ions as well as Pb^+ and Pb^{2+}. A completely phenomenological fit to the charge state data is provided by the expression

$$\overline{Q}_p = 0.38(T_{BP}/1000) + 0.6$$

where T_{BP} is the boiling point temperature of the cathode material in degrees kelvin. There are some exceptions to this predictor, and it should be used with caution. The MEVVA charge state distribution has been discussed in detail in Ref. 54.

5 OTHER VERSIONS

A somewhat different embodiment of the concept has been developed by Humphries and coworkers at the University of New Mexico [33–37]. Burkhart has described a novel version [62, 63] of the New Mexico-type sources in which an array of six metal vapor vacuum arcs has been used; see Figure 16.15. Use of multiple sources in parallel has several advantages, including limiting of the arc current per cathode, spatially averaging the plasma uniformity so as to allow extraction over a much larger area, and averaging-out of the shot-to-shot and temporal plasma density fluctuations. In the source shown in Figure 16.15, the circular array of six separate metal vapor vacuum arc plasma sources on the right side of the figure injects plasma into the annular, magnetically insulated, extractor configuration on the left side.

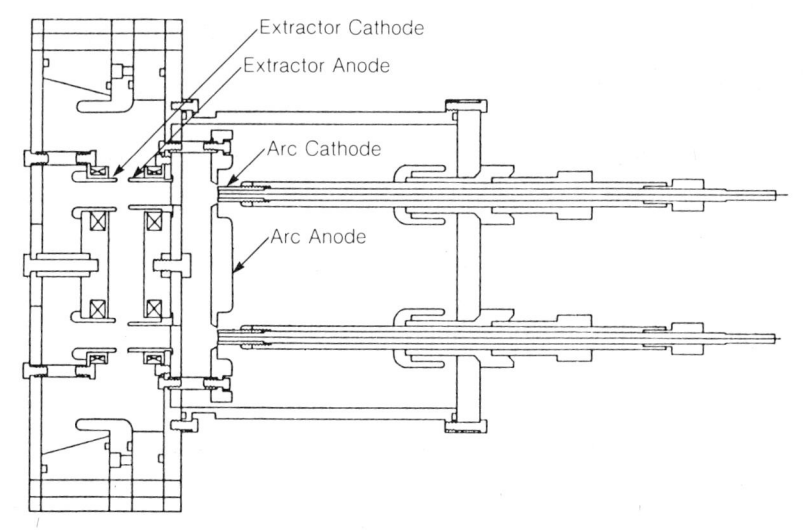

Figure 16.15 Sixfold array of metal vapor vacuum arc injectors of the New Mexico design, with magnetically insulated ion extraction. From Burkhart [62].

Thus, the beam is an annular beam, with a trace memory of the hexapole structure due to the six discrete plasma sources. The source has a particularly large extraction area of 30 cm^2. In a later variant, the magnetically insulated extractor configuration was replaced with a three-electrode, accel–decel design [63]. C$^+$ ions at a current of over 1 A at 15 kV, in a beam pulse of over 500-μs duration, were produced. A particularly significant innovation introduced into these kinds of ion sources, developed by the group at the University of New Mexico, is the incorporation of *grid-controlled extraction* of the ion beam, by means of which a particularly quiescent beam can be produced. This technique has been described in detail by Humphries, Burkhart, and Len, in Chapter 19.

Adler and Picraux have described their version of a metal vapor arc source [38], indicated schematically in Figure 16.16. The pulse length they have employed is in the microsecond range, and the source takes on some of the flavor of a spark source. This source has operated at a repetition rate of 32 Hz for pulses 6 μs long, and with cathodes of copper, titanium, and carbon. At an extraction voltage of 80 kV a peak ion current of 2 A was obtained. A most interesting feature of this source was the implied high charge state distribution of the ions produced. The ion charge state spectrum was not measured directly, but from measurements of the depth profile of titanium implanted into aluminum, the presence of titanium species from Ti$^+$ up to as high as Ti^{5+} was inferred. It is a characteristic of vacuum spark sources that the ions produced are, in general, quite highly stripped.

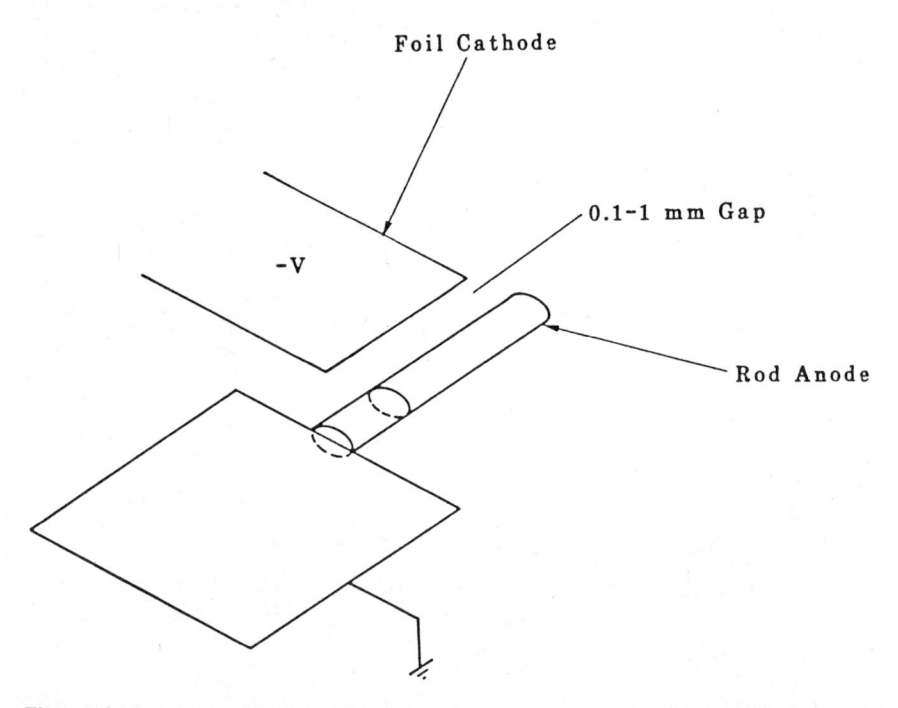

Figure 16.16 Schematic of the short-pulse metal vapor arc source described by Adler and Picraux [38].

6 FUTURE DIRECTIONS

The metal vapor vacuum arc source is a new addition to the gamut of ion sources available to the experimenter. The distinguishing characteristics of the source are the very high currents that can be produced and the fact that the species are metallic; also, the charge state species produced are in general multiply ionized. A great deal of further development remains to be done, and the source will take on new features and performance characteristics as this evolution proceeds. At the present time, the source is in its infancy, and both the design and the performance will mature with age—as sometimes happens.

REFERENCES

1. L. Valyi, *Atom and Ion Sources*, Wiley, New York, 1977.
2. J. Ishikawa and T. Takagi, *Jpn. J. Appl. Phys.* **22,** 534 (1983).
3. J. Ishikawa, Y. Takeiri, and T. Takagi, *Rev. Sci. Instrum.* **55,** 449 (1984).
4. A. Rockett, S. A. Barnett, and J. E. Greene, *J. Vac. Sci. Technol.* **B2,** 306 (1984).
5. A. Warwick, *IEEE Trans. Nucl. Sci.* **NS-32,** 1809 (1987).
6. R. K. Feeney, W. E. Sayle II, and J. W. Hooper, *Rev. Sci. Instrum.* **47,** 964 (1976).
7. I. C. Lyon and B. Peart, *J. Phys. E: Sci. Instrum.* **17,** 920 (1984).
8. B. Gavin, S. Abbott, R. MacGill, R. Sorensen, J. Staples, and R. Thatcher, *IEEE Trans. Nucl. Sci.* **NS-28,** 2684 (1981).
9. Y. Saito, Y. Mitsuoka, and S. Suganomata, *Rev. Sci. Instrum.* **55,** 1760 (1984).
10. R. W. Sorensen and H. E. Mendenhall, *Trans. AIEE, Pt. III* **45,** 1102 (1926).
11. J. D. Cobine, *Elec. Eng.* **81,** 13 (1962).
12. J. M. Lafferty (Ed.), *Vacuum Arcs—Theory and Application*, Wiley, New York, 1980.
13. B. Edlen, *Physica* **13,** 544 (1947).
14. W. A. Cilliers, R. V. Datla, and H. R. Griem, *Phys. Rev. A* **12,** 1408 (1975).
15. A. A. Plyutto, *Sov. Phys. JETP* **12** (6), 1106, (1961).
16. S. Takagi, S. Ohtani, K. Kadota, and J. Fujita, *Nucl. Instrum. Methods* **213,** 539 (1983).
17. A. A. Plyutto, V. N. Ryzhhov, and A. T. Kapin, *Sov. Phys. JETP* **20** (2), 328 (1965).
18. W. D. Davis and H. C. Miller, *J. Appl. Phys.* **40,** 2212 (1969).
19. C. W. Kimblin, *Proc. IEEE* **59,** 546 (1971).
20. C. W. Kimblin, *J. Appl. Phys.* **44,** 3074 (1973).
21. J. E. Daalder and P. G. E. Wielders, Proceedings 12th Int. Conf. on Phenomena in Ionized Gases, Eindhoven, Amsterdam: North-Holland, Pt. 1, p. 232.
22. I. Jorde, J. Kulsetas, and W. G. J. Rondeel, Proceedings, 12th Int. Conf. on Phenomena in Ionized Gases, Eindhoven, Amsterdam: North-Holland, Pt. 1, p. 240.
23. J. W. Robinson and M. Ham, *IEEE Trans. Plasma Sci.* **PS-3,** 222 (1975).
24. J. E. Daalder, *J. Phys. D: Appl. Phys.* **8,** 1647 (1975).
25. J. L. Hirshfield, *IEEE Trans. Nucl. Sci.* **NS-23,** 1006 (1976).
26. J. E. Daalder, *J. Phys. D: Appl. Phys.* **9,** 2379 (1976).
27. R. K. Wakerling and A. Guthrie (Eds.), *Electromagnetic Separation of Isotopes in Commercial Quantities*, National Nuclear Energy Series, USAEC (Washington, D. C.) 1951, p. 324.
28. E. I. Revutskii, G. M. Skoromnyi, Yu. F. Kulygin, and I. I. Goncharenko, *Proceedings, Sov. Conf. on Charged-Particle Accelerators, Moscow, 9–16 October 1968*, Vol. 1, A. A. Vasilev (Ed.), USAEC, p. 447.

29. I. G. Brown, J. E. Galvin, and R. A. MacGill, *Appl. Phys. Lett.* **47,** 358 (1985).
30. I. G. Brown, *IEEE Trans. Nucl. Sci.* **NS-32,** 1723 (1985).
31. I. G. Brown, J. E. Galvin, B. F. Gavin, and R. A. MacGill, *Rev. Sci. Instrum.* **57,** 1069 (1986).
32. I. G. Brown, J. E. Galvin, R. Keller, P. Spaedtke, R. W. Mueller, and J. Bolle, *Nucl. Instrum. Methods A* **245,** 217 (1986).
33. S. Humphries, Jr., M. Savage, and D. M. Woodall, *Appl. Phys. Lett.* **47,** 468 (1985).
34. C. Burkhart, S. Coffey, G. Cooper, S. Humphries, Jr., L. K. Len, A. D. Logan, M. Savage, and D. M. Woodall, *Nucl. Instrum. Methods B* **10/11,** 792 (1985).
35. S. Humphries, Jr., C. Burkhart, S. Coffey, G. Cooper, L. K. Len, M. Savage, D. M. Woodall, H. Rutkowski, H. Oona, and R. Shurter, *J. Appl. Phys.* **59,** 1790 (1986).
36. L. K. Len, C. Burkhart, G. W. Cooper, S. Humphries, Jr., M. Savage, and D. M. Woodall, *IEEE Trans. Plasma Sci.* **PS-14,** 256 (1986).
37. S. Humphries, Jr., and C. Burkhart, *Particle Accelerators* **20,** 211 (1987).
38. R. J. Adler and S. T. Picraux, *Nucl. Instrum.* **6,** 123 (1985).
39. N. Vidyardhi and R. S. N. Rau, *J. Phys. E: Sci. Instrum.* **6,** 33 (1973).
40. S. Kamakshaiah and R. S. N. Rau, *J. Phys. D: Appl. Phys.* **8,** 1426 (1975).
41. S. Kamakshaiah and R. S. N. Rau, *IEEE Trans. Plasma Sci.* **PS-5,** 164 (1977).
42. R. L. Boxman, *IEEE Trans. Electron Devices* **ED-24,** 122 (1977).
43. E. Hantzsche, Proceedings 13th Int. Conf. Phenomena in Ionized Gases (Berlin), Vol. 3, 1977, p. 121.
44. *IEEE Trans. Plasma Sci.* **PS-11** (Sep. 1983), special issue on vacuum discharge plasmas.
45. *IEEE Trans. Plasma Sci.* **PS-13** (Oct. 1985), special issue on vacuum discharge plasmas.
46. I. G. Brown, J. E. Galvin, R. A. MacGill, and R. T. Wright, 1987 Particle Accelerator Conference, Washington, DC, March 1987.
47. A. S. Gilmour, Jr., and D. L. Lockwood, *Proc. IEEE* **60,** 977 (1972).
48. Aremco Products, Inc., PO Box 429, Ossining, NY 10562.
49. Sauereisen Cements Co., Pittsburgh, PA 15238.
50. I. G. Brown, *IEEE Trans. Plasma Sci.* **PS-15,** 346, (1987).
51. I. G. Brown, J. E. Galvin, R. A. MacGill, and R. T. Wright, *Appl. Phys. Lett.* **49,** 1019 (1986).
52. See, e.g., Proceedings of the Third International EBIS Workshop, Cornell University, May 1985, V. O. Kostroun and R. W. Schmieder (Eds.).
53. B. Wang, Cornell University, private communication.
54. I. G. Brown, B. Feinberg, and J. E. Galvin, *J. Appl. Phys.*, **63,** 4889 (1988).
55. T. S. Green, *Rep. Prog. Phys.* **37,** 1257 (1974).
56. T. S. Green, *IEEE Trans. Nucl. Sci.* **NS-23,** 918 (1976).
57. R. Keller, Chapter 3, this book.
58. GSI 1986 Annual Report, Report GSI-87-1, March 1987, p. 322.
59. J. V. R. Heberlein and D. R. Porto, *IEEE Trans. Plasma Sci.* **PS-3,** 222 (1975).
60. V. M. Lunev, V. G. Padalka, and V. M. Khoroshikh, *Sov. Phys. Tech. Phys.* **22**(7), 858 (1977).
61. I. G. Brown, J. E. Galvin, R. A. MacGill, and R. T. Wright, *Rev. Sci. Instrum.* **58,** 1589 (1987).
62. C. Burkhart, Ph.D. Thesis, University of New Mexico, June 1987.
63. C. Burkhart and S. Humphries, Jr., in *Proceedings NATO Workshop on High Brightness Accelerators, Pitlochry, Scotland, 1986*, Plenum, New York, 1988.

17

Negative Ion Sources

K. N. Leung
Lawrence Berkeley Laboratory
University of California
Berkeley, California

Negative ions, notably H⁻ ions, are widely used in cyclotrons and tandem accelerators, in fueling storage rings of high-energy accelerators, and in generating energetic neutral beams for heating fusion plasmas. Negative ions can be formed by double charge exchange processes or by direct extraction from a negative ion source. In general, two distinct types of negative ion sources can be identified: (1) surface sources, in which the negative ions are generated by particle collisions with low work function surfaces, and (2) volume sources, in which the negative ions are produced by electron–molecule and electron–ion collision processes in the volume of a discharge plasma. This chapter reviews the development and some of the latest technology of these two types of negative ion sources. It can be seen that in just twenty years, steady-state negative ion beam current (such as H⁻) has been improved from several milliampere to more than 1 A.

1 SURFACE-PRODUCED NEGATIVE ION SOURCES

1.1 Sputtering-type Negative Ion Sources

The cesium beam sputter source developed by Middleton and Adams has been used to generate a wide variety of elemental and molecular ions [1, 2]. Figure 17.1 is a schematic drawing of the sputter source. Positive cesium ions from a surface ionization source are used to sputter the interior of a hollow conical target. Negative ions formed are extracted out of the rear aperture of the sputter target, and on emerging they are accelerated back to ground potential. The positive cesium ion energy is typically 20 to 30 keV and currents are limited to 1 to 2 mA. Negative

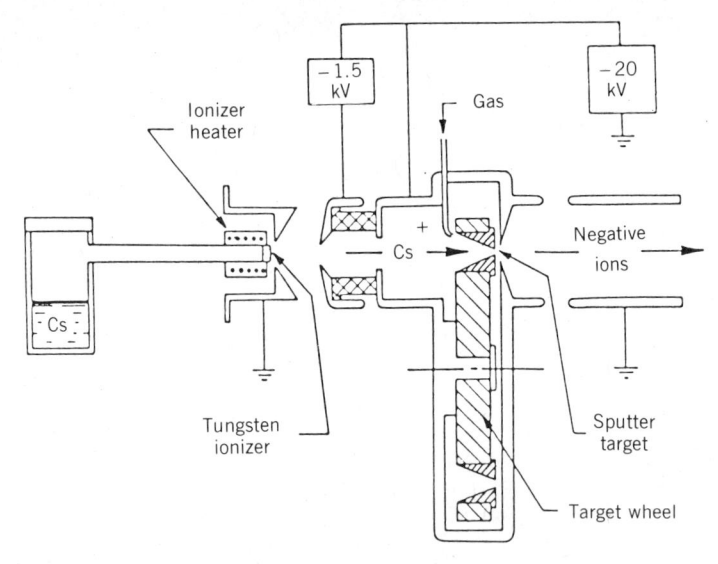

Figure 17.1 A schematic drawing of the cesium beam sputter source developed by Middleton and Adams [1, 2]. With permission from Plenum Press.

ion current generated from this source is usually in the range 0.1 to 10 μA. Negative ion species can be changed rapidly, requiring only a rotation of the target wheel.

Figure 17.2 shows another sputter type negative ion source (the so-called "inverted" sputter source) developed by Middleton in 1976 [1, 3]. An annular ionizer is used in this source geometry to generate the cesium sputter beams and the negative ions are extracted through the hole in the ionizer.

1.2 Plasma Surface Conversion Negative Ion Sources

There are two major types of plasma surface conversion negative ion sources. One type uses a Penning discharge geometry, has radial extraction, and employs a supporting gas to which cesium vapor is added. The sputter sources developed independently at the University of Aarhus [4], the University of Wisconsin [5], and in Novosibirsk [6, 7] fall into this category. The second type employs the multicusp plasma generator as the discharge chamber [8]. This source was originally developed at Lawrence Berkeley Laboratory for neutral beam application. Recently, it has been used as an H$^-$ source for particle accelerators at Los Alamos National Laboratory and at KEK in Japan [9, 10].

Figure 17.3 is a schematic drawing showing the principle of the Aarhus source. A Penning discharge generates a plasma of support gas and cesium. The function of the cesium is twofold: first, it acts as a sputtering agent, and second, it lowers the work function of the target surface so as to enhance the negative ion formation. The plasma forms a sheath in front of the spherical surface of the sputter cathode, which is biased at approximately -1 kV with respect to the anode. Positive ions

1 Surface-Produced Negative Ion Sources

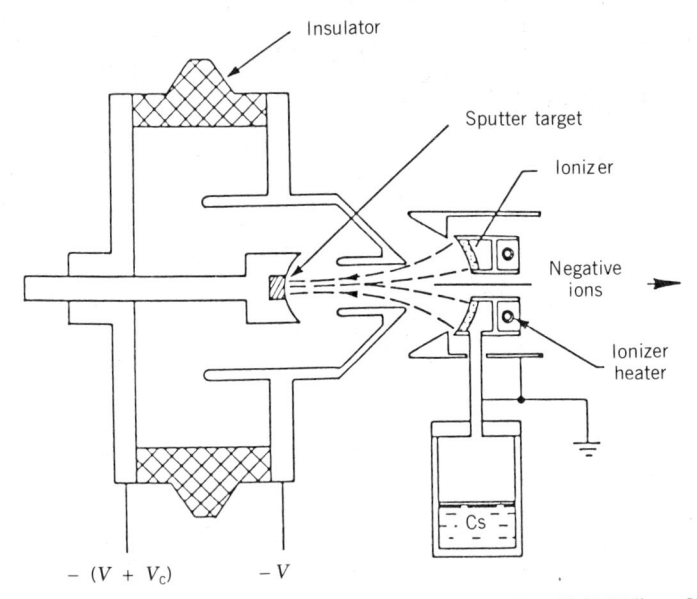

Figure 17.2 A schematic drawing of the inverted sputter source developed by Middleton [1, 3]. With permission from Plenum Press.

are accelerated from the plasma across the sheath and impinge on the surface of the sputter cathode. Negative ions formed on the cathode surface are accelerated back across the plasma boundary toward the exit aperture.

In Novosibirsk, a multislit H^- ion source with one-dimensional focusing was developed in 1979 [11]. Figure 17.4 shows a cross-sectional view of the source,

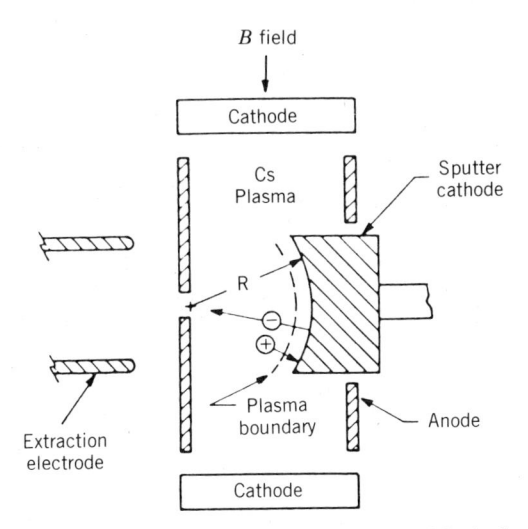

Figure 17.3 A schematic drawing showing the operating principles of the Aarhus negative ion source [1, 4]. With permission from Plenum Press.

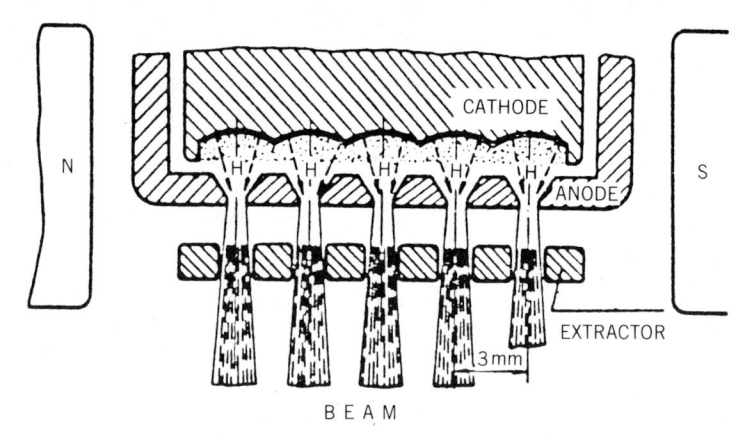

Figure 17.4 A schematic diagram of the Novosibirsk multislit H$^-$ ion source [11].

which can produce pulsed beam of H$^-$ ions with a current of 4 A. In 1981, Bel'chenko in Novosibirsk developed a honeycomb multiaperture source with focusing of H$^-$ ions from spherical dimples on a cathode surface to the circular emission holes in the anode [12]. From this source, a pulsed beam of H$^-$ ions with a current in excess of 11 A was produced and accelerated to 25 keV.

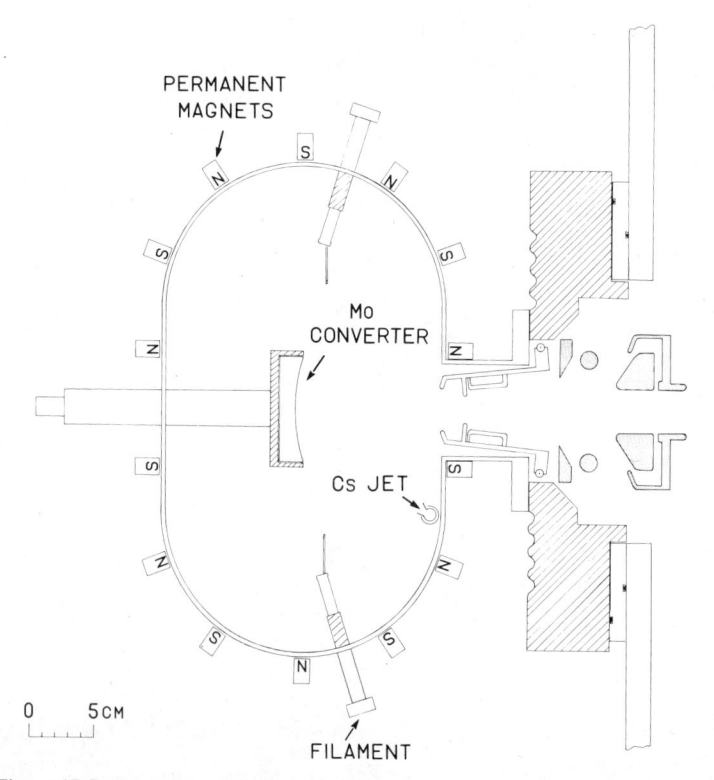

Figure 17.5 Schematic diagram of the LBL self-extraction negative ion source [13].

Figure 17.5 shows a schematic diagram of the large Lawrence Berkeley Laboratory surface-conversion H^- source. This source has been operated successfully to generate a steady-state H^- beam of current greater than 1 A [13]. The source is essentially a large multicusp plasma generator. To produce negative ions, a water-cooled, concave, molybdenum converter is inserted into the plasma. By biasing the converter negatively (~ 200 V) with respect to the plasma, positive ions from the plasma are accelerated across the sheath and strike the converter surface. Negative ions that are formed at the converter are then accelerated back through the sheath by the same potential. The bias voltage on the converter thus becomes the negative ion extraction potential. The converter surface is curved to geometrically direct the negative ions through the plasma to a position at the exit aperture which is located between two ceramic magnet columns. The B-field in the exit region is strong enough to reflect energetic primary electrons but it produces only a small lateral displacement on the trajectories of the negative ions.

In this type of surface negative ion source, cesium is introduced into the discharge so as to lower the converter work function and consequently enhance the H^- yield. Although this source was developed for H^- or D^- ions, other negative ions such as O^-, O_2^-, OH^-, B^-, and C^- have also been generated by this "self-extraction"-type source [14].

2 VOLUME-PRODUCED NEGATIVE ION SOURCES

2.1 Duoplasmatron, Magnetron, and Penning-type H^- Sources

In general, a hydrogen discharge contains not only positive ions (H^+, H_2^+, H_3^+) and electrons but also negative hydrogen ions. By using the photodetachment technique, Bacal et al. [15] have measured large concentrations of H^- ions in a dc hydrogen discharge plasma. In the past, attempts have been made to extract these H^- ions from the plasma of a duoplasmatron, a magnetron, or a Penning-type discharge source. Considerable progress has been made in increasing the extractable H^- current of these sources since the early 1960s.

The duoplasmatron, originally designed by Von Ardenne [16] for positive ion generation, was first adapted by Moak et al. [17] at Oak Ridge National Laboratory for H^- extraction. It was reported that 11 mA of H^- ions accompanied by 5 mA of electrons were extracted from a duoplasmatron ion source of the type shown in Figure 17.6. Later on, Wittkower et al. [18] investigated the geometry and conditions of a duoplasmatron which gave the best output and they obtained 70 mA of H^- ion accompanied by 65 mA of electrons.

An attempt was made by Collins and Gobbet [19] and independently by Lawrence et al. [20] to improve the H^- production efficiency by extracting from the edge of the duoplasmatron arc, where suitable conditions of high gas and electron density for negative ion formation should exist. It was found that the H^- output increased as the emission orifice in the anode was displaced relative to the axis of the intermediate electrode. A substantial reduction of the electron current in the extracted beam was also observed. With this new off-axis extraction config-

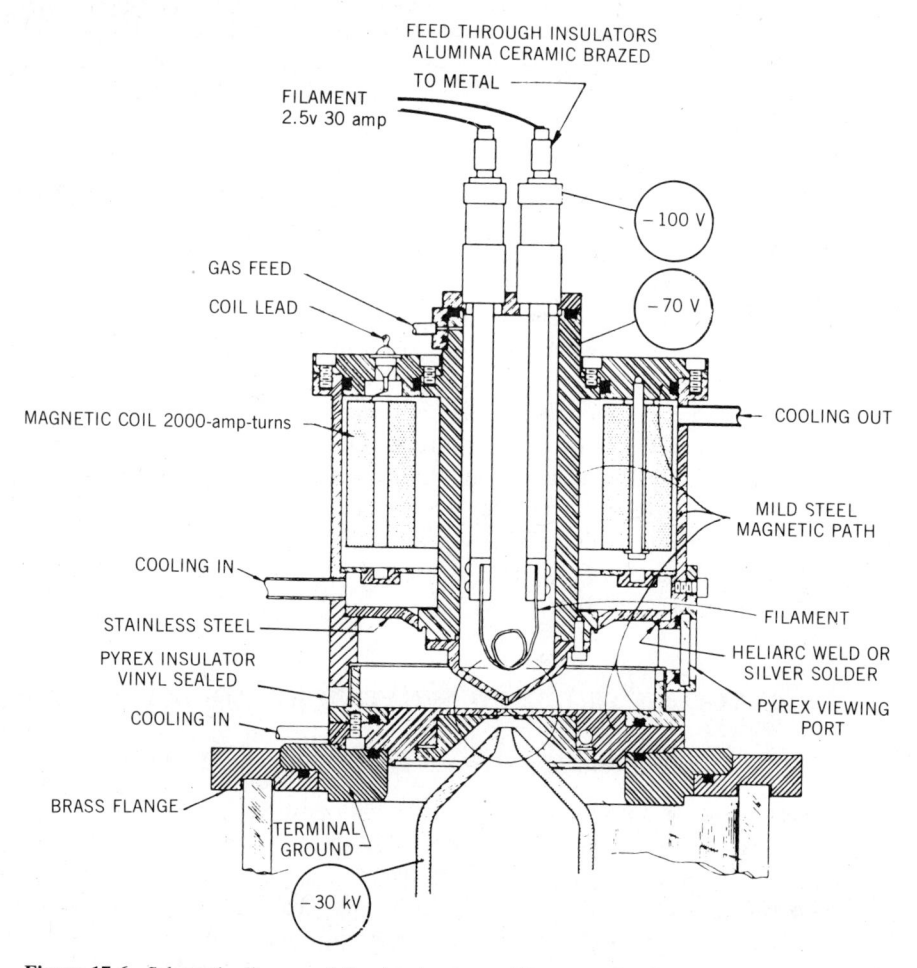

Figure 17.6 Schematic diagram of the duoplasmatron H⁻ source developed by Moak et al. [17]. With permission from AIP.

uration, the duoplasmatron source produced an H⁻ current of 80 mA with an electron load of 2 to 4 mA.

In 1972, Abroyan et al. [21] produced beams of H⁻ ions with current larger than 2 mA ($J^- \sim 120$ mA/cm^2) by operating a duoplasmatron in a high arc current (75 A), but pulsed mode (~ 100 ms). A displacement of 2 mm of the axis of the emission aperture from the axis of the discharge was found to be optimum with respect to the current density of H⁻ ions and the ratio of the ion current to electron current.

In 1972, a magnetron negative ion source, shown schematically in Figure 17.7, was developed in Novosibirsk by Bel'chenko et al. [22]. In this source, the discharge chamber has the shape of a racetrack and the plasma is kept circulating in the narrow space by the $E \times B$ drift motion. The H⁻ ions are extracted from

2 Volume-Produced Negative Ion Sources 361

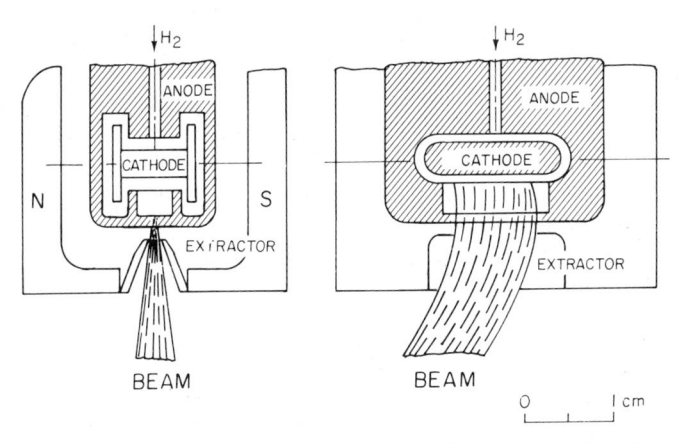

Figure 17.7 The magnetron negative hydrogen ion source developed by Bel'chenko et al. [22].

the plasma through the slit in the anode, elongated perpendicularly to the magnetic field. A pulsed beam of H⁻ ions with a current up to 22 mA and a high emission density of 220 mA/cm² was obtained with discharge currents as high as 100 A and an arc voltage of 500 V. The accompanying electron current was less than 100 mA and the source pressure was estimated to be about 1 Torr.

In 1963, Ehlers et al. [23] in Berkeley developed a direct extraction negative ion source which could be easily adapted for use in a cyclotron. This source (shown in Figure 17.8) is operated with a hot cathode Penning-type gas discharge. The

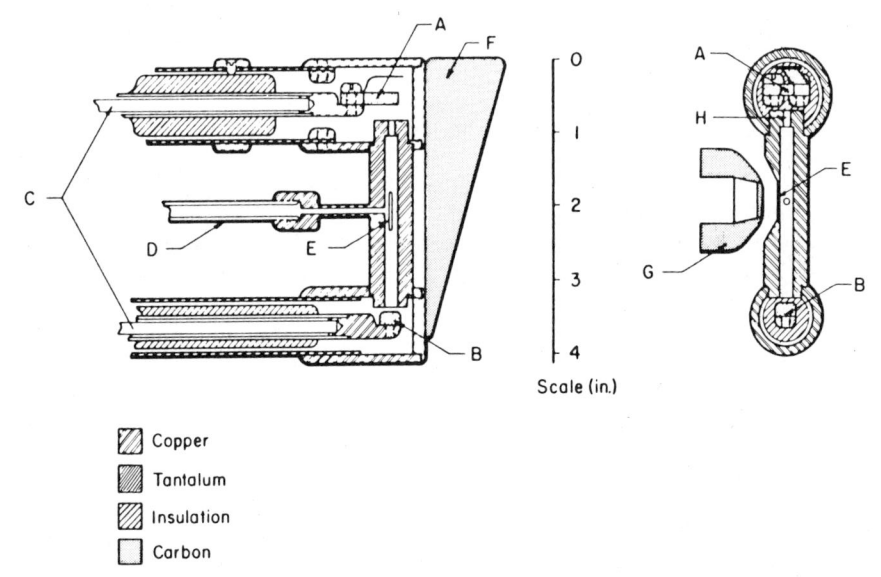

Figure 17.8 Cross-sectional drawing of a modified Penning H⁻ ion source [23]. A, heated filament; B, cold reflector cathode; C, water-cooled squirt tubes; D, gas feed lines; E, ion exit slit; F, trochoidal electron dump block; G, ion extraction electrode; H, arc-defining hole.

cathode is a 3.5-mm-thick tantalum plate, heated with a direct current of 380 A to the required temperature. A water-cooled tantalum electrode located directly opposite to the filament functions as the anti-cathode. The anode for the discharge is a copper cylinder, with an extraction slit of 1.17×12.5 mm^2 in its side wall. The magnetic field of 4000 G needed for the operation of the source is normally generated by the magnet of the cyclotron. A D$^-$ ion beam of 2 mA has been extracted from the arc in a direction normal to the magnetic field, with a discharge voltage of 300 V and a discharge current of 5 A. The gas consumption was approximately 23 sccm. Subsequently, this source was modified with a recess in the arc chamber where the extraction slit was located. With this new geometry, continuous H$^-$ current in excess of 5 mA ($J^- \sim 40$ mA/cm^2) was obtained.

In 1978, Gabovich et al. [24] increased the extracted H$^-$ ion current from the Ehler's type Penning Source to 15 mA merely by elongating the emission slit to 40 mm. In 1982, this reflex-type negative ion source was again modified in Berkeley [25] so that the anode was divided into sections: a top anode part and a wall anode part. By applying the appropriate bias voltage between these two anodes, 9.7 mA ($J^- \sim 100$ mA/cm^2) of H$^-$ and 4.1 mA ($J^- \sim 42$ mA/cm^2) of D$^-$ current were obtained in a dc operating mode. This increase in negative ion yield was attributed to the enhancement of anomalous plasma diffusion.

Recently, a Dudnikov type of Penning source (which requires cesium in normal operation) has been operated successfully in a pure hydrogen mode by using a low work function LaB$_6$ cathode [26]. It was found that the extracted H$^-$ current density was comparable to that of the cesium-mode operation; an H$^-$ current density of 350 mA/cm^2 has been obtained for an arc current of 55 A. The H$^-$ yield is closely related to the source geometry and the applied field. Experimental results have demonstrated that the majority of the H$^-$ ions extracted are formed by volume processes in this type of LaB$_6$ cathode source operation.

2.2 Multicusp Negative Ion Sources

In 1983, a novel method of extracting volume-produced H$^-$ ions directly from a multicusp source was reported by Leung et al. [27]. The multicusp plasma generator has demonstrated its ability to produce large volumes of uniform and quiescent plasmas with good gas and electrical efficiency [28]. A schematic diagram of the apparatus is shown in Figure 17.9. The source chamber (20 cm diam by 24 cm long) is surrounded externally by 10 columns of samarium–cobalt magnets which form a longitudinal line–cusp configuration for primary electron and plasma confinement. These magnet columns are connected at the end flange by four extra rows of magnets. The open end of the chamber is enclosed by a two-grid extraction system. A steady-state hydrogen plasma is produced by primary electrons emitted from tungsten filaments and the entire chamber wall served as the anode for the discharge.

When operated as a H$^-$ ion source, a water-cooled permanent magnet *filter* is included and it divides the entire source chamber into an arc discharge and an extraction region. This filter provides a limited region of transverse magnetic field

2 Volume-Produced Negative Ion Sources

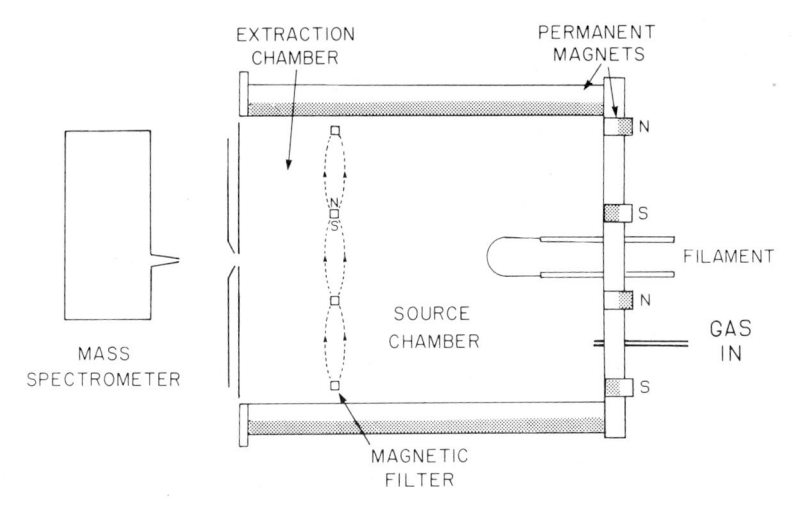

Figure 17.9 Schematic drawing of the LBL multicusp negative ion source [27].

which is strong enough to prevent the energetic primary electrons from entering the extraction zone. Excitation and ionization of the gas molecules are performed efficiently by the primaries in the discharge region. Both positive and negative ions, together with cold electrons, are able to penetrate the filter and they form a plasma with very low electron temperature in the extraction region. H^- ions are then produced in the extraction chamber and filter region via processes such as the dissociative attachment of vibrationally excited H_2 molecules [29], or the dissociative recombination of H_2^+ and H_3^+ ions [30, 31]. Initial results demonstrate that this filtered multicusp geometry will not only enhance the H^- ion yield but sizably reduce the extracted electron component.

It was demonstrated in 1984 that the filter-equipped multicusp source could provide high-quality H^- beams with sufficient current density (~ 40 mA/cm^2) to be useful for both neutral beam heating of fusion plasmas and accelerator application [32]. To produce this H^- current density, it was necessary to operate the source with a discharge current as high as 350 A.

Several methods to improve the efficiency of the multicusp H^- source have been investigated. By optimizing the extraction chamber length, a substantial improvement in the output has been achieved [33]. Experimental results have demonstrated that the H^- yield can be enhanced by choosing aluminum or copper as the chamber wall material [34] or by mixing hydrogen and xenon gases in the source discharge [35]. A substantial improvement in H^- yield also occurs when very low-energy electrons ($E \sim 1$ eV) are added into the filter or extraction regions [36].

Most recently, a small version of the filtered-multicusp source has been operated successfully to generate volume-produced H^- ions [37]. From this new source (Figure 17.10), H^- current density greater than 250 mA/cm^2 has been extracted from a 1-mm-diam aperture. The increase in H^- output is mainly due to an increase in the source plasma density.

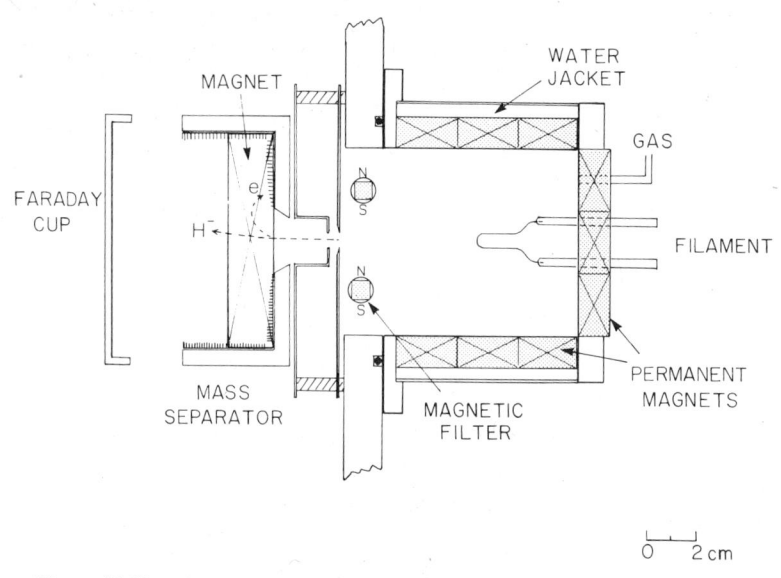

Figure 17.10 Schematic diagram of the LBL small multicusp H⁻ ion source [37].

Multicusp H⁻ sources and extraction systems are now being developed in various laboratories (Berkeley, Brookhaven, CERN, Culham, JAERI, Los Alamos, TRIUMF) to achieve the best electron suppression with no degradation of ion optics. In particular, large multicusp sources equipped with permanent magnet filters are now being used to provide high-current beams of H⁻ ions with multi-aperature extraction systems. It has been reported by Okumura et al. [38] that H⁻ ion beams of 1.26 A were extracted at 21 keV for 0.2 s from the large multicusp sources shown in Figure 17.11. The extractor consists of four grids, each having 209 apertures of 9 mm diam within a rectangular area of 12×26 cm². Small permanent magnets are installed in the second electrode so that a large fraction of the extracted electrons are deflected and captured by this electrode before they are accelerated to full energy. These results indicate that one might soon operate large area plasma sources to provide multiampere H⁻ beams in much the same manner as is now done to provide positive hydrogen ions for neutral beam systems.

2.3 Volume-produced Heavy Negative Ion Sources

The volume H⁻ sources described above can in principle also be used to generate negative ions of other gases such as oxygen, carbon monoxide, carbon dioxide, fluorine, chlorine, and sulfur hexafluoride which have high electron affinity energies. However, some of these gases are corrosive, attacking the cathodes of the ion sources, which sets a limit to the lifetime of the source operation. In some cases (for example, oxygen) it is almost impossible to sustain the discharge with hot tungsten filament cathodes. In this respect, a microwave ion source which operates without filaments can provide part of the solution.

2 Volume-Produced Negative Ion Sources

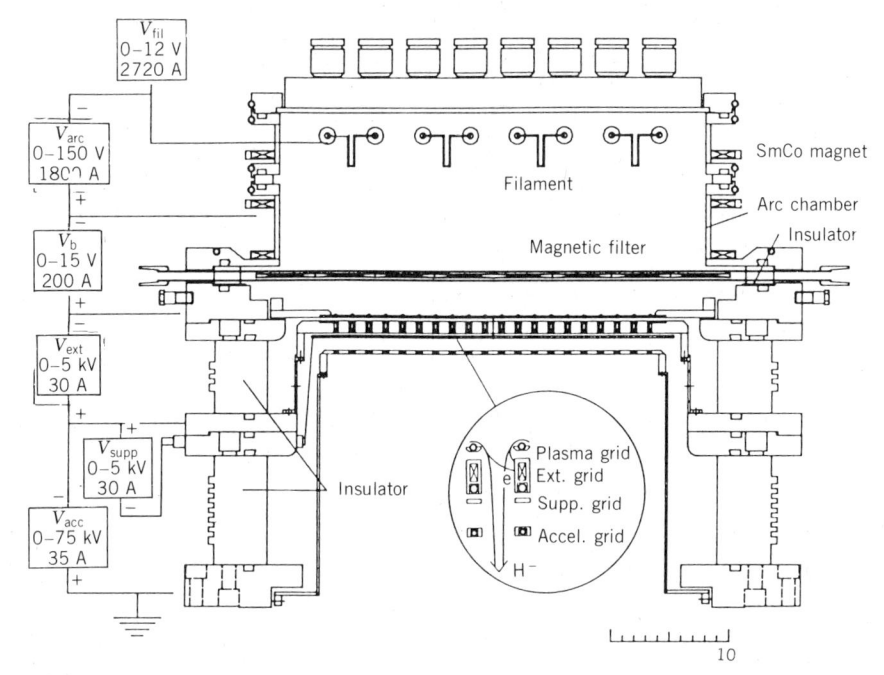

Figure 17.11 Cross-sectional view of the JAERI large volume H⁻ ion source together with the extraction system [38]. With permission from AIP.

Most recently, a compact microwave ion source [39] has been used to generate steady state beams of volume produced negative ions such as O⁻, C⁻, and Li⁻. This ion source, shown in Figure 17.12, is fabricated from a quartz tube with an ion extractor at one end and gas inlet at the other end. The tube is formed by joining a section of quartz tube with a 10-mm outside diameter to a section of a larger tube with a 27-mm outside diameter. The smaller tube is enclosed by a microwave cavity operating at a frequency of 2.45 GHz. Microwave power as high

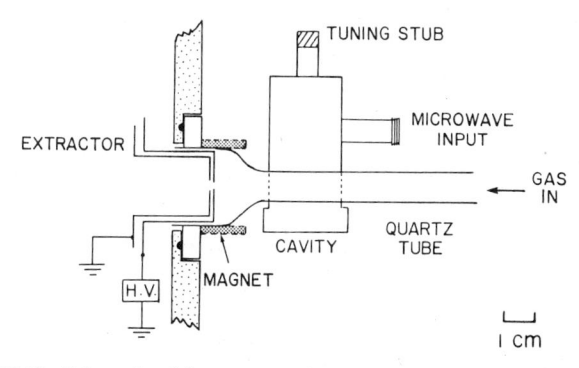

Figure 17.12 Schematic of the compact microwave ion source developed at LBL [39].

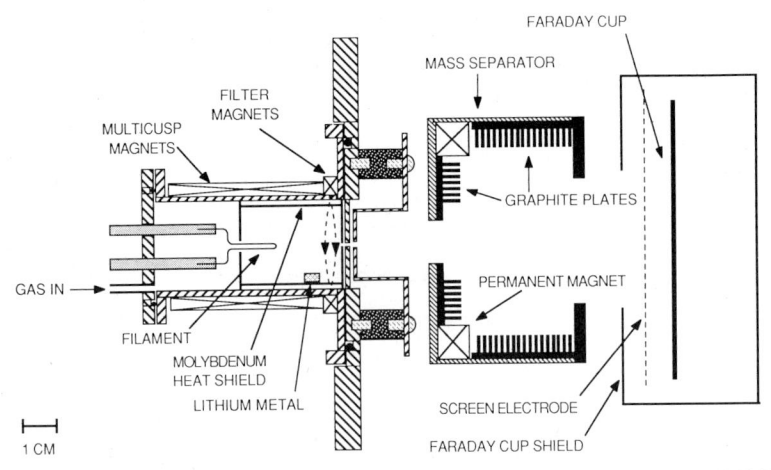

Figure 17.13 Schematic drawing of the multicusp Li⁻ ion source developed at LBL [40].

as 500 W can be coupled to the cavity via a coaxial cable. Cooling air is directed to the discharge tube through an opening in the body of the cavity. Ionization of the gas in the tube in initiated by a hand-held Tesla coil. A tuning stub and coupling slider are provided in the cavity to properly match the impedance of the discharge to that of the generator. In a manner similar to the multicusp H⁻ source, a permanent magnet magnetic filter is used to enhance the extracted negative ion yield. This filter is mounted externally and provides a transverse magnetic field in the region near the ion extractor.

To extract O⁻ ions, oxygen gas is used to form the discharge and the plasma electrode of the source is biased negative with respect to ground. The output signal from a mass spectrometer showed the dominant negative ion species to be O⁻. Li⁻ ions can be produced in the same source arrangement. In this case, the lithium is introduced into the source in solid form. The source is operated with argon gas and the Li vapor pressure gradually builds up due to discharge heating of the lithium crucible.

Attempts have been made to produce C⁻ ions by using gases such as CO, CO_2, or CH_4; however, no C⁻ ion signal has been detected over a range of different source operating conditions. By replacing the first or plasma electrode of the extractor with a negatively biased graphite disk, a C⁻ ion signal is observed when the source is operated with an argon plasma. In this arrangement, some carbon atoms are sputtered from the graphite electrode by the background Ar⁺ ions. They then react with the plasma electrons to form C⁻ ions.

Li⁻ ion beams have applications in fusion plasma diagnostics and neutral beam heating. Li⁻ can be formed by double charge exchange of Li⁺ in a vapor target, surface production on a cesiated converter surface, or volume production by dissociative attachment of vibrationally excited Li_2 molecules. Volume production of Li⁻ is preferred due to the lower negative ion temperature, which results in better beam emittance for these ion sources.

Volume-produced Li⁻ ions have been extracted from a small multicusp ion

2 *Volume-Produced Negative Ion Sources* 367

(a)

(b)

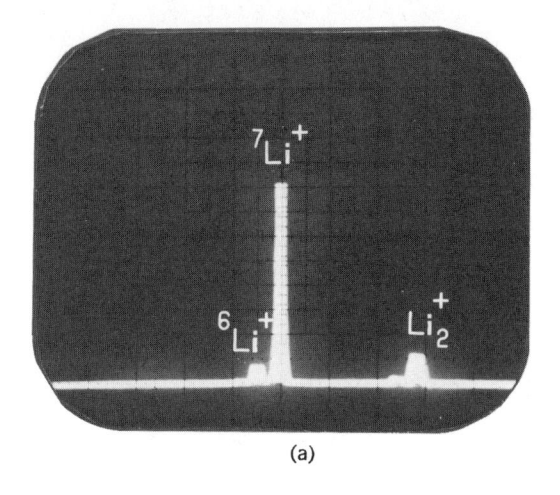

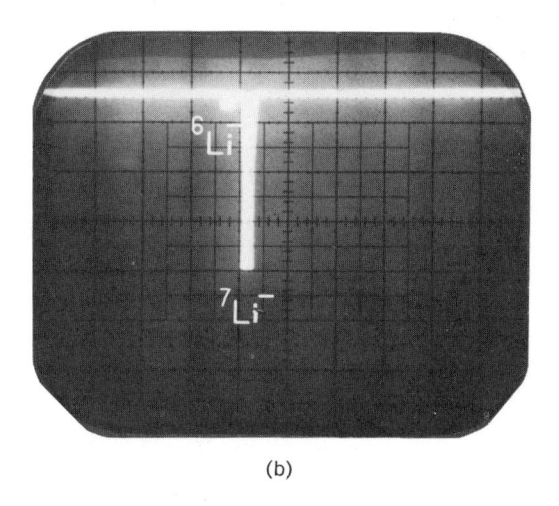

Figure 17.14 Mass spectrometer output signals showing (*a*) the positive ion species, and (*b*) the negative ion species in the extracted ion beam obtained from the small multicusp source shown in Figure 17.13. The oscilloscope sensitivity was 0.5 V/div for the positive species and 0.1 V/div for the negative ion species.

source using a permanent magnet filter [40]. A schematic diagram of the apparatus is shown in Figure 17.13. Argon is used as a supporting gas to initiate the discharge. Lithium is introduced as a vapor due to discharge heating and evaporation of a solid lithium sample placed inside the heat shield. Positive and negative extracted ion species have been analyzed by a magnetic mass spectrometer. The positive ion species are shown in Figure 17.14a. The positive ion beam is composed of both atomic and molecular lithium ions. Ar^+ is also present but is too small to be seen on this scale. Figure 17.14b shows the negative ion species in the extracted beam. Only the two lithium isotopes, $^6Li^-$ and $^7Li^-$, are seen with no impurities detected.

A steady-state Li$^-$ current of 14.9 μA ($J^- = 1.9$ mA/cm^2) has been obtained from this source with an arc voltage of 40 V and an arc current of 4 A. Since this is the first measurement of current density for a volume Li$^-$ ion source, substantial improvements in current density are possible with higher arc power, optimized lithium density in the ion source, and an improved magnetic filter geometry.

REFERENCES

1. R. Middleton, *Treatise on Heavy-Ion Science*, Vol. 7, D. Allan Bromley (Ed.), Plenum, New York, p. 53.
2. R. Middleton and C. T. Adams, *Nucl. Instrum. Methods* **118**, 329 (1974).
3. R. Middleton, *IEEE Trans. Nucl. Sci.* **NS-23**, 1098 (1976).
4. H. H. Andersen and P. Tykesson, *IEEE Trans. Nucl. Sci.* **NS-22**, 1632 (1975).
5. H. Vernon Smith, Jr., and H. T. Richards, *Nucl. Instrum. Methods* **125**, 497 (1975).
6. G. I. Dimov, Proceedings of the Second Symposium on Ion Sources and Formation of Ion Beams, Berkeley, 1974.
7. E. D. Bender, G. I. Dimov, and M. E. Kishinevsky, Proceedings of the National Seminar on Secondary Ion-Ion Emission, Kharkov, 1975.
8. K. W. Ehlers and K. N. Leung, *Rev. Sci. Instrum.* **51**, 721 (1980).
9. R. L. York and Ralph R. Stevens, Jr., AIP Conf. Proc. No. 111, 1984, p. 410.
10. A. Takagi, Y. Mori, K. Ikegami, and S. Fukumoto, *IEEE Trans. Nucl. Sci.* **NS-32**, 1782 (1985).
11. Yu. I. Belchenko and V. G. Dudnikov, *J. Phys. Colloque C7* **40**, 501 (1979).
12. Yu. I. Belchenko and V. G. Dudnikov, Proceedings of the XVth International Conf. on Phenomena in Ionized Gases, Minsk, 1981.
13. K. N. Leung and K. W. Ehlers, *Rev. Sci. Instrum.* **53**, 803 (1982).
14. K. N. Leung and K. W. Ehlers, Proceedings of the Third International Symposium on the Production and Neutralization of Negative Ions and Beams, Upton, NY, 1983.
15. M. Bacal and G. W. Hamilton, *Phys. Rev. Lett.* **42**, 1538 (1979).
16. M. Ardenne, *Atomkernenergie* **1**, 121 (1956).
17. C. D. Moak, H. E. Banta, J. N. Thurston, J. W. Johnson, and R. F. King, *Rev. Sci. Instrum.* **30**, 694 (1959).
18. A. B. Wittkower, R. P. Bastide, N. B. Brooks, and P. H. Rose, *Phys. Lett.* **3**, 336 (1963).
19. L. E. Collins and R. H. Gobbet, *Nucl. Instrum. Methods* **35**, 277 (1965).
20. G. P. Lawrence et al., *Nucl. Instrum. Methods* **32**, 357 (1965).
21. M. A. Abroyan, G. A. Nalivaiko, and S. G. Tsepakin, *Zh. Tekh. Fiz.* **42**, 876 (1972).
22. Yu. I. Bel'chenko, G. I. Dimov, and V. G. Dudnikov, *Investiya of USSR Academy of Science Ser. Fix.* **37**, 2573 (1973).
23. K. W. Ehlers, B. F. Gavin, and E. L. Hubbard, *Nucl. Instrum. Methods* **22**, 87 (1963).
24. M. D. Gabovich, Yu. N. Kozyrev, A. P. Naida, L. S. Simonenko, and N. A. Soloshenko, *Pis'ma ZhETF* **4**, 378 (1978).
25. K. Jimbo, K. W. Ehlers, K. N. Leung, and R. V. Pyle, *Nucl. Instrum. Methods A* **248**, 282 (1986).
26. K. N. Leung et al., *Rev. Sci Instrum.* **58**, 235 (1987).
27. K. N. Leung, K. W. Ehlers, and M. Bacal, *Rev. Sci. Instrum.* **54**, 56 (1983).
28. R. Limpaecher and K. R. MacKenzie, *Rev. Sci. Instrum.* **44**, 726 (1973).

References

29. M. Allen and S. F. Wong, *Phys. Rev. Lett.* **41**, 1795 (1978).
30. B. Peart and K. T. Dolder, *J. Phys. B* **8**, 1570 (1975).
31. B. Peart, R. A. Forrest, and K. Dolder, *J. Phys. B* **12**, 3441 (1979).
32. R. L. York, Ralph R. Stevens, Jr., K. N. Leung, and K. W. Ehlers, *Rev. Sci. Instrum.* **55**, 681 (1984).
33. K. N. Leung, K. W. Ehlers, and R. V. Pyle, *Rev. Sci. Instrum.* **56**, 364 (1985).
34. K. N. Leung, K. W. Ehlers, and R. V. Pyle, *Appl. Phys. Lett.* **47**, 227 (1985).
35. K. N. Leung, K. W. Ehlers, and R. V. Pyle, *Rev. Sci. Instrum.* **56**, 2097 (1985).
36. K. N. Leung, K. W. Ehlers, and R. V. Pyle, *Rev. Sci. Instrum.* **57**, 321 (1986).
37. K. N. Leung et al. *Rev. Sci. Instrum.*, in press.
38. Y. Okumura et al., Proceedings of the 4th International Symposium on the Production and Neutralization of Negative Ions and Beams, Upton, NY, 1986.
39. S. R. Walther, K. N. Leung, K. W. Ehlers, and W. B. Kunkel, *Nucl. Instrum. Methods B* **21**, 215 (1987).
40. S. R. Walther, K. N. Leung, and W. B. Kunkel, *Appl. Phys. Lett.* **51**, 566 (1987).

18

Light-Ion Sources for Inertial Confinement Fusion

R. A. Gerber
Sandia National Laboratories
Albuquerque, New Mexico

Light-ion drivers offer a potentially efficient and low-cost method to compress and heat an Inertial Confinement Fusion (ICF) target [1]. To produce a significant thermonuclear burn of the fuel, focused ion-beam intensities of 100 TW/cm^2 are required. The ion sources for these pulsed power drivers should produce a single-ion species and should be capable of providing current density levels up to 5 kA/cm^2, in pulse widths of 10 to 20 ns, at voltages up to several tens of megavolts. Most ion sources used in the past have produced multiple-ion species, including protons and heavier ion species such as carbon and oxygen. In the last few years there has been a substantial research effort to produce single-species sources. In this chapter the various types of light-ion sources presently being used and those under development will be discussed.

1 BACKGROUND

The major effort for light-ion fusion in the United States is now centered around the Particle Beam Fusion Accelerator II (PBFA II) [2] at Sandia National Laboratories; however, much of the early pioneering work related to ion diodes was done at Cornell University, Massachusetts Institute of Technology, and Naval Research Laboratory. An excellent review paper on intense pulsed ion beams, including the various types of ion diodes and their operation, is given by Humphries [3].

Efficient generation of intense ion beams relies on methods to suppress electron flow and thus increase the portion of the current carried by ions. One way to accomplish this is with an Applied-B diode [4–6]. In this diode a transverse magnetic field in the vacuum anode–cathode (A-K) gap inhibits electron flow and, by forming a virtual cathode tied to the applied magnetic field, can allow ion currents to flow that may be substantially in excess of the Child-Langmuir limit for the given physical anode–cathode separation.

Based on scaling studies [7], an Applied-B diode with a lithium ion source was selected for use on PBFA II. Figure 18.1 shows a sketch of an Applied-B diode for PBFA II. Power is fed to the diode from both the top and bottom, as indicated. This diode is designed to accelerate 4 MA of 30-MV lithium ions and focus them radially onto a 3- to 6-mm-radius target located in the center of the diode. Magnetic field coils located in both the cathode and anode structures can provide a 5-T insulating magnetic field. The ion source is located on the inner surface of the 30-cm-diameter anode, forming an active area of 800 cm^2. As shown in the figure, the anode surface is shaped so that the ions will focus on the target. Electrons emitted from the cathode tips form a virtual cathode at some location in the 1- to 2-cm anode–cathode gap.

Although most ion-diode experiments on smaller accelerators have been performed with proton beams in the past, lithium ions were selected for PBFA II

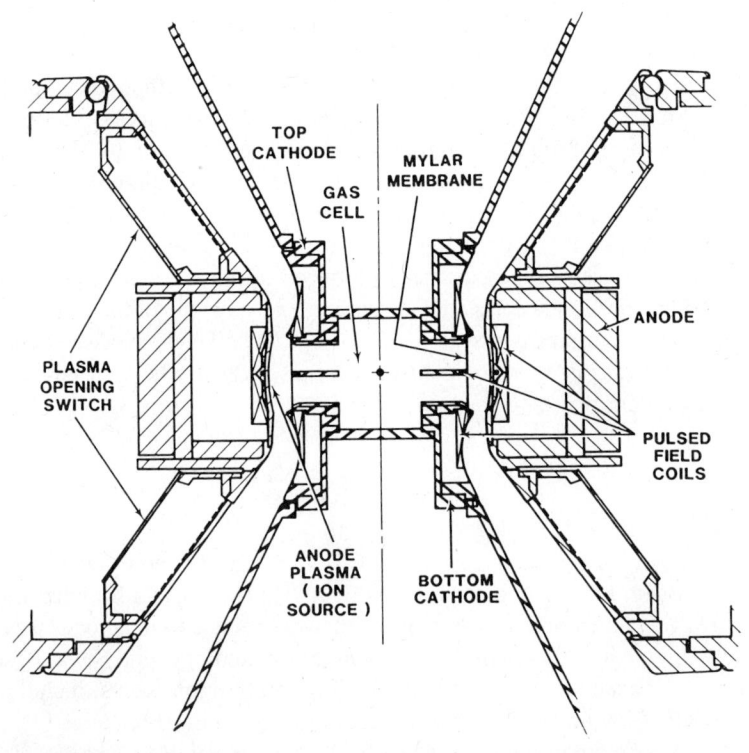

Figure 18.1 Cross-sectional view of the cylindrically symmetric PBFA-II magnetically insulated ion diode. Lithium ions are drawn from the ion source, accelerated by the anode–cathode gap potential, and focused onto an ICF target located at the center. From *J. Vac. Sci. Technol. A* **4**, 772 (1986). Reprinted by permission of K. W. Bieg.

for several reasons [8]. The use of lithium enables a large anode–cathode spacing, resulting in less impedance change during the pulse due to electrode plasma expansion. Lithium ions experience less magnetic deflection than protons in the diode, so the focusing is less sensitive to beam energy spread, allowing ions to reach the target for a longer portion of the pulse. Also, since the ionization potential of Li^+ is low (5.4 eV), while that of Li^{2+} is very large (75.6 eV), high Li^+ source purity should be achievable without significant contamination from Li^{2+}. Ions heavier than Li^+ would require either operating voltages greater than 30 MV, or the ions would have to be accelerated in a multiply charged state, to effectively couple to the ICF target.

The source requirements for the PBFA-II Applied-B diode are as follows: (1) It must be capable of providing 5 kA/cm^2 of Li^+, corresponding to a total ion current of nearly 4 MA. (2) The source should be spatially very thin, on the order of 1 mm or less, so that the ion source will conform to the surface curvature of the anode to allow the beam to be focused on the target. (3) The beam should be greater than 90% singly charged lithium, since this affects the efficiency of the total system. (4) The source must operate under the vacuum conditions of the PBFA-II diode section, which are typically on the order of a few times 10^{-5} Torr. (5) The source must be able to supply ions either immediately or shortly after (i.e., 1 to 2 ns) the 15-ns power pulse arrives at the diode. There has been an extensive research and development effort since 1983 to produce a pure lithium source. In addition to a review of existing ion sources, this chapter will also include results and directions of the lithium source research.

There are basically two types of ion sources being used for intense ion beam generation: (1) passive sources, which are produced directly or indirectly by the diode voltage itself, and (2) active sources, in which a preformed plasma is generated before the arrival of the power pulse.

2 PASSIVE SOURCES

2.1 Flashover Sources

The most popular and widely used ion source for Applied-B ion diodes is the flashover source. This source is a completely passive one in which a plasma is formed on a dielectric-filled anode during the operation of the diode. Although the plasma formation processes have been studied extensively [9–14], the exact mechanisms are still not fully understood. It is speculated [13] that electron leakage to the anode contributes to surface breakdown by generating electric fields with large tangential components at a dielectric surface. The impact of both primary and secondary electrons on the dielectric surface desorbs neutrals that can subsequently be ionized by avalanche electrons. Since many processes are involved in producing a plasma, there is a delay time between the start of the voltage pulse and the onset of the ion current.

Flashover anode sources seem to depend on having a dielectric–metal interface at the surface. They can be made in a variety of ways: embedding strips of polyethylene [9], covering the surface with velvet cloth or lapidary polishing pads [10], or covering the surface with acrylic plastic in which metal pins are inserted [15]. However, the most common procedure for fabricating a flashover source is to machine grooves in the anode surface and fill these grooves with a dielectric material [10, 14]. In general, these grooves are 0.5 to 2.0 mm wide, each spaced about 1 mm apart. For a proton-producing diode, the dielectric materials have been either epoxy ($C_8H_{11}O$ or $C_{21}H_{24}O_4$), polystyrene (CH), or carnauba wax (CH_2). Holographic interferometry and spectroscopy of flashover plasma produced in the diode of an 800-kV, 300-kA, 100-ns accelerator have shown the plasma originates near the edges of the grooves and then quickly covers the dielectric surfaces. The plasma density was determined to be about 5×10^{16} cm^{-3} and the temperature to be about 5 eV [10]. Electron density measurements by Pal and Hammer [12] at lower power levels (400 kV, 19 kA) indicated electron densities of 2×10^{15} cm^{-3} and an electron temperature of about 1 eV. They also found that greater than 10^{16} cm^{-3} neutrals were injected into the diode during the flashover process. Both the high-power and the low-power experiments support the explanation of the flashover plasma formation process previously discussed.

Early diode experiments at low power levels indicated that flashover sources produced a predominantly proton source. However, as the power levels increased and better diagnostics evolved, it was found that the ion beam was composed of protons, various charge states of carbon and oxygen, and surface impurities. The beam produced on PBFA I at current levels of 6 MA contained approximately 50% protons [16].

To obtain proper coupling to an ICF target and for efficient operation of an ion accelerator, single-species light-ion sources must be developed. A series of experiments were completed by Bieg et al. [14] to investigate the surface flashover mechanism and the effects of various dielectric materials, vacuum conditions, and surface preparation on the ion species. The experiments were conducted on a 500-kV Nereus accelerator [17] using a magnetically insulated extraction diode [13]. Figure 18.2 shows the experimental setup for these experiments. This diode produces a nonfocusing annular beam that is accessible to many diagnostics. Since ion-species determination was a major goal, the diode chamber was equipped with a Thomson parabola mass analyzer. There were also facilities to use rf glow-discharge cleaning of the anode surface. Initial experiments determined the ion species from polymer dielectrics commonly used for flashover sources. Table 18.1 shows the percentage of various ions in the beam without any surface preparation. The results show that the proton to carbon ratio is related to the composition of the bulk material. By substituting deuterated polystyrene for CH, it was determined that about one-third of the beam was due to surface contamination. The results shown in Table 18.1 demonstrate that a single-species beam cannot be obtained from room temperature flashover sources. Glow-discharge cleaning was investigated on a standard epoxy-filled anode in an attempt to increase the proton fractions. These results are shown in Table 18.2. By glow-discharge cleaning with hydrogen until the Nereus accelerator was pulsed, it was possible to increase the proton content to 70%, while considerably reducing the carbon fraction. Presum-

2 Passive Sources 375

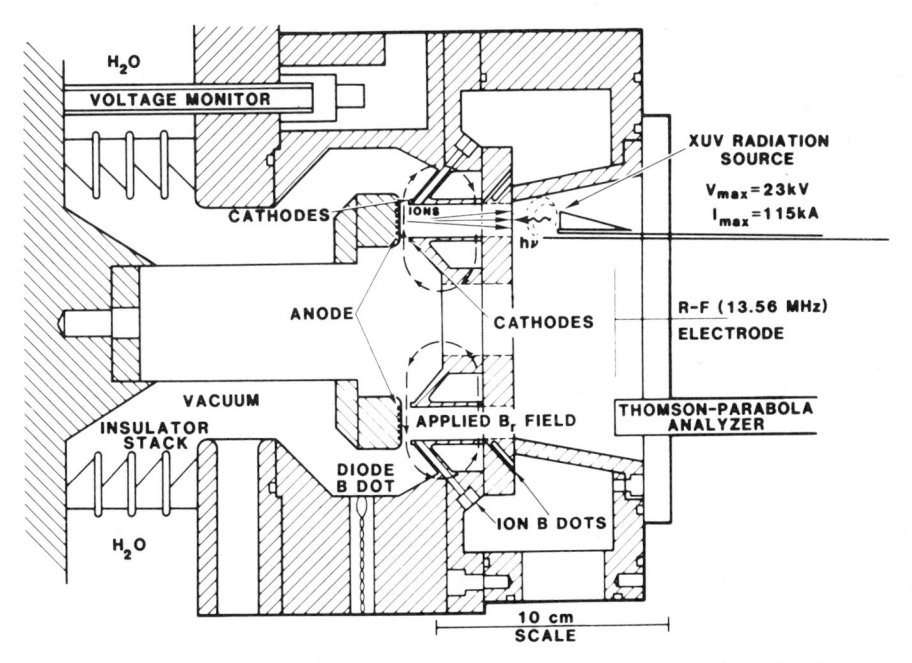

Figure 18.2 Nereus magnetically insulated ion diode. Ions are extracted from the surface flashover source as a hollow, annular beam along the diode axis. From *J. Vac. Sci. Technol. A* **4**, 772 (1986). Reprinted by permission of K. W. Bieg.

ably this is due to the formation of methane. The data in Table 18.2 also show that if the glow discharge was terminated three minutes before the machine was fired, the surface returned to its original condition.

A near-term lithium flashover source was required for initial pulse power tests on PBFA II. This nonoptimized source was needed to gain experience with Li-ion diodes and to develop nonprotonic diagnostics. In another series of experiments a

TABLE 18.1 Ion Species Distribution from a Flashover Source Using Anodes Filled with Various Polymer Dielectrics[a]

Material	H$^+$	D$^+$	C$^{+,2+,3+}$	O$^{+,2+}$	Misc.
Epoxy (C$_8$H$_{11}$O)	28	—	62	7	3
Polystyrene (CH)	20	—	66	6	8
Carnauba wax (CH$_2$)	67	—	29	1	3
Deuterated polystyrene (CD)	8	15	52	14	11

[a] Numbers give the percentage of each species in the beam. From *J. Vac. Sci. Technol. A* **3**, 1234 (1985). Reprinted by permission of K. W. Bieg.

TABLE 18.2 Effect of Glow Discharges on Ion Species from Flashover Sources Using Epoxy-Filled Anodes[a]

Treatment	H^+	$C^{+,2+,3+}$	$O^{+,2+}$	Misc. (Al, etc.)
Epoxy	28	62	7	3
H_2 glow, active	70	10	5	15
H_2 glow, 1-min delay	54	34	6	6
H_2 glow, 3-min delay	26	62	3	9
Ar glow, active	45	13	14	27
O_2 glow, active	36	17	5	42
O_2 Glow, 1- to 10-min delay	23	23	22	32

[a]Numbers give the percentage of each species in the beam. From *J. Vac. Sci. Technol. A* **3**, 1234 (1985). Reprinted by permission of K. W. Bieg.

number of lithium-bearing anode dielectric materials, including LiF, LiI, LiNO$_3$, and Li$_3$N, were investigated [18] on the same apparatus as shown in Figure 18.2. This well-controlled, well-diagnosed experiment compared the ion purity obtained from the various lithium compounds. This experiment also employed methods to prepare clean surfaces, such as glow-discharge cleaning, vacuum baking, and in situ source preparation by vacuum evaporation. The major ion-species diagnostic was a Thomson parabola ion analyzer. The lithium fluoride- and lithium nitride-filled anodes were prepared by evaporating a thin ($\simeq 3$ μm) coating on an epoxy-filled, grooved anode. Lithium nitrate and lithium iodide were melted directly into the anode grooves.

Table 18.3 shows the percentages of the various ions in the beam for the various configurations. An uncleaned LiF-overcoated, epoxy-filled anode gave a lithium percentage of 28%. Glow-discharge cleaning of this surface increased the lithium content to greater than 50%, while reducing the carbon and fluorine content. The heated but uncleaned LiNO$_3$ source gave purities similar to that previously obtained on the Proto-I accelerator [19]. Previous experiments using LiF produced a higher percentage lithium beam at current levels of 40–50 A/cm^2 [20]. The source that

TABLE 18.3 Ion Species Distribution from Lithium Flashover Sources with and without In-Diode Cleaning[a]

Source; Treatment	Li^+	H^+	$C^{+,2+}$	F^+	N^+	O^+	Misc.
LiF/epoxy; uncleaned	28	27	18	18	—	—	9
LiF/epoxy; glow-discharge cleaned	52	20	3	9	—	3	13
LiNO$_3$; 473 K	19	24	26	—	6	17	8
LiNO$_3$; 473 K + glow-discharge cleaned	52	1	1	—	13	26	7
LiI; uncleaned	39	36	12	—	—	2	11
LiI; 473 K + glow-discharge cleaned	85	5	—	—	—	3	7

[a]Numbers give the percentage of each species in the beam. From *J. Vac. Sci. Technol. A* **4**, 772 (1986). Reprinted by permission of K. W. Bieg.

gave the highest percentage Li$^+$ beam was heated LiI, using a glow discharge. As described in Ref. 18, Li$^+$ ion percentages from LiF and Li$_3$N of 68 and 85%, respectively, are achieved if the materials are evaporated on the suface after the diode is evacuated. However, it would be very difficult to prepare these surfaces on PBFA II. Both LiI and LiNO$_3$ are extremely hygroscopic, so LiF was chosen as the first lithium source to be fielded on PBFA II. To date several uncleaned LiF-overcoated anodes have been used on PBFA II and have produced total ion beam energies about one-third of that obtained from an epoxy-filled source. The Nereus experiments pointed out some inherent limitations of lithium flashover sources: Cleaned lithium-bearing sources produced less than one-third of the ion current obtained from uncleaned hydrocarbon sources, and there were long turn-on delays between the voltage pulse and the onset of ion current.

In summary, flashover sources are simple to use and convenient for diode studies but in general produce multiple-ion species both from the substrate material and from surface contamination. Another major disadvantage of these sources is an inherent time delay between the application of the accelerating voltage and the onset of ion current. This delay time, typically several nanoseconds, is associated with the complicated plasma formation mechanism.

2.2 Cryogenic-Anode Sources

Several years ago researchers from the Tokyo Institute of Technology proposed a cryogenically refrigerated anode source in which various source materials could be frozen on the surface [21]. Although this source is basically a flashover source, there are advantages in that high-purity beams of Xe, Kr, O, Ar, N, Ne, H, or D could be produced by cooling the anode with liquid helium, and H and/or D beams could be produced without carbon contamination for ices of H$_2$O or D$_2$O using liquid nitrogen coolant. In addition, these sources could be repetitively pulsed [22]. The sources are prepared in situ by cooling the anode under vacuum, then injecting the desired gas into the system, allowing the gas to freeze on the anode surface. Early experiments, using a low-energy accelerator and anode surfaces of either H$_2$O or D$_2$O ice, demonstrated the viability of these sources to produce 7 A/cm^2 of H$^+$ and 5 A/cm^2 of D$^+$, with a diode voltage of 80 kV [21].

The researchers at the Tokyo Institute of Technology have continued to improve their experimental facilities. At the present time they have an extraction-type Applied-B diode that can deliver a 500-keV ion beam with a total current of 560 A. The machine can operate repetitively at 0.2 Hz, and the anode is cooled by either a refrigerator that can cool to 20 K or a liquid helium system [23]. In one set of experiments [24] various species of ions were produced from the various ices: H$^+$ from H$_2$ at 5 K, Ar$^+$ from Ar at 30 K, C$^+$ and O$^+$ from CO$_2$ at 80 K, and N$^+$ from N$_2$ at 20 K. The measured ion current density of these sources in all cases exceeded the Child-Langmuir values for the applied voltage and anode–cathode spacing. By comparing the time of flight information of H$^+$ from H$_2$ with a polymer-filled flashover source at room temperature, it was established that an extremely pure H$^+$ beam was produced. Recently [23] nitrogen sources have been formed from N_2 ice at 20 K. A Thomson parabola mass analyzer has been added. A nearly 100% pure N^+ beam has been produced with no detectable proton or carbon impurities.

In addition to measuring the species purity of various sources, the group at Tokyo Institute of Technology has contributed greatly to our understanding of flashover sources [25, 26]. They have measured the electron density in the anode-cathode region, the time dependence of the emitted light intensity, the source uniformity, the anode–plasma expansion, and the dependence of turn-on time on temperature. Significant results are that the plasma density increases with increasing anode temperature and that turn-on time decreases with increasing temperature. Both of these effects could be explained by more neutrals being desorbed at higher temperatures, and thus more plasma being formed by avalanche ionization.

In summary, cryogenic-anode sources can be used to generate pure beams of a variety of species at high-current densities. In addition to light-ion ICF applications, these sources could be used for such diverse applications as exciting gas lasers or annealing semiconductors [22].

2.3 EHD-Driven Liquid Sources

One of the most promising lithium sources for PBFA II involves high-field evaporation of Li^+ ions from a liquid lithium surface. Theory [27] suggests that this electrohydrodynamically (EHD) driven source might satisfy source requirements for PBFA II, that is, pure, uniform, large-area sources capable of delivering 5 kA/cm^2, that will produce ions a few nanoseconds after the accelerating voltage is applied.

Figure 18.3 is a sketch that indicates the process by which ions are produced. When an electric field is applied to the surface of a liquid, electrohydrodynamic instabilities occur if the electric field stress exceeds the effects of surface tension (Figure 18.3a). The instability manifests itself by fluid motion that results in a series of cusplike protrusions on the liquid surface, as shown in Figure 18.3b. The temporal and spatial characteristics of the instability are governed by the growth time τ and wavelength λ of the most rapidly growing mode, and depend strongly on the value of the applied electric field. The electric field is greatly enhanced at the cusp tips. When the enhanced fields at the tips of the cusps exceed about 100 MV/cm [28], high-field evaporation of ions will occur, as in the case of single liquid metal ion sources (LMIS), described elsewhere in this book by Swanson and Bell. The results of the linear analysis show that wavelengths, which are related to emitter spacing, will be less than 1 μm and linear growth times will be less than 1 ns for applied electric fields expected on PBFA II (10 to 20 MV/cm). A full nonlinear analysis of the liquid surfaces predicts that ion emission will occur in less than three linear growth times (3 ns) for an applied field of 20 MV/cm [29], starting with an initial wave amplitude of 1.0 nm. Space-charge effects from multiple emitters, as well as source divergence, have also been calculated [30]. The results show that for an applied electric field of 10 MV/cm, the ion beam

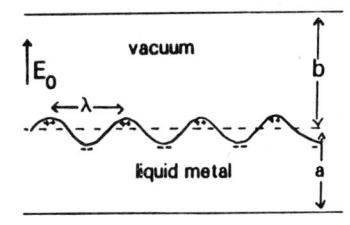

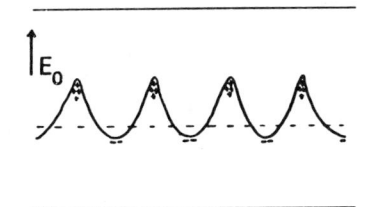

Figure 18.3 A sketch indicating the formation of an EHD-driven liquid ion source. (a) Electrohydrodynamically driven instabilities on a surface that exist when the electric field stress exceeds the effects of surface tension. Shown is the wavelength of the dominant instability. (b) As the instability grows, cusps are formed. When the enhanced electric field at the cusps reaches about 100 MV/cm, high-field evaporation of ions will occur. This sketch reprinted by permission of A. L. Pregenzer.

divergence should be less than 6 mrad and space-charge effects from neighboring emitters should not significantly affect the current from an individual emitter. The conclusion of the theoretical work to date is that a passive EHD-driven liquid source should produce 5 kA/cm^2 of Li$^+$, have a turn-on time ≤ 3 ns, and contribute less than 6 mrad to the divergence of the extracted beam.

The theory has also been extended to include liquid lithium nitrate, LiNO$_3$, which is a low melting point ionic salt. Although single-point EHD ion emission has not been studied for LiNO$_3$, EHD emission has been observed from lithium borate [31], that is also an insulator. The lithium atomic ion purity from this steady-state, single-emitter source was 80%.

An experiment was performed to determine if multiple-cusp formation would occur over a large area and to compare both wavelengths and growth times to the EHD theory [32]. In this experiment, electric fields were applied to the surface of various liquids. Low fields were used so that dominant wavelengths would be in the centimeter range and thus could be measured easily. The results demonstrate multiple-cusp formation over a surface area of 600 cm^2. The wavelengths for both water and ethanol, measured as a function of applied electric field, are in excellent agreement with the linear theory, and the growth times are consistent with the nonlinear analysis, giving us more confidence in EHD-driven, large-area sources.

Before ion-diode experiments could begin, a way had to be found to provide a vertical liquid anode surface (See Figure 18.1). One successful technique that has been developed is to suspend the liquid in a porous or fritted material. The porous material can be presoaked or can be filled in situ by capillary action from a reservoir located near the bottom of the anode [33].

As part of the effort to develop an EHD-driven ion source for PBFA II, experiments are being conducted with liquid lithium- and liquid lithium nitrate-filled anodes on a low-energy PI-110A accelerator [34]. This accelerator operates at a peak voltage of 1.5 MV, a peak current of 30 kA, and a pulse width of 24 ns. The accelerator was fitted with an extraction geometry Applied-B diode very similar to the one shown in Figure 18.2. Although applied electric fields of 10 MV/cm could not be achieved on this accelerator, these experiments allowed a test of methods to produce heated liquid-filled anodes. Since the maximum applied field that could be achieved was 2 MV/cm, anodes with surface textures were used to provide field enhancement.

The anodes used in these experiments were all annular in shape. The active area of the anode was 42 cm^2. This is the area of the anode face immediately across the diode from the 10.4-cm-diameter outer cathode and the 7.4-cm-diameter inner cathode. There were two basic anode designs used during these experiments, as shown in Figure 18.4.

The first type of anode (Figure 18.4a) was developed by the Energy Technology Engineering Center (ETEC). For this anode, a copper substrate for the liquid lithium was prepared by machining a spiral groove, having a pitch of approximately 100 μm and a peak-to-valley height of approximately 100 μm into the substrate. If partially filled with liquid, the geometry of this design provides a field enhancement of approximately three, thus increasing the probability of forming an EHD instability on the time scale of the accelerator pulse. A layer of molten lithium was spread over the heated substrate.

In the second type of anode (Figure 18.4b) the liquid was contained in a porous stainless-steel insert. The insert was 1.5 mm thick and 50% dense, and could be saturated with either liquid lithium or liquid lithium nitrate. The saturated stainless-steel insert could be mounted in either a stainless-steel housing or a copper housing.

The third type of anode used was a flashover source with epoxy-filled grooves. This type of anode has been used with many Applied-B diodes, including the

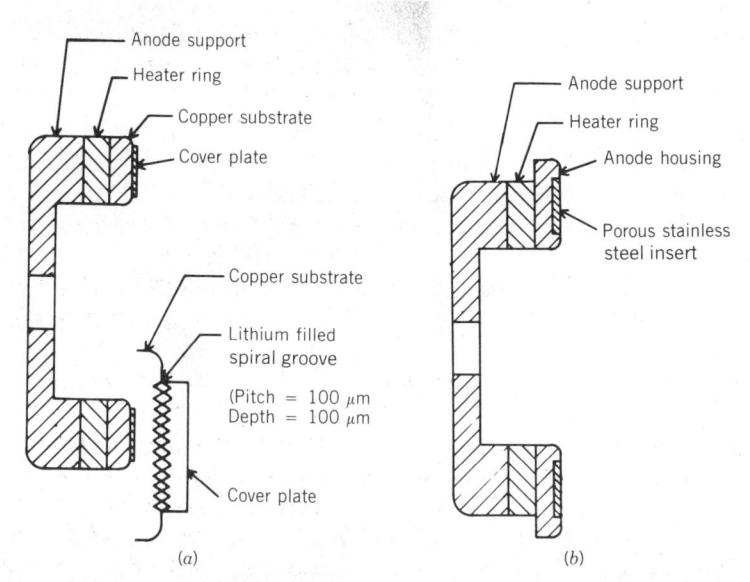

Figure 18.4 Two types of cylindrical anode configurations used with the extraction diode on a PI-110A accelerator. As described in the text, (a) shows the details of the ETEC-produced heated anode and (b) shows a heated anode in which the liquid is suspended in a porous metal insert. From *IEEE Trans. Plasma Sci.*, **PS-15**, 339 (1987). Reprinted by permission of IEEE and P. F. McKay © 1987 IEEE.

previously described Nereus experiments [14, 18]. The epoxy-filled anode was used to check accelerator performance and was used as a baseline with which to compare the results from the other anodes.

The primary purpose of the first experiments was to measure the ion species obtained from each of the various anode configurations. In addition, it was anticipated that differences in the electrical signals between epoxy-filled anodes and anodes filled with lithium-based substances would aid in the understanding of the ion production mechanism.

A summary of the ion species obtained with the various anodes is shown in Table 18.4. All the lithium-based anodes gave very high concentrations of lithium ions (69 to 96%), and the epoxy-filled anodes yielded species normally observed from anodes of this type. For the ETEC anodes, an attempt was made to do experiments both below and above the lithium melting temperature of 453 K. Based on visual observation of the surface just prior to the shots, it is doubtful that the surface was liquid, even at temperatures above 453 K. It is believed that the lithium had reacted with the background gas or the substrate and was a solid during the voltage pulse. In the case of the lithium nitrate anodes, both shots were performed above the $LiNO_3$ melting temperature of 533 K. The reason for the observed difference in the lithium content is unknown. The high lithium ion content using liquid lithium nitrate agrees with the preliminary results of a previous experiment [33]. The highest lithium ion percentage, as measured on the Thomson parabola analyzer, was produced by the lithium-filled porous stainless-steel anode, in which the lithium was a solid. This source produced >90% lithium ions. However, the data from these experiments did not allow one to distinguish a mechanism for ion production.

In a follow-up experiment [35] the beam composition and the ion currents were measured for both lithium and lithium nitrate as a function of temperature, from room temperature to above the melting point. The results showed the following: (1) The total ion charge, that is, the time integral of the ion current, was independent of temperature. (2) The composition of the beam was not temperature depen-

TABLE 18.4 Ion Beam Compositions for the Experiments Using Various Anode Configurations[a]

Type	Anode Housing Material	Temp.	Li^+	H^+	$C^{+,2+}$	$N^{+,2+}$	$O^{+,2+}$
Epoxy	Al	293 K		71	13	12	2
ETEC 4	Cu	505 K	82			4	12
ETEC 5	Cu	442 K	82	1	2	2	12
$LiNO_3$	SS	544 K	90			2	7
$LiNO_3$	Cu	564 K	69			4	27
Li	SS	414 K	96				3

[a]Shown are anode type, anode housing material (SS = stainless steel), temperature, and the percentages of the total charge of the ion beam carried by each species as inferred from the Thomson analyzer in which only the numbers of particle tracks are counted. From *IEEE Trans. Plasma Sci.* **PS-15**, 339 (1987). Reprinted by permission of IEEE and P. F. McKay © 1987 IEEE.

dent. For LiNO$_3$-filled anodes the average species concentration was 73% Li$^+$, 21% O$^+$, and 3% N$^+$. Lithium-filled anodes produced 95% Li$^+$ and 2% C$^+$. (3) The ion beam was space-charge limited: $J/J_{CL} \simeq 3$. Experiments were also done using anodes in which 1 to 5 μm of LiF was deposited on a porous stainless-steel anode. These anodes produced an 80% Li$^+$ source. Experiments using this field-enhanced source are now being done on PBFA II.

The results of the experiments on the PI-110A are very exciting in that a high purity Li$^+$ beam was produced. Various ion-production mechanisms have been proposed [34], but at this time it is not clear what mechanisms are responsible. Experiments are continuing to determine if the ion production is related to the EHD mechanism.

Passive EHD-driven liquid sources have the potential of providing a pure, uniform, lithium-ion beam. Initial experiments on PBFA II were unsuccessful because the existing diode (Figure 18.1) was incompatible with heated anodes. The anode field coils could not survive the high temperature required at the anode. Water-cooled coils have been obtained, and experiments are now in progress. Figure 18.5 shows a photograph of a nonfocusing EHD anode for PBFA II. The

Figure 18.5 A photograph of an EHD-type anode for PBFA II. A reservoir in the bottom contains solid LiNO$_3$. As the anode is heated, the liquid melts and saturates the porous material by capillary action, thus performing a vertical liquid surface from which ions can be extracted. The anode has six heater elements on the outside of the cylinder, as shown. Photograph used with permission from K. W. Bieg.

active area is porous stainless steel that extends into a reservoir located in the lower section of the anode. The reservoir will be filled with solid LiNO$_3$ prior to installation in the PBFA-II center section. After the diode is evacuated, the anode will be heated above the melting temperature of LiNO$_3$ (>533 K) so that the LiNO$_3$ will melt, and then by capillary action will wick the porous material to provide the vertical liquid surface.

3 ACTIVE SOURCES

An active ion source is one in which an anode plasma is produced by an external power source prior to the arrival of the main acceleration pulse. Potential advantages of active sources over the commonly used flashover sources are that [36] (1) turn-on time delay is reduced between the onset of the current pulse and the voltage pulse; (2) the source species and purity levels can be controlled; (3) the number of neutral particles, which cause premature diode closure, can be minimized; (4) the spatial uniformity of the source can be controlled; and (5) source damage can be minimized, allowing multiple pulses or even repetitive operation.

The majority of the early research on active sources was directed at producing volumetric plasma sources to uniformly fill the anode–cathode region in ion diodes, such as AMPFION [37]. The required plasma density for this diode was about 10^{15} cm^{-3}, with uniformity requirements of about 10%. One of the first plasma sources used was a magnetically driven carbon-plasma gun, driven by pulsed, low-energy capacitor banks [38, 39]. These guns produced from 10^{17} to 10^{18} multiply charged carbon ions, moving with a velocity of about 10 cm/μs. The plasma is formed in these basically coaxial devices by a surface breakdown over the graphite-coated insulator. For the original AMPFION diode experiments [37], six of these plasma guns produced densities on the order of 3 to 6 × 10^{14} cm^{-3}. The ions produced from these guns consisted of 40% C$^+$, 40% C^{2+}, and 15% H$^+$. A major difficulty with using these guns for plasma prefill was the nonuniform plasma fill. Although these carbon plasma guns did not provide an adequate plasma for an ion diode, they have been used extensively to reduce prepulse in ion diode experiments [40] and as plasma sources for plasma-opening switches [41]. To produce a more uniform plasma for plasma-filled diodes, a source with an array of graphite-coated spark gaps in the form of a flashboard was developed that consisted of approximately 10^3 discrete discharges [37]. A similar multiple-point plasma source was also used to produce a deuterium plasma, with a density of about 10^{14} cm^{-3}, by using deuterium-loaded titanium electrodes [42]. A major problem with these multiple sources is that current shunts through the expanding plasma, thereby limiting the plasma density.

Several other active sources have been used on various types of ion diodes, such as the nitrogen plasma fill of the reflex triode [43], the laser-produced plasma on the pinch reflex diode [44], and the pseudospark diode [45]. However, since the major emphasis for light-ion fusion is based on the Applied-B diodes, the remainder of the active sources described will be those with applicability to the Applied-B diode.

3.1 XUV-Driven Sources

Pulsed, high-current surface discharges were originally studied as intense UV photon sources for pumping lasers [46] and for x-ray laser research [47]. Sandia National Laboratories has had a research effort since 1984 to determine the characteristics of these sources and to determine if sufficient intensity levels can be obtained to preionize or clean anode surfaces. Scaling experiments [48] on a single XUV vacuum source, consisting of two copper electrodes separated by a distance of 5.5 mm and driven by a capacitor bank, demonstrated efficiencies (XUV energy/ input electrical energy) of from 1.5 to 3.2%. The peak intensity (energy) obtained was 80 MW (60 J). The UV light emitted was determined in four spectral regions 4 to 8 eV, 10 to 20 eV, 20 to 70 eV, and >70 eV. Most of the intensity was in the 10- to 70-eV region. Based on these results, an XUV system was designed and tested on the Applied-B diode of PBFA I [49, 50]. In the experiment, 16 discharges driven by 8 capacitor banks illuminated the entire epoxy-filled flashover anode with intensity levels of $\simeq 60$ kW/cm^2. It was found that XUV illumination had several effects on the diode's performance:

1. Normally, no current flowed in this diode until 5–20 ns after the arrival of the main accelerator power pulse. On shots with XUV illumination, diode and ion currents were observed in the diode during the ~ 200-kV accelerator prepulse, 20 ns prior to the arrival of the main accelerator pulse, indicating that the XUV sparks were generating plasma in or near the diode that was a source of ions to be accelerated as soon as any significant voltage arrived.
2. Shots with XUV illumination coupled more energy through the diode at early times in the main power pulse than shots without XUV. XUV shots, on the average, had conducted 10 mC more charge through the diode than non-XUV shots 10 ns after the voltage pulse rose past 1 MV. However, since a typical accelerator shot coupled >100 mC of charge through the diode at voltages above 1 MV during the pulse, the extra charge coupled due to the XUV illumination was not a significant effect.

A series of XUV-illumination experiments were also performed on the Nereus accelerator [51]. In these experiments, a LiF-coated flashover anode, which had been cleaned in an Ar/O$_2$ glow discharge, was illuminated at XUV levels of 60 kW/cm^2. The total fluence of XUV energy on the diode, prior to the accelerator pulse, was $\simeq 12$ mJ/cm^2. XUV illumination increased the peak ion current through the diode from 5 to 25 kA, roughly quadrupling the beam energy coupled through the diode. The turn-on time of the ion beam was reduced from ~ 25 ns without XUV to <10 ns with XUV illumination.

The extra ion current induced by XUV illumination contained very little lithium but consisted primarily of fluorine from the LiF, oxygen presumably deposited on the anode by the Ar/O$_2$ discharge, and hydrogen and carbon from incompletely cleaned surfaces in the diode. The fluorine content in the beam is not surprising, since both electron- and photon-stimulated desorption studies on LiF show that ground-state fluorine desorption is rapid, leaving the surface lithium enriched [52].

The XUV intensity levels used in diode experiments are sufficient to desorb neutrals and thereby shorten the flashover plasma production time. However, to melt and ionize lithium, intensities of $\simeq 1$ MW/cm^2 are required. Recent scaling studies [53], again on a single source, using a peaking capacitor switch arrangement to produce shorter pulses of XUV radiation, produced a source with a maximum intensity of 147 MW. A preliminary design of an XUV system for PBFA II using 12 arcs would produce intensity levels of 1 MW/cm^2 in a 300-ns (FWHM) pulse. However, at this time no experiments on PBFA II are planned.

Kitamura et al. [54] have recently reported experiments in which the effect of UV radiation on a sodium-overcoated anode in an Applied-B diode is studied. Sodium was chosen because of its low lattice energy, low work function, and low ionization potential; vaporization, photoelectron emission, and ionization would then be possible by low-energy photons. Sodium was coated on the anode by in situ vapor deposition. In a preliminary experiment a Xe-filled flash lamp provided the photons. An increase in the ion current was observed with UV radiation. In a later experiment a more powerful XUV source was employed. This experiment showed that intense ion beams were produced when XUV was used and that the beam was greater than 90% Na$^+$. Although there is a possibility that ions or plasma from the XUV source bombarded the surface, the authors inferred that the anode-plasma production was caused by photon irradiation. Additional experiments are ongoing to better understand this source; however, the results are very encouraging.

3.2 Gas-Breakdown Plasma Anode

Researchers at Cornell University have been developing a plasma-anode source over the last few years [36, 55, 56]. Results of source performance measurements on their 10^{10}-W, 1-μs, 150-kV LONGSHOT accelerator, fitted with an extraction-type magnetically insulated diode, have recently been reported. The source uses the induced voltage from a single-turn coil to produce an annular ring of plasma. The magnetically driven plasma stagnates against a preexisting applied magnetic field in the ion diode accelerating gap. The plasma thus serves as an anode from which ions are extracted. Figure 18.6 shows a sketch giving details of the diode and source.

The operation of the source follows this sequence of events: the slow insulating diode magnetic field is established first by a slow (~ 100-μs rise time) circuit. A magnetically driven fast valve is then opened to establish an annular gas fill in front of the fast coil. The gas is preionized by a circumferential array of spark gaps. The fast coil is then energized with a 1-μs rise time, inducing a voltage of $\simeq 17$ kV around the preionized annular plasma, which drives a current through it. The diamagnetic nature of the plasma causes it to be driven toward the diode

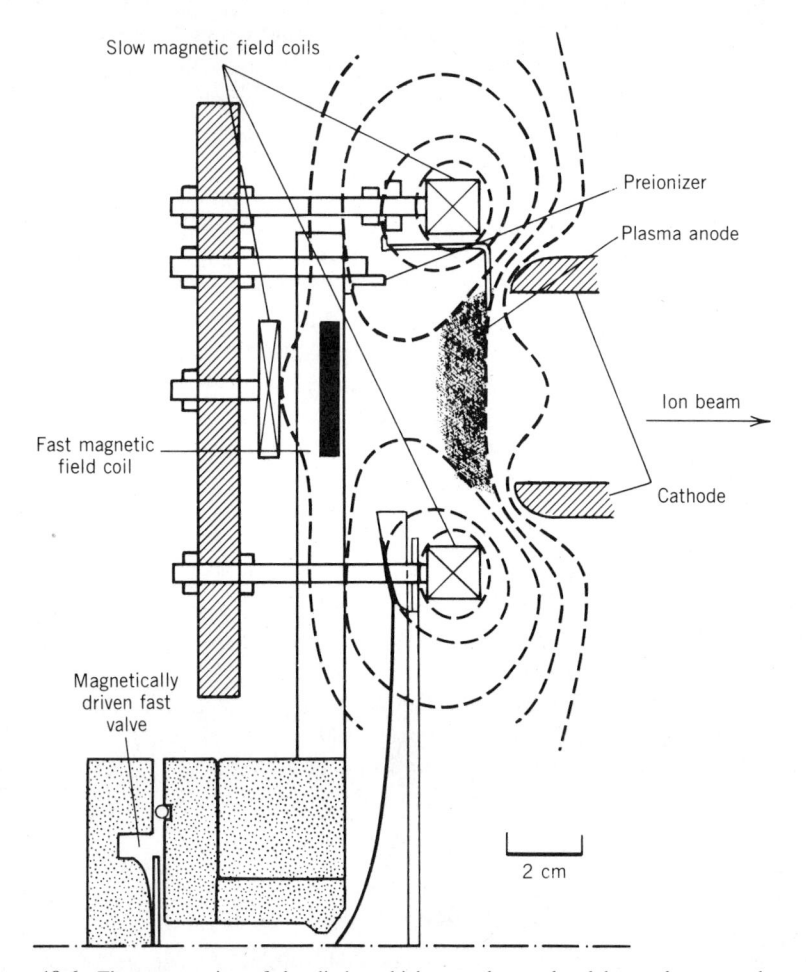

Figure 18.6 The construction of the diode, which uses the gas-breakdown plasma anode on the LONGSHOT accelerator at Cornell University. As described in the text, a fast magnetic field coil produces a moving plasma that stagnates against an applied magnetic field to produce a plasma anode at the approximate location shown in the figure. The electric field applied between the cathode and the plasma produces an annular ion beam. The figure also shows the magnetic field lines produced by the diode insulating slow magnetic field coils. Adapted from a figure in *J. Appl. Phys.* (1988). Reprinted by permission of D. A. Hammer.

accelerating gap, where it stagnates against the insulating field of the diode. The exact location of the plasma is determined by the relative magnitudes of the fast and slow magnetic fields at the time the high voltage is applied to the gap. The Cornell work improves upon earlier work by Humphries et al. [57]. The main difference in this earlier work is that plasma did not stagnate against a magnetic field, and consequently a lower plasma density was achieved.

The gas-breakdown plasma source is a versatile source and can operate under a variety of conditions. The ion species are determined by the type of gas injected.

By varying the timing of the various circuits (puff valve, preionizer, fast coil, and high-voltage pulse) as well as their charging voltage and the initial puff valve fill pressure, ion beam current densities from 20 to 200 A/cm^2 can be obtained. The impedance history of the diode can be varied from rising to flat to falling, depending upon the balance between the extracted beam current and the injected plasma flux [36].

Experiments using hydrogen gas have shown that a pure proton beam is produced at current levels >100 A/cm^2 over an area of about 400 cm^2 at 75 to 150 keV. The total ion current and energy are about twice as high as when a flashover source is used. Experiments using a nitrogen-gas puff have also produced beams at the 100 A/cm^2 level. Another very important aspect of this source is that more than 700 diode pulses were obtained with no major diode component requiring replacement.

The Cornell group is also working on an *e*xploding *m*etal *f*ilm *a*ctive *a*node *p*lasma *s*ource (EMFAAPS) [58]. In the initial experiments a thin (50 nm) aluminum film was deposited on an insulator on the anode surface. Using a plasma switch arrangement, a portion of the accelerator current was forced through the foil. The current vaporized the film and the vapor experienced electrical breakdown, producing a plasma. The time-of-flight ion data showed a prompt turn-on. The source produced 200 A/cm^2 of protons and 500 A/cm^2 of heavier gas ions which were adsorbed by the aluminum. This source, although it is not pure, apparently has reduced the turn-on time normally seen with flashover sources. With proper surface preparation to reduce hydrocarbons and absorbed gases, the source has the potential to provide a single-species ion source.

3.3 BOLVAPS/LIBORS Lithium Source

The active lithium ion source being developed for PBFA II is called BOLVAPS/LIBORS [59]. BOLVAPS, which stands for *b*oil-*o*ff *l*ithium *vap*or *s*ource, is a way to produce a thin lithium vapor layer near the anode surface by rapid ohmic heating of a thin-film laminate containing lithium. The vapor is then ionized by LIBORS (*l*aser *i*onization *b*ased *o*n *r*esonant *s*aturation) [60]. LIBORS is a laser process that produces a singly ionized lithium plasma. A conceptual sketch of this source for an Applied-B diode is shown in Figure 18.7. The surface of the metal anode consists of a dielectric onto which is sputtered a thin heater film of a refractory metal. A lithium-bearing film is then sputtered on the heater. A lithium–silver alloy was selected over pure lithium because lithium reacts very readily with the atmosphere, while LiAg is passive for a few hours. Producing the vapor near the anode is a two-step process. First, current from a slow (several milliseconds) capacitor bank is passed through the heater to melt the LiAg and drive off surface impurities ($\simeq 1100$ K). This slow-heating pulse is followed by a fast pulse (1 μs) that heats the LiAg to 1400 to 1600 K to produced a thin ($\simeq 1$ mm), high-density (10^{17} cm^{-3}) lithium vapor. As the vapor is established, light from a laser, tuned to the resonant transition of lithium (670.8 nm), ionizes the vapor by the LIBORS process. LIBORS is a single-photon collisional process that relies on saturating

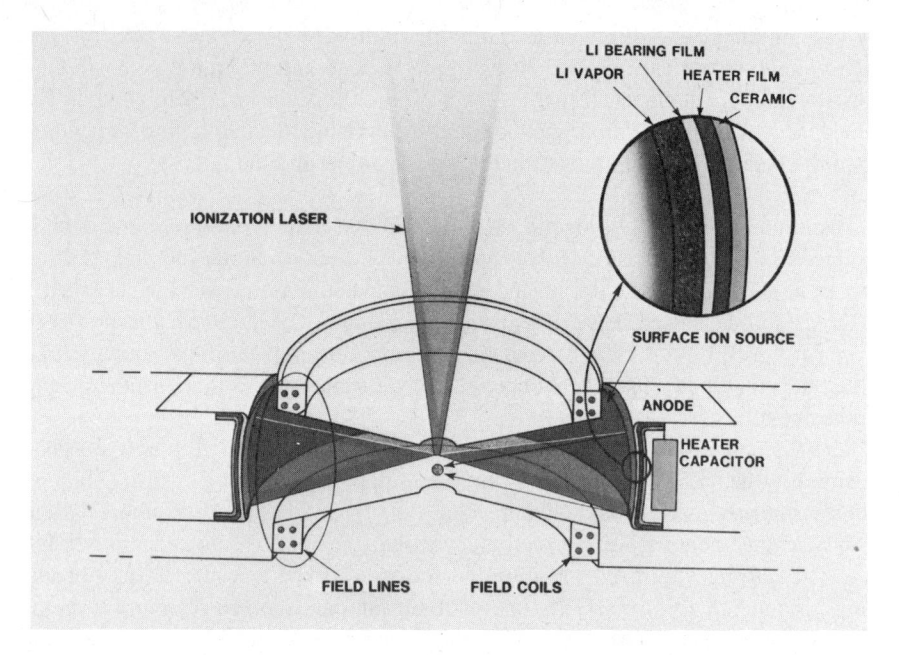

Figure 18.7 A conceptual diagram of the BOLVAPS/LIBORS ion source for an Applied-B diode. As described in the text, BOLVAPS provides a neutral lithium layer near the anode surface, which is then ionized by LIBORS, which is a laser process. This diagram used with permission from G. C. Tisone.

the resonant level. The ionization process has been studied extensively by Measures and coworkers [60–63]. In 1977, McIlrath and Lucatorto [64] achieved nearly 95% ionization in a column of lithium vapor with a 1-MW dye laser. Ionization is produced in a process by which a few electrons are first produced by photo-assisted Penning ionization or multiphoton ionization. These electrons then experience superelastic collisions with excited atoms, coverting laser energy into electron energy. Once the electron temperature reaches about 1 eV, rapid and complete ionization occurs. Because this is a single-photon process using a resonant line, only 10^4 or 10^5 W/cm^2 is required to ionize the vapor, compared to a few GW/cm^2 for multiphoton processes.

A laboratory experiment was performed to study the BOLVAPS/LIBORS source. A block diagram of the apparatus is shown in Figure 18.8. A tantalum heater foil 5 cm long and 2 cm wide was coated with a thin layer of LiAg alloy. Current from a slow (2.4 ms half-period) capacitor bank passed through the heater to raise the temperature to 1070 K, to melt the lithium alloy and drive off surface impurities. Late in the slow pulse (2 ms), a fast bank (8 μs half-period) heated the foil to 1300 K to produce the lithium vapor by evaporation.

A two-photon pumped fluorescence technique was developed [65] and used to measure the lithium neutral density as a function of time and as a function of distance above the substrate. As shown in Figure 18.8, a dye laser is tuned to

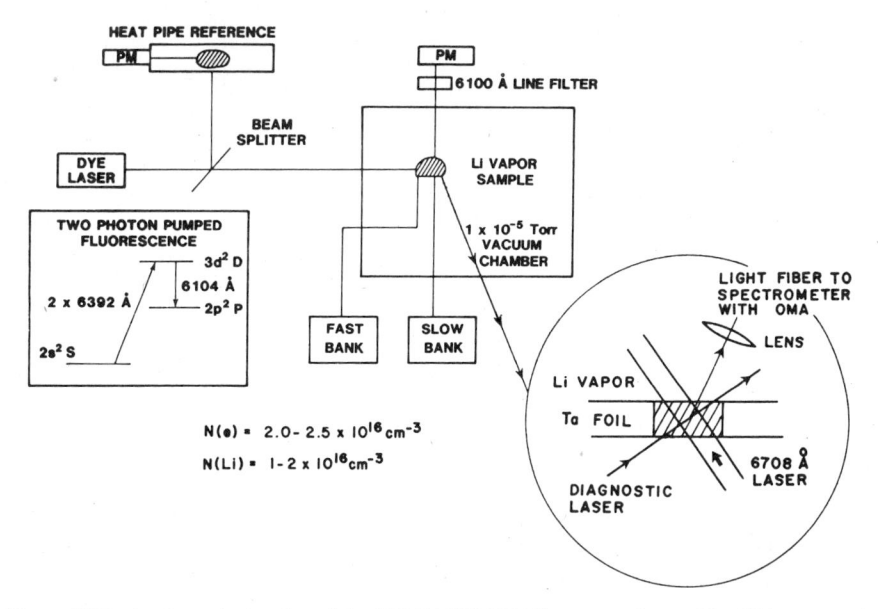

Figure 18.8 A schematic drawing of the BOLVAPS/LIBORS test experiment. The lithium vapor is produced by two-step ohmic heating of a LiAg alloy-coated tantalum foil. The detail shows the diagnostic laser and LIBORS laser beams crossing over the LiAg-coated tantalum foil. From *J. Appl. Phys.* **59**, 371 (1986). Reprinted by permission of P. L. Dreike.

twice the wavelength of the 2S to 3D transition and the fluorescence at 610.4 nm (3D to 2P) is observed. A lithium heat pipe is used as a reference. Figure 18.9 shows the lithium density as a function of distance above the substrate. A model described in Ref. 59 was fit to the data, and this comparison is also shown in the figure. Ionization of the BOLVAPS-produced neutral layer by the LIBORS process was also investigated using the experimental arrangement in Figure 18.8. The beam from a dye laser, tuned to the resonant transition at 670 nm, was focused with a cylindrical lens to a 0.1×1-cm beam and directed across the lithium vapor. The electron density was inferred by Stark broadening of the 460.3-nm atomic lithium line [66, 67]. For these measurements, the heater was driven to a higher current level, thereby producing more neutral density than shown in Figure 18.9. Neutral densities of 0.8 to 1.8×10^{16} cm^{-3} were measured, and the resulting electron density was 1.1 to 3.3×10^{16} cm^{-3}. Within experimental error, the BOLVAPS/LIBORS technique has thus shown that a nearly fully ionized plasma has been produced.

Higher densities (10^{17} cm^{-3}), in a thinner layer (~ 1 mm) than was obtained in this experiment, are required for PBFA II [33]. The results of the initial BOLVAPS/LIBORS experiment, a LIBORS kinetic model [68], and BOLVAPS evaporation model were used to specify the major source components. To maintain a thin neutral layer at high density, the evaporation pulse must be fast (1 μs) and heat the LiAg film to $\simeq 1600$ K. The laser system must produce in excess of 60

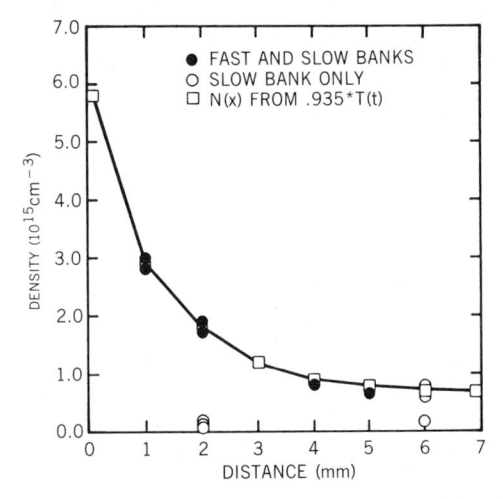

Figure 18.9 Lithium density versus distance 5 μs after the beginning of the "fast" heating pulse for experiments using both the fast and slow banks. The "slow bank only" points are for experiments without the fast bank. The fit to the measured densities is from a model described in Ref. 59. From *J. Appl. Phys.* **59,** 371 (1986). Reprinted by permission of P. L. Dreike.

MW at 670.8 nm to produce the necessary ionization. All the components necessary to implement BOLVAPS/LIBORS on PBFA II have been specified, and this source will be evaluated on PBFA II in 1988.

3.4 Other Research on Active Lithium Sources

Although BOLVAPS/LIBORS has been selected to be the first active lithium source to be evaluated on PBFA II, there are several potential sources for which the basic research has been completed. One class of such sources is laser produced plasmas. These sources rely on a single laser to both photoevaporate and ionize lithium from a lithium metal surface. The group at the Tokyo Institute of Technology [23, 69, 70] has used both a 20-ns, Q-switched ruby laser (λ = 693.4 nm) and a 1.8-μs, flashlamp-pumped dye laser tuned to the resonant transition of lithium (λ = 670.8 nm). They measured the electron density N as a function of the input laser intensity I. With the ruby laser they found that $N \propto I^{7.1}$, and with the dye laser, $N \propto I^{1.9}$. Plasma densities of 10^{16} cm^{-2} could be obtained at 40 MW/cm^2 with ruby, and the dye laser required intensities of only 1 MW/cm^2. This is a clear indication that the plasma in the latter case is in part produced by the LIBORS process, previously discussed. The Tokyo group estimates that densities of 10^{17} cm^{-3}, required for light-ion ICF sources, could be obtained with ≤400 mJ/cm^2 of resonant laser energy.

A series of experiments in which an ArF laser has been used to both photoevaporate and ionize lithium has recently been completed [71]. This technique should be efficient since an ArF photon (6.4 eV) has sufficient energy to ionize lithium by a single-photon process. In addition, the ionization cross section for this process is large. A proof-of-principle experiment was performed in which a low-energy laser (200 mJ, 20 ns) was used to irradiate lithium films. The electron density was inferred from Stark broadening of a lithium line. The results suggested that a lithium ion density of 10^{17} cm^{-3} could be achieved with less than 200 mJ/cm^2 of ArF radiation. These two sets of experiments demonstrate the clear advantage of using single-photon resonant processes over multiple-photon ionization for plasma production.

In addition to laser-produced sources, a potential PBFA-II source involving capillary discharges has been proposed and studied [72]. In this approach a fully ionized lithium plasma is produced in a pulsed capillary discharge. Following the pulse, the plasma recombines in an expansion section, producing a beam of fast-moving (cm/μs) neutral particles. This neutral beam would then pass through the applied magnetic field of the diode. The beam would then stagnate against the anode surface, producing a thin, fully ionized plasma. The plasma source for an Applied-B diode would consist of an array of sources of this type. Experiments on a single capillary-type source have shown that a thin (3 mm), high-density (10^{17} cm^{-3}) plasma can be produced by this method.

4 SUMMARY

A review of existing ion sources and those under development has been given. Although the generation of intense ion beams has been studied extensively since the early 1970s, the development of single-species ion sources is a relatively new research area. The major effort since 1983 has been directed toward developing a pure, high-current Li$^+$ source for the light-ion fusion acceleration, PBFA II.

The most common ion source used in the past for Applied-B diodes is the passive flashover source. The major disadvantages of this source are that multiple-ion species are produced and there is an inherent delay between the accelerating voltage pulse and the onset of the ion current. Results presented here show that modifications to the ion species distribution can be made by selecting the anode dielectric material and using surface preparation techniques, such as glow-discharge cleaning, vacuum bakeout, and in situ source preparation. XUV illumination of the surface has reduced turn-on delay of flashover sources. It is possible that by using a combination of techniques, such as in situ anode coating with ices (H$_2$, D$_2$, N$_2$, etc.), lithium compounds, or metals (Na), together with XUV illumination, the major disadvantages of flashover sources may be overcome.

A very promising passive lithium ion source is an EHD-driven liquid source in which ions are produced by high-field evaporation and/or ionization. Theory predicts that this source will satisfy all the requirements for PBFA II, if the applied electric fields are ≥ 10 MV/cm. Low-voltage experiments have demonstrated multiple-cusp formation over large surface areas. The measured wavelengths and growth times agree with theory. Diode experiments on a low energy accelerator are encouraging; however, a definite test of EHD-driven sources requires the high electric fields achievable on PBFA II. These experiments are now in preparation.

Active sources, in which an anode plasma is produced prior to the arrival of the accelerating pulse, have many advantages over passive sources because source parameters (ion species, density, and source thickness) can be varied to optimize the total system performance. The only active source for an Applied-B diode that has been characterized is the Gas-Breakdown Plasma Anode, developed by researchers at Cornell University. This source is capable of providing a pure ion beam from a variety of gases. To date, pure ion beams have been produced from both H_2 and N_2.

Several active lithium ion sources have been proposed for PBFA II. The results and status of the research on these sources have been described. A laser-based lithium plasma source, BOLVAP/LIBORS, has been selected to the be first active source to be evaluated on PBFA II.

Although this review has stressed the progress in lithium ion source development for PBFA II, many of the techniques of ion generation, such as EHD-driven liquid sources and BOLVAPS/LIBORS, can be extended to other metal-ion sources.

ACKNOWLEDGMENTS

The author thanks the following researchers at Sandia National Laboratories who critically reviewed this chapter and whose efforts provided many of the results presented here: K. W. Bieg, D. L. Cook, P. L. Dreike, J. Maenchen, P. F. McKay, J. A. Panitz, A. L. Pregenzer, K. R. Prestwich, J. K. Rice, M. A. Sweeney, G. C. Tisone, J. P. VanDevender, and J. R. Woodworth. The author also appreciates the information obtained from John Greenly and his colleagues at Cornell University, which is summarized in this chapter. Finally, the author expresses his gratitude to Janet Schow for typing and editing the manuscript.

REFERENCES

1. J. P. VanDevender and D. L. Cook, *Science* **232**, 837 (1986).
2. B. N. Turman et al., in *Proceedings of the 5th IEEE Pulsed Power Conference*, Arlington, VA, June 10–12, 1985, IEEE, New York, 1986, p. 155.
3. S. Humphries, Jr., *Nucl. Fusion* **20**, 1549 (1980).
4. F. Winterberg, *Physics of High Energy Density*, Academic, New York, 1971, p. 372.
5. R. N. Sudan and R. V. Lovelace, *Phys. Rev. Lett.* **31**, 1174 (1973).
6. P. Dreike, C. Eichenberger, S. Humphries, and R. Sudan, *J. Appl. Phys.* **47**, 85 (1976).
7. J. P. VanDevender, J. A. Swegle, D. J. Johnson, K. W. Bieg, E. J. T. Burns, J. W. Poukey, P. A. Miller, J. N. Olsen, and G. Yonas, *Laser and Particle Beams* **3**, 93 (1985).
8. J. P. VanDevender, in *Proceedings of the 5th International Conference on High-Power Particle Beams (Beams '83)*, San Francisco, CA, Sep. 12–14, 1983, R. J. Briggs and A. J. Toepfer (Eds.), p. 17.

References

9. J. Maenchen, L. Wiley, S. Humphries, Jr., E. Peleg, R. N. Sudan, and D. A. Hammer, *Phys. Fluids* **22** (3), 555 (1979).
10. D. J. Johnson, E. J. T. Burns, J. P. Quintenz, K. W. Bieg, A. V. Farnsworth, Jr., L. P. Mix, and M. A. Palmer, *J. Appl. Phys.* **52**, 168 (1981).
11. M. A. Sweeney, J. E. Brandenburg, R. A. Gerber, D. J. Johnson, J. M. Hoffman, P. A. Miller, J. P. Quintenz, S. A. Slutz, and K. W. Bieg, in *Proceedings of the 5th International Conference on High-Power Particle Beams (Beams '83)*, San Francisco, CA, Sep. 12–14, 1983, R. J. Briggs and A. J. Toepfer (Eds.), p. 207.
12. R. Pal and D. Hammer, *Phys. Rev. Lett.* **50** (10), 732 (1983).
13. D. J. Johnson, J. P. Quintenz, and M. A. Sweeney, *J. Appl. Phys.* **57**, 794 (1985).
14. K. W. Bieg, E. J. T. Burns, J. N. Olsen, and L. R. Dorrell, *J. Vac. Sci. Tech. A* **3**, 1234 (1985).
15. M. A. Greenspan, R. Pal, D. A. Hammer, and S. Humphries, Jr., *Appl. Phys. Lett.* **37**, 248 (1980).
16. P. L. Dreike, E. J. T. Burns, S. A. Slutz, J. T. Crow, D. J. Johnson, P. R. Johnson, R. J. Leeper, P. A. Miller, L. P. Mix, D. B. Seidel, and D. F. Wenger, *J. Appl. Phys.* **60**, 878 (1986).
17. K. R. Prestwich, *IEEE Trans. Nucl. Sci.* **NS-18**, 493 (1971).
18. K. W. Bieg, E. J. T. Burns, R. A. Gerber, J. N. Olsen, and K. P. Lamppa, *J. Vac. Sci. Tech. A* **4**, 772 (1986).
19. G. S. Mills, E. J. T. Burns, and D. J. Johnson, 1984 IEEE International Conference on Plasma Science, IEEE Publication No. 84CH1958-8, New York, 1984, p. 111.
20. J. M. Neri, D. A. Hammer, G. Ginet, and R. N. Sudan, *Appl. Phys. Lett.* **37**, 101 (1980).
21. K. Kasuya, K. Horioka, T. Takahashi, A. Urai, and M. Hijikawa, *Appl. Phys. Lett.* **39** (11), 887 (1981).
22. T. Takahashi, K. Horioka, M. Hijikawa, A. Urai, and K. Kasuya, *J. Appl. Phys.* **54** (8), 4269 (1983).
23. K. Kasuya, K. Horioka, H. Yoneda, and the PICA Group, in *Proceedings of the 6th IEEE Pulsed Power Conference, Arlington, VA, June 29–July 1, 1987*, p. 619 (IEEE Cat. No. 87CH2522-1).
24. K. Kasuya, K. Horioka, T. Takahashi, H. Yoneda, and H. Kuwabara, *IEEE Trans. Plasma Sci.* **PS-13**, 327 (1985).
25. H. Yoneda, K. Horioka, K. Ohbayashi, and K. Kasuya, *Appl. Phys. Lett.* **48**, 1196 (1986).
26. H. Yoneda, T. Urata, K. Ohbayashi, Y. Kim, K. Horioka, and K. Kasuya, 1987 IEEE International Conference on Plasma Science, IEEE Publication No. 87CH2451-3, New York, 1987, p. 100.
27. A. L. Pregenzer, *J. Appl. Phys.* **58**, 4509 (1985).
28. R. Gomer, *J. Appl. Phys.* **19**, 365 (1979).
29. A. L. Pregenzer and B. M. Marder, *J. Appl. Phys.* **60**, 3821 (1986).
30. A. L. Pregenzer and D. R. Kingham, 1987 IEEE International Conference on Plasma Science, IEEE Publication No. 87CH2451-3, New York, 1987, p. 81.
31. V. G. Dudnikov and A. L. Shabalin, *Sov. Tech. Phys. Lett.* **11** (7), 335 (1985).
32. A. L. Pregenzer and J. R. Woodworth, *J. Appl. Phys.*, to be published March 1, 1989.
33. R. A. Gerber, K. W. Bieg, E. J. T. Burns, P. L. Dreike, J. Maenchen, T. A. Mehlhorn, J. N. Olsen, A. L. Pregenzer, J. K. Rice, M. A. Sweeney, G. C. Tisone, and J. R. Woodworth, *IEEE Trans. Nucl. Sci.* **NS-32**, 1718 (1985).
34. P. F. McKay, R. A. Gerber, and A. L. Pregenzer, *IEEE Trans. Plasma Sci.* **PS-15**, 339 (1987).
35. P. F. McKay, 1987 IEEE International Conference on Plasma Science, IEEE Publication No. 87CH2451-3, New York, 1987, p. 6. Also P. F. McKay (private communications).
36. J. B. Greenly, M. Ueda, G. D. Rondeau, and D. A. Hammer, *J. Appl. Phys.* **63**, 1872 (1988).
37. C. W. Mendel, Jr. and G. S. Mills, *J. Appl. Phys.* **53**, 7265 (1982).

38. S. Humphries, Jr., C. W. Mendel, Jr., G. W. Kuswa, and S. A. Goldstein, *Rev. Sci. Instrum.* **50** (8), 993 (1979).
39. G. W. Mendel, Jr., D. M. Zagar, G. S. Mills, S. Humphries, Jr., and S. A. Goldstein, *Rev. Sci. Instrum.* **51**, 1641 (1980).
40. J. Maenchen et al., in *Proceedings of the 6th International Conference on High-Power Particle Beams (Beams '86)*, Kobe, Japan, June 9–12, 1986, p. 85.
41. B. V. Weber et al., in *Proceedings of the 6th International Conference on High-Power Particle Beams (Beams '86)*, Kobe, Japan, June 9–12, 1986, p. 882.
42. G. W. McClure and J. A. Webb, in *Proceedings of the 34th Annual Gaseous Electronics Conference*, 1981, p. 71.
43. V. M. Bistritsky, A. N. Didenko, A. V. Petrov, in *Proceedings of the 6th International Conference on High-Power Particle Beams (Beams '86)*, Kobe, Japan, June 9–12, 1986, p. 176.
44. N. Camarcat, A. Devin, C. Peugnet, and C. Patou, *Appl. Phys. Lett.* **45**, 513 (1984).
45. W. Bauer, A. Brandelik, A. Citron, H. Ehrier, E. Haltes, G. Melchior, K. Mittag, A. Rogner, and C. Schultheiss, in *Proceedings of the 6th International Conference on High-Power Particle Beams (Beams '86)*, Kobe, Japan, June 9–12, 1986, p. 192.
46. R. E. Beverly III, *Progress in Optics*, Vol. 16, North-Holland, New York, 1978, p. 350.
47. S. S. Glaros, S. L. Wong, and G. L. James, *Bull. Am. Phys. Soc.* **27**, 981 (1982).
48. J. R. Woodworth and P. F. McKay, *J. Appl. Phys.* **58** (9), 3364 (1985).
49. J. Maenchen et al., in *Proceedings of the 5th IEEE Pulsed Power Conference*, Arlington, VA, June 10–12, 1985, p. 102.
50. J. R. Woodworth, J. E. Maenchen, and P. F. McKay, in *Proceedings of the 5th IEEE Pulsed Power Conference*, Arlington, VA, June 10–12, 1985, p. 563.
51. J. R. Woodworth, E. J. T. Burns, D. L. Hanson, W. H. Jaramillo, and D. A. Pattison, *IEEE Trans. Plasma Sci.* **PS-15**, 384 (1987).
52. T. A. Green, G. M. Loubriel, D. M. Richards, N. H. Tolk, and R. F. Haglund, Jr., *Phys. Rev. B* **35**, 781 (1987).
53. J. R. Woodworth and W. H. Jaramillo, 1987 IEEE International Conference on Plasma Science, IEEE Publication No. 87CH2451-3, New York, 1987, p. 81.
54. A. Kitamura, K. Mitsuhashi, and S. Yano, in *Proceedings of the 6th International Conference on High-Power Particle Beams (Beams '86)*, Kobe, Japan, June 9–12, 1986, p. 204.
55. M. Ueda, J. B. Greenly, and G. D. Rondeau, *Bull. Am. Phys. Soc.* **28**, 1148 (1983).
56. M. Ueda, Ph.D. Thesis, Cornell University, 1986.
57. S. Humphries, Jr., R. J. M. Anderson, J. R. Freeman, and J. Greenly, *Rev. Sci. Instrum.*, **52** (2), 162 (1982).
58. G. D. Rondeau, C. Peugnet, J. B. Greenly, D. A. Hammer, B. R. Kusse, E. Pampellone, and R. N. Sudan, in *Proceedings of the 6th International Conference on High-Power Particle Beams (Beams '86)*, Kobe, Japan, June 9–12, 1986, p. 180.
59. P. L. Dreike and G. C. Tisone, *J. Appl. Phys.* **59**, 371 (1986).
60. R. M. Measures, N. Drewell, and P. Cardinal, *Appl. Opt.* **18**, 1824 (1979).
61. R. M. Measures, N. Drewell, and P. Cardinal, *J. Appl. Phys.* **50**, 2662 (1979).
62. R. M. Measures and P. G. Cardinal, *Phys. Rev. A* **23**, 804 (1981).
63. R. M. Measures, P. G. Cardinal, and G. W. Schinn, *J. Appl. Phys.* **52**, 1269 (1981).
64. T. J. McIlrath and T. B. Lucatorto, *Phys. Rev. Lett.* **38**, 1390 (1977).
65. G. C. Tisone, P. J. Hargis, and P. L. Dreike, Conference on Lasers and Electrooptics, Baltimore, MD, May 21–24, 1985, p. 202.
66. H. R. Griem, *Spectral Line Broadening by Plasmas*, Academic, New York, 1974, p. 170.
67. N. Konjevic and J. R. Roberts, *J. Chem. Ref. Data* **5**, 206 (1976).

68. J. K. Rice, R. A. Gerber, and G. C. Tisone, in *American Institute of Physics Conference Proceedings*, No. 146, AIP, New York, 1986, p. 654.
69. K. Horioka, H. Tamura, K. Ishitoya, K. Hashidate, J. Miyagowa, and K. Kasuya, in *Proceedings of the 6th International Conference on High-Power Particle Beams (Beams '86)*, Kobe, Japan, June 9–12, 1986, p. 235.
70. K. Horioka, Y. Hashidate, A. Kurosawa, H. Yoneda, and K. Kasuya, 1987 IEEE International Conference on Plasma Science, IEEE Publication No. 87CH2451-3, New York, 1987, p. 80.
71. S. Lin, K. Y. Tang, and R. O. Hunter, Jr., *Bull. Am. Phys. Soc.* **31** (9), 1570 (1986).
72. S. A. Goldstein and B. Hilko, 1986 IEEE International Conference on Plasma Science, IEEE Publication No. 86CH2317-6, New York, 1986, p. 31.

19

Ion Sources for Pulsed High-Brightness Beams

S. Humphries, Jr., C. Burkhart, and L. K. Len
Institute for Accelerator and Plasma Beam Technology
University of New Mexico
Albuquerque, New Mexico

High-brightness ion beams carry a large power flux and can be focused to a small area. Generation of such beams demands ion sources that can simultaneously meet the often conflicting requirements of high current density and low particle orbit divergence. High-flux sources, such as pulsed plasma guns or metal vapor vacuum arcs, usually produce plasmas with strong spatial and temporal flux variations. These variations distort the optics of conventional ion extraction gaps, degrading the beam emittance.

Methods to interface high-flux plasmas to extraction gaps are discussed in this chapter. The approach involves the use of electrostatically biased metal grids to control the plasma surface at the entrance to the extraction gap. In contrast to most ion extractors that rely on the formation of a plasma meniscus [1], the grid-controlled extractor decouples the beam optics from the properties of the plasma source. Decoupling allows the generation of high-current, low-divergence beams over large areas. Grid control also facilitates the generation of pulsed beams with fast current risetimes.

Definitions of emittance and brightness are reviewed in Section 1 along with required parameters of sources for injection into ion induction linacs [2] for Accelerator Inertial Fusion [3]. Section 2 introduces the physical principles of grid-controlled ion extractors, and Section 3 discusses the principle limitations to their use. Section 4 summarizes work performed in our laboratory on vacuum arc plasma sources. Experimental results on grid control and figures on available plasma flux are described in Sections 5 and 6. Measurements of extracted beam brightness and confirmation of single species, single charge state extraction for a variety of materials are presented in Section 7.

1 EMITTANCE AND BRIGHTNESS

The ion source sets a fundamental limit on the parallelism of particle orbits in the output beam of an ion accelerator. The quantities *emittance* and *brightness* quantify the concept of beam parallelism [4, 5]. Emittance can be defined in terms of a phase space density plot such as that of Figure 19.1a. Transverse ion orbits in the x direction at a particular axial location are represented by points on the plot. The distribution of points is usually indicated by isodensity contours. If motions in the x, y, and z directions are decoupled and all focusing forces are smooth, the phase space area of the beam distribution is preserved during acceleration.

Figure 19.1b illustrates how the phase space area determines the focal properties of the beam. Assume that a beam with phase space distribution boundary dimensions Δx and Δp_x is expanded and then focused by a lens of focal length f and diameter d. At the focal point, the distribution has phase space dimensions δ

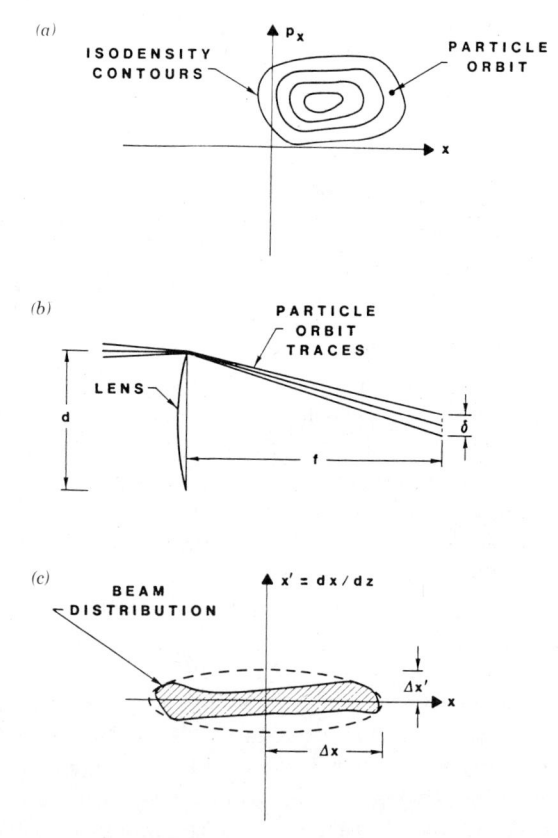

Figure 19.1 Transverse beam distributions. (*a*) Phase space plot of orbit parameters at a given time and axial location. (*b*) Focusing of a beam viewed in configuration space. (*c*) Trace space plot with emittance definitions.

and $(d/f)p_z$. Conservation of phase space area implies that

$$\delta = (\Delta x \Delta p_x / p_z)(2f/d) \tag{1}$$

where δ is the half-width of the focal spot. Equation (1) shows that smaller phase area implies a better focus.

In accelerator experiments, it is common to measure the angles of particle orbits, $x' = dx/dz = p_x/p_z$. Therefore, transverse beam distributions are usually represented in terms x and x' (trace space), as in Figure 19.1c. Such a representation is called an emittance plot. As shown, the distribution boundary is often defined in terms of a circumscribed ellipse. This reflects the fact that the peripheral orbits that can propagate through an extended linear focusing system with a prescribed transverse boundary trace out an ellipse in phase space. All particle distributions that fit within the boundary ellipse can be transported. The trace space boundary of an irregular distribution is defined in terms of the minimum area ellipse that circumscribes the distribution (see Figure 19.1c). Emittance in the x direction is defined as

$$\epsilon_x = \Delta x \Delta x' \tag{2}$$

Equation (1) can be rewritten in terms of emittance as

$$\delta = 2\epsilon_x (f/\text{number}). \tag{3}$$

where the lens f/number equals f/d. In some instances, the emittance is defined as the total area of the distribution ellipse, $\pi \Delta x \Delta x'$. To avoid confusion, the following notation is recommended: display of a factor of π in the units is a flag to indicate that the emittance value refers to trace space area divided by π. In other words, $\epsilon = 5\pi$-m-mrad implies that $\Delta x \Delta x' = 5 \times 10^{-3}$ m-rad.

Clearly, emittance is not conserved during acceleration. Conservation of phase space area implies that the normalized emittance,

$$\epsilon_{nx} = \gamma(v_z/c)\epsilon_x = \beta\gamma\epsilon_x \tag{4}$$

(proportional to $\Delta x \Delta p_x$) is conserved. The quantity γ is the standard relativistic beam energy parameter; for most ion beams, γ is close to unity.

Charged particle beam applications such as Accelerator Inertial Fusion call for a high current density focused to a small target (Figure 19.1b). In terms of the focal spot half-width and total beam current, the current density at the target is $j_i = I/\pi\delta^2$. If the beam emittances in the x and y directions are equal, the focused current density can be rewritten

$$j_i = I/(\pi\epsilon)^2 (f/\text{number})^2. \tag{5}$$

Similarly, the peak flux available from a beam can be characterized by the quantity brightness, B, defined as

$$B = I/(\pi\epsilon)^2 \quad (\text{A}/\text{m}^2 - \text{steradian}) \tag{6}$$

A related quantity, the normalized brightness, is conserved during particle acceleration,

$$B_n = I/(\pi\epsilon_n)^2 = B/(\beta\gamma)^2 \tag{7}$$

Normalized brightness is a useful quantity to compare charged particle sources for applications such as inertial fusion and free electron lasers since it signifies an inherent beam quality independent of the final beam energy. Emittance and current measurements of the output beam can be projected backward to derive the normalized source brightness.

The induction linear ion accelerator approach to Accelerator Inertial Fusion makes severe demands on plasma sources in terms of brightness, reproducibility, species purity, and flux rise time. For near-term experiments with multiple beams [6–9] relatively high current densities (20–40 mA/cm^2) of intermediate mass ions ($A = 20$–30) are necessary. In far-term devices, heavy ions in the mass range $A = 100$–200 will be used. The area per beam is in the range 10–20 cm^2; proposed configurations have 16 independent beams. The plasma source must reach an operating equilibrium in a few microseconds and supply plasma for a pulselength of 5–10 μs. The rise time of extracted beam current should be 0.1 μs or less. Ideally, there should be no plasma or neutral prefill of high-gradient accelerating structures prior to the extraction pulse. During the pulse, the source must maintain a high degree of spatial and temporal uniformity for good beam optics. Ideally, the extracted beam should be in a single ionization state since it is unlikely that charge-state separation can be performed on large-diameter intense ion beams without significant emittance growth. To provide the power density necessary to reach fusion ignition, the normalized emittance for a 300-mA beam should be less than $\epsilon_n = 4 \times 10^{-7}\pi$-m-rad to satisfy focusing demands. This figure corresponds to a normalized brightness exceeding 10^{11} A/(m-rad)^2. The brightness figure implies that the transverse energy of ions extracted from a source at 20 mA/cm^2 should be less than a few electronvolts.

2 PRINCIPLES OF GRID-CONTROLLED ION EXTRACTORS

A conventional ion extractor [10, 11] is illustrated in Figure 19.2a. A quiescent source (such as a thermionic cathode emitter in a gas background) generates a plasma. The available ion flux from the plasma is given by the Bohm expression [12]. The expanding plasma penetrates through one or more apertures to a high-voltage extraction gap. In the gap, the applied voltage separates ions from electrons

2 Principles of Grid-Controlled Ion Extractors

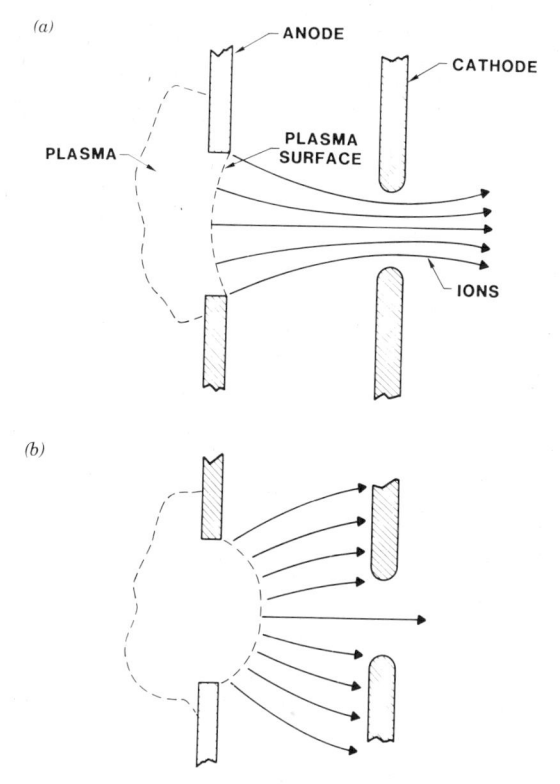

Figure 19.2 Conventional ion extractor. (a) Plasma ion flux matched to space-charge-limited flow in acceleration gap. (b) Excess plasma flux.

and accelerates them to high energy. The condition that ion emission takes place from a surface of zero electric field implies that the flux of ions across the acceleration gap is space-charge limited. In consequence, the shape of the plasma emission surface is determined by a balance between the particle fluxes given by the Bohm and Child [13] expressions. At an optimum match, the plasma surface defines a concave shape inside the extraction aperture [1]. The resulting focusing compensates space-charge expansion, leading to a parallel beam at the cathode aperture, as shown in Figure 19.2a.

If there is a mismatch between the properties of the plasma source and extraction gap, the optics of the output beam is affected. Figure 19.2b illustrates a situation with increased plasma flux compared to Figure 19.2a. To achieve flux balance, the plasma bulges into the extraction gap, increasing the emission area and decreasing the gap width. The result is a defocused beam. Conversely, reduced plasma flux causes the surface to recede into the extraction aperture, leading to overfocusing. Beam properties can be adversely affected by temporal variations of the plasma flux, spatial variations of the flux between apertures, and changes in the extraction voltage.

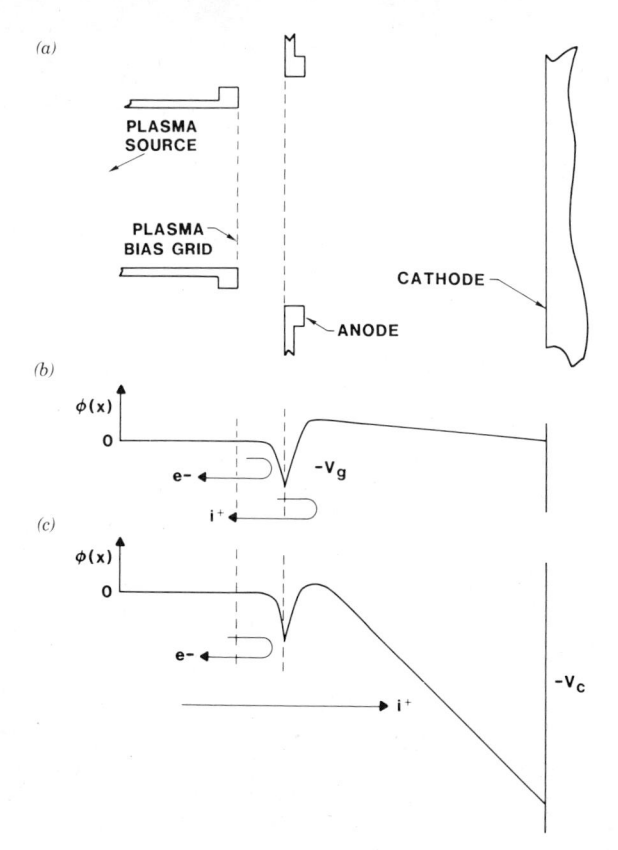

Figure 19.3 Principle of the grid-controlled plasma anode. (*a*) Grid geometry. (*b*) Particle flow and variation of electrostatic potential with zero extraction voltage. (*c*) Particle flow and variation of electrostatic potential with an applied extraction voltage.

The grid-controlled ion extractor, illustrated in Figure 19.3a, solves the matching problem by utilizing electrostatic confinement of the source plasma [14, 15]. The anode is a fine wire mesh maintained at a negative potential with respect to the plasma source. The potential difference is high compared to the kinetic energy of most of the plasma electrons. The electrostatic potential of the plasma is constrained by contact with the walls of the plasma expansion chamber and a plasma grid. In subsequent discussions, the two grids of Figure 19.3a are referred to as the anode grid and the plasma grid. The effect of the biased grids on particle flow from the plasma is illustrated in Figures 19.3b and c. Plasma electrons are prevented from reaching the anode mesh; ions are separated from the electrons in a thin sheath near the anode grid. Some ions are collected by the anode. Most of the ions pass through the grid to the extraction gap.

In the absence of an extraction voltage (Figure 19.3b), the space charge of the ions leads to the formation of a virtual anode. In this case, the electrostatic potential in the extraction gap rises to a value equal to the directed ion energy. The length scale for the creation of a virtual anode is comparable to the extraction sheath width. Ion orbits are reversed in direction; the ions either return to the plasma or reflex about the anode until they are collected. When an extraction voltage is applied (Figure 19.3c), the potential barrier in the extraction gap is lowered, allowing some ions to cross to the cathode. The ion current density is given by the Child limit corrected for the initial directed ion energy.

The main role of the biased grids is separation of positive particles from negative particles so that only ions enter the extraction gap. The anode in a grid-controlled extractor acts much like a thermionic ion source [16]. Although ions are continually injected into the extraction gap, the vacuum ion flow is negligible in the absence of an extraction voltage. When a voltage is applied, ion flux rises almost instantaneously to the space-charge-controlled level. Grid-controlled ion extraction has the following relative advantages:

1. Extracted beam flux is not affected by spatial or temporal source variations. This characteristic is particularly important for a multibeam accelerator with multiple plasma sources.
2. There is no plasma prefill of the extraction gap prior to and between pulsed extraction of ions beams.
3. Current can be initiated or extinguished on rapid time scales.

3 LIMITING FACTORS

There are a number of factors that constrain the voltage between the anode and plasma grids and may limit the current density and pulselength of grid-controlled ion extractors. The basic requirement for successful operation is that the bias voltage prevents the expanding plasma from entering the extraction gap by repelling the plasma electrons. Clearly, the magnitude of the voltage must exceed the directed kinetic energy (expressed in eV) of the bulk of the plasma electrons. Furthermore, the dimension of openings in the mesh must be smaller than the scale length of the plasma ion extraction sheath [17]. If the latter condition is not met, plasma can protrude through the mesh openings allowing free plasma flow over the reduced effective aperture size.

To prevent plasma flow into the extraction gap, the grid voltage must be considerably higher than the quantity (kT_e/e), where T_e is the plasma electron temperature. This condition follows from the fact that the thermal flux of electrons in a plasma is much higher than the flux of ions. If the ions have a directed energy E_i, then the ratio of electron to ion flux in a plasma of infinite extent is approximately [17]

$$j_e/j_i \cong (kT_e/E_i)^{1/2} (m_i/m_e)^{1/2} \qquad (8)$$

The flux of plasma into the extraction gap is governed by the directed motion of the ions. As long as one electron can overcome the electric field of the biased anode grid for every ion that passes, the plasma flux is not significantly impeded. The implication is that the electron flux must be reduced by a factor that exceeds the ratio of Eq. (8). From electrostatic probe theory [17], the fraction of electrons in a Maxwellian distribution that can pass through a grid biased to a voltage $-V_g$ with respect to the plasma is

$$f_e \cong \exp\left(-eV_g/kT_e\right) \quad (9)$$

Combining Eqs. (8) and (9) leads to the following constraint on the anode grid bias voltage for electrostatic confinement of the expanding plasma:

$$eV_g < (-kT_e) \ln\left[(kT_e m_i/E_i m_e)^{1/2}\right] \quad (10)$$

As an example, assume that $E_i = 10$ eV and $kT_e = 3$ eV. These parameters typify carbon plasmas used for 40-mA/cm² extraction experiments. Equation (10) predicts that the minimum grid voltage for plasma flow reduction is $V_g = -17$ V. Figure 19.4 shows a semilog plot of the measured ion flux downstream from a biased emission grid as a function of the bias voltage. The graph shows an initial rise in the flux with applied bias voltage resulting from acceleration of the ions. The ion flux dropped at the voltage level predicted by Eq. (10). The decrease in ion flux at high grid voltage implies an electron temperature of 14 eV rather than the 3-eV figure characteristic of the bulk of the electron distribution. This comes

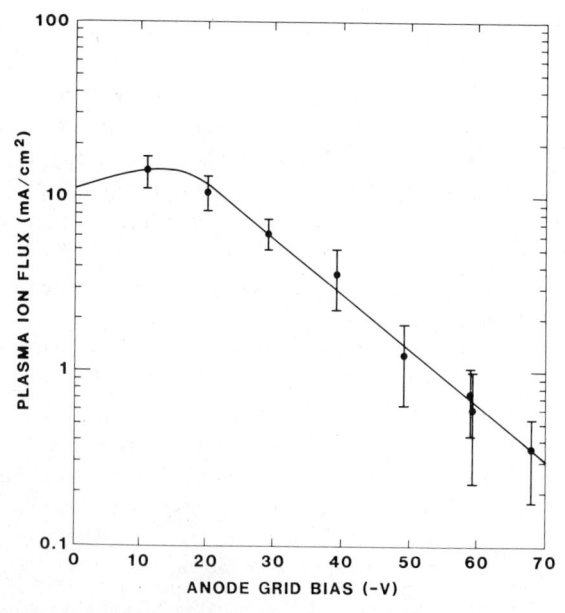

Figure 19.4 Average directed plasma ion flux downstream from anode grid as a function of bias voltage V_g. No extraction voltage. Carbon plasma, extractor 15 cm from arc source.

about because the residual ion flux measurement is sensitive to the high-energy tail of the electron distribution. Figure 19.4 indicates that the electron distribution had a non-Maxwellian tail. In general, we found anode grid voltages in the range $\leqslant -100$ V were necessary to reduce the plasma flow to a negligible level.

The sheath width for extraction of ions from a plasma is given by electrostatic probe theory as

$$\lambda = \left(\frac{4\epsilon_0}{9}\right)^{1/2} \left(\frac{2q}{m_i}\right)^{1/4} \frac{|V_g|^{3/4}}{j_i^{1/2}} \tag{11}$$

where V_g is the anode grid bias and j_i is the available ion current density from the plasma. If the sheath width does not exceed the halfwidth of the anode mesh opening, plasma can leak through the mesh, as shown in Figure 19.5a. The anode grid must be composed of a fine mesh to control high-density plasmas at moderate bias voltages. In our experiments, we use a stainless steel mesh with an opening halfwidth of $\delta/2 = 0.011$ cm. In comparison, Eq. (11) predicts a sheath width of 0.031 cm for an aluminum plasma with $j_i = 20$ mA/cm^2 and $V_g = -150$ V. For a high-brightness beam, the sheath width should exceed the width of the grid openings by a substantial margin. If this is not the case, the extraction optics is affected. Emission surface irregularities lead to beam divergence on the length

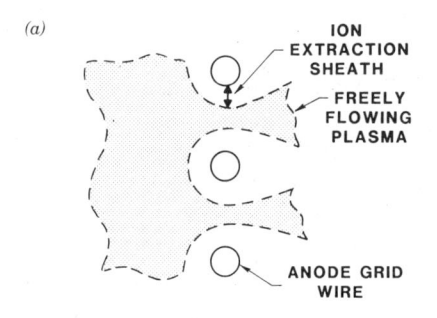

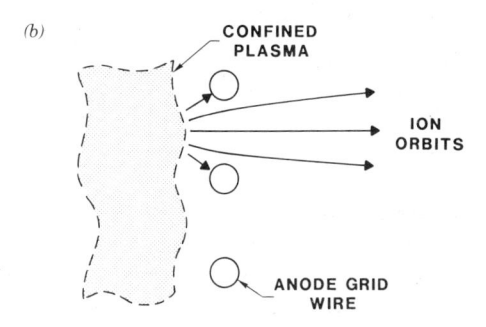

Figure 19.5 Plasma effects at the anode grid. (*a*) Plasma leakage. (*b*) Emission surface irregularity.

scale of a mesh cell, as shown in Figure 19.5b. The dependence of microscopic beam divergence on current density and grid bias is presently unknown. Complex two-dimensional calculations or numerical simulations must be carried out to determine the self-consistent emission surface shape.

The bias voltage and grid geometry also set limits on beam emittance through facet lens focusing [18]. This phenomenon arises from the distorted electric fields at the apertures of a grid. The relative effect can easily be estimated. Assuming zero electric field on the downstream side of the anode (space-charge-limited emission), the facet lens focal length is

$$f = -4\lambda \tag{12}$$

where λ is given by Eq. (11). The deflection angle for ions passing near grid wires spaced a distance δ apart (see Figure 19.5) is $\delta d/8\lambda$. Therefore, the maximum transverse energy gained by ions passing through the grid is

$$\Delta E_\perp \leqslant e|V_g|\left(\delta/8\lambda\right)^2 \tag{13}$$

Again, a large sheath width (compared to the mesh opening) is advantageous. As an example of facet lens effects, consider aluminum ions with $j_i = 20$ mA/cm² and an anode grid bias of $V_g = -150$ V. Equation (12) implies a transverse energy of $\Delta E_\perp = 1.3$ eV, within the range required for Accelerator Inertial Fusion.

Effects of facet lens defocusing and emission surface distortion are reduced when V_g is high. The upper limit on bias voltage is set by discharges between the anode and plasma grids. The breakdown level depends on the plasma density and pulselength. We have found that grid breakdown follows the net ion energy deposited in the anode grid wires. Although we have successfully used bias voltages as high as 300 V for short pulse plasmas, the intergrid voltage is limited to about 100 V for plasma pulses in the range $\geqslant 100$ μs. The frequency of breakdowns increases at high plasma density; the upper density limit for a carbon plasma is about 7×10^{11} cm^{-3}. Breakdown problems are exacerbated when the plasma is supplied by a vacuum arc source with a large gap. A broad arc gap leads to flux bursts and elevated plasma electron temperature. We achieved good operation with an arc gap of 0.8 mm. Although the vacuum arc fulfills our present requirements for Accelerator Inertial Fusion, other plasma sources may have advantages for grid-controlled extractors. The ideal plasma source should be relatively stable, have a high directed ion energy, and have a low-temperature Maxwellian electron distribution.

4 ARC PLASMA SOURCES FOR PULSED ION BEAMS

We used metal vapor vacuum arcs in most of our grid extraction experiments [19, 20]. They have the advantages of ease of operation, high available flux, reproducibility, and access to a wide variety of ion species [21–30]. With regard to Accel-

4 Arc Plasma Sources for Pulsed Ion Beams

erator Inertial Fusion experiments [31], vacuum arcs can produce ions of interest for both near-term experiments (Al^+, C^+) and reactor drivers (U^+, Bi^+). A review of vacuum arc physics and technology is given in Chapter 16.

A source geometry suitable for low duty cycle operation is shown in Figure 19.6. Plasma is produced by a high-density arc between a beveled stainless steel anode and a cathode. The ion species can be changed simply by replacing the cathode insert. The cathode is connected through a ballast resistor to a pulse-forming network (typical pulselength, $\Delta t_p \sim 100~\mu s$) with a standing voltage. The pulse-forming network is charged to a voltage $\geqslant 1$ kV; arc current is typically 100 to 500 A. The source can be command triggered with low jitter by overvolting the gap with a series transformer to initiate explosive plasma emission from the cathode. This method requires high voltage (~ 50 kV) but is compatible with an ultraclean system. An alternative method, developed by Brown [28], involves striking a surface discharge on an insulator near the vacuum gap to generate a starting plasma. This method gives reliable ignition of the discharge with moderate trigger voltage ($\leqslant 20$ kV).

We have used metal vapor vacuum arcs to generate a wide variety of ion species including p^+ and d^+ using hydrated titanium cathodes. We found that the available current density was proportional to the arc current and to the width of the vacuum

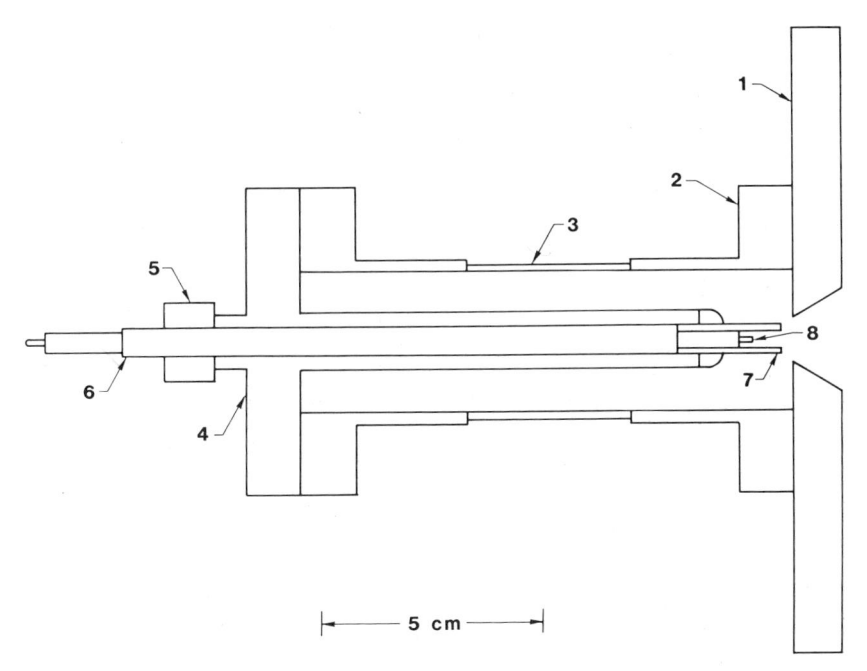

Figure 19.6 Geometry of arc plasma source. 1, Stainless steel arc anode and mounting flange; 2, stainless steel flange; 3, vacuum fitting with shielded glass insulator; 4, cathode support; 5, vacuum feedthrough with O-ring seal; 6, Teflon-insulated rigid coaxial tubing; 7, replaceable cathode; 8, insulator spark source.

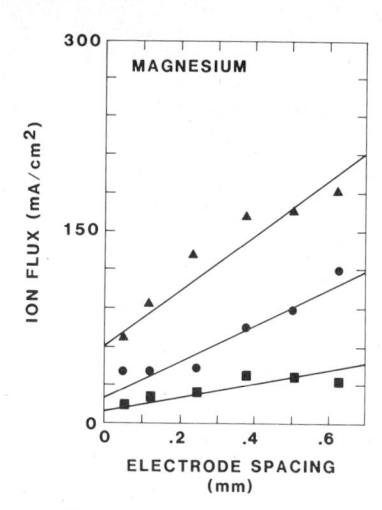

Figure 19.7 Ion flux 10 cm from a metal vapor vacuum arc as a function of arc current and gap spacing for a magnesium cathode. Arc current: 300 A (squares), 600 A (circles), 900 A (triangles).

gap between the electrodes. Figure 19.7 shows net flux measurements 10 cm from an arc with a magnesium cathode. Even at the relatively large distance, the available flux was high (>0.1 A/cm^2), exceeding the capability of most ion extractors. We also operated an array of six independent vacuum arcs as part of a project on advanced high-flux extractors [32]. The array supplied a flux of C$^+$ ions with available ion current density approaching 1 A/cm^2 at a 10-cm distance.

Time-of-flight measurements to measure the species content of the expanding plasmas were performed with a miniature pulsed accelerator. In low duty cycle operation, we observed that contaminants were initially present, but after a clearing period of 10–20 μs the plasma consisted entirely of material from the cathode. In experiments on grid-controlled ion extraction, the extractor voltage was delayed 40 μs from arc initiation to allow electrode cleaning. Within the measurement limits of the species detector, the extracted beams consisted of a single-ion species. The exception occurred when single arcs were operated at very high current (~ 1 kA). In this case, melted anode spots were formed and Fe$^+$ ions were detected. For most metal cathodes, a significant fraction (10–50%) of doubly charged ions were detected in an unconstrained, expanding plasma.

An oscillograph of arc current and plasma flux for an aluminum plasma source cathode is shown in Figure 19.8. The flux was recorded at a point 18.5 cm from the arc. The 20-μs delay time for arrival of plasma at the probe implies a directed ion energy of about 10 eV. The flux trace shows typical behavior for a pulsed metal vapor arc source; there was an initial burst of plasma at start-up followed by an equilibrium phase with random flux variations on a 5-μs time scale. The amplitude of flux variations varied from 10 to 100%, depending on the nature of the cathode material and the size of the gap. Soft materials like indium gave a relatively constant flux compared to hard materials like aluminum. High atomic number led to smoother flux because of velocity dispersion during the longer time-of-flight to the measurement position. Larger arc gaps typically resulted in higher arc voltages and a more erratic flux indicative of plasma instabilities.

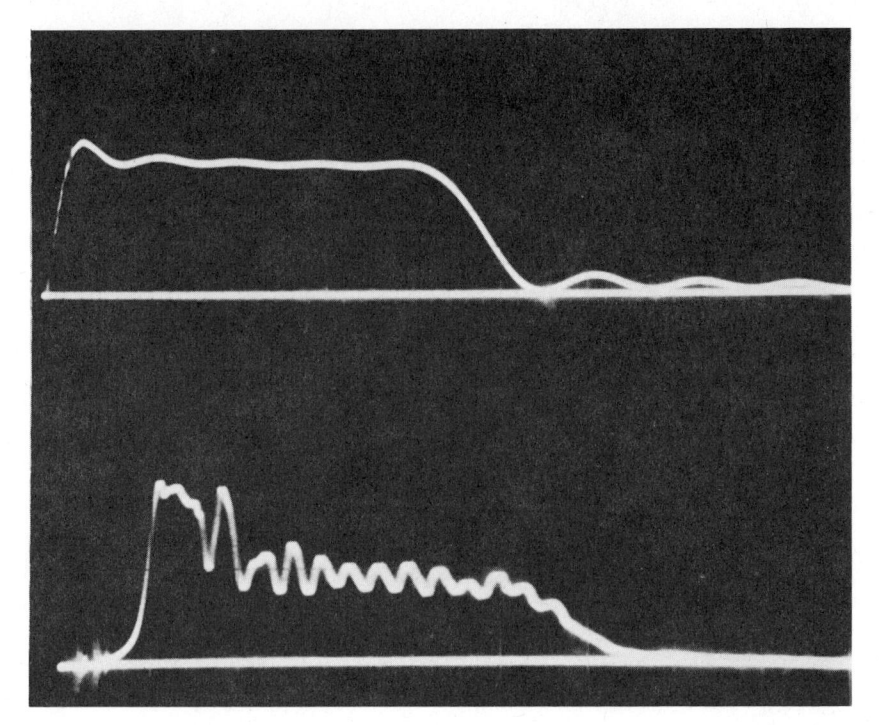

Figure 19.8 Plasma flux measured by ion saturation probe 15 cm from arc with a 3-kV charge on PFN. *Top*: Plasma arc driving current, 200 A/div. *Bottom*: Available plasma ion flux, 12.4 mA/cm^2/div, 20 μs/div.

5 OBSERVATION OF GRID-CONTROLLED EXTRACTION

The apparatus we used for grid-controlled ion extraction experiments [33–35] is shown in Figure 19.9. There was a 15-cm-long drift region between the arc source and the extractor. The drift region served three purposes:

1. The plasma flux was reduced to a level of about twice the space-charge-limited current density limit for the 30-kV extraction gap.
2. The extraction gap was effectively isolated from neutrals evolved from the arc.
3. Selection of forward directed ions from the expanding plasma led to an extracted beam with low transverse divergence.

The geometry of the extraction gap was chosen to model the control gap of the Los Alamos National Laboratory 2-MV Accelerator Inertial Fusion injector [8]. Extraction voltages up to 35 kV were applied across a 1.8-cm gap for pulses 5 to 10 μs in duration. Ion beams at 40 mA/cm^2 were extracted through a cathode mesh composed of 0.0025-cm-diameter wires with no observed gap breakdowns.

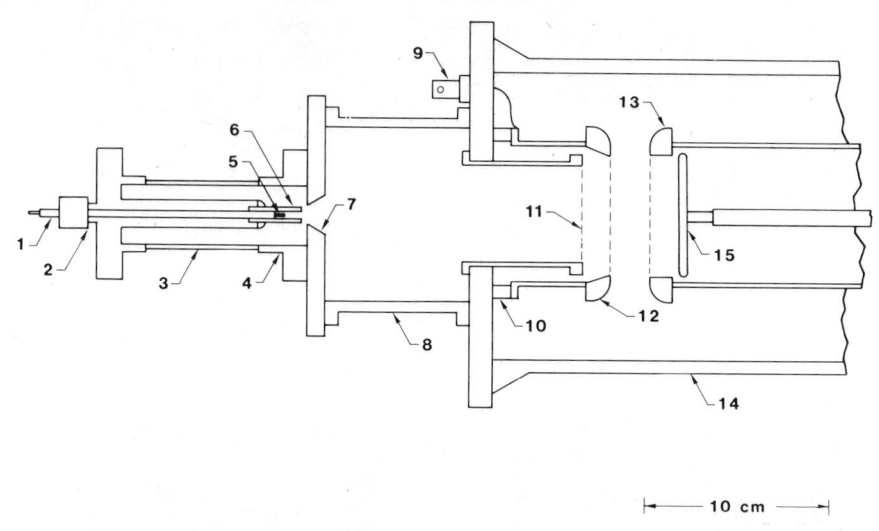

Figure 19.9 Apparatus for grid-controlled ion extraction experiments. 1, Rigid coaxial tubing feedthrough for initiator spark; 2, compression seal; 3, glass insulator; 4, stainless steel flange; 5, recessed initiator spark; 6, replaceable arc cathode; 7, arc anode, stainless steel; 8, plasma expansion chamber; 9, vacuum feedthrough for anode bias; 10, ceramic standoff; 11, plasma grid; 12, biased anode, extraction gap; 13, cathode, extraction gap; 14, glass vacuum chamber; 15, large area collector.

We performed experiments to verify that the extractor operated in a space-charge-limited mode with sufficient anode grid bias. Figure 19.10 shows results with an aluminum arc cathode. The top trace shows the voltage applied to the extraction gap, and the bottom trace shows the signal from a collector behind the cathode. The traces were taken with the same extraction voltage and plasma source properties, but with varying anode grid voltage. In Figure 19.10a, the grid voltage was low. There was a large spike in the collector signal when the extractor voltage was applied. This was caused by plasma prefill of the extraction gap. The signal following start-up was irregular. The extraction gap operated in the source-limited mode; therefore, the extracted current followed random variations in the plasma flux. The flux level was consistent with the measurements of available flux measured with a plasma probe. As the grid voltage was increased (Figures 19.10b and c), plasma prefill was reduced, and the signal became more regular. At a grid voltage of -100 V (Figure 19.8d), the ion extraction signal reached a constant value at a level well below the source limit. There was no evidence of plasma prefill. Furthermore, the flux of high-energy ions was relatively independent of fluctuations in the source.

Figure 19.11 shows similar results for an indium plasma. For low anode grid voltage, the collector current (top trace) reflects the effect of plasma prefill before application of the voltage. Note the negative signal level before the voltage pulse; this was caused by electron collection from plasma leaking across the extraction

5 Observation of Grid-Controlled Extraction

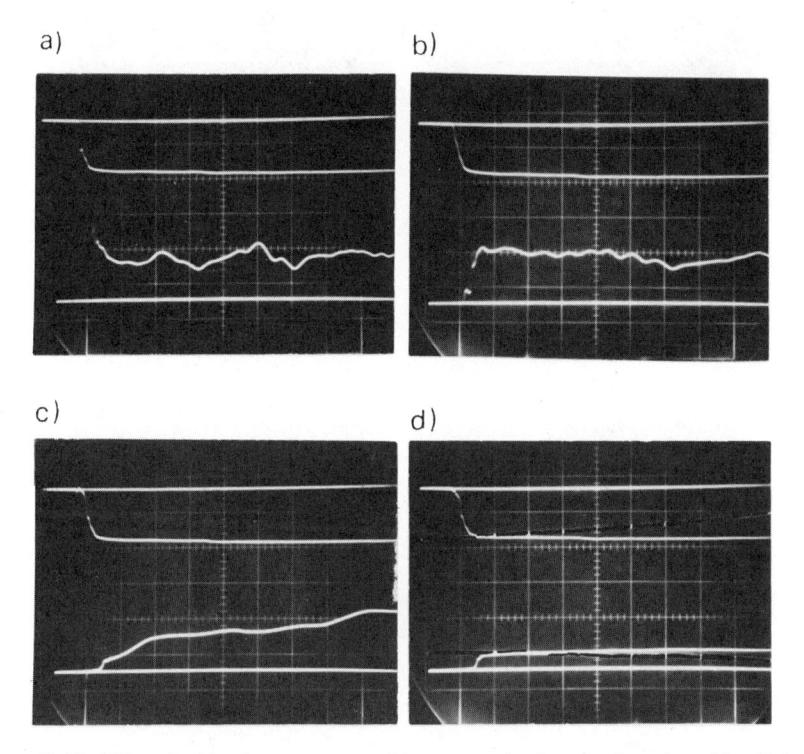

Figure 19.10 Effect of grid voltage on extracted ion current density, aluminum ions. *Top*: Extractor voltage, 4 kV/div. *Bottom*: Extracted ion current density, 6.2 mA/cm^2/div, 2 µs/div. (a) Grid voltage, −25 V; (b) grid voltage, −50 V; (c) grid voltage, −75 V; (d) grid voltage, −100 V.

gap. In Figure 19.11a, there was a gap breakdown associated with plasma prefill. Note further than the flux level in Figures 19.11b and c had little relation to the gap voltage. In contrast, at high anode grid voltage (Figure 19.11d) the ion flux was clearly controlled by the extraction voltage. The small spike at the beginning of the flux signal in Figures 19.11c and d, is a real feature corresponding to velocity bunching of the heavy indium ions traversing the 2-cm distance from the anode to the collector.

Control of ion flux by the extractor voltage at high grid voltage is demonstrated in the series of oscillographs of Figure 19.12. The properties of the aluminum ion source were maintained constant while the extractor voltage was raised. Note that all oscillographs are two-shot overlays. In the first two traces (Figures 19.12a and b), the space-charge-limited flux was well below the source limit; hence, the traces are almost identical. When the voltage was raised, the extraction gap flow became comparable to the source flow, and ion flux reflected random variations in the source flux. We have observed that the current density fluctuations at high current density are eliminated when the plasma flux is increased.

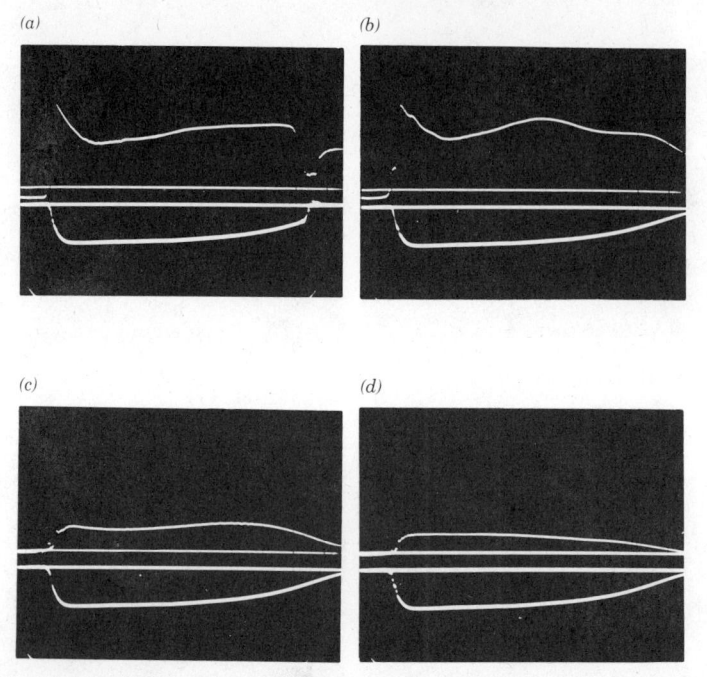

Figure 19.11 Effect of grid voltage on extracted ion current density. Indium ions. *Top*: Net ion current, 100 mA/div. *Bottom*: Extractor voltage, 10 kV/div. Grid voltage: (*a*) 0 V, (*b*) −50 V, (*c*) −100 V, (*d*) −150 V.

6 FLUX MEASUREMENTS

Estimates of flux requirements for near-term Acceleration Inertial Fusion experiments are in the range 10–20 mA/cm^2. Grid-controlled extractors have a demonstrated extracted flux in this range for a wide variety of ions. Measurements of extracted current density of Al$^+$ and In$^+$ ions at extraction voltages up to 30 kV are summarized in Figure 19.13. The measurements were performed using an apertured probe with an electron trap to eliminate contributions of neutralizing electrons and secondary electrons from the collector. The dashed lines are plotted with a slope corresponding to $V^{3/2}$. The ion flux closely follows the space-charge prediction for the 1.8-cm-wide extraction gap. The relative fluxes of indium and aluminum vary by a factor of two at the same voltage, in agreement with space-charge flow predictions.

Current density measurements for high-flux C$^+$ extraction are summarized in Figure 19.14. Current density approaching 50 mA/cm^2 was observed for 10-μs pulses with no problem of grid or extraction gap breakdown. The available plasma flux was approximately twice this figure, so that smooth space-charge-limited current waveforms were observed over the entire operational range. The dashed line is an absolute Child law prediction. The deviation at low extraction voltage is consistent with the 100-V directed energy of the ions entering the gap.

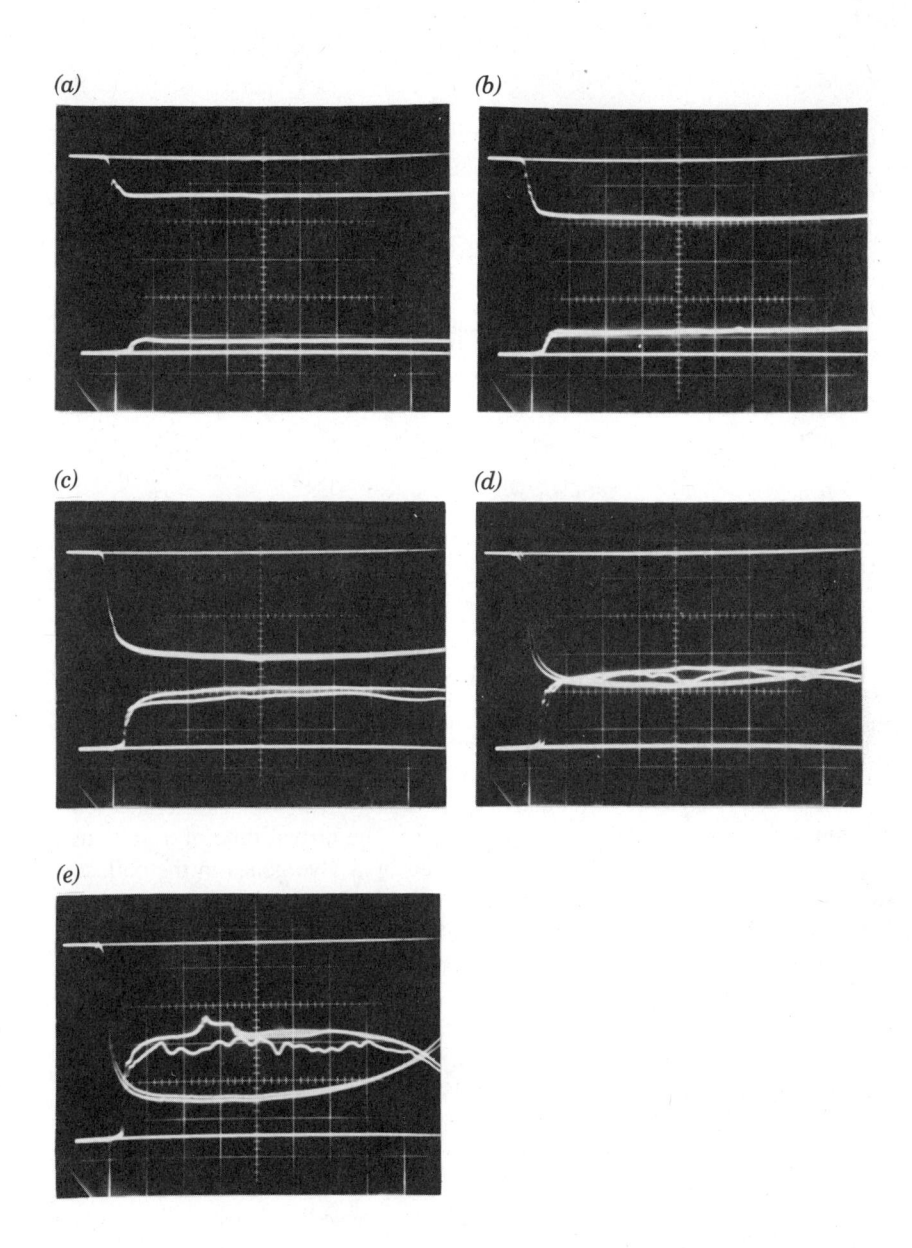

Figure 19.12 Ion current density as a function of extraction voltage, aluminum ions. Grid voltage: −150 V. *Top*: Extractor voltage, 4 kV/div. *Bottom*: Extracted ion current density, 3.2 mA/cm^2/div, 2 μs/div. Two-shot overlays on all traces. Peak extractor voltage: (*a*) 4 kV, (*b*) 6 kV, (*c*) 11 kV, (*d*) 14 kV, (*e*) 17 kV.

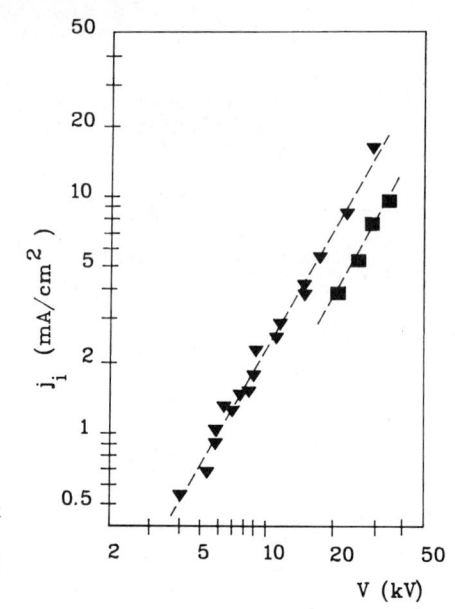

Figure 19.13 Log–log plot of average ion flux versus extraction gap voltage for Al$^+$ and In$^+$. Grid voltage: -150 V. 1.8-cm gap width. Dashed lines have $V^{3/2}$ slope.

7 MEASUREMENTS OF BEAM SPECIES AND EMITTANCE

The flux of ion beams from a low-energy ion extractor is quite sensitive to variations of injector voltage. This fact can be used to advantage for measurements of species in pulsed ion beams. If the rise time of the extractor voltage is comparable to the ion transit time, under most circumstances the arrival time of a particular ion species at a downstream detector is marked by a divergence in the collected current density. The characteristic signature of velocity bunching, which can be predicted from the accelerating voltage waveform, is useful for identifying species and charge states. A typical time-of-flight probe signal for aluminum ions from an extraction gap operated in the space-charge-limited mode is shown in Figure 19.15. The extractor voltage signal is displayed on the top; the probe signal is on the bottom. Note the substantial delay and the clean aluminum signal. The trace indicates that the beam is predominantly Al$^+$. If doubly ionized aluminum and light ion contaminants were present, their signals would be separated enough in time to allow easy identification. Similar results were obtained with indium. Carbon ion measurements were performed with a detector 70 cm from the extraction gap. Excellent signals were obtained with delay times exceeding 1 μs. There was no indication of multiply charged ions.

The generation of pure beams of singly charged ions from a vacuum arc is an unexpected result. Our plasma measurements and those of others indicate a high percentage of multiply charged ions in freely expanding vacuum arc plasmas. The absence of multiply charged ions from grid-controlled extractors may result from the fact that there is a stationary, electrostatically confined plasma between the arc

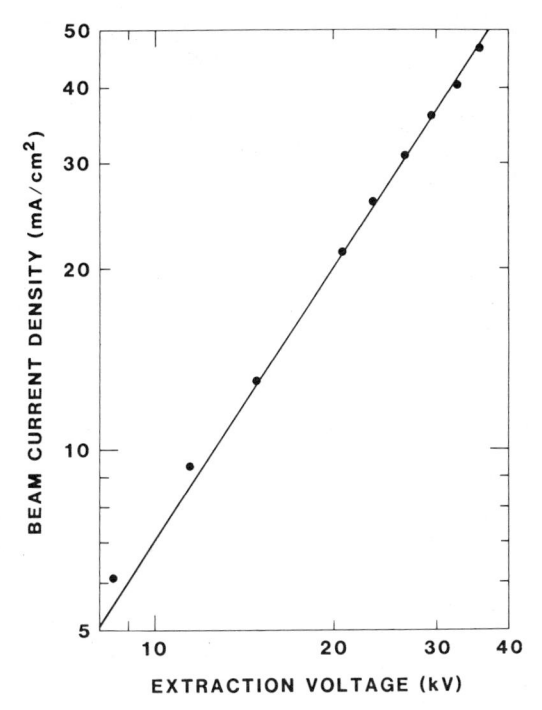

Figure 19.14 Log–log plot of average ion flux versus extraction gap voltage for high-current C$^+$ extraction. Anode grid voltage: -100 V, 1.49-cm gap. Solid line is absolute Child law prediction.

source and the extraction point. The measured electron temperature implies that the plasma potential is in the range 30–40 eV. If singly and doubly charged ions from the source have about the same kinetic energy, the multiply charged ions are less likely to enter the intervening plasma. Charge state filtering is useful for generating beams with a well-defined charge state. On the other hand, the effect may be an impediment for the generation of multiply charged ion beams.

We performed measurements of the beam divergence by replacing the cathode mesh by a solid plate with small apertures. The image of the projected ion orbits was detected with a channel-electron-multiplier array. Typical photographic output with the detector located 37 cm from the aperture plate is shown in Figure 19.16. The data of Figure 19.16a were taken with a relatively high voltage between the plasma source and grid (-250 V) and with a high extraction voltage (27 kV). The beams formed well-defined circles with an edge diameter of about 0.7 cm. The time-integrated spots had relatively small diameter, indicating a high degree of stability in the plasma extraction surface adjacent to the anode grid. The measured radius of the spots, corrected for space-charge expansion, implies a beam envelope divergence angle of $\Delta r' = 6.7$ mrad. The divergence angle corresponds to a peak ion transverse energy of 1.2 eV. If it is assumed that the angular divergence is representative of the distribution across the 5-cm diameter of the active

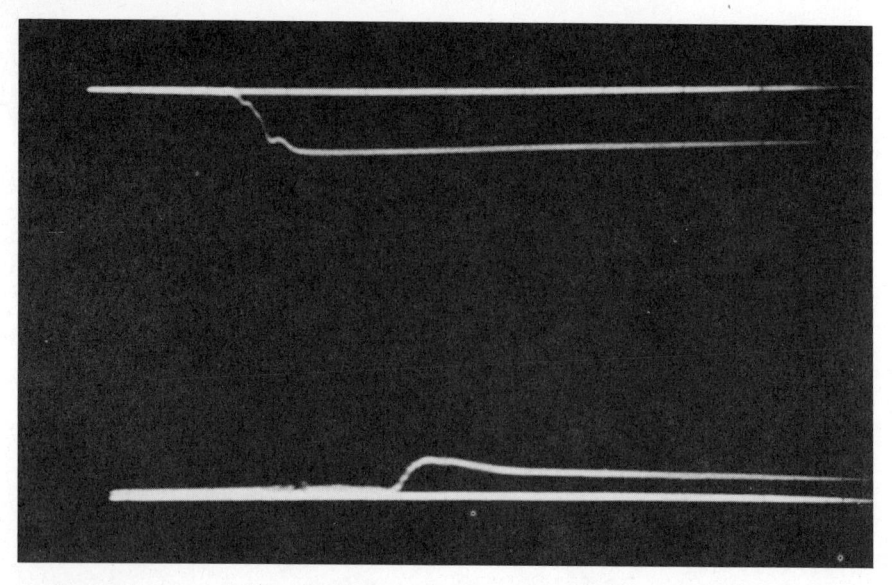

Figure 19.15 Time-of-flight detector, aluminum ions. Distance from cathode mesh to probe: 13 cm. *Top*: Extractor voltage, 4 kV/div. *Bottom*: Detector output (uncalibrated), 1 μs/div.

anode (r_0), then the emittance of the extracted beam can be estimated (neglecting nonlinear optical effects in the extraction gap). The normalized emittance is given by Eqs. (2) and (4) as 2.5×10^{-7} π-m-rad.

We observed that the emittance exhibited a variation with grid voltage (between the plasma source and gap anode) consistent with the predictions of Eq. (13). Figure 19.16b shows output from the channel-electron-multiplier array detector for the same extractor voltage as Figure 19.16a but with the anode grid voltage lowered to -100 V. For the plasma density in question, the extraction sheath width was predicted to be comparable to the mesh spacing, resulting in stronger facet lens focusing and a distorted plasma surface. Note the broadened central spot and the prominent beam halos in Figure 19.16b.

Recent measurements on high-density (>20 mA/cm^2) C$^+$ beams indicate that the formation of a virtual anode may substantially increase the emittance. Under conditions of source-limited extraction, the measured emittance was consistent with previous results. Upon transition to space-charge-limited extraction, the beam emittance increased by a factor of five. The transition from source limited to space-charge-limited flow has been induced by increasing the plasma flux as well as by decreasing the extraction voltage. The effect upon emittance is similar in both cases. During the transition to space-charge-limited flow, the point of ion extraction is moved from the surface of the plasma upstream of the anode grid to the downstream virtual anode. Ion deflection by nonuniform space charge in the virtual anode is suspected as the source of the increased transverse energy. Models examined to date have been inadequate to explain the observed effects. Additional

8 Conclusions 417

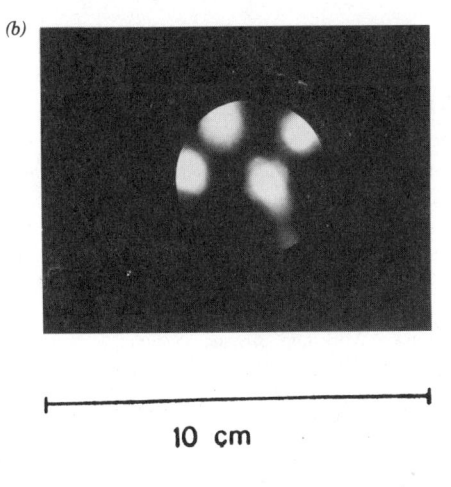

Figure 19.16 Time-integrated photographic output from CEMA beam divergence detector. Average extraction gap voltage: 27 kV. Al$^+$ ions, 15 mA/cm^2. (*a*) Voltage between anode grid and plasma source, -250 V. (*b*) Grid voltage, -100 V.

modeling of the virtual anode is required to determine the emittance growth mechanism.

8 CONCLUSIONS

Grid-controlled ion extractors supplied by metal vapor vacuum arcs may meet the near-term and far-term requirements of injectors for Accelerator Inertial Fusion. The physical principles of grid-controlled ion extractors are, for the most part, well understood. Additional numerical work is required to understand the shape of the plasma extraction sheath at the grid anode and its effect on ion optics. Intermediate mass ion current density in the range $\leqslant 50$ mA/cm^2 can be achieved for pulsed beams. At reduced current density, the grid-controlled vacuum arc source could be an attractive continuous-duty-cycle source of low-emittance beams. Such sources are simple to construct and operate and do not cause significant gas loading of downstream acceleration structures. The electrostatic confinement of extraction plasmas and reduction of plasma prefill accomplished with grids is an attractive feature for pulsed beam systems. The method may have application to improvement of the performance of modulated ion beam extractors.

ACKNOWLEDGMENTS

We thank D. Keefe of Lawrence Berkeley Laboratory, whose suggestion of the use of electrostatic grids to prevent plasma prefill in injectors provided much of the motivation for this work. We gratefully acknowledge the invaluable technical guidance we received from the late K. Riepe of Los Alamos National Laboratory. This work was supported by Los Alamos National Laboratory and the U.S. Department of Energy.

REFERENCES

1. See, for instance, C. Lejeune, in *Applied Charged Particle Optics—Very-High-Density Beams*, A. Septier (Ed.), Academic, New York, 1983, p. 207.
2. A. Faltens, E. Hoyer, and D. Keefe, Proceedings, 4th Int'l. Conf. High Power Electron and Ion Beam Research and Tech., H. J. Doucet and J. M. Buzzi (Eds.), Ecole Polytechnique, 1981, p. 751; A. Faltens, E. Hoyer, D. Keefe, and L. J. Laslett, *IEEE Trans. Nucl. Sci.* **NS-26,** 3106 (1979).
3. R. C. Arnold (Ed.), Proceedings, Heavy Ion Fusion Workshop, Argonne National Laboratory, ANL-79-41, 1979; W. B. Herrmannsfeldt (Ed.), Proceedings, Heavy Ion Fusion Workshop, Lawrence Berkeley Laboratory, LBL-10301, 1980.
4. See, for instance, J. D. Lawson, *The Physics of Charged Particle Beams*, Clarendon, Oxford, 1977.
5. See, for instance, C. Lejeune and J. Aubert, in *Applied Charged Particle Optics*, Part A, A. Septier (Ed.), Academic, New York, 1980.
6. D. L. Judd (Ed.), Multiple Beam Experiment (MBE), Conceptual Design and Program Description, Lawrence Berkeley Laboratory, PUB-5123, 1984.
7. E. O. Ballard, E. A. Meyer, H. L. Rutkowski, R. P. Shurter, and F. W. Van Haaften, *IEEE Trans. Nucl. Sci.* **NS-32,** 1788 (1985).
8. H. L. Rutkowski, H. Oona, E. A. Meyer, R. P. Shurter, and L. S. Englehart, *IEEE Trans. Nucl. Sci.* **NS-32,** 1742 (1985).
9. R. T. Avery, C. S. Chavis, T. J. Fessenden, D. E. Gough, T. F. Henderson, D. Keefe, J. R. Meneghetti, C. D. Pike, D. L. Vanacek, and A. T. Warwick, *IEEE Trans. Nucl. Sci.* **NS-32,** 3187 (1985).
10. See, for instance, R. G. Wilson and G. R. Brewer, *Ion Beams with Applications to Ion Implantation*, Wiley, New York, 1973, p. 178.
11. See, for instance, K. W. Ehlers, W. R. Baker, K. H. Berkner, W. S. Cooper, W. B. Kunkel, R. V. Pyle, and J. W. Stearns, in Proceedings, Symp. Ion Sources and Formation of Ion Beams, C. P. Pezzotti (Ed.), Lawrence Berkeley Laboratory, LBL-3399, 1974, p. I-5.
12. D. Bohm, *Characteristics of Electric Discharges in Magnetic Fields*, McGraw-Hill, New York, 1949.

13. C. D. Child, *Phys. Rev.* **32,** 492 (1911).
14. J. T. Crow, A. T. Forester, and D. M. Goebel, *IEEE Trans. Plasma Sci.* **PS-6,** 535 (1978).
15. W. L. Johnson, G. B. Johnson, and J. T. Verdeyen, *J. Appl. Phys.* **47,** 4442 (1976).
16. W. Chupp, A. Faltens, E. Hartwig, E. Hoyer, D. Keefe, C. Kim, M. Lampel, E. Lofgren, R. Nemetz, S. S. Rosenblum, J. Shiloh, M. Tiefenback, and D. Vanacek, *IEEE Trans. Nucl. Sci.* **NS-28,** 3389 (1981).
17. See, for instance, F. F. Chen, in *Plasma Diagnostic Techniques*, R. H. Huddleston and S. L. Leonard (Eds.), Academic, New York, 1965.
18. See, for instance, K. J. Hansen and R. Lauer, in *Focusing of Charged Particles*, A. Septier (Ed.), Academic, New York, 1967, p. 296.
19. C. Burkhart, S. Coffey, G. Cooper, S. Humphries, Jr., L. K. Len, A. D. Logan, M. Savage, and D. M. Woodall, *Nucl. Instrum. Methods B* **10/11,** 792 (1985).
20. L. K. Len, C. Burkhart, G. W. Cooper, S. Humphries, Jr., M. Savage, and D. M. Woodall, *IEEE Trans. Plasma Sci.* **PS-14,** 256 (1986).
21. A. A. Plyutto, V. N. Ryzhkov, and A. T. Kapin, *Sov. Phys. JETP* **20,** 328 (1965).
22. W. D. Davis and H. C. Miller, *J. Appl. Phys.* **40,** 2212 (1969).
23. C. W. Kimblin, *Proc. IEEE* **59,** 546 (1971); *J. Appl. Phys.* **44,** 3074 (1973).
24. V. M. Lunev, V. D. Ovcharenko, and V. M. Khoroshikh, *Sov. Phys. Tech. Phys.* **22,** 855 (1978).
25. H. C. Miller, *J. Appl. Phys.* **52,** 4523 (1981).
26. J. M. Lafferty (Ed.), *Vacuum Arcs: Theory and Applications*, Wiley, New York, 1980.
27. R. J. Adler and S. T. Picraux, *Nucl. Instrum. Methods B* **6,** 123 (1985).
28. I. G. Brown, J. E. Galvin, and R. A. MacGill, *Appl. Phys. Lett.* **47,** 358 (1985).
29. I. G. Brown, *IEEE Trans. Nucl. Sci.* **NS-32,** 1723 (1985).
30. J. V. R. Heberlein and D. R. Porto, *IEEE Trans. Plasma Sci.* **PS-11,** 152 (1983).
31. H. L. Rutkowski, H. Oona, E. A. Meyer, R. P. Shurter, L. S. Englehardt, and S. Humphries, Jr., *IEEE Trans. Nucl. Sci.* **NS-32,** 1742 (1985).
32. C. Burkhart and S. Humphries, Jr., Proceedings, NATO Workshop on High Brightness Accelerators, Pitlochry, Scotland, 1986, (Plenum, NY, 1988).
33. S. Humphries, Jr., C. Burkhart, S. Coffey, G. Cooper, L. K. Len, M. Savage, and D. M. Woodall, *J. Appl. Phys.* **59,** 1790 (1986).
34. S. Humphries, Jr. and C. Burkhart, *Particle Accelerators* **20,** 211 (1987).
35. L. K. Len, S. Humphries, Jr., and C. Burkhart, AIP Conference Proceedings No. 152, International Symposium on Heavy Ion Fusion, Washington, DC, May 1968.

Appendixes

APPENDIX 1 Physical Constants

c	Speed of light in free space	2.9979×10^8	m/s
ϵ_0	Permittivity of free space	8.8542×10^{-12}	F/m
μ_0	Permeability of free space	$4\pi \times 10^{-7}$	H/m
e	Electron charge	1.6022×10^{-19}	C
m_e	Electron mass	9.1094×10^{-31}	kg
m_p	Proton mass	1.6726×10^{-27}	kg
m_u	Atomic mass unit, amu	1.6605×10^{-27}	kg
h	Planck's constant	6.6261×10^{-34}	J · s
σ	Stefan-Boltzmann constant	5.6705×10^{-8}	J/m² · s · (K)⁴
k	Boltzmann's constant	1.3806×10^{-23}	J/K
N_A	Avogadro's number	6.0221×10^{23}	molecules/mole
n_0	Loschmidt's number	2.6868×10^{19}	molecules/cm³
a_0	Bohr radius (radius of hydrogen atom)	0.5292×10^{-8}	cm
	Wavelength associated with 1 eV	12,399	Å
	Energy associated with 1 eV	1.6022×10^{-19}	J
e/k	Temperature associated with 1 eV	11,605	K

$$1 \text{ amu} = 931.49 \text{ MeV}$$

$$m_e = 0.5110 \text{ MeV}$$

$$m_p/m_e = 1836.15$$

$$\pi = 3.14159265$$

$$e = 2.71828183$$

Molecular density at 20 °C and 1 mTorr pressure = 3.3×10^{13} molecules/cm³

APPENDIX 2 Some Plasma Parameters*

Electron plasma frequency	$f_{pe} = 8.98 \times 10^3 \, n_e^{1/2}$	Hz
Ion plasma frequency	$f_{pi} = 2.10 \times 10^2 \, QA^{-1/2} n_i^{1/2}$	Hz
Electron cyclotron frequency	$f_{ce} = 2.80 \times 10^6 \, B$	Hz
Ion cyclotron frequency	$f_{ci} = 1.52 \times 10^3 \left(\dfrac{Q}{A}\right) B$	Hz
Electron mean thermal velocity	$\bar{v}_{Te} = 6.69 \times 10^7 \, T_e^{1/2}$	cm/s
Ion mean thermal velocity	$\bar{v}_{Ti} = 1.57 \times 10^6 \left(\dfrac{T_i}{A}\right)^{1/2}$	cm/s
Ion sound speed	$c_s = 9.79 \times 10^5 \, (\gamma Q T_e / A)^{1/2}$	cm/s
Electron cyclotron radius	$r_e = 3.81 \times T_e^{1/2}/B$	cm
Ion cyclotron radius	$r_i = 1.64 \times 10^2 \, (A T_i)^{1/2}/(QB)$	cm
Debye length	$\lambda_D = 7.43 \times 10^2 \left(\dfrac{T_e}{n_e}\right)^{1/2}$	cm

*In these expressions n_e and n_i are the electron and ion densities in cm^{-3}, T_e and T_i are the electron and ion temperatures in eV, B is the magnetic field strength in Gauss, Q is the ion charge state, A is the ion mass in amu, and γ is the specific heat ratio. The cyclotron radii are given in terms of mean thermal velocities, $r = \bar{v}_T/\omega_c$.

APPENDIX 3 Table of the Elements

Element	Symbol	Z	A	MP (°C)	BP (°C)	Density (g/cm^3)	Electrical Resistivity ($\mu\Omega \cdot$ cm)	Ionization Potential (eV)
Hydrogen	H	1	1	-259	-253			13.60
Helium	He	2	4	-272	-269			24.59
Lithium	Li	3	6.9	180	1,350	0.534	8.5	5.39
Beryllium	Be	4	9	1,290	2,970	1.85	4.0	9.32
Boron	B	5	10.8	2,300	2,550(s)	2.34	4E12	8.30
Carbon	C	6	12	3,550	4,800	2.26	1375	11.26
Nitrogen	N	7	14	-210	-196			14.55
Oxygen	O	8	16	-218	-183			13.62
Fluorine	F	9	19	-220	-188			17.42
Neon	Ne	10	20.2	-249	-246			21.56
Sodium	Na	11	23	98	890	0.97	4.2	5.14
Magnesium	Mg	12	24.3	649	1,090	1.74	4.45	7.65
Aluminum	Al	13	27	660	2,470	2.70	2.65	5.98
Silicon	Si	14	28.1	1,410	2,400	2.33		8.15
Phosphorous	P	15	31	44	280	1.8-2.7	1E17	10.48
Sulfur	S	16	32.1	115	445	2.07	2E23	10.36
Chlorine	Cl	17	35.5	-100	-35			13.02
Argon	Ar	18	39.9	-189	-186			15.76
Potassium	K	19	39.1	64	760	0.862	6.15	4.34
Calcium	Ca	20	40.1	840	1,480	1.55	4.0	6.11
Scandium	Sc	21	45	1,540	2,830	3.00	51	6.56
Titanium	Ti	22	47.9	1,660	3,290	4.54	42	6.84
Vanadium	V	23	50.9	1,890	3,380	6.11	25	6.74
Chromium	Cr	24	52	1,860	2,670	7.18	13	6.76
Manganese	Mn	25	54.9	1,245	1,960	7.4	185	7.43
Iron	Fe	26	55.8	1,535	2,750	7.87	9.71	7.90
Cobalt	Co	27	58.9	1,495	2,870	8.92	6.25	7.86
Nickel	Ni	28	58.7	1,453	2,800	8.90	6.84	7.63
Copper	Cu	29	63.5	1,083	2,570	8.95	1.68	7.73
Zinc	Zn	30	65.4	420	907	7.13	5.9	9.39
Gallium	Ga	31	69.7	30	2,400	5.91	17.4	6.00
Germanium	Ge	32	72.6	937	2,850	5.32		7.88
Arsenic	As	33	74.9	817[a]	613(s)	5.73	33	9.81
Selenium	Se	34	79	217	685	4.79	12	9.75
Bromine	Br	35	79.9	-7	59	3.12		11.85
Krypton	Kr	36	83.8	-157	-152			14.00
Rubidium	Rb	37	85.5	39	687	1.53	12.5	4.18
Strontium	Sr	38	87.6	770	1,380	2.54	23	5.69
Yttrium	Y	39	88.9	1,520	3,340	4.46	60	6.53
Zirconium	Zr	40	91.2	1,852	4,380	6.51	40	6.95
Niobium	Nb	41	92.9	2,470	4,740	8.57	12.5	6.88
Molybdenum	Mo	42	95.9	2,620	4,610	10.22	5.2	7.10
Technetium	Tc	43	98.9	2,172	4,880	11.5		7.28
Ruthenium	Ru	44	101.1	2,310	4,200	12.45	7.2	7.36
Rhodium	Rh	45	102.9	1,966	3,730	12.41	4.51	7.45
Palladium	Pd	46	106.4	1,554	2,950	12.02	10.5	8.33
Silver	Ag	47	107.9	962	2,210	10.50	1.59	7.58
Cadmium	Cd	48	112.4	321	765	8.65	7.0	8.99
Indium	In	49	114.8	157	2,080	7.31	8.37	5.78
Tin	Sn	50	118.7	232	2,300	7.31	11	7.34
Antimony	Sb	51	121.7	631	1,700	6.62	40	7.84
Tellurium	Te	52	127.6	450	990	6.24	4.4E5	9.01

APPENDIX 3 (*Continued*)

Element	Symbol	Z	A	MP (°C)	BP (°C)	Density (g/cm³)	Electrical Resistivity (μΩ·cm)	Ionization Potential (eV)
Iodine	I	53	126.9	113	184	4.93	1E15	10.45
Xenon	Xe	54	131.3	-112	-107			12.13
Cesium	Cs	55	132.9	28	680	1.89	20	3.89
Barium	Ba	56	137.3	725	1,700	3.5	50	5.21
Lanthanum	La	57	138.9	920	3,450	6.15	80	5.61
Cerium	Ce	58	140.1	798	3,460	6.77	75	5.57
Praseodymium	Pr	59	140.9	931	3,510	6.77	68	5.42
Neodymium	Nd	60	144.2	1,020	3,070	7.0	64	5.45
Promethium	Pm	61	145	1,080	2,460			5.55
Samarium	Sm	62	150.4	1,075	1,800	7.54	105	5.60
Europium	Eu	63	152	822	1,600	5.25	90	5.64
Gadolinium	Gd	64	157.2	1,311	3,250	7.90	140	6.16
Terbium	Tb	65	158.9	1,360	3,200	8.23	115	5.98
Dysprosium	Dy	66	162.5	1,410	2,600	8.55	90	5.93
Holmium	Ho	67	164.9	1,470	2,720	8.78	82	6.02
Erbium	Er	68	167.3	1,530	2,900	9.05	86	6.10
Thulium	Tm	69	168.9	1,545	1,950	9.32	70	6.18
Ytterbium	Yb	70	173	820	1,200	6.96	25	6.22
Lutetium	Lu	71	175	1,660	3,400	9.84	58	6.15
Hafnium	Hf	72	178.5	2,225	4,600	13.2	35	6.80
Tantalum	Ta	73	180.9	3,000	5,425	16.65	13	7.88
Tungsten	W	74	183.8	3,410	5,660	19.3	5.6	7.98
Rhenium	Re	75	186.2	3,180	5,650	21.02	19.3	7.87
Osmium	Os	76	190.2	3,040	5,030	22.57	9.5	8.75
Iridium	Ir	77	192.2	2,410	4,430	22.65	5.3	9.05
Platinum	Pt	78	195.1	1,772	3,900	21.45	10.6	8.96
Gold	Au	79	197	1,064	2,810	19.3	2.21	9.23
Mercury	Hg	80	200.6	-39	357	13.55	98	10.44
Thallium	Tl	81	204.4	303	1,460	11.85	18	6.11
Lead	Pb	82	207.2	327	1,750	11.35	20.6	7.41
Bismuth	Bi	83	209	271	1,560	9.75	107	7.29
Polonium	Po	84	210	254	965			8.43
Astatine	At	85	210	302	337			
Radon	Rn	86	222	-71	-62			10.75
Francium	Fr	87	223	27	677			
Radium	Ra	88	226	700	1,200			5.28
Actinium	Ac	89	227	1,050	3,200			6.90
Thorium	Th	90	232	1,750	4,790	11.72	13	6.20
Protactinium	Pa	91	231					
Uranium	U	92	238	1,132	3,900	18.95	30	6.20
Neptunium	Np	93	237	640	3,900			
Plutonium	Pu	94	244	641	3,300			
Americium	Am	95	243	994	2,600			
Curium	Cm	96	247	1,340				
Berkelium	Bk	97	247					
Californium	Cf	98	251					
Einsteinium	Es	99	254					
Fermium	Fm	100	257					
Mendelevium	Md	101	258					
Nobelium	No	102	259					
Lawrencium	Lr	103	260					

Note: MP, BP, melting and boiling point temperatures; (s), sublimes; a, at a pressure of 28 atmospheres. Electrical resistivity is at or near 20 °C.

APPENDIX 4 Ionization Potentials for Multiply Charged Ions of All the Elements*

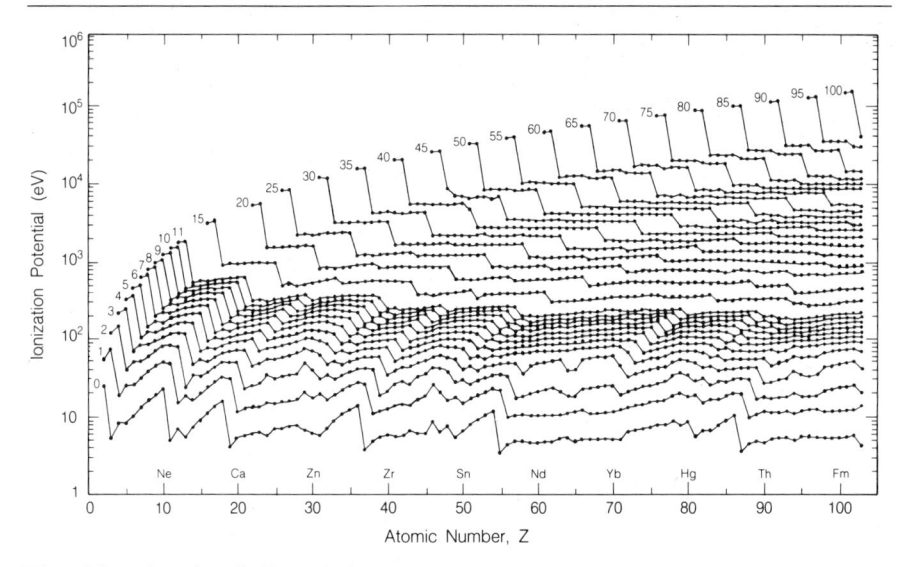

*Plotted from data given by T. A. Carlson, C. W. Nestor, Jr., N. Wasserman, and J. D. McDowell, *Atomic Data* **2**, 63 (1970).

APPENDIX 5 Work Functions of the Elements*

IA	IIA	IIIB	IVB	VB	VIB	VIIB	VIII			IB	IIB	IIIA	IVA	VA	VIA	VIIA
3 Li 2.9	4 Be 4.98											5 B 4.45	6 C 5.0			
11 Na 2.75	12 Mg 3.66											13 Al 4.28	14 Si 4.85	15 P ⋯	16 S ⋯	
19 K 2.30	20 Ca 2.87	21 Sc 3.5	22 Ti 4.33	23 V 4.3	24 Cr 4.5	25 Mn 4.1	26 Fe 4.5	27 Co 5.0	28 Ni 5.15	29 Cu 4.65	30 Zn 4.33	31 Ga 4.2	32 Ge 5.0	33 As 3.75	34 Se 5.9	Br
37 Rb 2.16	38 Sr 2.59	39 Y 3.1	40 Zr 4.05	41 Nb 4.3	42 Mo 4.6	43 Te ⋯	44 Ru 4.71	45 Rh 4.98	46 Pd 5.12	47 Ag 4.26	48 Cd 4.22	49 In 4.12	50 Sn 4.42	51 Sb 4.55	52 Te 4.95	I
55 Cs 2.14	56 Ba 2.7	57 La 3.5	72 Hf 3.9	73 Ta 4.25	74 W 4.55	75 Re 4.96	76 Os 4.83	77 Ir 5.27	78 Pt 5.65	79 Au 5.1	80 Hg 4.49	81 Tl 3.84	82 Pb 4.25	83 Bi 4.22	84 Po ⋯	At
87 Fr ⋯	88 Ra ⋯	89 Ac ⋯	90 Th 3.4	91 Pa ⋯	92 U 3.63											
			58 Ce 2.9	59 Pr ⋯	60 Nd 3.2	61 Pm ⋯	62 Sm 2.7	63 Eu 2.5	64 Gd 3.1	65 Tb 3.0	66 Dy ⋯	67 Ho ⋯	68 Er ⋯	69 Tm ⋯	70 Yb ⋯	71 Lu 3.3
			90 Th 3.4	91 Pa ⋯	92 U 3.63	93 Np ⋯										

*Work function ϕ_{exp} in electron volts. Data are for polycrystalline specimens. From H. B. Michaelson, *J. Appl. Phys.* **48**, 4729 (1977).

APPENDIX 6 Vapor Pressure Curves of the Elements*

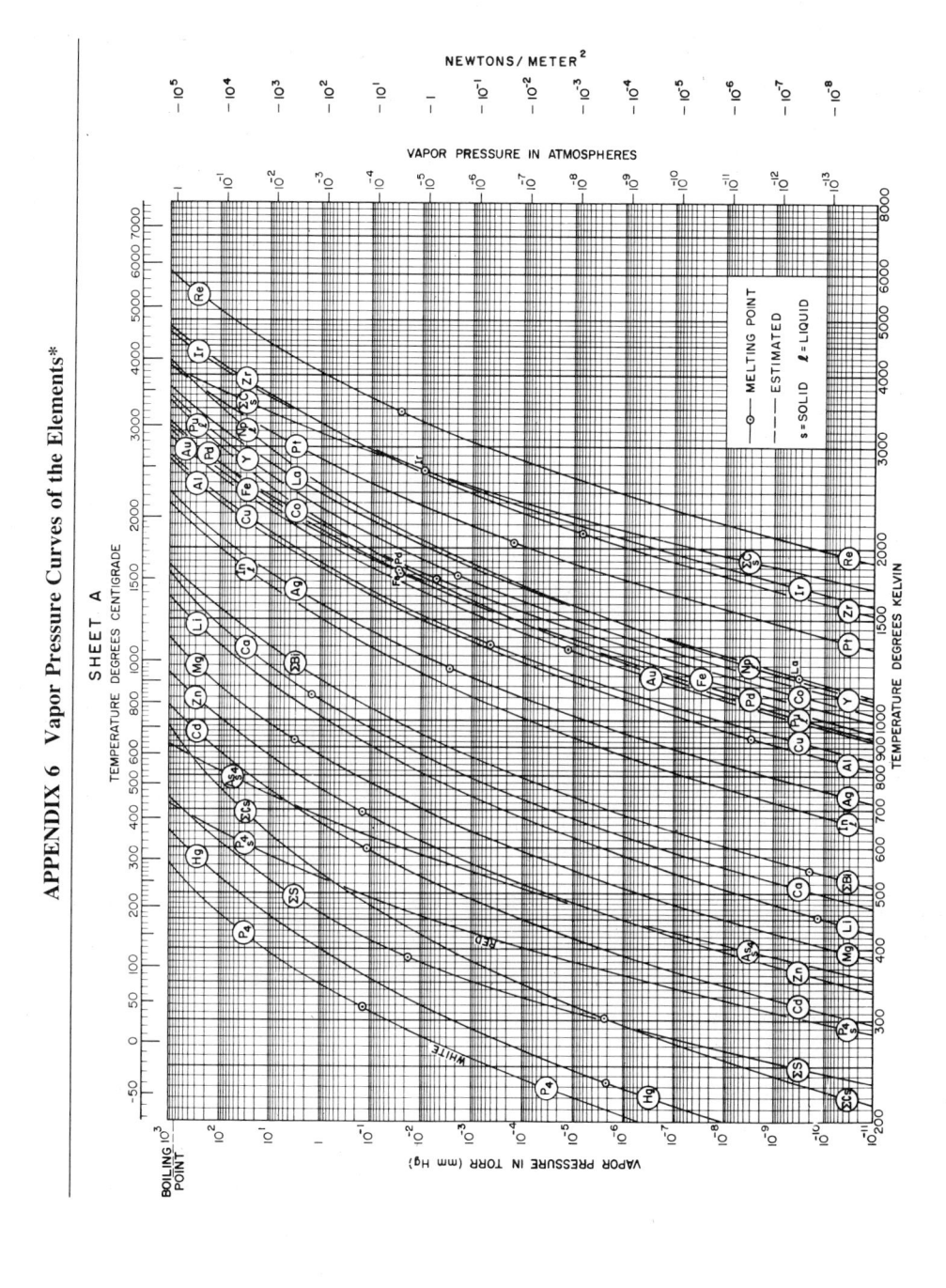

SHEET A

APPENDIX 6 (Continued)

SHEET B

SHEET C

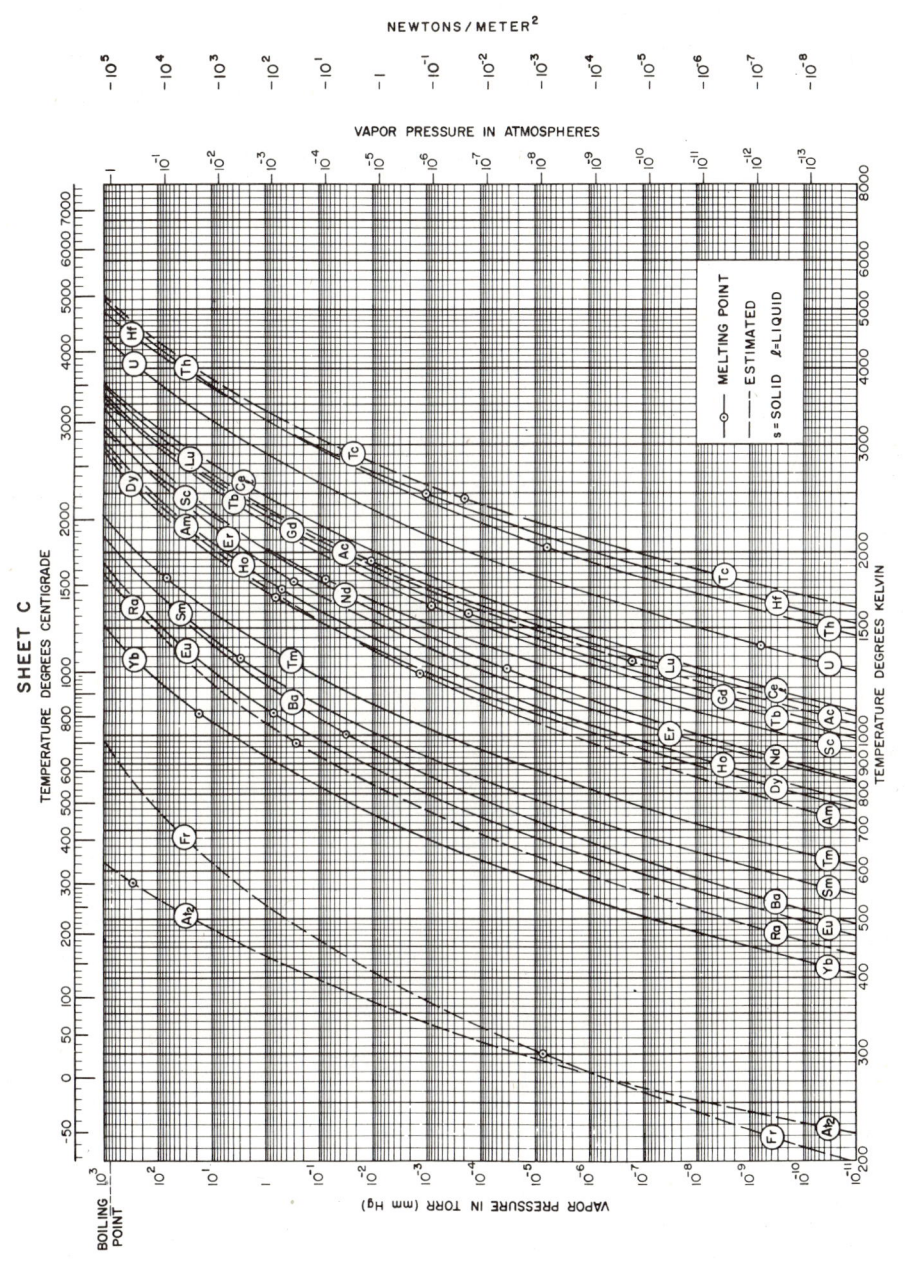

*From R. E. Honig and D. A. Kramer, *RCA Rev.* **30**, 285 (1969); © 1969 RCA Corp.

Name Index

Abbott, S., 171, 183, 331
Abdulmanov, W. G., 271
Abroyan, M. A., 360
Adams, C. T., 355
Adler, R. J., 332, 352, 406
Ahmed, H., 259
Aitken, D., 187, 199, 201, 202
Albert, J., 31
Alexeff, I., 210
Alford, D. B., 147
Allen, M., 363
Allison, P., 32, 38, 80, 95, 96, 120
Allison, S. K., 145
Al'ludov, Yu. K., 310
Alpert, V. A., 245, 269, 281
Alvarez, I., 301
Amboss, K., 259
Amemiya, H., 231
Ames, F., 299
Anan'in, O. B., 301, 303, 304
Anastasevitch, V. S., 76
Andersen, H. H., 163, 356
Anderson, C. E., 170, 179
Anderson, C. L., 313, 314
Anderson, K., 313, 327
Anderson, O. A., 76, 83
Anderson, R. J., 299, 301, 308, 309, 386
Antaya, T. A., 210
Ardenne, M., 359
Ariener, J., 168, 259, 260, 267, 269, 270, 271, 281
Armstrong, D. D., 29, 155
Arnold, R. C., 397
Arnush, D., 157
Asmussen, J., 146, 157, 239, 240
Attal, P., 171

Auberty, J., 398
Avery, R. T., 400

Bacal, M., 359, 362
Backouche, L., 171
Backus, J., 168, 169
Bainbridge, K. T., 145
Bajard, M., 176
Baker, W. R., 400
Baldin, A. M., 303
Ballard, E. O., 400
Banta, H. E., 359
Barabash, L. Z., 301
Bardet, R., 208
Barnett, C. F., 171
Barnett, S. A., 142, 331
Barr, D., 319, 320
Bartoli, C., 314, 326
Basile, R., 171
Basova, T. A., 310
Bastide, R. P., 359
Bauer, W., 383
Baumann, H., 171
Bay, H. L., 163
Bayless, J. R., 157
Bayly, A. R., 313, 327
Bechtold, V., 210, 214, 219
Becker, R., 114, 211, 255, 269, 270
Belchenko, Yu. I., 357, 358, 360
Bell, A. E., 313, 314, 317, 318, 319, 323, 325, 326
Benassayag, G., 317, 320
Bender, E. D., 356
Bennett, J. R. J., 168, 170, 171
Berkner, K. B., 48, 400
Bernas, R., 38, 179, 187

Name Index

Bernhardi, K., 208
Bethge, K., 171, 175
Beuscher, H., 210
Beverly, R. E., 384
Bex. L., 171, 176
Beznogikh, Yu. D., 303
Bieg, K. W., 372, 373, 374, 376, 379, 392
Bieth, C., 171
Bijkerk, F., 303
Billquist, P., 207, 210
Bistritsky, V. M., 383
Bittencourt, J. A., 7
Blewett, J. P., 145
Bliman, S., 208
Blommers, J., 314, 326
Blosser, H. G., 210
Bochev, B., 273
Bock, R., 152
Boers, J. E., 110
Bohm, D., 175, 400
Bolle, J., 332
Bolson, A., 145
Bonch-Osmolovskiy, A. G., 257
Bonnai, J. F., 42
Bourg, F., 210, 211, 223
Bourgarel, M. P., 171
Boxman, R. L., 332
Bozack, M. J., 325, 326
Brady, J. E., 314
Brainard, J. P., 42, 154
Brandelik, A., 383
Brandenberg, J. E., 373
Brewer, G. R., 24, 107, 400
Briand, P., 208, 211
Briggs, R. J., 288
Brillouin, L., 74
Broers, A. N., 259
Broodbent, D., 170
Brooks, N. B., 359
Brown, I. G., 1, 7, 48, 137, 230, 269, 270, 331, 332, 334, 339, 343, 347, 349, 351, 406, 407
Brown, K. L., 109
Brown, S. C., 7
Bruck, M., 164
Bruneteau, J., 308
Budker, G. I., 257
Bulirsch, R., 116
Burkchalter, P. G., 300
Burkhart, C., 332, 349, 351, 352, 397, 406, 408, 409
Burns, E. J. T., 372, 373, 374, 376, 379, 384
Busch, H., 55
Buzzi, J. M., 397
Bykovskii, Yu. A., 300, 301, 303, 304, 305, 310

Cabrespine, A., 171
Calbick, C. J., 29, 60
Caldecourt, L. R., 190
Campbell, I. E., 164
Cap, F. F., 152
Cardin, P., 171
Cardinal, P., 387, 388
Carey, D. C., 109
Carlson, T. A., 19, 425
Carlston, C. F., 162
Carmacat, N., 383
Carruth, M. R., 301
Carter, G., 179
Catlin, L. L., 155
Chan-tung, N., 214
Chavet, I., 38, 176, 187
Chavis, C. S., 400
Chen, F. F., 7, 403, 404
Chevalier, A., 214
Child, C. D., 25, 151, 401
Choate, L. M., 270
Chupp, W., 403
Cilliers, W. A., 331, 332
Citron, A., 383
Clampitt, R., 314
Clark, D. J., 176, 210
Clark, W. M. 313, 314, 318, 326
Clift, B. E., 207, 210
Cobine, J. D., 215, 331
Coffey, S., 332, 351, 406, 409
Colliex, C., 314
Colligon, J. S., 179
Collins, L. E., 359
Comeaux, A., 162
Compton, K. T., 25, 151
Conrad, J. R., 290
Consoli, T., 182
Cook, C., 230
Cook, D. L., 371, 392
Cooper, G., 332, 349, 351, 406, 409
Cooper, W. S., 36, 48, 110, 400
Cordey, J. G., 152
Cortois, A., 270, 271
Coupland, J. R., 29
Crandall, D. H., 212, 301
Crewe, A. C., 69
Crow, G., 313, 325, 327
Crow, J. T., 374, 402
Csanky, G., 313
Cutler, P. H., 317

Daalder, J. R., 332
Dahimene, M. 146, 240

Name Index

Dahlbacka, G. H., 300
Daley, H. L., 315, 317
D'Angelo, N., 145
Datla, R. V., 331, 332
Davis, R. C., 158
Davis, W. D., 332, 349, 406
Davisson, C. J., 29, 60
Dawton, R. H. 163, 187
Dayton, I. E., 64
d'Cruz, C., 320
Dearnaley, G., 152
Debernardi, J., 211
Debye, P., 12
Degtyarenko, N. N., 300
Degtyarev, V. G., 310
DeMichelis, C., 301
Demirchanov, R. A., 152, 158, 281, 294
Devin, A., 383
Devons, S., 170
Didenko, A. N., 383
Dimov, G. I., 356, 360
DiVergilio, W. F., 157
Dolder, K. T., 363
Done, K. C. W., 190
Donets, E. D., 21, 245, 246, 259, 260, 262, 269, 270, 271, 272, 274, 275, 276, 277, 281
Dorrell, L. R., 373, 374
Doschek, G. A., 300
Dote, T., 231
Doucet, H. J., 397
Dousson, S., 214
Dreike, P., 371, 374, 379, 387, 388, 392
Drewell, N., 387, 388
Druaux, J., 42, 179
Dudnikov, V. G., 357, 358, 360, 379
Dugar-Zhabov, V., 208
Dunn, D. A., 76, 77
Dupas, L., 208
Dworetsky, S., 141

Eastwood, J. W., 109
Edlen, B., 331
Ehler, A. W., 300
Ehlers, K. W., 24, 48, 155, 170, 356, 359, 361, 362, 363, 365, 400
Ehret, H. P., 210, 219
Ehrier, H., 383
Eichenberger, C., 371
Ektessabi, A., 281, 282
Eldridge, O., 213
Elesin, V. F., 300
El-Kareh, A. B., 62
El-Kareh, J. C. B., 62
Ellsworth, C. E., 176

Emig, H., 36, 163
Emigh, C. R., 29
Engemann, J., 148
Englehardt, A. G., 308
Englehart, L. S., 400, 407
Evans, W., 207, 210

Fabre, E., 308
Faltens, A., 397, 403
Farenik, V. I., 168
Farnsworth, A. V., Jr., 373
Faure, J., 254, 269, 270, 271
Feeney, R. K., 145, 331
Feibelman, W. A., 300
Feinberg, B., 254, 269, 270, 343, 349, 351
Feldman, U., 300
Fessenden, T. J., 400
Finn, T., 300
Flood, W. S., 176
Fois, M., 42
Forester, A. T., 402
Forrest, R. A., 363
Fosnight, V. V., 157
Francis, R. J., 190
Franczak, B., 108
Freeman, J. H., 152, 162, 187, 188, 190, 203, 205
Freeman, J. R., 386
Freisinger. J., 157
Friedrich, L., 210, 219
Früchtenicht, J. F., 307
Frodl, R., 211
Froehlich, H., 152, 158
Fujita, J., 331
Fukumoto, S., 356

Gabovich, M. D., 76, 168, 175, 176, 182, 362
Galvin, J. E., 332, 334, 343, 349, 351, 406, 407
Gamo, K., 314, 318
Gard, G. A., 203
Gastineau, B., 270, 271
Gautherin, G., 142, 168, 182
Gavin, B. F., 163, 167, 170, 171, 176, 179, 183, 331, 332, 361
Gayraud, R., 171
Geller, R., 2, 207, 208, 210, 211, 214, 223, 230
Gerber, R. A., 371, 373, 376, 379, 389
Gerritsen, H. C., 303
Gesley, M. A., 327
Gilles, J. P., 148
Gilmour, A. S., Jr., 334
Ginet, G., 376
Glaros, S. S., 384
Glaser, W., 221
Gobbet, R. H., 359

Goble, C., 172, 182
Goebel, D. M., 402
Goede, A., 155
Goede, H., 157
Goldstein, C., 267, 269, 270, 281
Goldstein, H., 91
Goldstein, S. A., 383, 391
Golovanevski, K., 208
Golvbev, A. A., 301
Gomer, R., 316, 378
Goncharenko, I. I., 332
Gorbachev, S. K., 171
Goretskii, V. P., 80
Gormezano, C., 208
Gough, D. E., 400
Gough, R. A., 176
Grandchamp, J. P., 148
Gray, L. G., 301, 308, 309
Green, T. A., 385
Green, T. S., 29, 47, 76, 77, 155, 172, 176, 182, 346
Greene, J. E., 142, 331
Greenly, J. B., 383, 385, 386, 387, 392
Greenspan, M. A., 374
Griem, H. R., 331, 332, 389
Grigorev, Yu P., 176
Grivet, P., 64, 65
Grotzschel, R., 320
Grunder, H. A., 170
Guharay, S. K., 42
Gusev, V. P., 301, 303, 304, 305
Guthrie, A., 175, 332
Gutkin, T. I., 152, 158
Guvarov, A. I., 303
Guyard, J., 32

Hadinger, G., 176
Hadinger-Espi, G., 176
Haglund, R. F., Jr., 385
Halback, K., 36, 110
Hale, J., 224
Haltes, E., 383
Ham, M., 332
Hamilton, G. W., 76, 160, 359
Hamm, R. W., 36, 50, 270
Hammer, D. A., 373, 374, 376, 383, 387
Hammond, D. P., 29
Hanawa, T., 138
Hand, L. N., 65
Hansen, K. J., 406
Hanson, D. L., 384
Hantzsche, E., 332
Haq, F. U., 145
Harbour, P., 78
Hargis, P. J., 388

Harrison, E. R., 72
Hartwig, E., 403
Harwood, L. H., 210
Hasan, M. A., 142
Hashidate, K., 390
Haught, A. E., 308
Haushahn, G., 179
Haydarov, R. T., 301
Haydon, S. C., 7
Heberlein, J. V. R., 347, 406
Heil, H., 170
Heinz, O., 145, 230
Henderson, C. M., 145
Henderson, T. F., 400
Hendricks, C. D., 315
Herrmannsfeldt, W. B., 110, 397
Hess, W., 208
Hijikawa, M. 377
Hilko, B., 391
Hill, R. B., 182
Hinkel, H., 197
Hiroshim, H., 325
Hirsch, E. H., 145
Hirsh, M. N., 7
Hirshfield, J. L., 332
Hockney, R. W., 109
Hoffman, J. M., 373
Hofmann, I., 101
Hofmann, K., 40, 42
Hoh, F., 179
Hoisington, R. W. R., 170
Holley, W. R., 176
Holmes, A. J. T., 24, 47, 53, 59, 76, 77, 80, 95, 104, 152
Holtkamp, D. B., 95
Honig, R. E., 429
Hooper, E. B., 48, 76, 83, 168
Hooper, J. W., 145, 331
Hopwood, J., 146
Hora, H., 301
Horioka, K., 377, 378, 390
Hornsby, J. S., 110
Hosoki, S., 314, 326
Hoyer, E., 397, 403
Hubbard, E. L., 361
Huckel, W., 12
Hughes, J. L., 301
Hughes, R. H., 299, 301, 308, 309
Humphries, S. Jr., 68, 332, 349, 351, 352, 371, 373, 374, 383, 386, 397, 406, 407, 408, 409
Hundley, J. L., 145
Hunter, R. O., 390
Hurst, G. S., 299
Huston, N. E., 169

Name Index

Ichimiya, T., 231
Iinuma, K., 145
Ikegami, K., 356
Iluyshchenko, V. I., 245, 269, 281
Inman, M., 24, 152
Inoue, N., 42
Ioffe, M. S., 170, 171
Irons, F. E., 300
Iselin, Ch., 109
Ishida, S., 42
Ishikawa, J., 24, 42, 239, 281, 282, 286, 287, 290, 291, 293, 331
Ishitani, T., 314, 326, 327
Ishitoya, K., 390
Ismail, H., 182
Itoh, H., 234
Ivannikov, R. I., 179
Iwaki, M., 242
Iwao, H., 281, 282
Izumi, K., 242

Jacoby, W., 163
Jacquot, B., 208, 210, 211, 214, 223, 230
Jacquot, C., 208, 230
Jaeger, E. F., 110
James, G. L., 384
Janes, G. S., 254
Jaramillo, W. H., 384, 385
Jayaram, R., 235
Jeffries, D. K., 313, 314, 327
Jergenson, J. B., 313, 314, 318, 326
Jimbo, K., 362
Johnson, D. J., 372, 373, 374, 376
Johnson, G. B., 402
Johnson, J. W., 359
Johnson, P. G., 145
Johnson, P. R., 374
Johnson, W. L., 402
Joho, W., 109
Jones, C. M., 210
Jones, D. A., 301
Jones, E. W., 145
Jones, M. E., 83
Jones, R. J., 170, 171, 175, 179
Jongen, Y., 207, 210, 214, 215
Jorde, I., 332
Jouffrey, B., 317
Judd, D. L., 400
Judd, R. A., 36
Jullens, R. A., 313, 314

Kadish, A., 83
Kadota, K., 331
Kalnins, G. J., 210

Kamakshaiah, S., 332, 347
Kamegai, M., 145
Kamke, D., 145
Kanayama, T., 325
Kane, E. L., 301
Kaneko, O., 42
Kang, N. K., 317
Kanomata, I. 208, 230, 231, 233, 235
Kanter, M. 176
Kapchinskij, I. M., 99, 100
Kapin, A. T., 332, 349, 406
Kartashov, S. V., 259, 277
Kasuya, K., 377, 378, 390
Katsubo, L. P. 76
Katsurai, M., 301
Kaufman, H. R., 154, 155
Kazes, E., 317
Keefe, D., 42, 397, 400, 403, 418
Keller, R., 23, 32, 36, 39, 40, 42, 45, 50, 101, 120, 151, 152, 156, 161, 163, 332, 346
Kelley, G. G., 158
Kenefick, R. A., 270
Kessi, O, 148
Kahn, M., 141
Khoroshikh, V. M., 349
Kidd, P. W., 157
Kilpatrick, W. D., 42
Kim, C., 403
Kim, Y., 378
Kimblin, C. W., 332, 347, 349, 406
Kimura, S., 199
King, R. F., 359
Kingham, D. R., 317, 318, 319, 322, 378
Kirschner, J., 139
Kishinevsky, M. E., 356
Kitamura, A., 385
Klabunde, J., 77, 80, 101
Klein, H., 211, 269, 270
Kleinod, M., 269, 270
Kluge, H. J., 299
Knall, J., 142
Knauer, W., 321
Knight, R. D., 308
Koch, J., 187, 188
Koike, H., 208, 230, 231, 234, 235
Kolosov, I. V., 301, 303, 304, 305, 310
Komuro, M., 325
Konjevic, N., 389
Kopecky, V., 230
Korschinek, G., 301, 306
Koshkarev, D. G., 301
Kostroun, V. O., 271, 339
Kosyrev, Yu. P., 301, 304, 305
Kovpik, O., 171

Kozharin, A. A., 168
Kozyrev, Yu. N., 362
Kozyrev, Yu. P., 300, 303, 305
Kramer, D. A., 429
Kramer, S. D., 299
Krechet, K. I., 301
Kreindel, Yu. E., 168
Krohn, V. E., 314, 326
Kroll, T., 101
Kruger, C. H., Jr., 7
Kubena, R. L., 313, 314
Kubena, R. S., 327
Kulsetas, J., 332
Kulygin, Yu. F., 332
Kunhardt, E. E., 7
Kunkel, W. B., x, 142, 147, 365, 367, 400
Kunori, M., 199
Kuroda, T., 42
Kurosawa, A., 390
Kursanov, U. V., 152, 158, 281, 294
Kusano, N., 281, 282
Kusse, B. R., 387
Kuswa, G. W., 383
Kutner, V. B., 176, 179, 305
Kutsarova, T., 273
Kuwabara, H., 377
Kuznetsov, V. I., 179
Kuznetsova, I. P., 176

Lafferty, J. M., 331, 332, 406
Lagrange, J. M., 171
Laman, H., 308
Lampel, M., 403
Lamppa, K. P., 376
Langmuir, I., 25, 151, 294
Lapitskii, Yu. I., 301
Lapostolle, P., 32, 93
Laptev, I. D., 310
Laslett, L. J., 397
Latyshev, S. V., 301
Lauer, R., 406
Lawrence, G. P., 359
Lawson, J. D., 58, 176, 398
Lee, H., 83
Leeper, R. J., 374
Leffel, C. S., Jr., 138
Leible, K., 180
Lejeune, C., 31, 142, 148, 168, 397, 398, 401
Lemons, D. S., 83
Len, L. K., 332, 349, 351, 352, 397, 409
Leung, K. N., 24, 147, 155, 161, 240, 355, 356, 359, 362, 363, 365, 367
Levine, M. A., 269, 270
Levi-Setti, R., 313, 325, 327

Limpaecher, R., 155, 362
Lin, S., 390
Livingston, R. S., 171
Lockwood, D. L., 334
Loeb, H., 157
Lofgren, E., 403
Logan, A. D., 332, 351
Lorenz, D., 230
Lossy, R., 148
Lotz, W., 20, 211, 246
Loubriel, G. M., 385
Lovelace, R. V., 305, 371
Lucatorto, T. B., 388
Ludwig, P., 211
Luessen, L. H., 7
Lunev, V. M., 349, 406
Luther-Davis, B., 301
Lynch, F., 207, 210
Lyneis, C., 207, 210, 215, 219, 224
Lyon, I. C., 331

McClure, G. W., 383
McDowell, J. D., 425
McGaffey, R. W., 110
MacGill, R. A., 171, 183, 331, 332, 334, 349, 406, 407
McIlrath, T. J., 388
McKay, P. F., 379, 381, 382, 384, 392
McKenzie, K. R., 155, 362
Maenchen, J., 373, 379, 383, 384, 392
Magnusson, G. D., 162
Magyary, S. B., 36, 110
Mahadevan, P., 162
Mahoney, J. F., 315, 317
Mahoney, L., 146
Mahony, C., 320, 327
Mair, G. L. R., 316
Makarov, L. G., 303
Makov, B. N., 170, 171, 174, 176
Manka, C. K., 301
Marder, B. M., 378
Marrs, R. E., 269, 270
Marti, F., 210
Masada, H., 327
Mathews, H. G., 210
Matsui, T., 314, 318
Mayer, H. P., 321
Mayer-Boericke, C., 210
Measures, R. M., 387, 388
Mehlhorn, T. A., 379
Meiners, L. G., 147
Melchior, G., 383
Melin, G., 208
Melngailis, J., 326

Name Index

Menat, M., 176
Mendel, C. W., 383
Mendenhall, H. E., 331
Meneghetti, J. R., 400
Meriwether, J. R., 176
Meszaros, P., 110
Meyer, E. A., 155, 400, 407
Meyer, F. W., 210, 224, 301
Michaelson, H. B., 425
Middleton, R., 355, 356
Mikhailovskii, A. B., 288
Miki, K., 110
Miller, H. C., 332, 349, 406
Miller, P. A., 372, 373, 374
Miller, P. D., 210
Miller, V. E., 170
Millikan, R. A., 331
Mills, C. B., 171
Mills, G. S., 376, 383
Mineev, F. I., 171
Minehara, E., 207, 210
Miskovsky, N. M., 317
Mitchner, M., 7
Mitsuhashi, K. 385
Mitsuoka, Y., 331
Mittag, K., 383
Mix, L. P., 373, 374
Miyagowa, J., 390
Mizutani, Y., 59
Moak, C. D., 359
Moellenbeck, J., 210, 219
Moeller, W., 145
Monchinskii, A., 303
Moore, R. D., 315, 317
Morgan, O. B., 158
Mori, Y., 356
Morikawa, J., 42
Morozov, P. M., 170, 171, 174, 176
Morris, D., 176
Motley, R. W., 145
Mott-Smith, J. M., 294
Mowat, J. R., 161
Mozley, R. F., 64
Mueller, A., 211, 214
Mueller, D. W., 29
Mueller, M., 176, 177, 180, 181
Mueller, R. W., 332
Musil, J., 230

Nahring, F., 320
Naida, A. P., 80, 362
Nakashima, S., 242
Nalivaiko, G. A., 360
Namba, S., 314, 318

Nanobashvili, S., 230
Nasser, E., 7
Nassibian, G. N., 170
Nelson, R. S., 152
Nemetz, R., 403
Neri, J. M., 376
Nestor, C. W., Jr., 19, 425
Neuberger, B. S., 83
Nevolin, V. N., 310
Nihei, N., 42
Nishimura, H., 42
Noehmayer, F., 42, 156
Novick, R., 141
Novikov, I. K., 303

Oberson, R., 42
Oblizin, A. N., 305
O'Hagan, J. B., 42, 154
Ohara, Y., 42, 45, 59, 157
Ohbayaski, K., 378
Ohmori, T., 301
Ohtani, S., 271, 331
Oka, Y., 42
Okamoto, Y., 230
Okudaira, S., 233
Okumura, Y., 59, 364
Olivo, M., 42
Olney, R. D., 313
Olsen, J. N., 372, 373, 374, 376, 379
Olson, R. E., 161
Omura, Y., 242
Oona, H., 332, 351, 400, 407
Orloff, J., 320, 326
Ose, Y., 110
Osher, J. E., 160
Oskam, H. J., 7
Ota, K., 42
Ovsyannikov, V. P., 259, 260, 270, 272, 273, 274, 277
Owren, H. W., 147, 240

Pack, J. L., 308
Padalka, V. G., 349
Pal, R., 373, 374
Palmer, M. A., 373
Pampellone, E., 387
Panitz, J. A., 392
Panofsky, W. K. H., 65
Pardo, R., 207, 210
Pasyuk, A. S., 168, 171, 176, 177, 179, 301, 303, 304, 305
Patou, C., 383
Pattison, D. A., 384
Payne, M. G., 299

Peacock, N. J., 300
Pearlman, T. S., 300
Peart, B., 331, 363
Pederson, D. O., 301
Peklenkov, V. D., 301, 303, 304, 305
Peleg, E., 373
Penning, F. M., 167
Perel, J., 315, 317
Peterson, J., 230
Petroff, P. M., 320
Petrov, A. V., 383
Petty, C. C., 148
Peugnet, C., 383, 387
Phaneuf, R. A., 300, 301
Phelps, A. V., 308
Picraux, S. T., 332, 352
Pierce, J. R., 27, 59, 61
Pigarov, Y. D., 174
Pike, C. D., 400
Pikin, A. I., 259, 270
Piosczyk, B., 36, 157
Pirart, C., 210
Plyutto, A. A., 331, 332, 349, 406
Polk, D. H., 308
Pontonnier, M., 210, 211
Popov, S. N., 176
Porto, D. R., 347, 406
Postma, H., 208
Poukey, J. W., 372
Pourrezaei, K., 320
Prangere, F., 182
Pregenzer, A. L., 378, 379, 392
Prestwich, K. R., 374, 392
Prewett, P. D., 313, 316, 320, 327
Pyle, R. V., 48, 362, 363, 400

Quintenz, J. P., 373

Raspopin, A. M., 303
Rau, R. S. N., 332, 347
Reaves, R. T., 145
Reich, J., 210
Reinhard, D. K., 146
Rensch, D. B., 313
Revutskii, E. I., 332
Rice, J. K., 379, 389, 392
Richards, D. M., 385
Richards, H. T., 356
Richter, R. M., 170
Riepe, K., 418
Ringo, G. R., 314, 326
Ripin, B. H., 141
Riviere, A. C., 29

Roberts, J. R., 389
Robinson, J. W., 332
Robinson, R. S., 154, 155
Rockett, A., 142, 331
Rogers, J. P., 317
Rogner, A., 383
Rondeau, G. D., 383, 385, 387
Rondeel, W. G. J., 332
Root, J., 146, 157, 239
Roppel, T., 146
Rose, P. H., 359
Rosenblum, S. S., 403
Rosenfeld, J. P., 301, 308
Rostomashvili, G., 230
Rothacker, F., 109
Rozhkov, A. M., 168
Rudenauer, F. G., 320
Ruster, W., 299
Rutkowski, H. L., 155, 332, 351, 400, 407
Ryckewaert, G., 210
Rynn, N., 145
Ryzhhov, V. N., 332, 349, 406

Sacherer, F. J., 32
Safanov, S., 208
Saitoh, T., 234
Saitoh, Y., 331
Sakudo, N., 208, 229, 230, 231, 233, 234, 235, 237, 238, 242
Sakurai, K., 42
Salzborn, E., 211, 214
Sano, F., 24, 281, 282, 291, 293
Sato, T., 42
Satoh, Y., 145
Satou, S., 242
Savage, M., 332, 349, 351, 406, 409
Sayle, II, W. E., 145, 331
Schinn, G. W., 388
Schlachter, A. S., 161
Schmiedbergen, P., 230
Schmieder, R. W., 269, 270, 339
Schmitt, N., 110
Schneider, J. D., 29, 155
Schoenenberg, M. H., 42
Schoenlein, A., 77, 80, 101
Schroeer, J. M., 141
Schulte, H., 163, 172, 176
Schultheiss, C., 383
Schweickert, H., 210, 219
Schwind, G. A., 313, 314, 318, 319, 326
Seidel, D. B., 374
Seidel, M., 288, 290
Sekiguchi, T., 301

Name Index

Self, S. A., 24, 76, 77, 118, 288
Seliger, R. L., 42, 153, 313, 314, 318, 326
Sellmair, J., 301, 306
Semenyushkin, I. N., 303
Septier, A., 34, 64, 65, 107, 182
Sessler, A. M., 301
Shabalin, A. L., 379
Shapiro, V. D., 288, 290, 291
Sharkov, B. Yu., 301
Sherman, J. D., 32, 38, 80, 95, 120
Sherwood, B. A., 155
Sherwood, E. M., 164
Shiloh, J., 403
Shimada, M., 234
Shoemaker, F. C., 64
Shohet, J. L., 7
Shoji, F., 138
Shubaly, M. R., 36, 42, 50, 110, 160, 161
Shukuri, S., 327
Shumshurov, A. V., 301
Shurter, R., 332, 351, 400, 407
Sidenius, G., 162, 205
Sil'nov, S. M., 300
Simonenko, L. S., 362
Singh, B., 155
Skilbreid, O., 205
Skoromnyi, G. M., 332
Skripal, L. P., 281, 294
Slutz, S. A., 373, 374
Smith, D. K., 148
Smith, H. V., 38, 80, 356
Smith, J., 110
Smith, W. W., 141
Smithells, C. J., 164
Soloshenko, N. A., 76, 362
Sorensen, R., 171, 183, 331
Sorensen, R. W., 331
Sortais, P., 211
Spaedtke, P., 36, 40, 42, 101, 107, 111, 119, 121, 156, 163, 332
Spitzer, L., 305
Stanku, V., 179
Staples, J., 171, 183, 331
Stearns, J. W., 161, 400
Steffen, K. G., 65
Steiger, W., 320
Stenzel, R. L., 141
Stephen, J., 152
Stetsenko, S. G., 305
Stevens, E. H., 313, 314
Stevens, R. R., 356, 363
Steyaert, J., 210
Stix, T. H., 230

Stoer, J., 116
Stover, G., 177
Struckmeier, J., 101, 109
Stuhlinger, E., 152
Sucov, E. W., 308
Sudan, R. N., 305, 371, 373, 376, 387
Sudraud, P., 314, 317
Suganomata, S., 331
Sugiura, H., 142
Sujatha, N., 317
Sundgren, J. E., 142
Sunka, P., 288, 290
Suvorov, K. G., 305
Suzuki, K., 233
Swanson L W. 313, 314, 317, 318, 319, 320, 321, 322, 323, 325, 326, 327
Swatik, D. S., 315
Sweeney, M. A., 373, 379, 392
Swegle, J. A., 372
Szuszczewicz, E., 230

Tabata, O., 199
Takagi, A., 356
Takagi, T., 24, 42, 110, 164, 239, 281, 282, 286, 287, 290, 291, 293, 331
Takahashi, T., 377
Takayama, S., 314, 326
Takebe, M., 145
Takeiri, Y., 42, 239, 331
Tamagawa, H., 210, 230
Tamura, H., 314, 326, 327
Tang, K. Y., 390
Tanque, H., 325
Tauth, T., 176
Taylor, G. I., 315
Tekawa, M. M., 170
Temple, W., 203
Thatcher, R., 171, 183, 331
Thompson, E., 59
Thompson, S. P., 314, 316, 326
Thoneman, P. C., 148
Thurston, J. N., 359
Tiefenback, M., 403
Tisone, G. C., 379, 387, 388, 389, 392
Tokiguchi, K., 208, 230, 231, 234, 235, 236
Tolk, N. H., 141, 385
Tonon, G. F., 301
Torii, Y., 234, 314
Tretyakov, Yu. P., 171, 177, 179
Tsepakin, S. G., 360
Tsintsadze, N., 230
Tsuji, H., 281, 282
Tuggle, D. W., 327

Turman, B. N., 371
Twai, T., 271
Tykesson, P., 356

Uchida, T., 42
Ueda, M., 383, 385
Ukegawa, T., 314
Umemura, K., 314, 326
Urai, A., 377
Urata, T., 378
Utlaut, M. W., 313, 314, 318, 326, 327
Utterback, N. G., 307
Uzienko, D. A., 305

Vadeev, V. P., 269, 270
Vahrenkamp, R. P., 42, 153
Valles, J. R., 307
Valyi, L., 168, 331
Vanacek, D. L., 400, 403
van Brug, H., 303
van der Walle, J., 314
van der Wiel, M. J., 303
VanDevender, J. P., 371, 372, 373, 392
van Haaften, F. W., 400
van Steenbergen, A., 95
Varga, I. K., 145
Vasseur, P., 308
Venkatesan, T., 319, 320
Verdeyen, J. T., 402
Vidyardhi, N., 332
Viehboeck, F., 164
Vladimirskij, V. V., 99, 100
Vlasov, V. V., 168
von Ardenne, M., 152, 158
von Rohden, H., 314, 326
Vorobev, E. D., 168, 179

Wada, Y., 327
Wagner, A., 319, 320
Wakerling, R. K., 175, 332
Wallmeroth, K., 299
Walsh, T. R., 99
Walther, S. R., 147, 240, 365, 367
Wang, B., 339
Wang, V., 313, 314
Wang, Y. L., 325, 327
Ward, J. W., 313, 314, 325, 327

Warwick, A., 331, 400
Wasserman, N., 425
Watanabe, I., 234
Waterson, M., 207, 210
Waugh, A. R., 313, 319, 327
Webb, J. A., 383
Weber, B. V., 383
Weber, C., 110
Weeden, A. E., 161
Weijiang, Z., 177
Weiss, M., 32
Wenger, D. F., 374
Whealton, J. H., 110
Whitson, J. C., 110
Wielders, P. G. E., 332
Wiesemann, K., 208
Wiley, L., 373
Williams, N., 199
Willmann, P. A., 76, 83
Wilson, R. G., 24, 107, 400
Winter, H., 20, 164, 168, 172, 175, 211
Winterberg, F., 371
Wittkower, A. B., 359
Wolf, B. H., 20, 163, 168, 169, 172, 175, 176, 180
Wong, S. F., 363
Wong, S. L., 384
Woodall, D. M., 332, 349, 351, 406, 409
Woodworth, J. R., 379, 384, 385, 392
Wright, L., 83
Wright, R. T., 334, 349

Yahiku, A. T., 315, 317
Yamada, I., 281, 282, 314
Yano, S., 385
Ye, X. M., 301
Yonas, G., 372
Yoneda, H., 377, 378, 390
York, R. L., 356, 363
Yoshida, K., 242
Young, J. P., 299
Young, Y. L., 313, 327

Zacek, F., 230
Zagar, D. M., 383
Zajec, E., 170, 177
Zinov'ev, Yu. P., 303
Zucker, Z., 170, 171

Subject Index

Aarhus ion source, 356
ABEL ion source, 173
Aberrations, 109
Accelerating column, 122, 127
Accelerating gap, 59, 122, 127
ADAM ion source, 170
Aluminosilicates, 145
AMPFION ion source, 383
Angular relaxation time, 13
Aperture-lens (electrode-lens) effect, 27, 29, 60
Applied-B diode, 305, 351, 371, 374
Arc plasma, 8
Atomic oven, 142, 179, 223, 226
AXCEL code, 110
AXCEL-GSI code, 25, 111

Basic ion sources, 3, 137
Bayard-Alpert gauges, 137
Beam:
    aberrations, 109
    brightness, 34, 92, 313, 325, 398
    current, normalized, 151
    divergence, 25, 31
    enevelope, 98
    formation, 23
    halo, 35
    optics, 54
    quality parameters, 30
    transport, 2, 53
    transport channels, 69, 88
BEAM code, 110
Beam-plasma discharge, 281, 287
Beam-plasma ion source, 4, 281
Bending magnet, 66
Bernoulli term, 317
Beta-eucriptite, 145

Biot-Savart law, 70
Blewett-Jones ion source, 145
Boltzmann constant, 8
Boltzmann distribution, 78, 86
BOLVAPS/LIBORS ion source, 387
Brightness, 34, 92, 313, 325, 398
Brillouin flow, 73, 74, 90, 250, 261
Broad beam ion source, 148
Busch's Theorem, 55

Calutron ion source, 187
Canonical angular momentum, 56, 91
Capillary discharge, 391
Cathode simulation, 116
Cathode spots, 332
Charge conservation equation, 114
Charge exchange, 214
Charge neutrality, 12, 87, 119
Child-Langmuir law, 25, 29, 34, 39, 116, 151, 372, 377, 412
CHORDIS ion source, 156, 162, 163
CIRCE plasma device, 208
Collisions, 13, 14
Compact microwave ion sources, 239, 365
Compact PIG ion source, 178
Computer modeling, 3, 38, 107
Courant-Snyder parameters, 98
Critical density, 214, 230
Cross section, 14
    charge exchange, 214
    electron impact ionization, 211, 214, 246, 249, 272
CRYEBIS ion source, 259
Cryogenic-anode ion sources, 377
Cyclotron frequency, 16
Cyclotron radius, 15

Debye length, 12, 305. *See also* Sheath
Discretization, 133
Dispenser source, 145
Dissociative attachment and recombination, 363
Distribution functions, 8, 287
Divergence, 25, 31
Drifting beam, 101
Drift space, 9, 69, 121
Drift tubes, 253, 286
Duopigatron ion source 40, 158
Duoplasmatron ion source, 158, 163

EBIS, 4, 212, 245, 281, 309
EBIS ion sources:
   Cornell EBIS, 271
   CRYEBIS, 270
   CRYEBIS-2, 271
   DIONE, 271
   Frankfurt EBIS, 270
   IEL, 269
   IEL-2, 269
   Kansas EBIS, 271
   KRION, 270
   KRION-2, 270
   KRION-3, 271
   LBL EBIS, 270
   LLNL EBIS, 271
   NICE EBIS, 271
   Novosibirsk EBIS, 271
   SILFEC, 269
   SNLL EBIS, 271
   Texas EBIS, 270
   TOFEBIS, 270
ECR (Electron Cyclotron Resonance) ion source, 1, 4, 19, 132, 145, 207, 309
   CAPRICE ion source, 217
   CP-ECR ion source, 218
   ECR (RIKEN), 218
   ECR (Texas A&M), 218
   ECR (Uppsala), 218
   ECREVETTE, 210
   ECREVIS, 221
   HISKA ion source, 217
   ISIS ion source, 217
   LBL-ECR ion source, 218, 219
   LISKA ion source, 217
   MAFIOS, 208
   MicroMAFIOS, 210
   MiniMAFIOS, 210
   OCTOPUS ion source, 217
   ORNL-ECR ion source, 218
   PHECR ion source, 218
   RT-ECR ion source, 218
   SC-ECR ion source, 218
   SuperMAFIOS, 216

   TripleMAFIOS, 209
ECR surface, 213
EHD (electrohydrodynamic) instability, 378
Einzel lens 50, 61
Electron-bombardment ion sources, 139
Electron cyclotron frequency, 16
Electron cyclotron radius, 15
Electron cyclotron resonance, 213, 229
Electron density, 7
Electron plasma frequency, 14
Electron volt, 8
Electrostatic aperture lens, 60
Electrostatic confinement, 402
Electrostatic electron reflector, 199, 249
Electrostatic lens, 59, 60
Electrostatic quadrupole, 64
Elements, table of, 423
EMFAAPS ion source, 387
Emittance, 30, 33, 92, 120, 182, 398
   ellipse, 31, 97, 98, 108, 398
   ellipse parameters, 98, 109
   growth, 100
   measurement, 95, 416
Energy distribution function, 9
Energy flow, 171
Energy relaxation time, 13
Envelope equations, 100
Envelope model, 108
Expansion cup, 158
Extraction slit, 46, 182, 187, 233, 235, 237

Facet lens, 406. *See also* Aperture-lens effect
Feed materials, 179, 203
Field evaporation, 316
Field ionization, 316, 319
Field ion microscope, 324
Finite Difference Method (FDM), 115
Finite element Method (FEM), 115
Flashover ion source, 373
Focusing of ion beams, 59, 313
Freeman ion source, 3, 187, 238

Gas flow rate, 174
Gas reflux, 175, 216
Gaussian optics, 53
Gauss-Seidel method, 116
Glaser lens, 221
Grid-controlled extraction, 5, 352, 397, 402
Gyroradius, 15
Gyrotropic plasma, 15

Halo, 35
Hamilton's equations, 92
Hexapole magnetic field, 208
High current gaseous ion sources, 3, 151

Subject Index

Hill's equation, 97
Hydrodynamic instability, 288

Inertial Confinement Fusion (ICF), 371, 374, 397
ION code, 110
Ion confinement time, 175, 215
Ion cyclotron radius, 16, 66, 74
Ion extraction, 2, 23
Ion implantation, semiconductor, 188, 190, 235
Ion-ion collision time, 14
Ionization, 17
   electron impact, 152, 211, 214, 246, 272
Ionization factor, 246
Ionization potential, 18, 423, 425
Ionization time, 254
Ion plasma frequency, 15, 74
Isotope separators, 187, 190

Kapchinskij-Vladimirskij (K–V) distribution, 32, 99
Kaufman ion source, 148, 155
Kilpatrick limit, 43
KOBRA3 code, 110, 119
KRION-2 ion source, 259

Lagrangian, 91
Laminar flow, 70
Langmuir probe, 232
Lanthanum hexaboride, 259, 362
Laplace equation, 116
Laser ion source, 4, 299
Lens:
   electrostatic, 50, 59, 60
   magnetic, 62
   quadrupole, 63, 65
   solenoidal, 62
Light ion sources for inertial confinement fusion, 5, 371
Lintott Series III Freeman ion source, 194
Liouville's Theorem, 34, 54, 91, 98
Liquid metal ion source, 4, 313
LOIS ion source, 302
LONGSHOT ion source, 385
Lorentz force, 71, 112
Lotz formula, semi-empirical ionization cross section, 20, 246, 249

MAFIOS ion source 208
Magnetic:
   bottle, 199
   bucket, 19, 155
   filter, 362, 367
   homogenizer, 270
   lens, 59, 62
   multipole, 19, 141, 152, 213

   quadrupole, 63
Magnetically insulated diode, 305, 351, 371, 374
Magnetron negative ion source, 360
Matrix method, 58
MATS III ion source, 160
Maxwellian distribution function, 8, 9, 10, 17, 94, 292, 307, 404
Mean free path, 14
Mean particle energy, 9
Meniscus, 24, 397, 401
Metal ion sources 142, 179, 222, 224, 242, 313, 331
Metal vapor vacuum arc ion source (MEVVA) 5, 48, 145, 331, 406
Microdroplets, 320, 327
MicroMAFIOS ion source, 210
MicroMEVVA ion source, 145, 339
Microwave ion source, 4, 145, 157, 229, 364
Microwave oven, 147
Microwaves, off-resonance, 231
MiniMAFIOS ion source, 210
Minimum-B magnetic field, 152, 209, 213
MIRKO code, 108
Monocusp ion source, 154
Multiaperture extraction, 45
Multiaperture lens, 233
Multicusp ion source, 155
Multicusp magnetic field, 146, 157, 208, 356, 359
Multiple ionization, 19, 246
Multiply charged ions, 154, 161, 167, 175, 207, 211, 245, 256, 300, 306, 308, 343, 349

Negative ion sources, 5, 86, 118, 355
Nereus ion source, 375
Neutrality condition, 12, 87, 119
Neutral particle reflux, 175, 216
Newton-Raphson iteration technique, 118
Non-laminar beams, 91
Nonlinear optics, 100
Normalized beam current, 27
Nova Freeman ion source, 196
Nuclear properties, 274

Off-resonance microwave plasma, 230
Overdense plasma, 214
Oxide-coated cathode, 141

Paraxial ray equation, 53, 54, 58, 73
PARMILA code, 109
PARMTEK code, 109
PARMTRA code, 109
Particle flow, 17
Particle-in-cell codes, 109

Subject Index

Paschen law, 215
PBFA II, 371
Penning ion source, 153, 158, 357, 361
Penning mode:
  high pressure, 168
  low pressure, 168
Percentage ionization, 8
Perveance, 27, 71, 73, 102
Photodetachment, 359
Photoionization, 299
Physical constants, 421
PI 9000 Freeman ion source, 194, 200
Picket fence magnetic field, 155, *See also*, Magnetic, mulitpole; Multicusp magnetic field
PICOHISKA ion source, 210
Pierce contours, 28
Pierce electrodes, 27, 182
PIG (Philips, or Penning, Ionization Gauge) ion source, 3, 18, 163, 167
Plasma:
  charge neutrality condition, 12, 79, 119
  density, 7
  frequency, 14, 213
  generators, 151
  instabilities, 13, 172, 175
  magnetic field effects, 15
  map, 10
  oscillations, 176
  parameters, 7, 422
  pressure, 16, 174
  sheath, 12
  simulation, 118
  temperature, 9
Poisson equation, 76, 112, 115, 119, 292
PRE-ISIS ion source, 210
Primary electrons, 18, 167, 169, 296
  velocity distribution, 172, 287
Pulsed high-brightness ion beams 5, 397

Quadrupole, electrostatic, 64
Quadrupole doublet, 65
Quasilinear relaxation, 290
Quasineutrality, 12

Rare earth metals, 145
Rate coefficient, 212
Ray tracing, 109, 119
  codes, 112
Reflex discharge, 18, 156, 158, 362
Relativistic parameter, 33
Relaxation time, 13
Resistivity, 423
RF ion source, 148, 157, 230
RFQ (Radio Frequency Quadrupole), 208

RIG-20 ion source, 158

Screening distance, *see* Sheath
Secondary electrons, 18, 296
Sector magnet, 67
Self-extraction source, 359
Self fields, 70, 73
Self's law, 121
Sheath, 11, 12, 305, 405, 234
Shifted Maxwellian, 17
Single-stage discharge, 153
SLAC code, 110
SNOW code, 110
Space charge, 54, 132, 151
  blowup, 24, 72
  neutralization, 71, 75, 121, 282, 292
Spark source, 352
Spodumene, 145
Sputtering, 163, 168, 179, 203, 355
Superexcited states, 277
SuperMAFIOS ion source, 207, 209
Surface breakdown, 373
Surface ionization, 145
Synchrotrom, zero-gradient, 69

Taylor cone, 315, 321
Thermal equilibrium, 8
Thomson parabola ion analyzer, 376, 378
Time-of-flight ion analyzer, 262, 268, 272, 304, 349, 408, 414
Transfer matrix, 59, 108
Transport channel, 61
TRANSPORT code, 109
Transverse energy, 120
TripleMAFIOS ion source, 209

U^{92+}, 19, 246, 257
Upper hybrid frequency, 288
Uranium plasma, 334

Vacuum arc, 331
Vacuum spark, 331, 352
Vapor pressure, 427
Velocity distribution function, 8, 173
Virtual anode, 403
Virtual source size, 324

Wien mass filter, 314
WOLF code, 110
Work functions, 426

X-ray spectroscopy, 277
XUV-driven ion sources, 384

Zeolite, 145